Lecture Notes in Computer Science 6508

Commenced Publication in 1973
Founding and Former Series Editors:
Gerhard Goos, Juris Hartmanis, and Jan van Leeuwen

Editorial Board

David Hutchison
Lancaster University, UK

Takeo Kanade
Carnegie Mellon University, Pittsburgh, PA, USA

Josef Kittler
University of Surrey, Guildford, UK

Jon M. Kleinberg
Cornell University, Ithaca, NY, USA

Alfred Kobsa
University of California, Irvine, CA, USA

Friedemann Mattern
ETH Zurich, Switzerland

John C. Mitchell
Stanford University, CA, USA

Moni Naor
Weizmann Institute of Science, Rehovot, Israel

Oscar Nierstrasz
University of Bern, Switzerland

C. Pandu Rangan
Indian Institute of Technology, Madras, India

Bernhard Steffen
TU Dortmund University, Germany

Madhu Sudan
Microsoft Research, Cambridge, MA, USA

Demetri Terzopoulos
University of California, Los Angeles, CA, USA

Doug Tygar
University of California, Berkeley, CA, USA

Gerhard Weikum
Max Planck Institute for Informatics, Saarbruecken, Germany

Lecture Notes in Computer Science 6503

Weili Wu
Ovidiu Daescu (Eds.)

Combinatorial Optimization and Applications

4th International Conference, COCOA 2010
Kailua-Kona, HI, USA, December 18-20, 2010
Proceedings, Part I

 Springer

Volume Editors

Weili Wu
University of Texas at Dallas
Department of Computer Science
Richardson, TX 75083, USA
E-mail: weiliwu@utdallas.edu

Ovidiu Daescu
University of Texas at Dallas
Department of Computer Science
Richardson, TX 75080, USA
E-mail: daescu@utdallas.edu

Library of Congress Control Number: 2010939794

CR Subject Classification (1998): F.2, G.2.1, G.2, C.2, E.1, I.3.5

LNCS Sublibrary: SL 1 – Theoretical Computer Science and General Issues

ISSN   0302-9743
ISBN-10   3-642-17457-4 Springer Berlin Heidelberg New York
ISBN-13   978-3-642-17457-5 Springer Berlin Heidelberg New York

springer.com

© Springer-Verlag Berlin Heidelberg 2010
Printed in Germany

Typesetting: Camera-ready by author, data conversion by Scientific Publishing Services, Chennai, India
Printed on acid-free paper   06/3180

Preface

The 4th Annual International Conference on Combinatorial Optimization and Applications (COCOA 2010) took place in Big Island, Hawaii, USA, December 18–20, 2010. Past COCOA conferences were held in Xi'an, China (2007), Newfoundland, Canada (2008) and Huangshan, China (2009).

COCOA 2010 provided a forum for researchers working in the areas of combinatorial optimization and its applications. In addition to theoretical results, the conference also included recent works on experimental and applied research of general algorithmic interest. The Program Committee received 108 submissions from more than 23 countries and regions, including Australia, Austria, Canada, China, Denmark, France, Germany, Hong Kong, India, Italy, Japan, Korea, Mexico, New Zealand, Poland, Slovak Republic, Spain, Sweden, Switzerland, Taiwan, UK, USA, Vietnam, etc.

Among the 108 submissions, 49 regular papers were selected for presentation at the conference and are included in this volume. Some of these papers will be selected for publication in a special issue of the *Journal of Combinatorial Optimization*, a special issue of *Theoretical Computer Science*, a special issue of *Optimization Letters*, and a special issue of *Discrete Mathematics, Algorithms and Applications* under the standard refereeing procedure.

We thank all authors for submitting their papers to the conference. We are grateful to all members of the Program Committee and all external referees for their work within demanding time constraints. We thank the Organizing Committee for their contribution to making the conference a success. We also thank Jiaofei Zhong and Donghyun Kim for helping us create and update the conference website and maintain the Springer Online Conference Service system and Shawon Rahman for helping in local arrangements.

Finally, we thank the conference sponsors and supporting organizations for their support and assistance. They are the University of Texas at Dallas, the University of Hawaii at Hilo, and the National Science Foundation of USA.

December 2010   Weili Wu
Ovidiu Daescu

Organization

COCOA 2010 was organized by the department of Computer Science, the University of Texas at Dallas, in cooperation with the University of Hawaii at Hilo.

Executive Committee

General Co-chairs	Ding-Zhu Du (University of Texas at Dallas, USA)
	Panos M. Pardalos (University of Florida, USA)
	Bhavani Thuraisingham (University of Texas at Dallas, USA)
PC Co-chairs	Ovidiu Daescu (University of Texas at Dallas, USA)
	Weili Wu (University of Texas at Dallas, USA)
Local Chair	Shawon Rahman (University of Hawaii at Hilo, Hawaii)

Program Committee

Farid Alizadeh	Rutgers University, USA
Mikhail (Mike) J. Atallah	Purdue University, USA
Giorgio Ausiello	Università di Roma, Italy
Piotr Berman	Penn State University, USA
Vladimir Boginski	University of Florida, USA
Annalisa De Bonis	Università di Salerno, Italy
Sergiy Butenko	Texas A&M University, USA
Gruia Calinescu	Illinois Institute of Technology, USA
Gerard Jennhwa Chang	National Taiwan University, Taiwan
Zhi-Zhong Chen	Tokyo Denki University, Japan
Chuangyin Dang	City University of Hong Kong, Hong Kong
Vladimir Deineko	The University of Warwick, UK
Zhenhua Duan	Xidian University, China
Omer Egecioglu	University of California, Santa Barbara, USA
Dan Hirschberg	University of California, USA
Tsan-sheng Hsu	Academia Sinica, Taiwan
Hejiao Huang	Harbin Institute of Technology, China
Wonjun Lee	Korea University, South Korea
Asaf Levin	The Technion, Israel
Yingshu Li	Georgia State University, USA
Guohui Lin	University of Alberta, Canada
Liying Kang	Shanghai University, China
Naoki Katoh	Kyoto University, Japan
Ilias S. Kotsireas	Wilfrid Laurier University, Canada

Anastasia Kurdia	Smith College, USA
Mitsunori Ogihara	University of Miami, USA
Jack Snoeyink	The University of North Carolina at Chapel Hill, USA
Ileana Streinu	Smith College, USA
Vitaly Strusevich	University of Greenwich, UK
Zhiyi Tan	Zhejiang University, China
Doreen Anne Thomas	University of Melbourne, Australia
Alexey A. Tuzhilin	Moscow State University, Russia
Amy Wang	Tsinghua University, China
Caoan Wang	Memorial University of Newfoundland, Canada
Feng Wang	Arizona State University, USA
Lusheng Wang	City University of Hong Kong, Hong Kong
Wei Wang	Xi'an Jiaotong University, China
Weifan Wang	Zhejiang Normal University, China
Chih-Wei Yi	National Chiao Tong University, Taiwan
Alex Zelikovsky	George State University, USA
Cun-Quan Zhang	West Virginia University, USA
Huaming Zhang	University of Alabama in Huntsville, USA
Louxin Zhang	National University of Singapore, Singapore
Xiao Zhou	Tohoku University, Japan

Referees

Ferdinando Cicalese	Salvatore La Torre	Gaolin Milledge
Paolo D'Arco	Yuan-Shin Lee	Seth Pettie
Gianluca De Marco	Chung-Shou Liao	J.K.V. Willson
Natallia Katenka	Hongliang Lu	Wei Zhang
Donghyun Kim	Hsueh-I Lu	Zhao Zhang
Stefan Langerman	Martin Milanič	Jiaofei Zhong

Table of Contents – Part I

Termination of Multipartite Graph Series Arising from Complex
Network Modelling 1
  Matthieu Latapy, Thi Ha Duong Phan, Christophe Crespelle, and
  Thanh Qui Nguyen

Simple Cuts Are Fast and Good: Optimum Right-Angled Cuts in Solid
Grids 11
  Andreas Emil Feldmann, Shantanu Das, and Peter Widmayer

Evacuation of Rectilinear Polygons 21
  Sándor Fekete, Chris Gray, and Alexander Kröller

A Fast Algorithm for Powerful Alliances in Trees 31
  Ararat Harutyunyan

NP-Completeness of Spreading Colored Points 41
  Ovidiu Daescu, Wenqi Ju, and Jun Luo

Construction of Mixed Covering Arrays of Variable Strength Using a
Tabu Search Approach 51
  Loreto Gonzalez-Hernandez, Nelson Rangel-Valdez, and
  Jose Torres-Jimenez

Feasibility-Based Bounds Tightening via Fixed Points 65
  Pietro Belotti, Sonia Cafieri, Jon Lee, and Leo Liberti

A Characterisation of Stable Sets in Games with Transitive
Preference 77
  Takashi Matsuhisa

Linear Coherent Bi-cluster Discovery via Beam Detection and Sample
Set Clustering 85
  Yi Shi, Maryam Hasan, Zhipeng Cai, Guohui Lin, and
  Dale Schuurmans

An Iterative Algorithm of Computing the Transitive Closure of a Union
of Parameterized Affine Integer Tuple Relations 104
  Bielecki Wlodzimierz, Klimek Tomasz, Palkowski Marek, and
  Anna Beletska

Bases of Primitive Nonpowerful Sign Patterns 114
  Guanglong Yu, Zhengke Miao, and Jinlong Shu

Extended Dynamic Subgraph Statistics Using h-Index Parameterized
Data Structures... 128
    David Eppstein, Michael T. Goodrich, Darren Strash, and
    Lowell Trott

Discrete Optimization with Polynomially Detectable Boundaries and
Restricted Level Sets 142
    Yakov Zinder, Julia Memar, and Gaurav Singh

Finding Strong Bridges and Strong Articulation Points in Linear
Time 157
    Giuseppe F. Italiano, Luigi Laura, and Federico Santaroni

Robust Optimization of Graph Partitioning and Critical Node
Detection in Analyzing Networks.................................. 170
    Neng Fan and Panos M. Pardalos

An Efficient Algorithm for Chinese Postman Walk on Bi-directed
de Bruijn Graphs 184
    Vamsi Kundeti, Sanguthevar Rajasekaran, and Heiu Dinh

On the Hardness and Inapproximability of Optimization Problems on
Power Law Graphs 197
    Yilin Shen, Dung T. Nguyen, and My T. Thai

Cyclic Vertex Connectivity of Star Graphs 212
    Zhihua Yu, Qinghai Liu, and Zhao Zhang

The Number of Shortest Paths in the (n, k)-Star Graphs 222
    Eddie Cheng, Ke Qiu, and Zhi Zhang Shen

Complexity of Determining the Most Vital Elements for the 1-median
and 1-center Location Problems.................................. 237
    Cristina Bazgan, Sonia Toubaline, and Daniel Vanderpooten

PTAS for Minimum Connected Dominating Set with Routing Cost
Constraint in Wireless Sensor Networks 252
    Hongwei Du, Qiang Ye, Jioafei Zhong, Yuexuan Wang,
    Wonjun Lee, and Haesun Park

A Primal-Dual Approximation Algorithm for the Asymmetric
Prize-Collecting TSP 260
    Viet Hung Nguyen

Computing Toolpaths for 5-Axis NC Machines 270
    Danny Z. Chen and Ewa Misiołek

A Trichotomy Theorem for the Approximate Counting of
Complex-Weighted Bounded-Degree Boolean CSPs 285
    Tomoyuki Yamakami

A Randomized Algorithm for Weighted Approximation of Points by a
Step Function 300
  Jin-Yi Liu

Approximating Multilinear Monomial Coefficients and Maximum
Multilinear Monomials in Multivariate Polynomials.................. 309
  Zhixiang Chen and Bin Fu

The Union of Colorful Simplices Spanned by a Colored Point Set....... 324
  André Schulz and Csaba D. Tóth

Compact Visibility Representation of 4-Connected Planc Graphs 339
  Xin He, Jiun-Jie Wang, and Huaming Zhang

Some Variations on Constrained Minimum Enclosing Circle Problem ... 354
  *Arindam Karmakar, Sandip Das, Subhas C. Nandy, and
  Binay K. Bhattacharya*

Searching for an Axis-Parallel Shoreline 369
  Elmar Langetepe

Bounded Length, 2-Edge Augmentation of Geometric Planar Graphs ... 385
  *Evangelos Kranakis, Danny Krizanc, Oscar Morales Ponce, and
  Ladislav Stacho*

Scheduling Packets with Values and Deadlines in Size-Bounded
Buffers 398
  Fei Li

Transporting Jobs through a Processing Center with Two Parallel
Machines 408
  Hans Kellerer, Alan J. Soper, and Vitaly A. Strusevich

Author Index 423

Table of Contents – Part II

Coverage with k-Transmitters in the Presence of Obstacles 1
  Brad Ballinger, Nadia Benbernou, Prosenjit Bose, Mirela
  Damian, Erik D. Demaine, Vida Dujmović, Robin Flatland,
  Ferran Hurtado, John Iacono, Anna Lubiw, Pat Morin,
  Vera Sacristán, Diane Souvaine, and Ryuhei Uehara

On Symbolic OBDD-Based Algorithms for the Minimum Spanning
Tree Problem 16
  Beate Bollig

Reducing the Maximum Latency of Selfish Ring Routing via Pairwise
Cooperations 31
  Xujin Chen, Xiaodong Hu, and Weidong Ma

Constrained Surface-Level Gateway Placement for Underwater Acoustic
Wireless Sensor Networks 46
  Deying Li, Zheng Li, Wenkai Ma, and Hong Chen

Time Optimal Algorithms for Black Hole Search in Rings 58
  Balasingham Balamohan, Paola Flocchini, Ali Miri, and
  Nicola Santoro

Strong Connectivity in Sensor Networks with Given Number of
Directional Antennae of Bounded Angle 72
  Stefan Dobrev, Evangelos Kranakis, Danny Krizanc,
  Jaroslav Opatrny, Oscar Morales Ponce, and
  Ladislav Stacho

A Constant-Factor Approximation Algorithm for the Link Building
Problem 87
  Martin Olsen, Anastasios Viglas, and Ilia Zvedeniouk

XML Reconstruction View Selection in XML Databases: Complexity
Analysis and Approximation Scheme 97
  Artem Chebotko and Bin Fu

Computational Study for Planar Connected Dominating Set Problem . . . 107
  Marjan Marzban, Qian-Ping Gu, and Xiaohua Jia

Bounds for Nonadaptive Group Tests to Estimate the Amount of
Defectives 117
  Peter Damaschke and Azam Sheikh Muhammad

A Search-Based Approach to the Railway Rolling Stock Allocation
Problem   131
    Tomoshi Otsuki, Hideyuki Aisu, and Toshiaki Tanaka

Approximation Algorithm for the Minimum Directed Tree Cover   144
    Viet Hung Nguyen

An Improved Approximation Algorithm for Spanning Star Forest in
Dense Graphs   160
    Jing He and Hongyu Liang

A New Result on $[k, k+1]$-Factors Containing Given Hamiltonian
Cycles   170
    Guizhen Liu, Xuejun Pan, and Jonathan Z. Sun

Yao Graphs Span Theta Graphs   181
    Mirela Damian and Kristin Raudonis

A Simpler Algorithm for the All Pairs Shortest Path Problem with
$O(n^2 \log n)$ Expected Time   195
    Tadao Takaoka and Mashitoh Hashim

New Min-Max Theorems for Weakly Chordal and Dually Chordal
Graphs   207
    Arthur H. Busch, Feodor F. Dragan, and R. Sritharan

A Simpler and More Efficient Algorithm for the Next-to-Shortest Path
Problem   219
    Bang Ye Wu

Fast Edge-Searching and Related Problems   228
    Boting Yang

Diameter-Constrained Steiner Tree   243
    Wei Ding, Guohui Lin, and Guoliang Xue

Minimizing the Maximum Duty for Connectivity in Multi-Interface
Networks   254
    Gianlorenzo D'Angelo, Gabriele Di Stefano, and Alfredo Navarra

A Divide-and-Conquer Algorithm for Computing a Most Reliable
Source on an Unreliable Ring-Embedded Tree   268
    Wei Ding and Guoliang Xue

Constrained Low-Interference Relay Node Deployment for Underwater
Acoustic Wireless Sensor Networks   281
    Deying Li, Zheng Li, Wenkai Ma, and Wenping Chen

Structured Overlay Network for File Distribution   292
    Hongbing Fan and Yu-Liang Wu

Optimal Balancing of Satellite Queues in Packet Transmission to
Ground Stations   303
    Evangelos Kranakis, Danny Krizanc, Ioannis Lambadaris,
    Lata Narayanan, and Jaroslav Opatrny

The Networked Common Goods Game.   317
    Jinsong Tan

A Novel Branching Strategy for Parameterized Graph Modification
Problems   332
    James Nastos and Yong Gao

Listing Triconnected Rooted Plane Graphs   347
    Bingbing Zhuang and Hiroshi Nagamochi

Bipartite Permutation Graphs Are Reconstructible   362
    Masashi Kiyomi, Toshiki Saitoh, and Ryuhei Uehara

A Transformation from PPTL to S1S   374
    Cong Tian and Zhenhua Duan

Exact and Parameterized Algorithms for Edge Dominating Set in
3-Degree Graphs   387
    Mingyu Xiao

Approximate Ellipsoid in the Streaming Model   401
    Asish Mukhopadhyay, Animesh Sarker, and Tom Switzer

Author Index   415

Termination of Multipartite Graph Series Arising from Complex Network Modelling

Matthieu Latapy[1], Thi Ha Duong Phan[2],
Christophe Crespelle[1], and Thanh Qui Nguyen[3]

[1] LIP6, CNRS and Université Pierre et Marie Curie (UPMC - Paris 6)
104 avenue du Président Kennedy, 75016 Paris, France
Firstname.Lastname@lip6.fr
[2] Institute of Mathematics, 18 Hoang Quoc Viet, Hanoi,
Vietnam and LIAFA, Université Paris 7, 2 place Jussieu, 75005 Paris, France
phanhaduong@math.ac.vn, phan@liafa.jussieu.fr
[3] University of CanTho, 1, Ly Tu Trong, Quan Ninh Kieu, CanTho, Vietnam
ntquy2004@yahoo.com

Abstract. An intense activity is nowadays devoted to the definition of models capturing the properties of complex networks. Among the most promising approaches, it has been proposed to model these graphs via their clique incidence bipartite graphs. However, this approach has, until now, severe limitations resulting from its incapacity to reproduce a key property of this object: the overlapping nature of cliques in complex networks. In order to get rid of these limitations we propose to encode the structure of clique overlaps in a network thanks to a process consisting in iteratively *factorising* the maximal bicliques between the upper level and the other levels of a multipartite graph. We show that the most natural definition of this factorising process leads to infinite series for some instances. Our main result is to design a restriction of this process that terminates for any arbitrary graph. Moreover, we show that the resulting multipartite graph has remarkable combinatorial properties and is closely related to another fundamental combinatorial object. Finally, we show that, in practice, this multipartite graph is computationally tractable and has a size that makes it suitable for complex network modelling.

1 Introduction

It appeared recently [10,1,3] that most real-world complex networks (like the internet topology, data exchanges, web graphs, social networks, or biological networks) have some non-trivial properties in common. In particular, they have a very low density, low average distance and diameter, an heterogeneous degree distribution, and a high local density (usually captured by the clustering coefficient [10]). Models of complex networks aim at reproducing these properties. Random [1] graphs with given numbers of vertices and edges [4] fit the density and

---

[1] In all the paper, *random* means uniformly chosen in a given class.

W. Wu and O. Daescu (Eds.): COCOA 2010, Part I, LNCS 6508, pp. 1–10, 2010.
© Springer-Verlag Berlin Heidelberg 2010

distance properties, but they have homogeneous degree distributions and low local density. Random graphs with prescribed distributions [9] and the preferential attachment model [2] fit the same requirement, with the degree distribution in addition, but they still have a low local density. As these models are very simple, formally and computationnaly tractable, and rather intuitive, there is nowadays a wide consensus on using them.

However, when one wants to capture the high local density in addition to previous properties, there is no clear solution. In particular, we are unable to construct a random graph with prescribed degree distribution and local density. As a consequence, many proposals have been made, e.g. [10,3,7,6] , each with its own advantages and drawbacks. Among the most promising approaches, [6,7] propose to model complex networks based on the properties of their clique incidence bipartite graph (see definition below). They show that generating bipartite graphs with prescribed degree distributions for bottom and top vertices and interpreting them as clique incidence graphs results in graphs fitting all the complex network properties listed above, including heterogeneous degree distribution and high local density.

However, the bipartite model suffers from severe limitations. In particular, it does not capture overlap between cliques, which is prevalent in practice. Indeed, as evidenced in [6,8], the neighbourhoods of vertices in the clique incidence bipartite graph of a real-world complex network generally have significant intersections: cliques strongly overlap and vertices belong to many cliques in common. On the opposite, when one generates a random bipartite graph with prescribed degree distributions, the obtained bipartite graph have much smaller neighbourhood intersections, almost always limited to at most one vertex (under reasonable assumptions on the degree distributions). Indeed, the process of generation based on the bipartite graph is equivalent to randomly choosing sets of vertices of the graph (with prescribed size distribution) that we all link together. Because of the constraints imposed on this size distribution by the low density of the graph , the probability of choosing several vertices in common between two such random sets tends to zero when the graph grows. As a consequence, the bipartite model fails in capturing the overlapping nature of cliques in complex networks. This leads in particular to graphs which have many more edges than the original ones (two cliques of size d lead to $d.(d-1)$ edges in the model graph, while the overlap between cliques make this number much smaller in the original graph).

Our contribution. Since the random generation process of the bipartite graph is not able to generate non-trivial neighbourhood intersections (that is having cardinality at least two), a natural direction to try to solve this problem consists in using a structure explicitly encoding these intersections. This can be done using a tripartite graph instead of a bipartite one: one may encode any bipartite graph $B = (\perp, \top, E)$ into a tripartite one $T = (\perp, \top, C, E')$ where C is the set of non-trivial maximal bicliques (complete bipartite graphs having at least two bottom vertices and two top vertices) of B and E' is obtained from E by adding the edges between any biclique c in C and all the vertices of B which belong to c and removing the edges between vertices of C. This process, which we call *factorisation*, can be iterated to encode any graph in a multipartite one where there are hopefully no non-trivial neighbourhood intersections.

In this paper, we show that this iterated factorising process do not end for some graphs. We then introduce variations of this base process and study them with regard to termination issue. Our main result is the design of such a process, which we call *clean factorisation*, that terminates on any arbitrary graph. In addition, we show that the multipartite graph on which terminates this process has remarkable combinatorial properties and is strongly related to a fundamental combinatorial object. Namely, its vertices are in bijection with the chains of the inf-semilattice of intersections of maximal cliques of the graph. Finally, we give an upper bound on the size and computation time of the graph on which terminates the iterated clean factorising process of G, under reasonable hypothesis on the degree distributions of the clique incidence bipartite graph of G; therefore showing that this multipartite graph can be used in practice for complex network modelling.

Outline of the paper. We first give a few notations and basic definitions useful in the whole paper. We then consider the most immediate generalisation of the bipartite decomposition (Section 2) and show that it leads to infinite decompositions in some cases. We propose a more restricted version in Section 3, which seems to converge but for which the question remains open. Finally, we propose another restricted version in Section 4 for which we prove that the decomposition scheme always terminates.

Notations and preliminary definitions. All graphs considered here are finite, undirected and simple (no loops and no multiple edges). A graph G having vertex set V and edge set E will be denoted by $G = (V, E)$. We also denote by $V(G)$ the vertex set of G. The edge between vertices x and y will be indifferently denoted by xy or yx.

A k-partite graph G is a graph whose vertex set is partitioned into k parts, with edges between vertices of different parts only (a bipartite graph is a 2-partite graph, a tripartite graph a 3-partite graph, etc): $G = (V_0, \ldots, V_{k-1}, E)$ with $E \subseteq \{uv \mid u \in V_i, v \in V_j, i \neq j\}$. The vertices of V_i, for any i, are called the *i-th level* of G, and the vertices of V_{k-1} are called its *upper vertices*.

$\mathcal{K}(G)$ denotes the set of maximal cliques of a graph G, and $N^G(x)$ the neighbourhood of a vertex x in G. When $G = (V_0, \ldots, V_{k-1}, E)$ is k-partite, we denote by $N_i^G(x)$, where $0 \leq i \leq k - 1$, the set of neighbours of x at level i: $N_i^G(x) = N^G(x) \cap V_i$. When the graph referred to is clear from the context, we omit it in the exponent. A *biclique* of a graph is a set of vertices of the graph inducing a complete bipartite graph. We denote $B(G)$ the clique incidence graph of $G = (V, E)$, *i.e.* its bipartite decomposition: $B(G) = (V, \mathcal{K}(G), E')$ where $E' = \{vc \mid c \in \mathcal{K}(G), v \in c\}$.

In all the paper, an operation will play a key role, we name it *factorisation* and define it generically as follows.

Definition 1 (factorisation). *Given a k-partite graph $G = (V_0, \ldots, V_{k-1}, E)$ with $k \geq 2$ and a set V_k' of subsets of $V(G)$, we define the factorisation of G with respect to V_k' as the $(k+1)$-partite graph $G' = (V_0, \ldots, V_k, (E \setminus E_-) \cup E_+)$ where:*

- *V_k is the set of maximal (with respect to inclusion) elements of V_k',*
- *$E_- = \{yz \mid \exists X \in V_k, y \in X \cap V_{k-1} \text{ and } z \in X \setminus V_{k-1}\}$, and*
- *$E_+ = \{Xy \mid X \in V_k \text{ and } y \in X\}$.*

When $V_k \neq \emptyset$, the factorisation is said to be effective.

In the rest of the paper, we will refine the notion of factorisation by using different sets V_k' on which is based the factorisation operation, and we will study termination of the graph series resulting from each of these refinements.

The converse operation of the factorisation operation is called *projection*.

Definition 2 (projection). *Given a k-partite graph $G = (V_0, \ldots, V_{k-1}, E)$ with $k \geq 3$, we define the projection of G as the $(k-1)$-partite graph $G' = (V_0, \ldots, V_{k-2}, (E \cap (\bigcup_{1 \leq i \leq k-2} V_i)^2) \cup A_+)$ where $A_+ = \{yz \mid \exists i, j \in [\![1, k-2]\!], i \neq j \text{ and } y \in V_i \text{ and } z \in V_j \text{ and } \exists t \in V_{k-1}, yt, zt \in E\}$ is the set of edges between any pair of vertices of $\bigcup_{1 \leq i \leq k-2} V_i$ having a common neighbour in V_{k-1}.*

It is worth to note that the projection is the converse of the factorisation operation independently from the set V_k' used in the definition of the factorisation.

2   Weak Factor Series

As explained before, our goal is to improve the bipartite model of [6,7] in order to be able to encode non-trivial clique overlaps, that is overlaps whose cardinality is at least two. Since these overlaps in the graph result from the neighbourhood overlaps of the upper vertices, the purpose of the new model we propose is to encode the graph into a multipartite one by recursively eliminating all non-trivial neighbourhood overlaps of the upper vertices. We first describe this process informally, then give its formal definition and exhibit an example for which it does not terminate.

Neighbourhood overlaps of the upper vertices in a bipartite graph $B = (V_0, V_1, E)$ may be encoded as follows. For any maximal[2] biclique C that involves at least two upper vertices and two other vertices, we introduce a new vertex x in a new level V_2, add all edges between x and the elements of C, and delete all the edges of C, as depicted on Figure 1. We obtain this way a tripartite graph $T = (V_0, V_1, V_2, E')$ which encodes B (one may obtain B from T by the projection operation) and which has no non-trivial neighbourhood overlaps in its first two level (V_0 and V_1).

---

[2] The reason why one would take the maximal bicliques is simply to try to encode all neighbourhood overlaps using a reduced number of new vertices. Notice that there are other ways to reduce even more the number of new vertices created, for example by taking a biclique cover of the edge set of B. This is however out of the scope of this paper.

Fig. 1. Example of multipartite decomposition of a graph. From left to right: the original graph; its bipartite decomposition; its tripartite decomposition; and its quadripartite decomposition, in which there is no non-trivial neighbourhood overlap anymore. In this case, the decomposition process terminates.

This process, which we call a *factorising step*, may be repeated on the tripartite graph T obtained (as well as on any multipartite graph) by considering the bipartite graph between the upper vertices and the other vertices of the tripartite (or multipartite) graph, see Figure 1. All k-partite graphs obtained along this iterative factorising process have no non-trivial neighbourhood overlap between the vertices of their $k - 1$ first levels. Then, the key question is to know whether the process terminates or not.

We will now formally define the factorising process and show that it may result in an infinite sequence of graphs. In the following sections, we will restrict the definition of the factorising step in order to always obtain a finite representation of the graph.

Definition 3 ($V_k^\bullet$ and weak factor graph). *Given a k-partite graph $G = (V_0, \ldots, V_{k-1}, E)$ with $k \geq 2$, we define the set $V_k^\bullet$ as:*

$$V_k^\bullet = \{\{x_1, \ldots, x_l\} \cup \bigcap_{1 \leq i \leq l} N(x_i) \mid l \geq 2, \forall i \in [\![1, l]\!], x_i \in V_{k-1} \text{ and } |\bigcap_{1 \leq i \leq l} N(x_i)| \geq 2\}.$$

The weak factor graph $G^\bullet$ of G is the factorisation of G with respect to $V_k^\bullet$.

The weak factorisation admits a converse operation, called projection, which is defined in Section 1. It implies that the factor graph of G, as well as its iterated factorisations, is an encoding of G.

The *weak factor series* defined below is the series of graphs produced by recursively repeating the weak factorising step.

Definition 4 (weak factor series $\mathcal{WFS}(G)$). *The weak factor series of a graph G is the series of graphs $\mathcal{WFS}(G) = (G_i)_{i \geq 1}$ in which $G_1 = B(G)$ is the clique incidence graph of G and, for all $i \geq 1$, G_{i+1} is the weak factor graph of G_i: $G_{i+1} = G_i^\bullet$. If for some $i \geq 1$ the weak factor operation is not effective then we say that the series is* finite.

Figure 1 gives an illustration for this definition. In this case, the weak factor series is finite. However, this is not true in general; see Figure 2. Intuitively, this is due to the fact that a vertex may be the base for an infinite number of

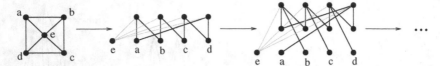

Fig. 2. An example graph for which the weak factorising process is infinite. From left to right: the original graph G, its bipartite decomposition $B(G)$, and its tripartite decomposition $B(G)^\bullet$. The shaded edges are the ones involving vertex e, which play a special role: all the vertices of the upper level of the decompositions are linked to e. The structure of the tripartite decomposition is very similar to the one of the bipartite decomposition, revealing that the process will not terminate.

factorising steps (like vertex e in the example of Figure 2). The aim of the next sections is to avoid this case by giving more restrictive definitions.

3   Factor Series

In the previous section, we have introduced weak factor series which appear to be the most immediate extension of bipartite decompositions of graphs. We showed that, unfortunately, weak factor series are not necessarily finite. In this section, we introduce a slightly more restricted definition that forbids the repeated use of a same vertex to produce infinitely many factorisations (as observed on the example of Figure 2). However, we have no proof that it necessarily gives finite series, which remains an open question.

Definition 5 (V_k° and factor graph). *Given a k-partite graph $G = (V_0, \ldots, V_{k-1}, E)$ with $k \geq 2$, we define the set V_k° as:*

$$V_k^\circ \ = \ \{X \in V_k^\bullet \text{ such that } | \bigcap_{y \in X \cap V_{k-1}} N_{k-2}(y)| \geq 2\}.$$

The factor graph *G° of G is the factorisation of G with respect to V_k°.*

This new definition results from the restriction of the weak factor definition by considering only sets $X \in V_k^\bullet$ such that the vertices of $X \cap V_{k-1}$ have at least two common neighbours at level $k - 2$. In this way, the creation of new vertices depends only on the edges between levels $k - 1$ and $k - 2$ (even though some other edges may be involved in the factorisation operation). Thus, a vertex will not be responsible for infinitely many creations of new vertices. This restriction also plays a key role in the convergence proof of the *clean factor series*, defined in next section. That is why we think it may be possible that it is sufficient to guarantee the convergence of the factor series, but we could not prove it with this sole hypothesis.

4   Clean Factor Series

In the two previous sections, we studied two multipartite decompositions of graphs. The first one is very natural but it does not lead to finite objects. The second one remains very general but we were unable to prove that it leads to finite object. As a first step towards this goal, we introduce here a more restricted definition for which we prove that the decomposition is finite. This new combinatorial object has many interesting features, and we consider it worth of study in itself. In particular, we prove that it is a decomposition of a well-known combinatorial object: the inf-semi-lattice of the intersections of maximal cliques of G. This correspondence allows to calculate quantities of graph G from elements of M. One of such results is an explicit formula (not presented here) giving the number of triangles in G, which is a very important parameter of complex networks.

The *clean factor graph* (defined below) is a proper restriction of the factor graph in which the vertices at level $k - 1$ used to create a new vertex at level k are required to have exactly the same neighbourhoods at all levels strictly below level $k - 2$, except at level 1. Intuitively, this requirement implies that the new factorisations push further the previous ones and are not simply a rewriting at a higher level of a factorisation previously done. The particular role of level 1 will allow us to differentiate vertices of the multipartite graph by assigning them sets of nodes at level 0. Let us now formally define the clean factor graph and its corresponding series.

Definition 6 (V_k^* and clean factor graph). *Given a k-partite graph $G = (V_0, \ldots, V_{k-1}, E)$ with $k \geq 4$, we define the set V_k^* as:*

$$V_k^* = \{X \in V_k^\circ \mid \forall x, y \in X \cap V_{k-1}, \forall p \in \{0\} \cup [\![2, k-3]\!], N_p(x) = N_p(y)\}.$$

The clean factor graph *G^* of G is the factorisation of G with respect to V_k^*.*

Definition 7 (clean factor series $\mathcal{CFS}(G)$). *The clean factor series of a graph G is the series of graphs $\mathcal{CFS}(G) = (G_i)_{i \geq 1}$ in which $G_1 = B(G)$ is the clique incidence graph of G, $G_2 = G_1^\circ$, $G_3 = G_2^\circ$ and, for all $i \geq 3$, G_{i+1} is the clean factor graph of G_i: $G_{i+1} = G_i^*$. If for some i the clean factor operation is not effective then we say that the series is* finite.

The rest of this section is devoted to proving the following theorem.

Theorem 1. *For any graph G, the clean factor series $(G_i)_{i \geq 1}$ is finite.*

Notation 1. *Let $(G_i)_{i \geq 1}$ be the clean factor series of G. For any $i \geq 1$, any $x \in V_i$ and any $j < i$, we denote by $V_j(x)$ the set $N_j^{G_i}(x)$ and by $V(x)$ the set $\bigcup_{0 \leq j < i} V_j(x)$.*

Remark 1. In the rest of the paper, when referring to Definition 6, it is worth keeping in mind that for $x \in V_{k-1}$ and $p \leq k - 3$, the sets $N_p(x)$ and $N_p(y)$ used in the definition are precisely the sets $V_p(x)$ and $V_p(y)$.

Definition 8. *We denote by $\mathcal{O}'$ the set $\{O \subseteq V(G) \mid \exists k \geq 2, \exists C_1, \ldots, C_k \in \mathcal{K}(G), (\forall j, l \in [\![1, k]\!], j \neq l \Rightarrow C_j \neq C_l) \text{ and } O = \bigcap_{1 \leq i \leq k} C_i\}$; and by $\mathcal{O}$ the set $\{O \in \mathcal{O}' \mid |O| \geq 2\}$. For any $O \in \mathcal{O}'$, we denote by $K(O)$ the set $\{C \in \mathcal{K}(G) \mid O \subseteq C\}$. We also denote by $\mathcal{C}$ the set $\{Y \subseteq \mathcal{K}(G) \mid \exists O \in \mathcal{O}', Y = K(O)\}$.*

It is clear from the definition that $\mathcal{O}'$ is closed under intersection, this is also the case for $\mathcal{C}$.

In all the G_i's of the clean factor series, vertices at level 0 correspond to vertices of G, vertices at level 1 correspond to the maximal cliques of G, that is for any $y \in V_1, V_0(y) \in \mathcal{K}(G)$. That is the reason why in the following we do not distinguish between the elements of $\mathcal{K}(G)$ and those of V_1. We will show that the vertices of V_2 correspond to the elements of $\mathcal{O}$. Indeed, $x \mapsto V_0(x)$ is a bijection from V_2 to $\mathcal{O}$. First, for any $x \in V_2$, by definition, $|V_0(x)| \geq 2$, then $V_0(x) = \bigcap_{y \in V_1(x)} V_0(y)$ belongs to $\mathcal{O}$. Let $O \in \mathcal{O}$. Let us show that $X = K(O) \cup \bigcap_{y \in K(O)} V_0(y)$ is a maximal element of V_2°. First note that $X \cap V_0 = O$ and then $|X \cap V_0| \geq 2$. Now, if you augment X with an element of $y \in V_1 \setminus K(O)$, since $y \notin K(O)$, $X \cap V_0$ will decrease. Thus X is maximal and there is a corresponding $x \in V_2$ such that $V_0(x) = O$. Furthermore, it is straightforward to see that the maximality of $V(x)$ implies that $V_1(x) = K(O)$. Which proves the uniqueness of the $x \in V_2$ such that $V_0(x) = O$.

Definition 9. *Let G be a graph and let $(G_i)_{i \geq 1}$ be its clean factor series. The characterising sequence $S(x) = (O_1(x), \ldots, O_{k-1}(x))$ of a vertex $x \in V_k$, with $k \geq 2$, is defined by:*

- $O_1(x) = V_0(x)$
- $\forall j \in [\![2, k-1]\!], O_j(x)$ *is the unique element[3] of $\mathcal{O}'$ such that $K(O_j(x)) = \bigcap_{y \in V_j(x)} V_1(y)$.*

Note that O_j is properly defined. Indeed, since $\mathcal{C}$ is closed under intersection, a simple recursion would show that for all $i \geq 3$ and for all $y \in V_i$, $V_1(y) = \bigcap_{z \in V_{i-1}} V_1(z) \in \mathcal{C}$.

Theorem 2 is our main combinatorial tool for proving the finiteness of the clean factor series (Theorem 1). Its proof is rather intricate, but it gives much more information than the finiteness of the series. By associating a sequence of sets to each vertex in levels greater than V_2 in the multipartite graph, we show that each such vertex corresponds to a chain of the inf-semi-lattice of the intersections of maximal cliques of G. The correspondence thereby highlighted between this very natural structure and the multipartite factorisation scheme we introduced is non-trivial and of great combinatorial interest.

Theorem 2. *Let G be a graph and $(G_i)_{i \geq 2}$ its clean factor series. We then have the following properties:*

1. *$\forall k \geq 2, \forall x \in V_k, O_1(x) \subsetneq \ldots \subsetneq O_{k-1}(x)$ and if $k = 3$, $O_2(x) \in \mathcal{O}$ and if $k \geq 4$, $(O_2(x), \ldots, O_{k-2}(x)) \in \mathcal{O}'^{k-3}$*
2. *$\forall k \geq 2, \forall x, y \in V_k, x \neq y \Rightarrow S(x) \neq S(y)$,*
3. *$\forall k \geq 2, \forall (O_1, \ldots, O_{k-1}) \in \mathcal{O}'^{k-1}, O_1 \subsetneq \ldots \subsetneq O_{k-1} \Rightarrow \exists x \in V_k, S(x) = (O_1, \ldots, O_{k-1})$.*

---

[3] By convention, $O_j(x) = V(G)$ when $\bigcap_{y \in V_j(x)} V_1(y) = \varnothing$.

For lack of space, we do not give the proof of Theorem 2. It can be made by recursion on k. The key of our proof is that we could characterise, for any $k \geq 3$, the vertices at level $k-1$ involved in the creation of a new vertex x at level k : roughly, they are those vertices y such that there exist $O_1, \ldots, O_{k-3}, O_m, O_M \in \mathcal{O}$ and $S_y = (O_1, \ldots, O_{k-3}, O_k - 2(y))$ is such that $O_m \subseteq O_{k-2}(y) \subseteq O_M$. Then, the characterising sequence of the created vertex x is $S(x) = (O_1, \ldots, O_{k-3}, O_m, O_M)$. Please refer to the webpages of the authors for a complete version of the paper including proof of Theorem 2.

Theorem 1 is a corollary of Theorem 2. Indeed, Theorem 2 states that the characterising sequence $(O_1(x), \ldots, O_{k-1}(x))$ of any node x at level k is such that $O_1(x) \subsetneq \ldots \subsetneq O_{k-1}(x)$. The strict inclusions imply that the length of the characterising sequence, which is equal to $k - 1$, cannot exceed the height h of the inclusion order of elements of $\mathcal{O}$. Since $h \leq n - 1$, necessarily V_{n+1} is empty. It follows that the clean factor series is finite and stops at rank at most n.

Size of the multipartite model. The size of the multipartite graph M obtained at termination of the clean factor series can be exponential in theory, as the number of maximal cliques itself may be exponential. But in practice, its size is quite reasonable and it can be computed efficiently. Theorem 3 below shows that under reasonable hypotheses, the size of M only linearly depends on the number of vertices of G, with a multiplicative constant reflecting the complexity of imbrication of maximal cliques.

Theorem 3. *If every vertex of G is involved in at most k maximal cliques and if every maximal clique of G contains at most c vertices, then $|V(M)| \leq 4 \times min(k\, 2^c\, c!\, , 2^k\, k!) \times n$.*

This upper bound can be obtained by bounding the number of sequences $O_1, \ldots, O_i$ in two different ways: either by considering sequences ending with a fixed set $O_i = A$, which are obtained by starting from set A and removing vertices one by one; or by considering sequences starting with a fixed set $O_1 = B$, which are obtained by starting from a maximal clique containing B and intersecting it by one more maximal clique containing B at each step.

In practice, parameters k and c are quite small, as they are often constrained by the context itself independently from the size of the graph. Then, the size of M is small. An important consequence is that, using algorithms enumerating the cliques or bi-cliques of a graph (see [5] for a recent survey), M can be computed efficiently, that is in low polynomial time, since the number of maximal cliques is small.

5   Perspectives

Many questions arise from our work. The first one is to find minimal restrictions of the factorising process that guarantee termination. On the other hand, for processes that do not always terminate, one may determine on which classes of graphs those processes terminate. Another question of interest is the termination speed, as well as the size of the obtained encoding: proving upper bounds with softer hypothesis would be desirable.

Finally, the use of multipartite decompositions as models of complex networks, in the spirit of the bipartite decomposition, asks for several questions. In this context, the key issue is to generate a random multipartite graph while preserving the properties of the original graph. To do so, one has to express the properties to preserve as functions of basic multipartite properties (like degrees, for instance) and to generate random multipartite graphs with these properties. This is a promising direction for complex network modelling, but much remains to be done.

Acknowledgements. We warmly thank Jean-Loup Guillaume, Stefanie Kosuch and Clémence Magnien for helpful discussions.

References

1. Albert, R., Barabási, A.-L.: Statistical mechanics of complex networks. Reviews of Modern Physics 74, 47 (2002)
2. Barabasi, A.-L., Albert, R.: Emergence of scaling in random networks. Science 286, 509–512 (1999)
3. Dorogovtsev, S.N., Mendes, J.F.F.: Evolution of networks. Advances in Physics 51 (2002)
4. Erdös, P., Rényi, A.: On random graphs I. Publications Mathematics Debrecen 6, 290–297 (1959)
5. Gély, A., Nourine, L., Sadi, B.: Enumeration aspects of maximal cliques and bicliques. Discrete Applied Mathematics 157(7), 1447–1459 (2009)
6. Guillaume, J.-L., Latapy, M.: Bipartite structure of *all* complex networks. Information Processing Letters (IPL) 90(5), 215–221 (2004)
7. Guillaume, J.-L., Latapy, M.: Bipartite graphs as models of complex networks. Physica A 371, 795–813 (2006)
8. Latapy, M., Magnien, C., Del Vecchio, N.: Basic notions for the analysis of large two-mode networks. Social Networks 30(1), 31–48 (2008)
9. Molloy, M., Reed, B.: A critical point for random graphs with a given degree sequence. In: Random Structures and Algorithms (1995)
10. Watts, D., Strogatz, S.: Collective dynamics of small-world networks. Nature 393, 440–442 (1998)

Simple Cuts Are Fast and Good:
Optimum Right-Angled Cuts in Solid Grids[*]

Andreas Emil Feldmann[1], Shantanu Das[2], and Peter Widmayer[1]

[1] Institute of Theoretical Computer Science, ETH Zürich, Switzerland
{feldmann,widmayer}@inf.ethz.ch
[2] Laboratoire d'Informatique Fondamentale, Aix-Marseille University, France
shantanu.das@lif.univ-mrs.fr

Abstract. We consider the problem of bisecting a graph, i.e. cutting it into two equally sized parts while minimising the number of cut edges. In its most general form the problem is known to be NP-hard. Several papers study the complexity of the problem when restricting the set of considered graphs. We attempt to study the effects of restricting the allowed cuts. We present an algorithm that bisects a solid grid, i.e. a connected subgraph of the infinite two-dimensional grid without holes, using only cuts that correspond to a straight line or a right angled corner. It was shown in [13] that an optimal bisection for solid grids with n vertices can be computed in $\mathcal{O}(n^5)$ time. Restricting the cuts in the proposed way we are able to improve the running time to $\mathcal{O}(n^4)$. We prove that these restricted cuts still yield good solutions to the original problem: The best restricted cut is a bicriteria approximation to an optimal bisection w.r.t. both the differences in the sizes of the partitions and the number of edges that are cut.

1 Introduction

The *graph partitioning* problem requires that the vertex set of a given graph is partitioned into a given number p of equal-sized subsets in such a way that the number of edges having end-points in distinct partitions is minimised. This problem has applications ranging from task allocation in parallel computers to automated design of electronic circuits. In our case the motivation stems from finite element simulations of human bone in order to diagnose osteoporosis [2]. In this setting the underlying graph structure is a three dimensional grid, which is why we chose two dimensional grids as a first step towards solving the problem for the former graphs. In general the problem seems to be extremely difficult to solve within a reasonable runtime. The *bisection* problem, which is the graph partitioning problem for $p = 2$, is already known to be NP-hard for general graphs, graphs of bounded degree, and d-regular graphs (see [5] for an overview).

---

[*] We gratefully acknowledge discussions with Peter Arbenz who introduced the human bone simulation problem to us, and the support of this work through the Swiss National Science Foundation under Grant No. 200021_125201/1, "Data Partitioning for Massively Parallel Computations in the Hypergraph Model".

W. Wu and O. Daescu (Eds.): COCOA 2010, Part I, LNCS 6508, pp. 11–20, 2010.
© Springer-Verlag Berlin Heidelberg 2010

For other graph classes polynomial-time bisection algorithms are known, such as for trees, solid grid graphs, or hypercubes [5]. For planar graphs and unit disk graphs the complexity of the bisection problem is unknown.

We feel, however, that not only the structure of the graph, but also the structure of the cuts decisively influences the complexity of the problem. Can simpler cuts allow for a more efficient computation while still returning valuable information? In this paper, we initiate the study of simple cuts, limiting ourselves to solid grid graphs, which are finite connected subgraphs of the infinite two-dimensional grid without holes. It was shown by Papadimitriou and Sideri [13] that an $\mathcal{O}(n^5)$ algorithm exists that computes the optimal bisection for solid grids. By restricting the cuts in solid grids to orthogonal lines with at most one (rectangular) bend, we improve the runtime to $\mathcal{O}(n^4)$. This runtime gain comes at a rather small loss in the quality of the partition: We get a bicriteria approximation to the optimal bisection in the sense that for any $\varepsilon > 0$ there is a $k \in [(1 - \varepsilon)n/2, (1 + \varepsilon)n/2]$ such that cutting k vertices from the grid using the restricted cuts is a $\mathcal{O}(1/\sqrt{\varepsilon})$-approximation to the bisection width. Since we use a dynamic programming approach we can find the approximate solution for any given ε.

Related Work. For the graph partitioning problem on arbitrary graphs, where the vertex set is to be cut into p sets, Räcke and Andreev [1] show that there can not exist a polynomial time approximation algorithm if one requires the sets to have equal size, unless P=NP. This implies that for the partitioning problem a bicriteria approximation is inevitable. Accordingly they present an algorithm that for any $\varepsilon > 0$ allows the size of each set in the partition to deviate from n/p by at most the factor $1 + \varepsilon$. If C^* denotes the number of edges cut by the optimal solution in which all sets have size at least $\lfloor n/p \rfloor$, they prove that the number of edges cut by the computed solution are at most $\mathcal{O}(\log^{1.5}(n)/\varepsilon^2) \cdot C^*$. In case the set sizes are allowed to deviate by a factor of 2 from n/p, Krauthgamer et al. [11] give a bicriteria approximation algorithm that computes a solution in which at most $\mathcal{O}(\sqrt{\log n \log p}) \cdot C^*$ many edges are cut.

The problem of bisecting a graph into two equally sized parts has received a lot of attention as it is regarded as a first step towards the more challenging problem of partitioning a graph into p sets. However, as mentioned before, the bisection problem was shown to be NP-hard in general [8]. Therefore there were many attempts at finding good approximate solutions or solving special cases in polynomial time. For instance Feige and Krauthgamer [7] give an algorithm with approximation ratio $\mathcal{O}(\log^{1.5} n)$ for general graphs. A different example of such an algorithm is given by Bui et al. [4] who present a method that finds the optimal bisection of almost all regular graphs whose bisection width is small. The authors note that these input instances are especially hard to solve using heuristics since in graphs with large bisection width any bisection is reasonably good.

Several special classes of graphs have been studied in this context and a listing of these can be found in the survey by Díaz et al. [5]. One class that is relevant to our present work are planar graphs. To the best of our knowledge it is still an open question whether bisecting planar graphs is NP-hard. Feige and

Krauthgamer [7] present an algorithm with a logarithmic approximation factor. Bui and Peck [3] give a fixed parameter algorithm for bisecting planar graphs. Their algorithm has a running time that is exponential in the number of edges cut by an optimal solution. Hence for planar graphs with logarithmically valued bisections there is a polynomial time algorithm. Díaz *et al.* [6] used the above work by Bui and Peck to develop a PTAS and a NCAS for planar graphs.

Another class of graphs that is relevant to our present work are trees. For these MacGregor [12] presents an algorithm that computes the optimal bisection in time $\mathcal{O}(n^3)$. Goldberg and Miller [9] present a parallel algorithm that runs on $\mathcal{O}(n^2)$ many processors and needs $\mathcal{O}(\log^2 n \log \log n)$ time to compute the bisection width of a tree. They also claim that with some modification the running time of the algorithm given by MacGregor can be reduced to $\mathcal{O}(n^2)$.

For the class of graphs studied in this paper, viz. solid grid graphs, the only known polynomial time algorithm is by Papadimitriou and Sideri [13]. In that paper the authors characterise the types of cuts that constitute an optimal bisection of a solid grid graph. In a preprocessing step they compute the relevant set of cuts in time $\mathcal{O}(n^4)$ and using a dynamic programming approach they then search through the computed set to find the optimal bisection in $\mathcal{O}(n^5)$ time. For grid graphs containing holes, the authors give a reduction from planar graphs with polynomial weights to grid graphs with holes. However, as mentioned before, the complexity for bisecting planar graphs is unknown.

The bisection of grids is related to the geometric problem of bisecting a polygon into parts of equal areas, which was shown to be NP-hard by Koutsoupias *et al.* [10]. They also give a $\mathcal{O}(n^5)$ algorithm for approximately bisecting a polygon of n sides, which is based on the algorithm for solid grid graphs given in [13].

Our Contributions. In an attempt to understand the effect of restricting the structure of the cuts on the computational complexity and the quality of the solution, we allow only subsets of the edges that form a straight line or a right-angled corner (see Figure 1 for an illustration). We present a dynamic programming algorithm that finds the optimal bisection restricted to this special class of cuts. Due to the restriction to these simple cuts only, we are able to approach the problem from a different angle than the authors of [13] did. Their approach was to see the solid grid as a sort of polygon. This is also reflected by the fact that their algorithm naturally carries over to polygons as shown in [10]. Our approach is to define a root in the grid and then use a bottom-up computation with respect to the root. Therefore some of our ideas are borrowed from the above mentioned work of MacGregor [12] on bisecting trees.

The advantage of restricting the types of cuts to simple cases is twofold. The first advantage is that the preprocessing time needed for this set of cuts is $\mathcal{O}(n^2)$ instead of $\mathcal{O}(n^4)$ when considering all relevant cuts. The second is that our algorithm needs time $\mathcal{O}(n^4)$ instead of $\mathcal{O}(n^5)$ as the algorithm to compute the optimal bisection with unrestricted cuts does. The downside however is that the result of the computation is an approximation to the optimal bisection with arbitrary cuts. Since we use a dynamic programming approach our algorithm does not only output a solution to the bisection problem but also for any other

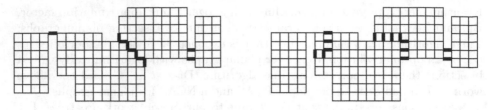

Fig. 1. An optimal k-cut (for $k = 110$) to the left and a k-cut using a restricted set of cut types to the right. The bold edges are those constituting the cut. Notice that the cut to the right has three more edges than the one to the left.

value $k \in \{0, ..., n\}$ such that one of the sets has size k and the other $n - k$. Hence for any $\varepsilon > 0$ our algorithm is able to output the best partition, i.e. the one with the smallest number of cut edges, of the vertices into two sets such that each has a size that deviates at most by the factor $1 + \varepsilon$ from $n/2$. At the same time, if C^* denotes the number of edges cut by an optimal solution in which both sets have size at least $\lfloor n/2 \rfloor$, we can guarantee that the number of edges cut by the partition computed by our algorithm is at most $\mathcal{O}(1/\sqrt{\varepsilon})$ times C^*.

We expect that the algorithm described in this paper can be generalised to other graph classes and cuts therein. Hence new insights into the structure of optimal cuts in different types of graphs can lead to more efficient algorithms to compute optimal or approximate bisections.

2   Restricting the Cut-Edges

Classically a cut of a graph G, with vertex set V and edge set E, is defined as a partition of V and its quality is measured by the number of cut-edges, i.e. those edges that have both endpoints in different sets of the partition (see e.g. [8]). We introduce a different definition: Given an arbitrary subset of edges $E' \subseteq E$ inducing a set of connected components G_1 to G_m when removed from G, any union $W = \bigcup_{i \in M} V_i$, where $M \subseteq \{1, .., m\}$, of the vertex sets of the respective components is said to be *cut-out* by E'. We will specify which vertices are cut-out by a set of edges when needed. A k-cut S is a set $S \subseteq 2^E$ of sets of edges for which the edges $E(S) := \{e \in s | s \in S\}$ cut-out a set of vertices W of cardinality k. The *cut-size* of a k-cut S is the sum $\sum_{s \in S} |s|$ of the sizes of the edge sets in S. An *optimal k-cut* is one that minimises the cut-size. Notice that this definition allows a k-cut to include an edge several times in different edge sets and that this edge will then also be counted several times for the cut-size. However in the most general form of the definition as stated above, any duplicate (i.e. redundant) edges may be removed from one of the edge sets to yield a different cut which cuts-out the same set of vertices and has a smaller cut-size. In particular this means that any optimal cut corresponding to the classical definition using a partition of V, can be transformed into an optimal k-cut S, for some k, corresponding to our definition by, for instance, simply including the set of cut-edges of the partition in the set S.

Definition 1. *Given a set $\widetilde{\Gamma} \subseteq 2^E$, a k-cut S is called $\widetilde{\Gamma}$-restricted if $S \subseteq \widetilde{\Gamma}$.*

We are interested in finding a set $\widetilde{\Gamma} \subseteq 2^E$ such that the optimal $\widetilde{\Gamma}$-restricted k-cut is a good approximation to the optimal k-cut. Notice that it may not be possible to cut-out k vertices if $\widetilde{\Gamma}$ is too restrictive. Hence we have to show that the set we use to restrict the k-cuts will always allow k vertices to be cut-out. Also notice that contrary to the non-restricted case it may happen that in an optimal $\widetilde{\Gamma}$-restricted k-cut S some edges are included in several edge sets in S.

Given a plane graph G, i.e. a planar graph that is embedded in the plane, we define its dual (multi-)graph $D = (F, E_D)$ in the usual way: The vertices in F are identified with the faces of G. Let f_∞ be the face corresponding to the exterior of G (i.e. the only face of infinite size). We define F_{in} to be the (possibly empty) set of *interior faces* (i.e. all faces except for f_∞). For every edge e in G there is an edge $e^* \in E_D$ between the faces that touch e. We say that the edges e^* and e *correspond* to one another. Accordingly we say that a subset of E and a subset of E_D correspond to one another if their respective edges correspond to one another.

Definition 2. *For any non-empty simple cycle $o \subseteq E_D$ in D we call the set $s \subseteq E$ of corresponding edges in G a segment. Let $\mathcal{E}_D \subseteq 2^{E_D}$ be the set of simple cycles in D and*

$$\mathcal{E} := \{s \subseteq E \mid \exists\, o \in \mathcal{E}_D : e \in s \Leftrightarrow e^* \in o\}$$

be the set of corresponding segments in G.

In any plane graph a partition of the vertices, i.e. a cut in the traditional sense, induces a set of cycles in the dual graph via the cut-edges, and vice versa. Therefore an optimal k-cut S in a plane graph, in which there are no redundant edges, is always a set of segments from the set $\mathcal{E}$. We will refer to any set of vertices cut-out by the corresponding edges $s \in \mathcal{E}$ of a cycle $o \in \mathcal{E}_D$ as being cut-out by o.

Any single cycle from $\mathcal{E}_D$ cuts-out two non-empty vertex sets that are true subsets of the whole vertex set V of the given graph, namely the vertices of the two components when the corresponding edges are removed. Let $V_i \subset V$ and $V_i' \subset V$, for $i \in \{1, 2\}$, be the respective non-empty vertex sets for two cycles $o, o' \in \mathcal{E}_D$. The segments $s, s' \in \mathcal{E}$ that correspond to the cycles o and o', respectively, are said to *cross* if $V_i \cap V_j' \neq \emptyset$ for all pairs $i, j \in \{1, 2\}$ (see Figure 2 for an illustration). Notice that o and o' share at least one interior face. We call these shared faces the *crossing points* of the segments s and s'. If o and o' share some interior faces but do not cross, we say that s and s' *touch*. In this case the shared interior faces are called *touching points*. Finally if the cycles o and o' touch and share at least one edge the segments s and s' are said to *overlap* and the shared edges of s and s' are called *overlapping edges*. Notice that the edges of a pair of crossing segments can always be seen as the edges of a pair of (different) touching segments, depending on which path is followed at the crossing points (making them touching points) to form the corresponding

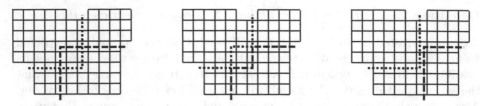

Fig. 2. From left to right: Two segments crossing, touching, and overlapping in a plane grid graph. The two segments are indicated by a dashed and a dotted line.

cycles. This means that for any k-cut S of a plane graph we can always find a k-cut of equal cut-size for which the segments do not cross. We will call such a k-cut *non-crossing*.

We call a finite connected subgraph G of the infinite two dimensional grid a *grid graph*. Since G is a plane graph it has a dual graph as defined above. A *hole* of G is an interior face which in the dual graph has a degree greater than 4. A grid graph is called *solid* if its set of faces does not contain any holes. We assume that the embedding of any given solid grid graph G has the property that the vertices are coordinates in $N \times N$ and an edge between two vertices v and w only exists if the Euclidean distance between v and w is 1.

It is fairly easy to see, and was also proven in [13], that for a solid grid graph any cycle in $\mathcal{E}_D$, corresponding to a segment from $\mathcal{E}$ of an optimal k-cut S, will always contain the exterior face. The reason is that otherwise the cycle can be "moved" in an arbitrary direction in the grid until its segment either overlaps with another segment or the cycle includes the exterior face. In both cases the resulting cut has a smaller cut-size since in the former case the overlapping edges can be removed from the cut, while in the latter case the boundary acts as a "natural barrier" so that some of the edges that were cut before are now no longer needed to cut-out the same number of vertices. Because of this we will restrict the used segments from $\mathcal{E}$ in a way such that the corresponding cycles in D all include the exterior face.

For any set of edges $q \in E$ (e.g. a segment) that corresponds to a path[1] $p \in E_D$ in the dual graph we define the length $l_q := |q|$. If $l_q \geq 2$ we say that q *ends* in a face f whenever there is exactly one edge in q that touches f and we say that q *embraces* a face f if there are more than one edge in q touching f. In the special case when $l_q = 1$ we say that q ends in a face its edge e touches if e touches two distinct faces, and we say that q embraces such a face if e touches only one face, i.e. if e is a bridge. If q ends in the two (distinct) faces f and f' we say that q *lies between* f and f'. We call an edge $\{v, w\} \in E$ in a grid graph G *horizontal* if v and w have the same y-coordinate. Accordingly we call it *vertical* if v and w have the same x-coordinate. A *bar* $b \subseteq E$ is a non-empty set of either only horizontal or only vertical edges that corresponds to a path p in D, such that if p is not a cycle then b does not embrace the exterior face. Notice that the latter means that if a bar b embraces f_∞ it is a segment. The *orientation* ω of

---

[1] In accordance with the usual terminology, any vertex on a path p has at most two incident edges in p.

b is *horizontal* if the edges in b are vertical, and accordingly it is *vertical* if the edges are horizontal.

Definition 3. *We call a segment* $s \in \mathcal{E}$ *a* straight *segment if it is a bar. A segment* $s \in \mathcal{E}$ *is a* corner *segment if there is an interior face* c, *a horizontal, and a vertical bar* b_h *and* b_v, *respectively, such that* $s = b_h \cup b_v$ *and both* b_h *and* b_v *lie between* c *and* f_∞. *We refer to* c *as the* corner *of* s *and we say that* b_h *and* b_v *are its respective horizontal and vertical bars. For the given solid grid graph* G *for which a* k-*cut is to be computed, we denote the set of corner and straight segments by* Γ.

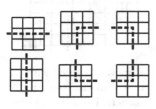

Fig. 3. A horizontal and a vertical straight segment, and a corner segment of orientation down-right, up-right, down-left, and up-left. The segments are indicated by dashed lines.

Notice that for a straight segment the orientation is well-defined since it is a bar. For a corner segment s we also define its *orientation* ω which can either be *down-right*, *down-left*, *up-right*, or *up-left*. Of these four types "down" and "up" refer to whether the vertical bar of s lies above or below its corner c, while "left" and "right" refer to whether its horizontal bar lies to the left or to the right of c. See Figure 3 for an illustration of all possible segments in the set Γ.

We now turn to the task of showing that for any solid grid graph it is possible to cut-out any number of vertices k from V by only using segments from Γ. Since we will make use of the following Lemma in a more general setting later on, we will prove a slightly more general statement.

Lemma 4. *For a solid grid graph* G *and any* $k \in \{0, ..., n\}$ *there is a non-crossing* Γ-*restricted* k-*cut* S *in which all corner segments have the same given orientation* ω.

If the set of segments was restricted so that only straight lines were allowed, it is easy to see that no guarantee as the one given in Lemma 4 could be given. On the other hand if in addition to straight and corner segments also segments with two or more corners were used, it is easy to observe that the number of segments to be considered would be asymptotically larger than the size of Γ. This would increase the running time of an algorithm based on the ideas presented in this paper to more than $\mathcal{O}(n^4)$. The use of Γ-restricted cuts is thus justified by the following theorem and the running time of our algorithm. The proof of the theorem is omitted since it involves techniques beyond the scope of the current paper.

Theorem 5. *Let* C^* *be the cut-size of an optimal* k-*cut in a solid grid* G *and* $\varepsilon \in]0, 1]$. *Then there exists a* Γ-*restricted non-crossing* k'-*cut, for some* $k' \in [(1 - \varepsilon)k, (1 + \varepsilon)k]$, *which has a cut-size of* $\mathcal{O}(1/\sqrt{\varepsilon}) \cdot C^*$.

Since the algorithm presented in this paper uses a dynamic programming approach, the output of the algorithm is a table containing the optimal non-crossing Γ-restricted k-cuts for all $k \in \{0, ..., n\}$. According to the above theorem we can hence look for the $k \in [(1 - \varepsilon)\frac{n}{2}, (1 + \varepsilon)\frac{n}{2}]$, for a desired $\varepsilon \in]0, 1]$, that has the smallest cut-size in the table and be sure that it is a $\mathcal{O}(1/\sqrt{\varepsilon})$ approximation of the optimal bisection. In the rest of this paper we will concentrate on how to compute an optimal non-crossing Γ-restricted k-cut.

3   Computing Optimal Non-crossing Γ-Restricted k-Cuts

To compute an optimal non-crossing Γ-restricted k-cut we use a dynamic programming approach. We fix a vertex r of the given solid grid graph G, which we will call the *root* of G, for which all other vertices $V \setminus \{r\}$ can be cut-out by a single segment $s_r \in \Gamma$. Notice that such a vertex always exists and r either has degree 1, in which case s_r is a straight segment, or degree 2, in which case it is a "convex corner" of the grid and s_r is a corner segment. (see Figure 4). With respect to r we can now define the non-empty set of vertices V_s of a segment $s \in \Gamma$ to be those vertices in G that are cut-out by s but do not include the root r, i.e. $V_s \subseteq V \setminus \{r\}$ and $V_s \neq \emptyset$. We will from now on only refer to V_s as the vertices cut-out by s. With respect to V_s we define $G_s = (V_s, E_s)$ to be the subgraph of G induced by the

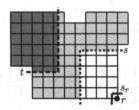

Fig. 4. A grid with a root r in the lower right corner. The set V_t is indicated by the dark grey area, and the set V_s is indicated by the dark and light grey areas together.

vertices in V_s and $n_s := |V_s|$ to be the number of vertices in G_s. Also we define $\Gamma_s := \{t \in \Gamma | V_t \subseteq V_s\}$ to be the set of segments that cut-out a subset of V_s. Notice that s is included in Γ_s and that no $t \in \Gamma_s$ will cross s. Also notice that $\Gamma_t \subseteq \Gamma_s$ for any $t \in \Gamma_s$, which also means that $\Gamma_t \subseteq \Gamma_s$ if and only if $V_t \subseteq V_s$. As a consequence the sets V_s and Γ_s can be used somewhat interchangeably and we will use the more convenient notion according to context.

For each segment $s \in \Gamma$ we can now define the vector $C_s \in \mathbb{N}^{n+1}$ for which the entry $C_s(k)$, for $k \in \{0, ..., n\}$, denotes the cut-size of the optimal non-crossing Γ_s-restricted k-cut. Lemma 4 guarantees that for any $k \leq n_s$ the entry $C_s(k)$ is well-defined since a non-crossing Γ-restricted k-cut in G_s, in which all corner segments have the same orientation as s, can easily be transformed to a Γ_s-restricted k-cut using the same method as in the proof of the Lemma. For any $k > n_s$ we set the entry $C_s(k)$ to infinity. To compute the vectors C_s we need the notion of *interference-free* subsets $S \subseteq \Gamma$ as defined below.

Definition 6. *A non-empty set of segments $S \subseteq \Gamma$ is called* interference-free *if $V_s \cap V_t = \emptyset$ for each pair of distinct segments $s, t \in S$. For such a set S we define $V_S = \bigcup_{s \in S} V_s$, $n_S = |V_S|$, $E_S = \bigcup_{s \in S} E_s$, and $\Gamma_S = \bigcup_{s \in S} \Gamma_s$.*

In addition to the vectors C_s we also define the vectors C_S for interference-free sets S so that $C_S(k)$ is the cut-size of the optimal non-crossing Γ_S-restricted k-cut. Again we set those entries $C_S(k)$ where $k > n_S$ to infinity, since Lemma 4 also guarantees that only for these values of k the entry of the vector is undefined. The following theorem shows how the vector C_S can be computed from the vectors C_s where $s \in S$. The definition of the vector C_S given below complies with the *min-convolution* of the vectors C_s as first defined by MacGregor [12,9].

Theorem 7. *For an interference-free set of segments $S \cup \{t\} \subseteq \Gamma$ the optimal non-crossing $\Gamma_{S \cup \{t\}}$-restricted k-cut can be defined recursively by*

$$C_{S \cup \{t\}}(k) = \begin{cases} C_t(k) & \text{if } S = \emptyset, \\ \min\{C_t(i) + C_S(k-i) \mid i \in \{\beta_1, ..., \beta_2\}\} & \text{if } k \leq n_t + n_S, \quad (1) \\ \infty & \text{else.} \end{cases}$$

where $\beta_1 := \max\{0, k - n_S\}$ and $\beta_2 := \min\{k, n_t\}$.

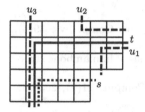

Fig. 5. The set of segments $S = \{u_1, u_2, u_3\}$ is not maximally interference-free since u_2 and u_3 can be exchanged with t such that $S^* = \{t, u_1\}$ is interference-free. The set S^* is maximally interference-free.

In order to compute the vectors C_s we will make use of the vectors C_{S^*} for special interference-free sets $S^* \subset \Gamma_s$. Namely we use those interference-free sets S^* for which no other interference-free set can be found that includes S^*, in the sense given in the definition below. See Figure 5 for an example.

Definition 8. *We call a non-empty set of segments $S^* \subseteq \Gamma_s \setminus \{s\}$ maximally interference-free with respect to $s \in \Gamma$ if it is interference-free and there does not exist a segment $t \in \Gamma_s \setminus \{s\}$ and a (possibly empty) subset $S' \subseteq S^*$ such that $V_u \subset V_t$ for all $u \in S'$ and the set $(S^* \setminus S') \cup \{t\}$ is interference-free. The set of maximally interference-free sets for a segment $s \in \Gamma$ is denoted by $\mathcal{I}_s$.*

The following theorem shows that we can make use of the maximally interference-free sets in order to compute the optimal Γ_s-restricted cuts. The corresponding algorithm will be based on a dynamic programming approach using the recursive equations given below and in Theorem 7.

Theorem 9. *Let $s \in \Gamma$ be a segment of length l_s and $k \in \{0, ..., n_s\}$. If $\mathcal{I}_s$ is non-empty, the optimal non-crossing Γ_s-restricted k-cut is*

$$C_s(k) = \min\{C_{S^*}(k), \ l_s + C_{S^*}(n_s - k) \mid S^* \in \mathcal{I}_s\}. \quad (2)$$

If $\mathcal{I}_s$ is empty then

$$C_s(k) = \begin{cases} 0 & \text{if } k = 0, \\ l_s & \text{if } k = 1, \quad (3) \\ \infty & \text{else.} \end{cases}$$

To show that an algorithm based on the above theorem is efficient we need the following statement.

Theorem 10. *For any $s \in \Gamma$ there are $\mathcal{O}(n)$ many members in $\mathcal{I}_s$. Also all sets of maximally interference-free sets can be constructed in time $\mathcal{O}(n^2)$ in a preprocessing step.*

Using this fact we can show that an algorithm based on Theorem 9 is efficient. Using an amortisation argument on the sizes n_t of the vertex sets cut-out by the segments $t \in S^*$ for a set $S^* \in \mathcal{I}_s$ in Theorem 7, we get a runtime of $\mathcal{O}(n^4)$.

Theorem 11. *An optimal non-crossing Γ-restricted k-cut of a solid grid graph G can be computed in time $\mathcal{O}(n^4)$ for any $k \in \{0, ..., n\}$.*

References

1. Andreev, K., Räcke, H.: Balanced graph partitioning. Theor. Comp. Sys. 39(6), 929–939 (2006)
2. Arbenz, P., Müller, R.: Microstructural finite element analysis of human bone structures. ERCIM News 74, 31–32 (2008)
3. Bui, T., Peck, A.: Partitioning planar graphs. SIAM J. Comput. 21(2), 203–215 (1992)
4. Bui, T.N., Chaudhuri, S., Leighton, F.T., Sipser, M.: Graph bisection algorithms with good average case behavior. Combinatorica 7(2), 171–191 (1987)
5. Díaz, J., Petit, J., Serna, M.J.: A survey of graph layout problems. ACM Comput. Surv. 34(3), 313–356 (2002)
6. Díaz, J., Serna, M.J., Torán, J.: Parallel approximation schemes for problems on planar graphs. Acta Informatica 33(4), 387–408 (1996)
7. Feige, U., Krauthgamer, R.: A polylogarithmic approximation of the minimum bisection. SIAM J. Comput. 31(4), 1090–1118 (2002)
8. Garey, M.R., Johnson, D.S., Stockmeyer, L.: Some simplified NP-complete graph problems. Theoretical Computer Science 1(3), 237–267 (1976)
9. Goldberg, M., Miller, Z.: A parallel algorithm for bisection width in trees. Computers & Mathematics with Applications 15(4), 259–266 (1988)
10. Koutsoupias, E., Papadimitriou, C.H., Sideri, M.: On the optimal bisection of a polygon. ORSA Journal on Computing 4(4), 435–438 (1992)
11. Krauthgamer, R., Naor, J., Schwartz, R.: Partitioning graphs into balanced components. In: Proceedings of the twentieth Annual ACM-SIAM Symposium on Discrete Algorithms (SODA), pp. 942–949. Society for Industrial and Applied Mathematics (2009)
12. MacGregor, R.M.: On partitioning a graph: a theoretical and empirical study. PhD thesis, University of California, Berkeley (1978)
13. Papadimitriou, C.H., Sideri, M.: The bisection width of grid graphs. In: Proc. of the First Annual ACM-SIAM Symposium on Discrete Algorithms (SODA), pp. 405–410 (1990)

Evacuation of Rectilinear Polygons*

Sándor Fekete, Chris Gray, and Alexander Kröller

Department of Computer Science, TU Braunschweig, Germany
{fekete,gray,kroeller}@ibr.cs.tu-bs.de

Abstract. We investigate the problem of creating fast evacuation plans for buildings that are modeled as grid polygons, possibly containing exponentially many cells. We study this problem in two contexts: the "confluent" context in which the routes to exits remain fixed over time, and the "non-confluent" context in which routes may change. Confluent evacuation plans are simpler to carry out, as they allocate contiguous regions to exits; non-confluent allocation can possibly create faster evacuation plans. We give results on the hardness of creating the evacuation plans and strongly polynomial algorithms for finding confluent evacuation plans when the building has two exits. We also give a pseudo-polynomial time algorithm for non-confluent evacuation plans. Finally, we show that the worst-case bound between confluent and non-confluent plans is $2 - O(\frac{1}{k})$.

1 Introduction

A proper evacuation plan is an important requirement for the health and safety of all people inside a building. When we optimize evacuation plans, our goal is to allow people to exit the building as quickly as possible. In the best case, each of a building's exits would serve an equal number of the building's inhabitants. However, there might be cases in which this can not happen. For instance, when there is a bottleneck between two exits, it might make sense for most of the building's inhabitants to stay on the side of the bottleneck closer to where they begin, even if this means that one exit is used more than the other.

In this research, we study the computation of evacuation plans for buildings that are modeled as grid polygons. We make the assumption that every grid square is occupied by exactly one person and that at most one person can occupy a grid square at any given time. This assumption arises from the inability of building designers to know exactly where people will be in the building in the moments before an evacuation, and this pessimistic view of the situation is the only sensible one to take. Also remember that in some cases, such as in airplanes, the situation in which nearly every bit of floor space is occupied before an evacuation is more common than the alternative.

Evacuation plans can be divided into two distinct types. In the first, signs are posted that direct every person passing them to a specific exit. In the second,

---

* This research was funded by the German Ministry for Education and Research (BMBF) under grant number 03NAPI4 "ADVEST", which fully funded Chris Gray.

W. Wu and O. Daescu (Eds.): COCOA 2010, Part I, LNCS 6508, pp. 21–30, 2010.

every person is assigned to a distinct exit that does not necessarily depend on the exits to which his or her neighbors are assigned. The first type of evacuation plan generates what is known as a *confluent flow* and the second generates a *non-confluent flow*. More precise definitions of these terms will be given later.

It is clear that in a non-confluent flow, people can be evacuated as quickly as in a confluent flow, and we show that in certain instances people can be evacuated significantly more quickly in a non-confluent flow than in a confluent flow. However, a confluent flow is significantly easier to carry out than a non-confluent one, so we also give results related to it. We first show that the problem of finding an optimal confluent flow belongs to the class of NP-complete problems if the polygon has "holes"—that is, if it represents a building with completely enclosed rooms or other spaces. We then give an algorithm with a running time linear in the description complexity of the region (which can be exponentially smaller than the number of cells) that computes an evacuation plan for buildings without holes that have two exits; a generalization to a constant number of exits is more complicated, but seems plausible. Finally, we show that the worst-case ratio between the evacuation times for confluent flows and non-confluent flows for k exits is $2 - O(\frac{1}{k})$.

1.1 Preliminaries

We are given a rectilinear polygon P on a grid. There exists, on the boundary of P, a number of special grid squares known as *exits*. We call the set of exits $\mathcal{E} = \{e_1, \ldots, e_k\}$. We assume that every grid square in P contains a person. A person can move vertically or horizontally into an empty grid square or an exit. The goal is to get each person to an exit as quickly as possible. When an exit borders more than one grid square, we specify the squares from which people can move into the exit.

The area of P is denoted by A and the number of vertices of P is denoted by n. Note that A can be exponential in n. The set of people that leave P through the exit e is called the *e-exit class*. We also write *exit class* to refer to the set of people who leave through an unspecified exit. The grid squares that are adjacent to the boundary of P are known as *boundary squares*.

There are two versions of the problem that we consider. We call these *confluent flows* and *non-confluent flows*. In the first, we add the restriction that every grid square has a unique successor. Thus, for every grid square s, people passing through s leave s in only one direction. This restriction implies that evacuation plans are determined by space only. It does not exist for non-confluent flows. It can be argued that informing people which exit to use is easier in the case of confluent flows since a sign can be placed in every grid square, informing the people who pass through it which exit to use. However, we show that non-confluent flows can lead to significantly faster evacuations.

The major difficulty in the problem comes from bottlenecks. We define a *k-bottleneck* to be a rectangular subpolygon B of P such that two parallel boundary edges of B are the same as two edges of P, where the distance between the common edges is k.

1.2   Related Work

The problem when restricted to confluent flows is similar in many ways to the unweighted Bounded Connected Partition (or 1-BCP) problem [12]. In this problem, one is given an unweighted, undirected graph $G = (V, E)$ and k distinguished vertices. The goal is to find k connected subsets of V, where each subset contains exactly one of the distinguished vertices and where each has the same cardinality. If we define the dual of the grid contained in our polygon to be the graph formed by connecting adjacent faces and let the distinguished vertices be the exits, then 1-BCP is clearly the same as evacuation restricted to confluent flows with k exits.

It was independently shown by Lovász [8] and Győri [7] that a solution to the 1-BCP problem can be found for every graph that is k-connected. However, their proofs are not algorithmic. Thus, there has been some work on finding partitions of size k for low values of k. For example, Suzuki *et al.* give an algorithm for 2-partitioning 2-connected graphs [14]. One algorithm claiming a general solution for k-partitioning k-connected graphs [10] is incorrect. Unfortunately, when P contains 1-bottlenecks, the graph obtained by finding the dual of the grid inside P is 1-connected. Therefore, the results of Lovász and Győri do not apply. Also, since the dual of the grid contained in our polygon can have size exponential in the complexity of the polygon, we would probably need to merge nodes and then assign weights to the nodes of the newly-constructed graph. The addition of weights, however, makes the BCP problem NP-complete, even in the restricted case in which the graph is a grid [3].

Another connection is to the problem of partitioning polygons into subpolygons that all have equal area. The confluent version of our problem, indeed, can be seen as the "discrete" version of that problem. The continuous version has been studied. One interesting result from this study is that finding such a decomposition while minimizing the lengths of the segments that do the partitioning is NP-hard even when the polygons are orthogonal [1]. However, polynomial algorithms exist for the continuous case when that restriction is removed [9].

Baumann and Skutella [2] consider evacuation problems modeled as earliest-arrival flows with multiple sources. They achieved a strongly-polynomial-time algorithm by showing that the function representing the number of people evacuated by a given time is submodular. Such a function can be optimized using the parametric search technique of Megiddo. Their approach is different from ours in that they are given an explicit representation of the flow network as input. We are not given this, and computing the flow network that is implicit in our input can take exponential time. Also, their algorithm takes polynomial time in the sum of the input and output sizes. However, the complexity of the output can be exponential in the input size.

Another, perhaps more surprising, related problem is machine scheduling. Viewed simply, confluent flows correspond to non-preemptive scheduling problems, while non-confluent flows correspond to preemptive scheduling problems. The NP-hardness result for optimal confluent flows is inspired by the hardness of scheduling jobs on non-preemptive machines [5], while the worst-case ratio

Fig. 1. The polygon P given a PARTITION instance of $\{11, 6, 9\}$. To keep the picture a manageable size, the elements have not been scaled and the left ends of the first and fifth rows are truncated.

between confluent and non-confluent flows is inspired by the list-scheduling approximation ratio [6].

2 Confluent Flows

As mentioned in the introduction, in a confluent flow, every grid square has the property that all people that pass through it use the same exit.

In this section, we present our results related to confluent flows. First, we show the NP-completeness of the problem of finding an optimal evacuation plan with confluent flows in a polygon with holes. This holds even for polygons with two exits. We then give a linear-time algorithm for polygons with two exits.

2.1 Hardness

Weak NP-hardness with two exits. We first sketch that the evacuation problem with confluent flows is NP-hard if we allow P to have holes. We reduce from the well-known NP-complete problem PARTITION; the idea is shown in Fig. 1.

Theorem 1. *The problem of finding an optimal confluent flow in a polygon with holes is NP-complete.*

Strong NP-hardness. Theorem 1 shows that the problem is *weakly NP-complete*. This means that the hardness of the problem depends on the areas of subpolygons being exponential in the complexity of the input. This implies that a pseudo-polynomial algorithm might exist.

However, we can show that if we allow $O(n)$ exits, the problem is *strongly NP-complete*. This means that the problem is still NP-complete when all of its numerical parameters are polynomially bounded in the size of the input.

Our reduction is from CUBIC PLANAR MONOTONE 1-IN-3 SATISFIABILITY (or CPM 1-IN-3 SAT for short). This is a variant of the SATISFIABILITY problem in which every clause contains exactly 3 literals, every variable is in exactly 3 clauses, the graph generated by the connections of variables to clauses is planar, every literal in every clause is non-negated, and a clause is satisfied if exactly one of its variables is true.

We use a reduction that is almost equivalent to the one showing that tiling a finite subset of the plane with right trominoes is NP-complete [11]. Details are contained in the full paper [4].

Theorem 2. *The problem of finding an optimal confluent flow in a polygon with holes and $O(n)$ exits is strongly NP-complete.*

2.2   Two Exits

When the polygon P has no holes and only two exits, e_1 and e_2, we can find an optimal confluent flow in $O(n)$ time. We first present an algorithm that takes cubic time that can be modified fairly easily into an algorithm that takes linear time.

Naïve algorithm. Notice that the case in which P has two exits is simpler than the case in which P has more exits because of the fact that both exit classes must each have one contiguous connection to the boundary of P.

We begin with a decomposition of P into rectangles. This decomposition is the overlay of two simpler decompositions: the vertical and horizontal decompositions. The vertical decomposition of a rectilinear polygon P is the partition of P into rectangles by the addition of only vertical line segments. Similarly, the horizontal decomposition of P is the partition of P into rectangles by the addition of only horizontal line segments. We call the overlay ω and its dual graph ω^*. We add the vertical and horizontal line segments from the grid points on opposite sides of e_1 and e_2 to ω as well. We then do the following for every pair of rectangles r_1 and r_2 in ω that have at least one edge of the boundary of P. We first ensure that e_1 is between r_1 and r_2 and that e_2 is between r_2 and r_1. We also ensure that there are no bottlenecks of size 1 between r_1 and r_2. If either of these conditions are not met, we proceed to the next pair of rectangles. We then set the e_1-exit class to be all the grid squares along the boundary of P between r_1 and r_2. We then add all the grid squares surrounded by the e_1-exit class to the e_1-exit class. We call the area of the e_1-exit class A_1. We define the e_2-exit class similarly and call its area A_2. We call the larger of the two areas A_ℓ and its corresponding exit e_ℓ. If A_ℓ is less than $A/2$, we can divide the rest of the grid squares evenly among the exit classes—see Lemma 2 for details—and return the solution. Otherwise, we attempt to make the e_ℓ-exit class as small as possible (while staying above $A/2$) inside r_1 and r_2 and assign the rest of the grid squares to the other exit class. We maintain a variable that tracks the area of the smallest such exit class e_{min}. If we get through all the possible pairs r_1 and r_2 without finding a pair that we can return, we return e_{min}. The proof of the following summarizing lemma can be found in the full paper.

Lemma 1. *The algorithm presented above is correct and takes $O(n^3)$ time.*

Linear algorithm. The algorithm above has two steps that lead to it taking cubic time: the loop over all pairs of rectangles on the boundary of P and the computation of the minimum area for an exit class that has a connection to the boundary that begins in one of the rectangles and ends in the other. In this section, we sketch a more clever solution that avoids these problems; again, full details are described in the full paper.

We begin by observing that if we update the area, each time we change the starting and ending points of the connection of the e_1-exit class to the boundary of the polygon rather than computing it anew, the total time spent computing the area depends on the sum of the complexities of the updated areas.

We also observe that we loop over the rectangles of ω with at least one edge attached to the boundary of P. This means that we do not really need to compute the entire overlay ω—only the intersections of ω with the boundary of P. These intersections can be computed in $O(n)$ time by computing the vertical and horizontal decompositions of P separately. See Fig. 2 for the idea.

Theorem 3. *In the confluent setting, the above algorithm finds the optimal evacuation plan for a polygon P with two exits in $O(n)$ time, where n is the number of vertices in P.*

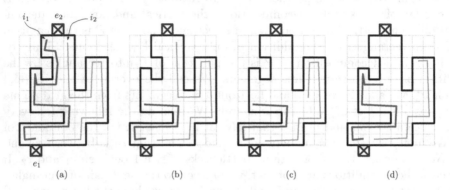

Fig. 2. An example of the linear-time algorithm. The pointers i_1 and i_2 denote the endpoints of the connection of the e_1-exit class to the boundary of the polygon.

If the overall number of grid pixels is exponential in the number of vertices of the polygon, particular care is necessary to ensure a small output complexity. Using a refined output encoding (which is described in the full paper), it is possible to note the following.

Lemma 2. *The above algorithm runs in $O(n)$ time and has output size that is linear in the input size.*

We conjecture that one can use algorithms similar to the naïve algorithm given above to compute the evacuation of any polygon with a constant number of exits, but the details become much more involved. We therefore leave this question to future work.

3   Non-confluent Flows

Compared with confluent flows, non-confluent flows are clearly a stronger model. We note that any confluent flow is a non-confluent flow, but not *vice versa*. We show that non-confluent flows can be as much as $2 - 2/(k + 1)$ times as fast as confluent flows by giving an example in which this is the case. We then argue that the ratio our example achieves is optimal.

3.1 Pseudo-polynomial Algorithm

In contrast to the case with confluent flows, for which we showed that finding an assignment of people to exits is strongly NP-complete when we are dealing with polygons with holes and $O(n)$ exits, we can show that, for non-confluent flows, a pseudo-polynomial algorithm exists.

The algorithm is based on the technique of using time-expanded networks to compute flows over time [13]. Therefore, we compute a flow network from the input polygon as follows. We create a source vertex s and a sink vertex t. For each grid square in P, we create two vertices—an *in* vertex and an *out* vertex. We connect the in vertex to the out vertex with an edge that has capacity 1 for every grid square. We then make, for some integer $T \geq 1$, T copies of the polygon $P_1, \ldots, P_T$, where each copy has these vertices and edges added. For every grid square of P_1, we connect s to the in vertex of the grid square with an edge that has capacity 1. We then connect the out vertex of every grid square in P_i to the in vertex of all its neighbors in P_{i+1} for all $1 \leq i \leq T - 1$. Again, the edges we use all have capacity 1. Finally, we connect the out vertex of every exit to t with an edge that has capacity 1. We call this flow network G.

It is fairly easy to see that if we are able to find a maximum flow of value A through G, then we are able to evacuate P in T time steps. However, we note that both T and $|G|$ can be exponential in the complexity of P, making this a pseudo-polynomial algorithm.

Theorem 4. *There exists a pseudo-polynomial algorithm to find an evacuation of a polygon with a non-confluent flow.*

3.2 Differences to Confluent Flows

The example that shows a large gap between confluent and non-confluent flows is a horizontal rectangle of width 1 with length $2k + mk$, for some integer $m \geq 1$. Attached to this rectangle are k vertical rectangles of width 1 and length mk—one at every other square for the first $2k$ squares. Between each vertical rectangle is an exit. Each exit can only be entered from the square to the left. See Figure 3 (a).

We can see that the example has an optimal confluent flow that requires $2mk + 3$ time steps: three to remove the people directly to the left of each exit and all the people below the exits, mk to remove the people in the vertical rectangles, and another mk for the people in the horizontal "tail" to go through the rightmost exit. On the other hand, in the optimal non-confluent flow, all exits can remain continuously busy. One way that this can happen is for m people from the horizontal rectangle to leave through successive exits, while people from the vertical rectangles are leaving through the other exits. Since the exits are continuously busy, the amount of time for all people to leave is $(k^2m + (2 + m)k)/k = mk + 2 + m$. The ratio between the confluent and non-confluent flows in this case is

$$\frac{2mk + 3}{mk + m + 2} = 2 - \frac{2m + 1}{mk + m + 2} \underset{m \to \infty}{\longrightarrow} 2 - \frac{2}{k + 1}.$$

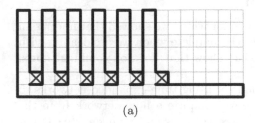
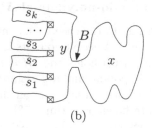

(a)   (b)

Fig. 3. (a) A polygon whose optimal non-confluent flow is nearly twice as fast as its optimal confluent flow. (b) The general shape of any polygon that realizes the maximum ratio between the confluent and non-confluent flows.

We now show that the ratio achieved in this example is tight. Our ratio is similar to (and inspired by) the upper bound for the list-scheduling approximation ratio [6] in machine scheduling.

Theorem 5. *The maximum ratio between the confluent flow and non-confluent flow in any grid polygon P is $2 - (2/(k+1))$.*

Proof. Let the ratio between the confluent flow and the non-confluent flow for a given polygon P be known as R_P. When calculating R_P, we assume that the confluent and non-confluent flows are calculated optimally for P.

We begin by observing that by reducing the size of the smallest bottleneck in a polygon P can only increase R_P. Suppose we have polygons P_a and P_b, where P_a has a minimum bottleneck size of at least 2, and P_b is the same as P_a, except that one grid square has been removed from the minimum bottleneck. The number of exit classes on one side of the smallest bottleneck in the confluent case can only decrease in P_b relative to P_a, while in the non-confluent case, they may stay the same. Thus, there may be an exit in the non-confluent solution that is used for longer than in the confluent solution. This implies that the number of steps that is required to evacuate the building in the confluent setting increases faster than the number of steps required under the non-confluent setting. Therefore, R_{P_a} is at least as large as R_{P_b}, and may be larger.

This implies that R_P is maximized when the size of the minimum bottleneck is minimized, so we can assume that the size of the minimum bottleneck is 1. We call the subpolygons on either side of the bottleneck P_1 and P_2.

Furthermore, we can easily see that increasing the difference between the number of exits in P_1 and P_2 can only increase R_P. This is because the number of people that must go through the bottleneck that separates P_1 and P_2 can only be increased by increasing this difference. Therefore, we can assume that all the exits are on one side of the bottleneck between P_1 and P_2. Without loss of generality, assume that all exits are in P_1.

Given this setup, we attempt to construct P so that as many exits as possible are used during as many time steps as possible in the non-confluent case. This implies that there must be some source of people in P_1. This is because only one person can go through the bottleneck between P_1 and P_2 per time step. Therefore, at most one person from P_2 can reach an exit per time step. However, by creating a supply of people in P_1, we allow the people from P_2 to queue in front of the exits.

So that the people from P_2 can queue in front of the exits, the route taken by the people in P_1 from the supply to the exit must not interfere with the paths of the people from P_2 to the exits. This means that the number of people in P_1 must be split and distributed to each exit.

Therefore, P has the form sketched in Figure 3(b). There is a bottleneck B. The number of people behind B is x, the amount of space for these people to queue in is y, and the supply of people for each exit e_i is s_i.

In both the confluent and non-confluent solutions, it takes time $2y/k$ to remove the y people in the queueing area. In the non-confluent solution, it is necessary that this is the first step performed. After this is done, the people from behind the bottleneck begin entering the queueing area. The people must therefore take turns exiting from the queueing area and exiting from the supplies that are attached to each exit. This implies that y is as small as possible (while satisfying $y \geq 2k$) and that $s_i \geq x$ for all $1 \leq i \leq k$. Using different values of s_i yields no advantage, so we assume that the value of s_i is some value s_x for all i.

The ratio between the confluent flow and non-confluent flow is thus

$$\frac{2y/k + 2s_x + 2x}{2y/k + 2s_x + 2x/k}$$

which is maximized according to our constraints when $s_x = x$ and when $y = 2k$. This gives a ratio of

$$\frac{2x + 2k/k}{x + x/k + 2k/k} = \frac{2x + 2}{x + x/k + 2} = 2 - \frac{2x + 2k}{xk + x + 2k} \xrightarrow[x \to \infty]{} 2 - \frac{2}{k + 1}$$

which is the claimed result. $\square$

4   Conclusions

We have discussed evacuations in grid polygons. We first showed that finding evacuations with confluent flows in polygons with holes is hard, even for polygons with only two exits. We then looked at algorithms to find evacuations with confluent flows. Finally, we showed that, while the difference between confluent and non-confluent flows is potentially significant, it is bounded.

Our work raises some questions that require further study. For simple polygons, there is evidence that a constant number of exits allows strongly polynomial solutions, even though some of the technical details are complicated. What is the complexity of finding an evacuation plan with a confluent flow when the number of exits is not constant? Next, can we find an fixed-parameter-tractable algorithm to find the confluent evacuation of polygons? Finally, can we find a polynomial algorithm that gives the optimal evacuation using non-confluent flows? Note that it is not even clear that the output size of such an algorithm is always polynomial.

Acknowledgments

We thank Estie Arkin, Michael Bender, Joe Mitchell, and Martin Skutella for helpful discussions; Martin Skutella is also part of ADVEST.

References

1. Bast, H., Hert, S.: The area partitioning problem. In: Proc. 12th Can. Conf. Comput. Geom. (CCCG 2000), Fredericton, NB, Canada, pp. 163–171 (2000)
2. Baumann, N., Skutella, M.: Earliest arrival flows with multiple sources. Math. Oper. Res. 34(2), 499–512 (2009); Journal version of 2006 FOCS article Solving evacuation problems efficiently
3. Becker, R., Lari, I., Lucertini, M., Simeone, B.: Max-min partitioning of grid graphs into connected components. Networks 32(2), 115–125 (1998)
4. Fekete, S.P., Gray, C., Kröller, A.: Evacuation of rectilinear polygons. Technical report (2010), http://arxiv.org/abs/1008.4420
5. Garey, M.R., Johnson, D.S.: Computers and Intractability: A Guide to the Theory of NP-Completeness. W. H. Freeman, New York (1979)
6. Graham, R.L.: Bounds on multiprocessing timing anomalies. SIAM Journal on Applied Mathematics 17, 416–429 (1969)
7. Györi, E.: On division of graphs to connected subgraphs. Combinatorics, Keszthely, 485–494 (1978)
8. Lovász, L.: A homology theory for spanning trees of a graph. Acta Mathematica Hungarica 30(3), 241–251 (1977)
9. Lumelsky, V.: Polygon area decomposition for multiple-robot workspace division. Int. Journal of Computational Geometry and Applications 8(4), 437–466 (1998)
10. Ma, J., Ma, S.: An $O(k^2 n^2)$ algorithm to find a k-partition in a k-connected graph. Journal of Computer Science and Technology 9(1), 86–91 (1994)
11. Moore, C., Robson, J.: Hard tiling problems with simple tiles. Discrete and Computational Geometry 26(4), 573–590 (2001)
12. Salgado, L.R., Wakabayashi, Y.: Approximation results on balanced connected partitions of graphs. Electr. Notes in Discrete Mathematics 18, 207–212 (2004)
13. Skutella, M.: An introduction to network flows over time. In: Research Trends in Combinatorial Optimization, pp. 451–482. Springer, Heidelberg (2009)
14. Suzuki, H., Takahashi, N., Nishizeki, T.: A linear algorithm for bipartition of biconnected graphs. Information Processing Letters 33(5), 227–231 (1990)

A Fast Algorithm for Powerful Alliances in Trees

Ararat Harutyunyan*

Department of Mathematics
Simon Fraser University
Burnaby, BC, V5A 1S6
Canada
aha43@sfu.ca

Abstract. Given a graph $G = (V, E)$ with a positive weight function w on the vertices of G, a global powerful alliance of G is a subset S of V such that for every vertex v at least half of the total weight in the closed neighborhood of v is contributed by the vertices of S. Finding the smallest such set in general graphs is NP-complete, even when the weights are all the same. In this paper, we give a linear time algorithm that finds the smallest global powerful alliance of any weighted tree $T = (V, E)$.

Keywords: Alliances, powerful alliances, weighted trees, algorithm.

1 Introduction

The study of alliances in graphs was first introduced by Hedetniemi, Hedetniemi and Kristiansen [9]. They introduced the concepts of defensive and offensive alliances, global offensive and global defensive alliances and studied alliance numbers of a class of graphs such as cycles, wheels, grids and complete graphs. The concept of alliances is similar to that of *unfriendly partitions*, where the problem is to partition $V(G)$ into classes such that each vertex has at least as many neighbors outside its class than its own (see for example [1] and [10]). Haynes et al. [7] studied the global defensive alliance numbers of different classes of graphs. They gave lower bounds for general graphs, bipartite graphs and trees, and upper bounds for general graphs and trees. Rodriquez-Velazquez and Sigarreta [15] studied the defensive alliance number and the global defensive alliance number of line graphs. A characterization of trees with equal domination and global strong defensive alliance numbers was given by Haynes, Hedetniemi and Henning [8]. Some bounds for the alliance numbers in trees are given in [6]. Rodriguez-Velazquez and Sigarreta [12] gave bounds for the defensive, offensive, global defensive, global offensive alliance numbers in terms of the algebraic connectivity, the spectral radius, and the Laplacian spectral radius of a graph. They also gave bounds on the global offensive alliance number of cubic graphs in [13] and the global offensive alliance number for general graphs in [14] and [11]. The concept of powerful alliances was introduced recently in [3].

* Research supported by FQRNT (Le Fonds québécois de la recherche sur la nature et les technologies) doctoral scholarship.

Given a simple graph $G = (V, E)$ and a vertex $v \in V$, the *open neighborhood* of v, $N(v)$, is defined as $N(v) = \{u : (u, v) \in E\}$. The *closed neighborhood* of v, denoted by $N[v]$, is $N[v] = N(v) \cup \{v\}$. Given a set $X \subset V$, the *boundary* of X, denoted by $\delta(X)$, is the set of vertices in $V - X$ that are adjacent to at least one member of X.

Definition 1. *A set $S \subset V$ is a* defensive alliance *if for every $v \in S$, $|N[v] \cap S| \geq |N[v] \cap (V - S)|$. For a weighted graph G, where each vertex v has a non-negative weight $w(v)$, a set $S \subset V$ is called a* weighted defensive alliance *if for every $v \in S$, $\sum_{u \in N[v] \cap S} w(u) \geq \sum_{u \in N[v] \cap (V-S)} w(u)$. A (weighted) defensive alliance S is called a* global (weighted) defensive alliance *if S is also a dominating set.*

Definition 2. *A set $S \subset V$ is an* offensive alliance *if for every $v \in \delta(S)$, $|N[v] \cap S| \geq |N[v] \cap (V - S)|$. For a weighted graph G, where each vertex v has a non-negative weight $w(v)$, a set $S \subset V$ is called a* weighted offensive alliance *if for every $v \in \delta(S)$, $\sum_{u \in N[v] \cap S} w(u) \geq \sum_{u \in N[v] \cap (V-S)} w(u)$. A (weighted) offensive alliance S is called a* global (weighted) offensive alliance *if S is also a dominating set.*

Definition 3. *A* global (weighted) powerful alliance *is a set $S \subset V$ such that S is both a global (weighted) offensive alliance and a global (weighted) defensive alliance.*

Definition 4. *The* global powerful alliance number *of G is the cardinality of a minimum size global (weighted) powerful alliance in G, and is denoted by $\gamma_p(G)$. A minimum size global powerful alliance is called a $\gamma_p(G)$-set.*

There are many applications of alliances. One is military defence. In a network, alliances can be used to protect important nodes. An alliance is also a model of suppliers and clients, where each supplier needs to have as many reserves as clients to be able to support them. More examples can be found in [9].

Balakrishnan et al. [2] studied the complexity of global alliances. They showed that the decision problems for global defensive and global offensive alliances are both NP-complete for general graphs. It is clear that the decision problems to find global defensive and global offensive alliances in weighted graphs are also NP-complete for general graphs.

The problem of finding global defensive, global offensive and global powerful alliances is only solved for trees. [5] gives a $O(|V|^3)$ dynamic programming algorithm that finds global defensive, global offensive and global powerful alliances of any weighted tree $T = (V, E)$.

In this short paper we give a $O(|V|)$ time algorithm that finds the global powerful alliance number of any weighted tree $T = (V, E)$. Combining our result with the obvious lower bound, we actually show that the problem of finding global powerful alliance number of any weighted tree $T = (V, E)$ is $\Theta(|V|)$.

In the next section, we present a linear time algorithm for minimum cardinality powerful alliance in weighed trees.

2 Powerful Alliances

In this section, we give a linear time algorithm that finds the minimum cardinality weighted global powerful alliance number of any tree. We assume all the weights are positive - the algorithm can be easily modified for the case where the weights are non-negative.

For a set $S \subset V$ define $w(S) := \sum_{u \in S} w(u)$.

It is clear that when the weight function is positive, the global powerful alliance problem can be formulated as follows.

Observation 1. *Let $G = (V, E)$ be a graph, and $w : V \to R^+\backslash\{0\}$ a weight function. Then a global powerful alliance in G is a set $S \subset V$ such that for all $v \in V$,*

$$\sum_{u \in N[v] \cap S} w(u) \geq \frac{w(N[v])}{2}.$$

Note that the condition that S is a dominating set is automatically guaranteed because the weights of all vertices are positive. We will use the above formulation in our algorithm.

Definition 5. *For $v \in V$, the* alliance condition *for v is the condition that $\sum_{u \in N[v] \cap S} w(u) \geq \frac{w(N[v])}{2}$.*

2.1 Satisfying the Alliance Condition for a Vertex

Throughout the algorithm, for every vertex v we need to find the smallest number of vertices necessary in the closed neighborhood of v to satisfy the alliance condition for v. To solve this problem, we use the following algorithm.

Given positive numbers $a_1, a_2, ..., a_n$, $FindMinSubset(a_1, a_2..., a_n)$ is the problem of finding the minimum k such that there is a k-subset of $\{a_1, a_2, ..., a_n\}$ the elements of which sum up to at least $\frac{1}{2}\sum_{i=1}^{n} a_i$. We give an algorithm that solves this problem in time $O(n)$. The algorithm also finds an instance of such a k-subset. Furthermore, out of all the optimal solutions it outputs a set with a maximum sum.

Algorithm FindMinSubset[A,n,T].
Input: An array A of size n of positive integers; a target value T.
Output: The least integer k such that some k elements of A add up to at least T.

1. If $n = 1$, return 1.
2. Set $i = \lceil \frac{n}{2} \rceil$.
3. Find the set A' of the i largest elements of A and compute their sum M.
4. If $M > T$ return $FindMinSubset[A', i, T]$; if $M = T$ return i; else return $i + FindMinSubset[A - A', n - i, T - M]$.

Lemma 1. *The above algorithm solves the problem $FindMinSubset(a_1, a_2, ...,$ $a_n)$ in time $O(n)$. Furthermore, if the solution is k, the algorithm finds the k largest elements.*

Proof. Let $S = \sum_{i=1}^{n} a_i$. It is clear that $FindMinSubset[A, n, T = \frac{S}{2}]$ where A is the array of the elements $(a_1, a_2, ..., a_n)$ will solve the desired problem. The analysis of the running time is as follows. Note that finding the largest k elements in an array of size n can be done in linear time for any k (see [4]). Therefore, in each iteration of $FindMinSubset$, Step 2 and Step 3 take linear time, say Cn. Since the input size is always going down by a factor of 2, we have that the running time $T(n)$ satisfies $T(n) \leq T(n/2) + Cn$. By induction, it is easily seen that $T(n) = O(n)$. It is clear that out of all the possible solutions, the algorithm picks the one with the largest weight.

We now describe the algorithm for weighted powerful alliances in trees.

2.2   An Overview of the Algorithm

We assume the tree is rooted. For each k, we order the vertices of depth k from left to right. By $C(v)$ we denote the set of children of v. We define $p(v)$ to be the parent of v.

In each iteration of the algorithm, we may label some vertices with "+", with "?p", or with "?c". The "+" vertices are going to be part of the powerful alliance. Under some conditions, we may also pick some of the vertices labelled with "?p" or with "?c".

Now, we give a brief intuition behind the algorithm. As noted above, when all the weights are positive, the global powerful alliance problem is simply finding the smallest set $S \subset V$ such that for every vertex $v \in V$, $w(N[v] \cap S) \geq \frac{w(N[v])}{2}$. Our algorithm is essentially a greedy algorithm. We root the tree at a vertex, and start exploring the neighborhoods of vertices starting from the bottom level of the tree. The vertices which have already been chosen to be included in the alliance set S are labelled with "+". For each vertex v, we find the smallest number of vertices in its closed neighborhood that need to be added to the vertices labelled "+" in v's closed neighborhood to satisfy the alliance condition for v. We do this using the algorithm FindMinSubset. In some cases we may get more than one optimal solution and it will matter which solution we pick (for example, if there is an optimal solution containing both v and $p(v)$ then this solution is preferable when we consider the neighborhood of $p(v)$). The complication arises when there is an optimal solution containing v, but not $p(v)$, and there is an optimal solution containing $p(v)$. If $w(v) > w(p(v))$, then it is not clear which solution is to be preferred because v is at least as good as $p(v)$ for satisfying the alliance condition for $p(v)$, but choosing $p(v)$ is preferable for satisfying the alliance condition for parent of $p(v)$, $p(p(v))$. In general, when we have to choose between a solution that contains v and one that contains $p(v)$ we give preference to the solution containing $p(v)$ unless: (i) we can immediately gain by choosing the solution with v due to choices made in previous iterations (ii)when satisfying the alliance condition for $p(v)$, it might theoretically be better to have chosen v. In case (ii), we label both v and $p(v)$ with "?", v with "?c", and $p(v)$ with "?p", and delay satisfying v's alliance condition for later iterations.

2.3   Labels and Sets

In the algorithm, we use four labels for vertices: "+", "?c", "?p" and "?c+". We assign a vertex v a label "+" when we can claim that v is contained in some minimum cardinality powerful alliance. Generally, we assign a vertex v a label "?c" when we have a choice of taking v or $p(v)$ to satisfy the alliance condition for v (but we can't choose both v and $p(v)$) and $w(v) > w(p(v))$. In this case, we also label $p(v)$ with "?p". For a vertex u, we define $D(u)$ to be the set of all "?c" children of u. Note that for every vertex u, eventually we must take u or $D(u)$ in our alliance, regardless whether these vertices are the best in the sense that they help satisfy the alliance condition for u. We change the label of a vertex v from "?c" to "?c+" if for satisfying the alliance condition for $p(v)$ it is better to choose v than $p(v)$. The reason we label it "?c+" and not "+" is that it may turn out that $p(p(v))$, the parent of $p(v)$, will be labelled "+", and and this may allow labelling $p(v)$ with "+" and unlabelling of v.

There are two occasions when we label a vertex v with "?p". The first is when a child of v is labelled "?c", as described above. The second case is when we see no optimal solution containing $p(v)$ that would satisfy v's alliance condition, but if $p(v)$ were later labelled "+" for another reason, then there would be an optimal solution containing v that would satisfy v's alliance condition. This solution would be preferable since it can decrease the number of vertices required to satisfy $p(v)$'s alliance condition.

We denote by $N^+[v]$ to be the set of all vertices in the closed neighborhood of v which are labelled with "+" at the current stage in the algorithm. For a vertex $v \in V$, we define $Findmin(v)$ to be the function that finds the smallest set of vertices in $N[v] \backslash N^+[v]$(i.e. the set of vertices not labelled with "+") that need to be added to the set of vertices in already labelled "+" in v's neighborhood, $N^+[v]$, to satisfy the alliance condition for v. If there is more than one such set, $Findmin(v)$ returns the set with maximum total weight. In the algorithm, by a *solution* for v we mean either the set $Findmin(v)$ or a set S such that $|S| = |Findmin(v)|$ and $S \cup N^+[v]$ satisfies the alliance condition for v. Note that it could be that $Findmin(v) = \emptyset$ since the alliance condition could already have been satisfied for v. Also, note that finding $Findmin(v)$ is done using the algorithm $FindMinSubset$ defined previously.

For a set of vertices X none of which are labelled "+", define $Findmin(v) : X$ to be the set $Findmin(v)$ under the assumption that the vertices of X are now labelled "+". We also have a special set X_v for every vertex v. This is the set of all children u of v labelled with "?p" with the property that $|Findmin(u) : \{u,v\}| < |Findmin(u)|$. This means that when trying to satisfy the alliance condition for v, if we see a solution that contains v and u, then we can safely label u and v with "+", regardless whether this solution contained $p(v)$ since we will gain a vertex when solve the alliance condition for u. Therefore, the X_v children of v can be used under some circumstances to satisfy the alliance condition for v. We have also two recursive functions, $Xcollect(v)$ and $Clear(v)$, that are used in the algorithm. $Xcollect(v)$ chooses all the X vertices in the subtree rooted at v, and finally settles the alliance condition for these vertices. It is used when we know that we can label v with a "+". $Clear(v)$ erases the labels of all the non X vertices that have label "?p" in the subtree rooted at v, and settles the alliance condition for them. Once we reach the root $r(T)$ of the tree, we no longer have the problem of deciding whether to give the parent of $r(T)$ a priority over $r(T)$ or its X_r children, since the parent does not exist. We can then settle the alliance conditions of all the "?" labelled vertices.

The functions $Clear(v)$ and $Xcollect(v)$ are as follows:

$Clear(v)$
For every child u of v labelled "?p" AND $u \notin X_v$,
      Remove the label "?p" (and "?c" if it exists) of u
      Replace all the labels of "?c+" from its children by "+".
      $Clear(u)$.
      $Findmin(u)$ and label the chosen vertices by "+".

$Xcollect(v)$
For every X_v child u of v
      Label u with "+".
      Remove "?p" label from u
      Remove all "?c+" labels from children of u.
      $Xcollect(u)$.
      $Findmin(u)$, and label chosen vertices with "+".

If vertex r is the root, we assume that $p(r) = \emptyset$ and $w(p(r)) = 0$. Also, we will say that $p(r)$ has label "+" so that we are in Case 1 or in Case 2.1.

2.4   The Algorithm

The algorithm has two cases: the vertex v under consideration is labelled "+", or "?p" or unlabelled (it can only receive the label "?c" during the iteration). Each case has two subcases depending on the label of $p(v)$. We often switch between the cases. For example, if the vertex v under consideration was unlabelled (Case 2) and gets a label "+", we jump to Case 1 and continue from there.

All X_v's are initially set to $X_v := \emptyset$.

Algorithm Weighted Powerful Alliances in Trees

for$(i = 0$ to $d)$ AND **for all** (vertices v at depth $d - i$) **do**

Algorithm 1. Algorithm for finding Minimum Cardinality Weighted Powerful Alliance in a tree T. **Case 1: v is labelled with "+".**

1: **Case 1.1: v has no "?p" children.**
2: **if** $p(v)$ is labelled "+" **then**
3:      $Findmin(v)$ and take any solution. Label all the picked vertices with "+".
4: **else if** $[p(v)$ is unlabelled or labelled "?p"$]$ **then**
5:      $Findmin(v)$
6:      **if** $\exists$ solution containing $p(v)$ **then**
7:          choose this solution. Label all picked vertices with "+".
8:          **if** v had label "?p" **then**
9:              remove this label, and remove the labels "?c" from all its children.
10:     **else**
11:         $Findmin(v)$
12:         Label picked vertices with "+"
        {Case1.2: v has at ≥ 1 "?p" child.}
13: Xcollect(v)
14: Clear(v)
15: Go back to Case 1.1

Algorithm 2. Case 2: v is labelled "?p" or unlabelled.

1: **Case 2.1: $p(v)$ is labelled "+".**
2: Let $D(v)$ be the set of children of v that have a label "?c".
3: $X_v := X_v \cup \{u \in C(v) : label(u) = \text{"?p"}, |Findmin(u) : \{u,v\}| < |Findmin(u)|\}$
4: Define $X_v' = X_v \cup v \cup p(v)$
5: $Findmin(v)$.
6: **if** $\exists$ solution $\ni v$ OR $\nexists$ solution $\supset D(v)$ OR $|Findmin(v) : \{X_v \cup v\}| < |Findmin(v)|$ **then**
7:      Label v with "+".
8:      Remove "?c" label's from v's children.
9:      Go to Case 1.
10: **else**
11:     Choose a solution $S \supset D(v)$. Label vertices of S with "+"
12:     Remove "?c" labels from vertices in $D(v)$, remove "?p" label from v
13:     Set $X_u := \emptyset$ for every vertex u in the subtree rooted at v
14:     Clear(v).

Algorithm 3. Case 2: v is labelled "?p" OR is unlabelled.

1: **Case 2.2:** $p(v)$ is labelled "?p" OR is unlabelled.
2: Let $D(v)$ be the set of children of v that have a label "?c".
3: $Findmin(v)$
4: $X_v := X_v \cup \{u \in C(v) : label(u) = $ "?p"$, |Findmin(u) : \{u, v\}| < |Findmin(u)|\}$
5: $X'_v = X_v \cup \{v\} \cup \{p(v)\}$
6: **if** $|Findmin(v) : \{X_v \cup v\}| \leq |Findmin(v)| - 2$ OR $\exists$ solution S such that $|S \cap X'_v| \geq 2$ OR $\exists$ solution S such that $S \supset \{v, p(v)\}$. **then**
7:     Label v with "+", remove its "?p" label. Remove "?c" labels from v's children. Go to Case 1.
8: **else if** $\exists$ a solution S such that $S \supset \{D(v) \cup p(v)\}$ **then**
9:     Let R be a solution $\ni p(v)$ with maximum total weight.
10:    **if** $X_v \cup \{v\} \cup R\backslash u \geq w(N[v])/2$ for some $u \in R$, $u \neq p(v)$ **then**
11:        Label v with "+", remove its "?p" label.
12:        Remove "?c" labels from v's children.
13:        Go to Case 1.
14:    **else if** $X_v \cup \{v\} \cup R\backslash p(v) < w(N[v])/2$ OR $w(v) \leq w(p(v))$ **then**
15:        Label $p(v)$ with "+".
16:        **if** $p(v)$ had label "?p" **then**
17:            remove it, and remove "?c" labels from $p(v)$'s children.
18:        Go to Case 2.1
19:    **else**
20:        Label $p(v)$ "?p" if it is not already. Add a label "?c" to v. Relabel "?c" vertices in $D(v)$ by "?c+".
21:        **if** $|Findmin(v) : X'_v| < |Findmin(v) : \{p(v)\}|$ **then**
22:            $X_{p(v)} := X_{p(v)} \cup \{v\}$. Label v with "?p", if v is unlabelled.
23:        Clear(v)
24: **else if** $\exists$ a solution S such that $S \supset D(v)$ **then**
25:    **if** $|Findmin(v) : \{X_v \cup v\}| < |Findmin(v)|$ **then**
26:        Label v with "+", remove its "?p" label.
27:        Remove "?c" labels from v's children.
28:        Go to Case 1.
29:    **else**
30:        Relabel "?c" labels of vertices in $D(v)$ by "?c+".
31:        **if** $|Findmin(v) : X'_v| < |Findmin(v) : \{p(v)\}|$ **then**
32:            $X_{p(v)} := X_{p(v)} \cup \{v\}$. Label v with "?p", if v is unlabelled
33:        **else**
34:            Label vertices of S with "+".
35:            Remove "?c+" labels from vertices in $D(v)$
36:        Clear(v).
37: **else**
38:    Label v with "+". Remove "?p" label from v. Remove "?c" labels from vertices in $D(v)$. Go to Case 1.

An illustration of the algorithm by an example (Fig. 1).

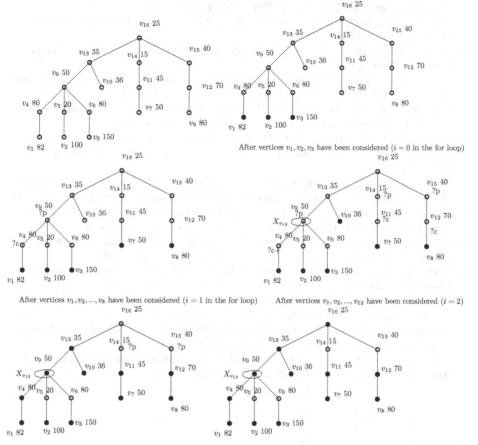

After vertices v_1, v_2, v_3 have been considered ($i = 0$ in the for loop)

After vertices $v_1, v_2, ..., v_8$ have been considered ($i = 1$ in the for loop)

After vertices $v_1, v_2, ..., v_{12}$ have been considered ($i = 2$)

After vertices $v_1, v_2, ..., v_{15}$ have been considered ($i = 3$)

After vertices $v_1, v_2, ..., v_{16}$ have been considered ($i = 4$)

Theorem 1. *Algorithm Powerful Alliances for Trees correctly computes the minimum cardinality weighted powerful alliance of a weighted tree $T = (V, E)$ in time $O(|V|)$.*

Due to space restrictions, we omit the proof.

Acknowledgements. I wish to thank Dr. Jacques Verstraete for helpful discussions, and Leonid Chindelevitch for indicating Lemma 1.

References

1. Aharoni, R., Milner, E.C., Prikry, K.: Unfriendly partitions of a graph. J. Combin. Theory Ser. B 50(1), 1–10 (1990)
2. Balakrishnan, H., Cami, A., Deo, N., Dutton, R.D.: On the complexity of finding optimal global alliances. J. Combinatorial Mathematics and Combinatorial Computing 58, 23–31 (2006)

3. Brigham, R.C., Dutton, R.D., Haynes, T.W., Hedetniemi, S.T.: Powerful alliances in graphs. Discrete Mathematics 309(8), 2140–2147 (2009)
4. Cormen, T.H., Leiserson, C.E., Rivest, R.L., Stein, C.: Introduction to Algorithms. McGraw-Hill, New York (2002)
5. Dean, B.C., Jamieson, L.: Weighted Alliances in Graphs. Congressus Numerantium 187, 76–82 (2007)
6. Harutyunyan, A.: Some bounds in alliances in trees. In: Cologne Twente Workshop on Graphs and Combinatorial Optimization (2010) (accepted)
7. Haynes, T.W., Hedetniemi, S.T., Henning, M.A.: Global defensive allliances in graphs. Electronic Journal of Combinatorics 10(1), R47 (2003)
8. Haynes, T.W., Hedetniemi, S.T., Henning, M.A.: A characterization of trees with equal domination and global strong alliance numbers. Utilitas Mathematica 66, 105–119 (2004)
9. Hedetniemi, S.M., Hedetniemi, S.T., Kristiansen, P.: Alliances in graphs. Journal of Combinatorial Mathematics and Combinatorial Computing 48, 157–177 (2004)
10. Milner, E.C., Shelah, S.: Graphs with no unfriendly partitions. In: A tribute to Paul Erdos, pp. 373–384. Cambridge Univ. Press, Cambridge (1990)
11. Rodriguez-Velazquez, J.A., Sigarreta, J.M.: On the global offensive alliance number of a graph. Discrete Applied Mathematics 157(2), 219–226 (2009)
12. Rodriguez-Velazquez, J.A., Sigarreta, J.M.: Spectal study of alliances in graphs. Discussiones Mathematicae Graph Theory 27(1), 143–157 (2007)
13. Rodriquez-Velazquez, J.A., Sigarreta, J.M.: Offensive alliances in cubic graphs. International Mathematical Forum 1(36), 1773–1782 (2006)
14. Rodriguez-Velazquez, J.A., Sigarreta, J.M.: Global Offensive Alliances in Graphs. Electronic Notes in Discrete Mathematics 25, 157–164 (2006)
15. Rodriguez-Velazquez, J.A., Sigarreta, J.M.: On defensive alliances and line graphs. Applied Mathematics Letters 19(12), 1345–1350 (2006)

NP-Completeness of Spreading Colored Points

Ovidiu Daescu[1], Wenqi Ju[2,3], and Jun Luo[3,*]

[1] Department of Computer Science, University of Texas at Dallas, USA
[2] Shenzhen Institutes of Advanced Technology, Chinese Academy of Sciences, China
[3] Institute of Computing Technology, Chinese Academy of Sciences, China
daescu@utdallas.edu, {wq.ju,jun.luo}@sub.siat.ac.cn

Abstract. There are n points in the plane and each point is painted by one of m colors where $m \leq n$. We want to select m different color points such that (1) the total edge length of resulting minimal spanning tree is as small as possible; or (2) the total edge length of resulting minimal spanning tree is as large as possible; or (3) the perimeter of the convex hull of m different color points is as small as possible. We prove NP-completeness for those three problems and give approximations algorithms for the third problem.

1 Introduction

Data from the real world are often imprecise. Measurement error, sampling error, network latency [5, 6], location privacy protection [1, 2] may lead to imprecise data. Imprecise data can be modeled by a continuous range such as square, line segment and circle [7, 8]. Imprecise data can also be modeled by discrete range such as point set. In discrete model, each point set is assigned one distinct color. Then the problem is converted to choosing one point from each colored point set such that the resulting geometric structure is optimal. The discrete model has applications in many areas such as Voronoi diagram [3], community system [9] and color-spanning object [3].

Related Work. Löffler and van Kreveld [7] address the problem of finding the convex hull of maximum/minimum area or maximum/minimum perimeter based on line segment or squares. The running time varies from $O(n \log n)$ to $O(n^{13})$.

Ju and Luo [8] propose an $O(n \log n)$ algorithm for the convex hull of the maximum area based on the imprecise data modeled by equal sized parallel line segment and an $O(n^4)$ algorithm for the convex hull of the maximum perimeter based on the imprecise data modeled by non-equal sized parallel line segment.

For discrete imprecise data model, Zhang et al. [13] use a brute force algorithm to solve the problem of the minimum diameter color-spanning set problem (MDCS). The running time of their algorithm is $O(n^k)$ if there are k colors and n points. Chen et al. [9] implement the algorithm in their geographical tagging system. Fleischer and Xu [10] show that MDCS can be solved in polynomial time

---

* This work was supported by 2008-China Shenzhen Inovation Technology Program (SY200806300211A).

W. Wu and O. Daescu (Eds.): COCOA 2010, Part I, LNCS 6508, pp. 41–50, 2010.

for L_1 and L_∞ metrics, while it is NP-hard for all other L_p metrics even in two dimensions and gave a constant factor approximation algorithm.

There are also other works on colored point sets problems [3, 4, 11].

Problem Definition. The problems we discuss in this paper are as follows:

Problem 1. **Min-MST.** There are n points in the plane and each point is painted by one of m colors where $m \leq n$. We want to select m different color points such that the total edge length of resulting minimal spanning tree is as small as possible.

Problem 2. **Max-MST.** There are n points in the plane and each point is painted by one of m colors where $m \leq n$. We want to select m different color points such that the total edge length of resulting minimal spanning tree is as large as possible.

Problem 3. **Min-Per.** There are n points in the plane and each point is painted by one of m colors where $m \leq n$. We want to select m different color points such that the perimeter of the convex hull of m different color points is as small as possible.

We will prove those three problems are NP-complete by reduction from 3-SAT problem and give approximation algorithms for the third problem.

2   Min-MST Is NP-Complete

First we show that this problem belongs to NP. Given an instance of the problem, we use as a certificate the m different color points chosen from n points. The verification algorithm compute the MST of those m points and check whether the length is at most L. This process can certainly be done in polynomial time.

We prove this problem is NP-hard by reduction from 3-SAT problem. We need several gadgets to represent variables and clauses of 3-SAT formula. The general idea is for a 3-SAT formula, we put some colored points on the plane such that the given 3-SAT formula has a true assignment if and only if the length of minimum MST of colored points equals some given value.

First we draw a point O with distinct color at $(0,0)$. For a 3-SAT formula ψ, suppose it has n variables $x_1, x_2, ..., x_n$ and m clauses. Let $x_{i,j,k}$ (or $\neg x_{i,j,k}$) be the variable x_i that appears at the j-th literal in ψ from left to right such that x_i (including $\neg x_i$) appears $k-1$ times already in ψ before it. For each variable x_i ($1 \leq i \leq n$), six points $p_i^{j'}$ ($1 \leq j' \leq 6$) are created on $p_i^1 = (400i - 300, 0)$, $p_i^2 = (400i - 200, 0)$, $p_i^3 = (400i - 100, 0)$, $p_i^4 = (400i, 0)$, $p_i^5 = (300i - 100, -100)$, $p_i^6 = (300i, -100)$. For each literal $x_{i,j,k}$ (or $\neg x_{i,j,k}$), eight additional points are created on $p_{i,j,k}^7 = (400i - 200, -100)$, $p_{i,j,k}^8 = (400i, -100)$, $p_{i,j,k}^9 = (0, 400j - 300)$, $p_{i,j,k}^{10} = (0, 400j - 200)$, $p_{i,j,k}^{11} = (0, 400j - 100)$, $p_{i,j,k}^{12} = (0, 400j)$, $p_{i,j,k}^{13} = (-\frac{100}{3m}, 400j - 200)$, $p_{i,j,k}^{14} = (-\frac{100}{3m}, 400j)$ respectively. Among those fourteen points, p_i^5 and p_i^6 have the same color, $p_{i,j,k}^7$ and $p_{i,j,k}^{13}$ have the

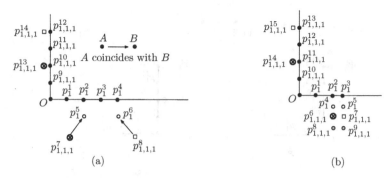

Fig. 1. Gadget for the first literal x_1 (a) for MIN-MST (b) for MAX-MST. Different symbols means different colors. All solid circles are with distinct colors.

same color, $p_{i,j,k}^8$ and $p_{i,j,k}^{14}$ have the same color, and all other points have different colors. Figure 1(a) shows gadget for the first literal x_1. According to above construction, we get a set P of $6n + 24m + 1$ points ($4n$ points are on x axis, $2n + 2 \times 3m = 2n + 6m$ points are on line $y = -100$, $4 \times 3m = 12m$ points are on y axis, $2 \times 3m = 6m$ points are on line $x = -\frac{100}{3m}$, one point is on point O) in the plane. Let T_P be the minimum MST over P. $4n + 12m + 1$ points $\subset P$ are on x and y axis and they are all with distinct colors. Therefore, all points on x and y axis and edges with length 100 connecting those points appear in T_P. Since p_i^5 and p_i^6 have the same color, $p_{i,j,k}^7$ and $p_{i,j,k}^{13}$ have the same color and $p_{i,j,k}^8$ and $p_{i,j,k}^{14}$ have the same color, we have to choose either $\{p_i^5, p_{i,j,k}^7, p_{i,j,k}^{14}\}$ or $\{p_i^6, p_{i,j,k}^8, p_{i,j,k}^{13}\}$ to get the minimum MST T_P with length $500n + 400 \times 3m + 3m \times \frac{100}{3m} = 500n + 1200m + 100$. Let choosing $\{p_i^5, p_{i,j,k}^7, p_{i,j,k}^{14}\}$ correspond x_i is *false* and choosing $\{p_i^6, p_{i,j,k}^8, p_{i,j,k}^{13}\}$ correspond x_i is *true* (see Figure 2).

Now, we illustrate how to represent the binary relation *or* in 3-SAT formula. We assume that the one clause is $x_{i1,j,k1} \vee x_{i2,j+1,k2} \vee x_{i3,j+2,k3}$. We create three *or* points $p_{i1,\lceil \frac{j}{3} \rceil}^{or}$, $p_{i2,\lceil \frac{j+1}{3} \rceil}^{or}$ and $p_{i3,\lceil \frac{j+2}{3} \rceil}^{or}$ with the same color (but different with all other colors) at $p_{i1,j,k1}^{13}$, $p_{i2,j+1,k2}^{13}$ and $p_{i3,j+2,k3}^{13}$ respectively. If one literal is in negation form $\neg x_{i,j,k}$, then we need to create the *or* point $p_{i,\lceil \frac{j}{3} \rceil}^{or}$ at $p_{i,j,k}^{14}$ instead of $p_{i,j,k}^{13}$. Figure 3 shows the gadget for the first clause $(x_1 \vee x_2 \vee \neg x_3)$ and Figure 4 shows the gadget for $\psi = (x_1 \vee x_2 \vee \neg x_3) \wedge (\neg x_1 \vee x_3 \vee x_4)$. Let T be the minimum MST over P plus $3m$ *or* points. For literal $x_{i,j,k}$ (or $\neg x_{i,j,k}$), if variable x_i is true (or false), then $\{p_i^6, p_{i,j,k}^8, p_{i,j,k}^{13}\}$ (or $\{p_i^5, p_{i,j,k}^7, p_{i,j,k}^{14}\}$) are chosen that means the *or* point $p_{i,\lceil \frac{j}{3} \rceil}^{or}$ can be chosen to construct T without adding any extra length comparing with T_P since $p_{i,\lceil \frac{j}{3} \rceil}^{or}$ is located at the same place as $p_{i,j,k}^{13}$ (or $p_{i,j,k}^{14}$). Therefore, if at least one literal is true in one clause, no extra length will be added to T_P for that clause. Otherwise, all three literals in one clause are false, then all three *or* points in that clause can not be covered by points in T_P. Since we have to choose one point from those three *or* points,

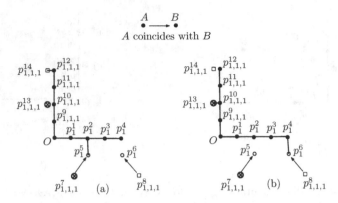

Fig. 2. Suppose x_1 is the first literal in the first clause of 3-SAT formula. (a) the minimum MST when x_1 is false. (b) the minimum MST when x_1 is true.

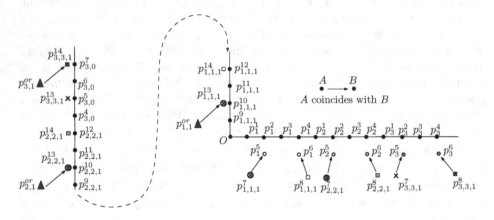

Fig. 3. Gadget for the clause $(x_1 \vee x_2 \vee \neg x_3)$ assuming it is the first clause

then the extra length of one will be added to the length of T_P. If m' clauses are not satisfiable, then the extra length of m' will be added to the length of T_P. Thus the lemma follows:

Lemma 1. *The 3-SAT formula ψ with n variables $x_1, x_2, ..., x_n$ and m clauses is satisfiable if and only if the length of T is equal to $500n + 1200m + 100$.*

3   Max-MST Is NP-Complete

First we show that this problem belongs to NP. Given an instance of the problem, we use as a certificate the m different color points chosen from n points. The verification algorithm compute the MST of those m points and check whether the length is at most L. This process can certainly be done in polynomial time.

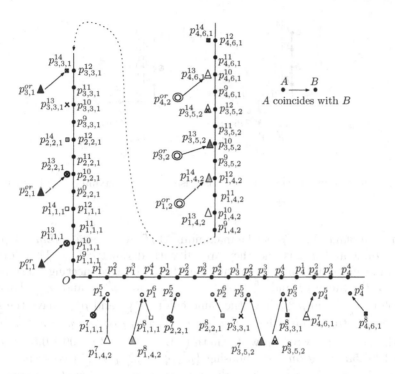

Fig. 4. Gadget for $\psi = (x_1 \vee x_2 \vee \neg x_3) \wedge (\neg x_1 \vee x_3 \vee x_4)$ for **Min-MST**. y axis is broken to save space.

We prove this problem is NP-hard by reduction from 3-SAT problem. We need several gadgets to represent variables and clauses of 3-SAT formula. The general idea is for a 3-SAT formula, we put some colored points on plane such that the given 3-SAT formula has true assignment if and only if the length of maximum MST of colored points equals some given value.

First we draw a point O with distinct color at $(0,0)$. For a 3-SAT formula ψ, suppose it has n variables $x_1, x_2, ..., x_n$ and m clauses. Let $x_{i,j,k}$ (or $\neg x_{i,j,k}$) be the variable x_i that appears at the j-th literal in ψ from left to right such that x_i (including $\neg x_i$) appears $k-1$ times already in ψ before it. For each variable x_i ($1 \leq i \leq n$), five points $p_i^{j'}$ ($1 \leq j' \leq 5$) are created on $p_i^1 = (201i - 101, 0)$, $p_i^2 = (201i - 1, 0)$, $p_i^3 = (201i, 0)$, $p_i^4 = (201i - 1, -10)$, $p_i^5 = (201i, -10)$. For each literal $x_{i,j,k}$ (or $\neg x_{i,j,k}$), ten additional points are created on $p_{i,j,k}^6 = (201i - 1, -9 - 2k)$, $p_{i,j,k}^7 = (201i, -9 - 2k)$, $p_{i,j,k}^8 = (201i - 1, -10 - 2k)$, $p_{i,j,k}^9 = (201i, -10 - 2k)$, $p_{i,j,k}^{10} = (0, 400j - 300)$, $p_{i,j,k}^{11} = (0, 400j - 200)$, $p_{i,j,k}^{12} = (0, 400j - 100)$, $p_{i,j,k}^{13} = (0, 400j)$, $p_{i,j,k}^{14} = (-0.5, 400j - 200)$, $p_{i,j,k}^{15} = (-0.5, 400j)$ respectively. Among those fifteen points, p_i^4 and p_i^5 have the same color, $p_{i,j,k}^6$ and $p_{i,j,k}^{14}$ have the same color, $p_{i,j,k}^7$ and $p_{i,j,k}^{15}$ have the same color, $p_{i,j,k}^8$ and $p_{i,j,k}^9$ have the same color, and all other points have different colors. Figure 1(b) shows gadget for the first literal x_1. According to above construction, we get a set P of $5n + 30m + 1$

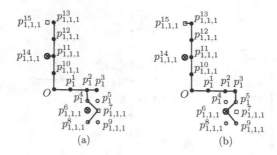

Fig. 5. (a) the maximum MST when x_1 is false. (b) the maximum MST when x_1 is true.

points in the plane. Let T_P be the maximum MST over P. $3n + 12m + 1$ points $\subset P$ are on x and y axis and they are all with distinct colors. Therefore, all points on x and y axis and edges with length 100 or 1 connecting those points appear in T_P. Since p_i^4 and p_i^5 have the same color, $p_{i,j,k}^6$ and $p_{i,j,k}^{14}$ have the same color, $p_{i,j,k}^7$ and $p_{i,j,k}^{15}$ have the same color, $p_{i,j,k}^8$ and $p_{i,j,k}^9$ have the same color, we have to choose either $\{p_i^4, p_{i,j,k}^7, p_{i,j,k}^8, p_{i,j,k}^{14}\}$ or $\{p_i^5, p_{i,j,k}^6, p_{i,j,k}^9, p_{i,j,k}^{15}\}$ to get the maximum MST T_P with length $(201 + 10)n + 3m(400 + 0.5 + 2\sqrt{2}) = 211n + 1201.5m + 6\sqrt{2}m$. Let choosing $\{p_i^4, p_{i,j,k}^7, p_{i,j,k}^8, p_{i,j,k}^{14}\}$ correspond x_i is *false* and choosing $\{p_i^5, p_{i,j,k}^6, p_{i,j,k}^9, p_{i,j,k}^{15}\}$ correspond x_i is *true* (see Figure 5).

For each literal $x_{i,j,k}$ (or $\neg x_{i,j,k}$), we create one *or* point $p_{i,\lceil \frac{j}{3} \rceil}^{or}$ at $p_{i,j,k}^{14}$ (or $p_{i,j,k}^{15}$) as in section 2. The remaining part of this section is similar to section 2.

Lemma 2. *The 3-SAT formula ψ with n variables $x_1, x_2, ..., x_n$ and m clauses is satisfiable if and only if the length of T is equal to $211n + 1201.5m + 6\sqrt{2}m$.*

4   Min-Per Problem Is NP-Complete

First we show that this problem belongs to NP. Given an instance of the problem, we use as a certificate the m different color points chosen from n points. The verification algorithm compute the perimeter of the convex hull of those m points and check whether the perimeter is at most p. This process can certainly be done in polynomial time.

We prove this problem is NP-hard by reduction from 3-SAT problem. We need several gadgets to represent variables and clauses of 3-SAT formula. For a given instance of 3-SAT problem, suppose it has k variables $x_1, x_2, ..., x_k$. First we draw a circle C. The general idea is for a 3-SAT formula, we put some colored points on and inside C such that the given 3-SAT formula has true assignment if and only if the perimeter of convex hull CH_{opt} with minimum perimeter equals some given value.

For two points a, b on C, let $\tilde{ab}$ be the arc of C from a to b in clockwise order and $\overline{ab}$ be the line segment between a, b. Let the length of $\overline{ab}$ be $|\overline{ab}|$. For

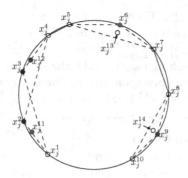

Fig. 6. The gadget for variable x_j. The colors of all empty circles are different from each other and from solid colored points. Solid line segments must appear on CH_{opt}. Dashed line segments mean they are candidates for edges of CH_{opt}.

each variable x_j, we put 10 points x_j^i where $i = 1, 2, ..., 10$ on circle C in such way: x_j^i where $i = 1, 2, ..., 10$ are arranged in clockwise order (see Figure 6). For those ten points, only x_j^2 and x_j^3 have the same color and all other points have different colors. Also $|\overline{x_j^1 x_j^2}| = |\overline{x_j^3 x_j^4}|$. Then we add two more points x_j^{11}, x_j^{12} where x_j^{11} is on line segment $\overline{x_j^1 x_j^2}$ and x_j^{12} is on line segment $\overline{x_j^3 x_j^4}$. x_j^{11} has the same color as the color of x_j^6 and x_j^{12} has the same color as the color of x_j^9. For different variables, the arcs they spanned do not overlap and they are arranged in clockwise order for increasing j. The relative distances between twelve points satisfy following equations:

redFormate changed:

$$|\overline{x_j^1 x_j^2}| + |\overline{x_j^2 x_j^4}| = |\overline{x_j^1 x_j^3}| + |\overline{x_j^3 x_j^4}| = p_1 |\overline{x_j^5 x_j^7}| = |\overline{x_j^8 x_j^{10}}| = p_2$$

$$|\overline{x_j^5 x_j^6}| = |\overline{x_j^6 x_j^7}| = |\overline{x_j^8 x_j^9}| = |\overline{x_j^9 x_j^{10}}| = p_3$$

$$|\overline{x_j^1 x_j^{11}}| + |\overline{x_j^{11} x_j^3}| - |\overline{x_j^1 x_j^3}| = |\overline{x_j^2 x_j^{12}}| + |\overline{x_j^{12} x_j^4}| - |\overline{x_j^2 x_j^4}| = \Delta p_1$$

$$2p_3 - p_2 = \Delta p_2; \Delta p_1 >> \Delta p_2$$

The inequality $\Delta p_1 >> \Delta p_2$ ensures the part of CH_{opt} on arc $\widehat{x_j^1 x_j^4}$ has to be $\overline{x_j^1 x_j^2} \cup \overline{x_j^2 x_j^4}$ or $\overline{x_j^1 x_j^3} \cup \overline{x_j^3 x_j^4}$. Suppose we choose $\overline{x_j^1 x_j^2} \cup \overline{x_j^2 x_j^4}$ to be the part of CH_{opt}. In order to cover the color of x_j^{12}, we have to choose $\overline{x_j^8 x_j^9} \cup \overline{x_j^9 x_j^{10}}$ for arc $\widehat{x_j^8 x_j^{10}}$. For the arc $\widehat{x_j^5 x_j^7}$, since the color of x_j^6 is the same as the color of x_j^{11} and it has been covered by $\overline{x_j^1 x_j^2} \cup \overline{x_j^2 x_j^4}$ already, we can just choose $\overline{x_j^5 x_j^7}$ for arc $\widehat{x_j^5 x_j^7}$. x_j^{11} acts as the negation of x_j^{12} that means if we choose one of them, the other one will not be chosen. Similarly, x_j^6 acts as the negation of x_j^9. Therefore, if the variable x_j is assigned to be 1, we choose $\overline{x_j^1 x_j^2} \cup \overline{x_j^2 x_j^4} \cup \overline{x_j^4 x_j^5} \cup \overline{x_j^5 x_j^7} \cup \overline{x_j^7 x_j^8} \cup \overline{x_j^8 x_j^9} \cup \overline{x_j^9 x_j^{10}}$ as part of CH_{opt} for the arc $\widehat{x_j^1 x_j^{10}}$. Otherwise if the variable x_j is assigned to be 0, we choose $\overline{x_j^1 x_j^3} \cup \overline{x_j^3 x_j^4} \cup \overline{x_j^4 x_j^5} \cup \overline{x_j^5 x_j^6} \cup \overline{x_j^6 x_j^7} \cup \overline{x_j^7 x_j^8} \cup \overline{x_j^8 x_j^{10}}$ as part of CH_{opt} for the arc $\widehat{x_j^1 x_j^{10}}$. Therefore, if we connect all k gadgets together to form CH_{opt}, the perimeter of CH_{opt} is $k \times (p_1 + 2p_2 + \Delta p_2) + p_4$ where $p_4 = 2 \sum_{j=1}^{k} (|\overline{x_j^4 x_j^5}| + |\overline{x_j^7 x_j^8}|) + \sum_{j=1}^{k-1} |\overline{x_j^{10} x_{j+1}^1}| + |\overline{x_k^{10} x_1^1}|$.

For three literals in each clause, we add three points with same color into gadgets constructed above corresponding to three literals. For example, if x_j appears in one clause, we put x_j^{13} inside the triangle $\triangle x_j^5 x_j^6 x_j^7$ such that $\Delta p_3 = |\overline{x_j^5 x_j^{13}}| + |\overline{x_j^{13} x_j^7}| - |\overline{x_j^5 x_j^7}| << \Delta p_2$. If $\neg x_j$ appears in one clause, we put x_j^{14} inside the triangle $\triangle x_j^8 x_j^9 x_j^{10}$ such that $\Delta p_3 = |\overline{x_j^8 x_j^{14}}| + |\overline{x_j^{14} x_j^{10}}| - |\overline{x_j^8 x_j^{10}}| << \Delta p_2$. We call x_j^6 and x_j^9 as *apex-point* and x_j^{13} and x_j^{14} as *or-point*. For three literals in one clause, if all three literals are 0, then corresponding *apex-point* will not be chosen as the vertex of CH_{opt}. In order to cover the color of *or-point*, one of three *or-points* has to be chosen as the vertex of CH_{opt}. Then one Δp_3 will be add into the perimeter of CH_{opt}. Therefore, for a given 3-SAT formula with k variables, if it has true assignment, then the perimeter of convex hull CH_{opt} with minimum perimeter over the gadgets constructed as above equals $k \times (p_1 + 2p_2 + \Delta p_2) + p_4$.

If the perimeter of convex hull CH_{opt} over the gadgets constructed as above equals $k \times (p_1 + 2p_2 + \Delta p_2) + p_4$, that means at least one literal of all clauses is true. Because otherwise, the perimeter of convex hull CH_{opt} will be $\geq k \times (p_1 + 2p_2 + \Delta p_2) + p_4 + \Delta p_3$. Therefore, the 3-SAT formula has true assignment.

Theorem 1. *There are n points in the plane and each point is painted by one of m colors where $m \leq n$. To select m different color points such that the perimeter of the convex hull of m different color points is as small as possible is NP-complete.*

5   Approximation Algorithms for Min-Per Problem

5.1 π-Approximation Algorithm

The algorithm is simple: for each point p, select $m-1$ points $p_1, ..., p_{m-1}$ such that the colors of $p, p_1, ..., p_{m-1}$ are different and the distance from p_i $(1 \leq i \leq m-1)$ to p is less than the distances from other points with the same color as p_i to p. For those m points, construct a convex hull CH_p and compute its perimeter. For all n point, we get n perimeters and the minimum one is what we want. The running time of this algorithm is $O(n(n + m \log m))$. Assume the perimeter we get is per_{app} and the optimal perimeter is p_{opt}. Now we prove $per_{app} \leq \pi per_{opt}$.

Suppose two vertices that decide the diameter of the optimal convex hull CH_{opt} are p_a and p_b (see Figure 7). Let $r = |\overline{p_a p_b}|$. We draw a circle C with center p_a and radius r. CH_{p_a} is constructed as above algorithm. CH_{p_a} is totally inside C since CH_{opt} is totally inside C that means there are at least one point from each color is inside C. Then perimeter of CH_{p_a} is $per_{p_a} \leq 2\pi r$. We know that $per_{app} \leq per_{p_a}$ and $per_{opt} \geq 2r$. Therefore $per_{app} \leq \pi per_{opt}$.

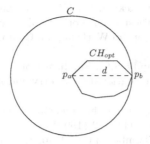

Fig. 7. Illustration of π-approximation algorithm

Theorem 2. *There is a π-approximation algorithm for problem 3 with running time $O(n^2 + nm\log m)$.*

5.2 $\sqrt{2}$-Approximation Algorithm

In [12], Abellanas et al. give an $O(\min\{n(n-m)^2, nm(n-m)\})$ algorithm for computing minimum perimeter axis-parallel rectangle R enclosing at least one point of each color. Let the perimeter of R be per_R and the diameter of R be d. If we construct CH_{app} over the points inside R, CH_{app} contains at least one point of each color and CH_{app} is totally inside R: Therefore $per_{app} \leq per_R$. There are some properties for R. There are at least one point on each edge of R. Colors of those four points are different. There are no other points inside R having the same color as those four points. Let R' be the smallest axis-parallel rectangle enclosing CH_{opt} and d' be the diameter of R'. Then $per_{opt} \geq 2d' \geq 2d$. We know $per_R \leq 2\sqrt{2}d$. Then $per_{app} \leq \sqrt{2}per_{opt}$.

Theorem 3. *There is a $\sqrt{2}$-approximation algorithm for problem 3 with running time $O(\min\{n(n-m)^2, nm(n-m)\})$.*

6 Conclusions

In this paper, we discussed three variations of spread colored points problems and proved NP-completeness of those three problems. For the third problem, we gave π and $\sqrt{2}$ approximation algorithms with $O(n^2 + nm\log m)$ and $O(\min\{n(n-m)^2, nm(n-m)\})$ running time respectively, where n is the number of points and m is the number of colors. In future work, it would be interesting to investigate more geometric problems for colored points. For example, maximizing the closest pair for spreading colored points.

References

1. Beresford, A.R., Stajano, F.: Location Privacy in pervasive Computing. IEEE Pervasive Computing 2(1), 46–55 (2003)
2. Gedik, B., Liu, L.: A customizable k-anonymity model for protecting location privacy. In: ICDCS, pp. 620–629 (2004)

3. Abellanas, M., Hurtado, F., Icking, C., Klein, R., Langetepe, E., Ma, L., Palop, B., Sacristan, V.: The farthest color Voronoi diagram and related problems. In: Proceedings of the 17th European Workshop on Computational Geometry (EWCG 2001), pp. 113–116 (2001)
4. Das, S., Goswami, P.P., Nandy, S.C.: Smallest Color-spaning Object Revisited. International Journal of Computational Geometry and Applications (IJCGA) 19(5), 457–478 (2009)
5. Pfoser, D., Jensen, C.: Capturing the uncertainty of moving-objects representations. In: Proc. SSDBM, pp. 111–132 (1999)
6. Sistla, P.A., Wolfson, O., Chamberlain, S., Dao, S.: Querying the uncertain position of moving objects. In: Temporal Databases: Research and Practice, pp. 310–337 (1998)
7. Löffler, M., van Kreveld, M.: Largest and Smallest Convex Hulls for Imprecise Points. Algorithmica 56(2), 235–269 (2008)
8. Ju, W., Luo, J.: New Algorithms for Computing Maximum Perimeter and Maximum Area of the Convex Hull of Imprecise Inputs Based On the Parallel Line Segment Model. In: CCCG 2009, pp. 1–4 (2009)
9. Chen, Y., Chen, S., Gu, Y., Hui, M., Li, F., Liu, C., Liu, L., Ooi, B.C., Yang, X., Zhang, D., Zhou, Y.: MarcoPolo: A community system for sharing and integrating travel information on maps. In: Proceedings of the 12th International Conference on Extending Database Technology (EDBT 2009), pp. 1148–1151 (2009)
10. Fleischer, R., Xu, X.: Computing Minimum Diameter Color-Spanning Sets. In: Lee, D.-T., Chen, D.Z., Ying, S. (eds.) FAW 2010. LNCS, vol. 6213. Springer, Heidelberg (2010)
11. Borgelt, M.G., van Kreveld, M., Löffler, M., Luo, J., Merrick, D., Silveira, R.I., Vahedi, M.: Planar bichromatic minimum spanning trees. J. Discrete Algorithms 7(4), 469–478 (2009)
12. Abellanas, M., Hurtado, F., Icking, C., Klein, R., Langetepe, E., Ma, L., Palop, B., Sacristan, V.: Smallest Color-Spanning Objects. In: Proc. 9th Annu. European Sympos. Algorithms, pp. 278–289 (2001)
13. Zhang, D., Chee, Y.M., Mondal, A., Tung, A.K.H., Kitsuregawa, M.: Keyword search in spatial databases: Towards searching by document. In: Proceedings of the 25th IEEE International Conference on Data Engineering (ICDE 2009), pp. 688–699 (2009)

Construction of Mixed Covering Arrays of Variable Strength Using a Tabu Search Approach

Loreto Gonzalez-Hernandez, Nelson Rangel-Valdez, and Jose Torres-Jimenez

Information Technology Laboratory, CINVESTAV-Tamaulipas
Km. 5.5 Carretera Cd. Victoria-Soto la Marina, 87130, Cd. Victoria Tamps., Mexico
agonzalez@tamps.cinvestav.mx, nrangel@cinvestav.mx, jtj@cinvestav.mx
http://www.tamps.cinvestav.mx/~jtj/

Abstract. The development of a new system involves extensive tests on the software functionality in order to identify possible failures. Also, a system already built requires a fine tuning of its configurable options to give the best performance in the environment it is going to work. Both cases require a finite set of tests that avoids testing all the possible combinations (which is time consuming); to this situation Mixed Covering Arrays (MCAs) are a feasible alternative. MCAs are combinatorial structures represented as matrices having a test case per row. MCAs are small, in comparison with brute force, and guarantees a level of interaction among the parameters involved (a difference with random testing). We present a Tabu Search (TS) algorithm to construct MCAs; the novelty in the algorithm is a mixture of three neighborhood functions. We also present a new benchmark for the MCAs problem. The experimental evidence showed that the TS algorithm improves the results obtained by other approaches reported in the literature, finding the optimal solution in some the solved cases.

Keywords: Mixed Covering Arrays, software testing, Tabu Search.

1 Introduction

Different areas of knowledge like artificial intelligence, machine learning or optimization, relies on the use of sets of data to measure the aptitude of a particular function to feedback a system so that its general performance is improved. A function is defined through a set of parameters and an expected output, based on this fact the test sets can be: a) a complete test set, the one that includes all the possible combinations of values of the parameters; b) a random test set, the one generated by a subset of the whole set of possible combination chosen at random; c) a t-wise test set, the one where a set of data is chosen guaranteeing a level of interaction between the values of the parameters.

The use of the complete test set would be ideal but the size of the set of data grows exponentially, this limits its use to functions where the number of parameters and their values is small. An alternative to this is the use of random

W. Wu and O. Daescu (Eds.): COCOA 2010, Part I, LNCS 6508, pp. 51–64, 2010.
© Springer-Verlag Berlin Heidelberg 2010

test sets, which are smaller in size and tend to proportionate a good sample of the complete set. However, based on recent studies that have shown that close to 100% of the failures are triggered with interactions among 6 parameters [14], tasks like testing or fine tuning software could find a better alternative in the t-wise test sets. A level of interaction t among parameters means that any combination of values among t parameters appears in the test set.

The problem of constructing t-wise test sets is known as the construction of Covering Arrays Problem (CAP). The CAP focuses on constructing a test set which contains the level of interaction indicated with the minimum number of tests. The solution of the CAP in general is hard, i.e. there is no efficient algorithm that solves it; the NP completeness of the problem of constructing CA was reported in [16] and [7].

Different approaches have been used to solve the CAP, some of them are: exact methods [1], greedy algorithms [22], genetic algorithms (GA) [21], tabu search (TS) [18], simulated annealing (SA) [6], ant colony optimization algorithm (ACO) [20], hill climbing (HC) [6], great deluge (GD) [3], SAT model [17] among others. Due to the complexity of the problem, approximated approaches are preferred instead of exact approaches, in order to provide good solutions in a reasonable time.

This paper focuses on a more general case of CAP: the problem of constructing Mixed Covering Arrays (MCAP), where the values of the different parameters involved in the problem are heterogeneous. We present a Tabu Search approach to construct Covering Arrays. What is novel in this approach is the use of a set of neighborhood functions with a probability to be selected, i.e., instead of using a single neighborhood function, the algorithm chooses between a set of predefined neighborhood functions according to a assigned probability to each function. In order to determine the best probability of selection for each neighborhood function, a fine tuning process was made over a set of discretized probabilities.

This document is organized as follows: section 2 summarizes the state of the art of the existing approaches that have been used to construct Covering Arrays (CAs) and Mixed Covering Arrays (MCAs); section 3 defines the problem of constructing Mixed Covering Arrays; section 4 details the Tabu Search algorithm proposed in this paper to solve MCAP; section 5 shows the results from the experiment where our TS is compared against the algorithms reported in the literature. The comparison was made based on the average size obtained by the approaches over an existing benchmark and a new benchmark proposed in this paper. Finally, section 6 presents the main contribution of this paper.

2   Related Work

There are several reported approaches for the construction of CAs [10], they are approximated methods, i.e. they do not guarantee that the provided solution is always optimal. A repository of CAs is available online[1], some of them are optimal or near to the optimal.

---

[1] http://www.tamps.cinvestav.mx/~jtj/authentication.php

The metaheuristics applied include genetic algorithms (GA), simulated annealing (SA), tabu search (TS), hill climbing (HC), ant colony optimization algorithm (ACO) and great deluge (GD). The greedy methods have been implemented in the algorithms Automatic Efficient Test Generator (AETG), Deterministic Density Algorithm (DDA), Test Case Generator (TCG) and In Parameter Order (IPO) which was subsequently extended to In Parameter Order General (IPOG). Other approaches like algebraic methods, constraint programming (CP) and EXACT (exhaustive search of combinatorial test suites) have also been applied. Details of these are given below:

Shiba et.al. [20] implemented GA, ACO and SA, Stardom [21] implemented SA, TS and GA, being SA which provided the best results. According to Stardom, Bryce [4] emphasizes that SA and TS have constructed many of the CA optimal or near to the optimal. Nurmela [18] also used TS for constructing CA and MCA and reported some upper bounds for them, Walker and Colbourn [12] employed TS using permutation vectors. Likewise Bryce and Colbourn [2] implemented an hybrid technique of greedy methods with the metaheuristics TS, HC, SA and GD.

Relating to the greedy algorithms, they generate one test at a time, some examples are AETG [5], DDA [7], IPO [16], IPOG [15], TConfig [24], IPOG-F [8] and TCG [23]. Cohen et.al. [6] implemented their own version of AETG and TCG. Other researches are focused on explaining the steps for constructing CA, one of these is proposed by William and Probert [25] who used algebraic methods and combinatorial theory. Besides the described methods, an algorithm that focuses on solving this problem of exact way was proposed by Hinch [11] which is based on the model CP, moreover, Yang and Zang [26] used a retrospective algorithm and incorporated it in a tool called EXACT.

The next section shows a formal definition for the problem of constructing Mixed Covering Arrays.

3   Definition of Mixed Covering Array

A mixed covering array $MCA(N; t, k, v_1 v_2 \ldots v_k)$ is a matrix M of size $N \times k$ where each column c_j, for $1 \leq j \leq k$, v_j can contain symbols from the set $\{0, 1, \ldots, v_j - 1\}$. Also, any combination of t columns $T = \{\tau_1, \tau_2, \ldots, \tau_t\}$ from M, known as t tuple, must contain all the possible combination of symbols derived from the set $\{\{0, \ldots, v_{\tau_1}\} \times \ldots \times \{0, \ldots, v_{\tau_t}\}\}$. The sets of symbols that each column can have is known as the alphabet and the cardinality of each set is known as the cardinality of the alphabet. The value t represents the level of interaction in the problem of constructing t-wise test sets, this parameter of MCA is known as the strength of the matrix.

The MCAP is defined as the construction of a $MCA(N; t, k, v_1 v_2 \ldots v_k)$ matrix with the minimum of rows. A short notation for the MCAP can be given using the exponential notation $MCA(N; t, k, v_1^{q_1} \ldots v_g^{q_w})$; the notation describes, that there are q_r parameters from the set $\{v_1, v_2, \ldots, v_k\}$ that takes v_s values. When the value of the minimum N is searched in the MCAP, i.e., the minimum size matrix for which an MCA exists for a set of given parameters, the notation is changed to $MCAN(t, k, v_1^{q_1} \ldots v_g^{q_w})$.

An example of an instance of MCAP can be given when considering the verification of a Switch WLAN in four different aspects: monitoring, management, maintenance and safety. The verification process in the switch involves four parameters, three of them have two possible values and one three. Table 1 shows the set of parameters of the Switch WLAN and their values.

Table 1. Parameters of a Switch WLAN, the firsth with three possible values and the rest with two

Monitoring	Management	Maintenance	Safety
PC	Load balancing	Interference	Denial of service
Access points	Connection	Barriers	Ad-hoc Networks
Sensors	–	–	–

A MCA for the Switch WLAN instance of the MCAP shown in Table 1 is given in Table 2. The complete test set for this instance of the MCAP would required $N = 3 \times 2 \times 2 \times 2 = 24$ test cases; instead, the constructed MCA only requires a set of data with $N = 6$ cases, with a level of interaction of 2. When in the tested system the number of its parameters grows, the difference between the number of cases of a MCA and the complete set is exponential even for moderate values of the strength t.

Table 2. Mapping of the $MCA(6; 2, 3^1 2^3)$ to the corresponding pair-wise test suite

Monitoring	Management	Maintenance	Safety
PC	Load balancing	Barriers	Ad-hoc Networks
PC	Connection	Interference	Denial of service
Access points	Load balancing	Interference	Ad-hoc Networks
Access points	Connection	Barriers	Denial of service
Sensors	Load balancing	Barriers	Denial of service
Sensors	Connection	Interference	Ad-hoc Networks

The next section presents the Tabu Search approach proposed in this paper to construct MCAs. Particularly, we present details about the neighborhood functions and the evaluation function developed for the strategy and their participation in the construction of MCAs.

4 Proposed Approach

The metaheuristic known as Tabu Search (TS) was first proposed by Glover and Laguna [9] and has been used to solve a large set of problems in artificial intelligence. The basic TS strategy improves a solution for a given problem by visiting neighbors that were created using heuristics over the solution and that do not belong to a tabu search list.

The key features of the approach presented in this paper are the followings: a) the routine to create the initial solution for the algorithm; b) the size of the tabu search list; c) the use of a mixture of neighborhood functions; d) the evaluation function; e) the process of fine tuning the probabilities of selection for each neighborhood function based on a complete test set of discretized probabilities. The approach will be referred as TSA from now on and in the following subsection details about each of the features will be given.

4.1   Creating the Initial Solution

Given an MCAP instance $MCA(N; t, k, v_1^{q_1} \ldots v_g^{q_w})$, a solution is represented as a bidimensional array M of size $N \times k$, where the columns are the parameters and the rows are the cases of the test set that is constructed. Each cell $m_{i,j}$ in the array accepts values from the set $\{0, 1, ..., v_j - 1\}$ where v_j is the cardinality of the alphabet of $j^{th} column$.

The process followed to generate the initial solution s_0 for TSA is iterative. Firstly, for the first t columns it generates v^t rows containing the combinations derived from $\{0, 1, ..., v_1\} \times ... \times \{0, 1, ..., v_j\}$, the alphabets of that columns; the same set of combinations is duplicated as many times as the number of wished rows N is completed. In the case that the missing rows is smaller than v^t, the combinations will be chosen in the order that they were initially generated until the total of N rows is completed. Once that the N rows for the first t columns were generated, the process is repeated for the next t columns and so on until the k columns of the initial matrix s_0 are filled. Table 3 shows an initial solution matrix s_0 generated according with the method described for $MCA(7; 2, 5, 3^1 2^4)$; the the first two columns have the symbols bolted that corresponds to the combinations derived from $\{0, 1, 2\} \times \{0, 1\}$.

Table 3. Initial solution s_0 for $MCA(7; 2, 5, 3^1 2^4)$

a	b	c	d	e
0	0	0	0	0
0	1	0	1	1
1	0	1	0	0
1	1	1	1	1
2	0	0	0	0
2	1	0	1	1
0	0	1	0	0

4.2   Tabu List Definitions

The Tabu Search approaches are characterized by the use of a list of movements; this list contains changes that can't be made by the algorithm for a certain period of time over the matrix M that is constructed, i.e. the neighborhood function can modify the solution if the modification involves changes that are not listed in the tabu list. However, an aspiration critera can be used to allow a change in a solution even if the change is in the tabu list; the aspiration critera consists in a improvement over the global solution, i.e. if the best known solution constructed by TSA so far can be improved by a change found in the tabu list then the change is allowed.

The TSA algorithm characterizes the tabu list through the following parameters: row, column, symbol, used neighborhood function and number of missings. Summarizing, when a neighborhood function $\mathcal{N}$ set the symbol v in the row i and column j of the matrix M and produces a number of missings m, the values $(\mathcal{N}, v, i, j, m)$ are added to the tabu list during a period of time $\mathcal{T}$.

The expiration time $\mathcal{T}$ is defined in this approach as the number of times that a neighborhood function is called. The following subsection shows the neighborhood functions used in TSA.

4.3 Neighborhood Functions

A neighbor is a solution created from another solution through heuristics. The approach proposed in this paper uses a combination of three different neighborhood functions to define the neighborhood of a solution. The three neighborhood functions modify the matrix M representing the solution for the $MCA(N; t, k, v_1^{q_1} \ldots v_g^{q_w})$.

In order to describe the neighborhood functions, three sets derived from an instance $MCA(N; t, k, v_1^{q_1} \ldots v_g^{q_w})$ of the MCAP will be defined as follows: a) the set $\mathcal{C} = \{c_1, c_2, ..., c_l\}$, where each of its elements c is a t tuple to be covered; b) the set $\mathcal{A}$, where each of its elements $\mathcal{A}_i$ is a set containing the combinations of symbols that must be covered in the t tuple $c_i \in \mathcal{C}$; and c) the set $\mathcal{R} = \{r_1, ..., r_N\}$, where each element $r_i \in \mathcal{R}$ will be a test set of the MCA that will be constructed. The cardinality $l = |\mathcal{C}|$ is given by the expression $\binom{k}{t}$. The cardinality $|\mathcal{A}_i|$ of each $\mathcal{A}_i \in \mathcal{A}$ is given by $|\{0, 1, ..., v_i - 1\}|$, where v_i is the cardinality of the alphabet of column i in the MCA that is constructed. The cardinality of the set $\mathcal{R}$ is N, the expected number of rows in the MCA. Table 4 contains the sets $\mathcal{C}, \mathcal{A}, \mathcal{R}$ derived from the $MCA(7; 2, 5, 3^1 2^4)$ instance shown in Table 3.

The neighborhood function $\mathcal{N}_1$ consists in two phases. In the first phase searches for a t tuple $c' \in \mathcal{C}$ such that it contains at least one symbol combination a' missing. To do that, the function $\mathcal{N}_1$ randomly chooses a t tuple $c_i \in \mathcal{C}$ to start with; then, it checks if c_i has a symbol combination $a \in \mathcal{A}_i$ not covered yet. If the neighborhood function $\mathcal{N}_1$ fails in its first try, it takes the next combination in order c_{i+1} if $i + 1 < \binom{k}{t}$ otherwise it takes c_1. This process continues until a non covered t tuple c' is found and one of its missing symbol combination is identified, denoted by a'.

Once that a non covered t tuple $c' \in \mathcal{C}$ is found and a missing symbol combination a' identified, the second phase of $\mathcal{N}_1$ starts. This phase searches for the best test case $r \in \mathcal{R}$ where the symbol combination a' can substitute the

Table 4. Example of the sets $\mathcal{C}, \mathcal{A}, \mathcal{R}$ derived from instance $MCA(7; 2, 5, 3^1 2^4)$ shown in Table 3

$\mathcal{C} \leftarrow$	$\{c_1 = (a,b), c_2 = (a,c), c_3 = (a,d), c_4 = (a,e), c_5 = (b,c),$
	$c_6 = (b,d), c_7 = (b,e), c_8 = (c,d), c_9 = (c,e), c_{10} = (d,e)\}$

$\mathcal{A} \leftarrow$	$\{\mathcal{A}_1 = \{(0,0), (0,1), (1,0), (1,1), (2,0), (2,1)\},$
	$\mathcal{A}_2 = \{(0,0), (0,1), (1,0), (1,1), (2,0), (2,1)\},$
	$\mathcal{A}_3 = \{(0,0), (0,1), (1,0), (1,1), (2,0), (2,1)\},$
	$\mathcal{A}_4 = \{(0,0), (0,1), (1,0), (1,1), (2,0), (2,1)\},$
	$\mathcal{A}_5 = \{(0,0), (0,1), (1,0), (1,1)\},$
	$\mathcal{A}_6 = \{(0,0), (0,1), (1,0), (1,1)\},$
	$\mathcal{A}_7 - \{(0,0), (0,1), (1,0), (1,1)\},$
	$\mathcal{A}_8 = \{(0,0), (0,1), (1,0), (1,1)\},$
	$\mathcal{A}_9 = \{(0,0), (0,1), (1,0), (1,1)\},$
	$\mathcal{A}_{10} = \{(0,0), (0,1), (1,0), (1,1)\}\}$

$\mathcal{R} \leftarrow$	$\{r_1 = \{0,0,0,0,0\},$
	$r_2 = \{0,1,0,1,1\},$
	$r_3 = \{1,0,1,0,0\},$
	$r_4 = \{1,1,1,1,1\},$
	$r_5 = \{2,0,0,0,0\},$
	$r_6 = \{2,1,0,1,1\},$
	$r_7 = \{0,0,1,0,0\}\}$

symbols defined by the non covered t tuple c' in that case. The test case r' will be the one that, when substituting the symbols described by c' for the symbol combination a', minimizes the total number of missing symbol combination in the MCA constructed. The number of evaluations of the objective function done by the neighborhood function $\mathcal{N}_1$ is $O(N)$, because in the worst case the function requires to change the symbol combination for c' in each of the N test cases.

The neighborhood function $\mathcal{N}_2$ works directly over the test set $\mathcal{R}$ that is being formed. This function randomly selects a column or parameter from the test set (which in our case will be a value $1 \leq j \leq k$). After that, for each different test case $r_i \in \mathcal{R}$, the function $\mathcal{N}_2$ changes the symbol at $r_{i,j}$, where j is the jth symbol in $r_i \in \mathcal{R}$, and evaluates the number of missing symbol combinations. In this neighborhood function, every possible change of symbol in $r_{i,j}$ is made. The number of evaluation functions performed in $\mathcal{N}_2$ are $O((v_j - 1) \cdot N)$, because there are $v_j - 1$ possible changes of symbols in column j and there are N different test cases. The neighborhood function $\mathcal{N}_2$ will choose to change the symbol at $r_{i,j}$ to v' iif changing the jth symbol in test case r_i for the symbol v' minimizes the number of missing combinations among all the other possible changes of symbols performed by the function.

Finally, the neighborhood function $\mathcal{N}_3$ is a generalization of the function $\mathcal{N}_2$ in the sense that it performs all the changes of symbols in the whole test set $\mathcal{R}$. Again, the change of symbol that minimizes the number of missing combinations will be the one chosen by this function to create the new neighborhood. The number of evaluations of the objective function performed by this neighborhood function is $O((v_j - 1) \cdot N \cdot k)$.

The following subsection defines the evaluation function used in this paper to implement the TSA algorithm.

4.4   Evaluation Function

In this paper, the objective function $f(M, \mathcal{C}, \mathcal{A})$ that will be minimized by TSA is the number of combination of symbols missing in the matrix MCA that is constructed. For a particular matrix M that represents a MCA, and sets $\mathcal{C}$ and $\mathcal{A}$, a formal definition for this function is shown in Equation 1.

$$f(M, \mathcal{C}, \mathcal{A}) : \sum_{\forall c \in \mathcal{C}} \sum_{\forall A_i \in \mathcal{A}} \sum_{\forall a \in A_i} g(M, c, a) .$$

$$\textbf{where}\quad g(M, c, a) = \begin{cases} 1 & \textbf{if } a \text{ in } c \text{ has not been covered yet in } M \\ 0 & \textbf{otherwise} \end{cases} \tag{1}$$

An example of the use of the evaluation function $f(M, \mathcal{C}, \mathcal{A})$ is shown in Table 5, where the number of missing symbol combinations in matrix M shown in Table 3 is counted. Table 5 shows in the first column the different combinations of symbols to be covered in the matrix. The rest of the columns shows the different t tuples in the matrix and the number of times that each combination of symbol is covered in M. A symbol $-$ represents that a combination of symbols must not be satisfied in a certain combination c. The results obtained from $f(M, \mathcal{C}, \mathcal{A})$ is shown at the end of the table, note that the matrix M still has 9 missing combinations making it a non MCA.

Table 5. Matrix $\mathcal{P}$ of symbol combinations covered in M (from Table 3) and results from evaluating M with $f(M, \mathcal{C}, \mathcal{A})$

$\mathcal{P}$	c_1	c_2	c_3	c_4	c_5	c_6	c_7	c_8	c_9	c_{10}
$(0,0)$	2	2	2	2	2	4	4	2	2	4
$(0,1)$	1	1	1	1	2	0	0	2	2	0
$(1,0)$	1	0	1	1	2	0	0	2	2	0
$(1,1)$	1	2	1	1	1	3	0	1	1	3
$(2,0)$	1	2	1	1	$-$	$-$	$-$	$-$	$-$	$-$
$(2,1)$	1	0	1	1	$-$	$-$	$-$	$-$	$-$	$-$
				$f(M, \mathcal{C}, \mathcal{A}) = 9$						

The cost of evaluating $f(M, \mathcal{C}, \mathcal{A})$ is $O(\binom{k}{t} \times N)$, because the operation requires to examine the N rows of the matrix M and the $\binom{k}{t}$ different t tuples. With the aim of improving the time of this calculation, we implemented the $\mathcal{P}$ matrix. This matrix is shown in Table 5 and it is of size $\binom{k}{t} \times v_{max}$, where $v_{max} = \prod_{i=1}^{i=t} v_i$ and v_i is the ith alphabet cardinality taken in decreasing order from the cardinalities of the columns of M. Each element $p_{ij} \in \mathcal{P}$ contains the number of times that the ith combination of symbols is found in the t tuple $c_j \in \mathcal{C}$; the value of p_{ij} is not taken into account if the ith combination of symbols must not be included in the t tuple c_j.

To avoid the expensive cost $O(\binom{k}{t} \times N)$ at every call of $f(M, \mathcal{C}, \mathcal{A})$, the matrix $\mathcal{P}$ is used for a partial recalculation of the cost of M, i.e., the cost of changing a symbol in a cell $m_{ij} \in M$ is determined and only the affected t tuples in $\mathcal{P}$ are updated, modifying the results from $f(M, \mathcal{C}, \mathcal{A})$ according to that changes. The cells in $\mathcal{P}$ that must be updated when changing a symbol from $m_{ij} \in M$ are the t tuples that involve the column j of the matrix M. On this way, the complexity taken for the update of $f(M, \mathcal{C}, \mathcal{A})$ is reduced to $O(\binom{k-1}{t-1} \times 2)$.

The next section presents the results obtained from the implementation of the TSA algorithm, following the details described in the present section. The algorithm was compared against an existing benchmark reported in the literature and a new benchmark proposed in this paper.

5   Experimental Evaluation

The algorithm TSA was implemented in C language and compiled with gcc. The instances have been run on a cluster using eight processing nodes, each with two dual-core Opteron Processors. The features of each node are: Processor 2 X Dual-Core AMD, Opteron Processor 2220, 4GB RAM Memory, Operating Systems Red Hat Enterprise Linux 4 64-bit and gcc 3.4 Compiler.

In order to show the performance of the TSA algorithm, three experiments were developed. The first experiment had as purpose to fine tune the probabilities of the neighborhood functions to be selected. The second experiment evaluated the performance of TSA over a new benchmark proposed in this paper. The results were compared against one of the best algorithms in the literature that constructs MCA of variable strength, the IPOG-F algorithm [8]. Finally, the third experiment was oriented to compared the performance of TSA according with results obtained by other approaches in a benchmark reported in the literature. The performance of the algorithms was compared firstly in the number of test cases generated by each approach and secondly in the time spent by them.

5.1   First Experiment: Fine-Tuning the Neighborhood Functions

In the TSA algorithm, every time that a new neighbor should be created, the used neighborhood function will be $\mathcal{N}_1$ with probability p_1, $\mathcal{N}_2$ with probability p_2 and $\mathcal{N}_3$ with probability p_3. In order to determine the values for p_1, p_2, p_3 which give the best performance for TSA, a complete test set of discrete probabilities was formed and tested.

The possible values for p_1, p_2, p_3 were set up to $\{0.1, 0.2, ..., 1.0\}$. The complete test set of values for the probabilities was defined according with the solution of the Equation 2, which results in a test set with 66 different combinations of probabilities.

$$p_1 + p_2 + p_3 = 1 . \tag{2}$$

Every combination of the probabilities was applied by TSA to construct the $MCA(30; 2, 19, 6^1 5^1 4^6 3^8 2^3)$ and each experiment was run 31 times. A summary of the performance of TSA with the probabilities that solved the 100% of the runs is shown in Table 6. The performance of the best combinations of probabilities over TSA is shown in column 4 in ascending order of the time spent by them to construct the MCA.

Table 6. Performance of TSA with the 6 best combinations of probabilities which solved the 100% of the runs to construct the $MCA(30; 2, 19, 6^1 5^1 4^6 3^8 2^3)$

p_1	p_2	p_3	Avg. time[sec]
0.4	0.6	0.0	0.11
0.2	0.8	0.0	0.15
0.3	0.7	0.0	0.16
0.6	0.3	0.1	0.17
0.8	0.2	0.0	0.18
0.1	0.9	0.0	0.18

Given the results shown in Table 6 the best configuration of probabilities was $p_1 = 0.6$, $p_2 = 0.4$, $p_3 = 0$ because it found the MCA in smaller time (in average). The values $p_1 = 0.6$, $p_2 = 0.4$, $p_3 = 0$ were kept fixed in the following two experiments.

5.2   Second Experiment: Solving a New Benchmark

This experiment compares the performance of the TSA algorithm against IPOG-F, one of the best heuristics found in the literature to build test sets of variable strength. For this purpose, a benchmark of 18 instances was designed. First column of Table 7 shows the MCA instances of this new benchmark; the interaction level of these instances range from $t = 2$ to $t = 6$ and the column size varies from 6 to 20. The theoretical optimal size of each MCA is shown in column 2. Given that TSA and IPOG-F are non deterministic strategies, each instance was solved 10 times by them and the average time (in seconds) and the minimum size of the MCA constructed were reported; this information is shown in columns 3 to 6 in Table 7.

According with the results shown in Table 7, the TSA requires more time to build MCAs than IPOG-F in average, however the extra time consumed by TSA allowed the construction of MCAs of smaller size. The TSA algorithm achieved the optimal solution in 15 from the 18 cases, while IPOG-F did it in only 4. In conclusion, TSA can construct MCAs of smaller size than IPOG-F.

Table 7. Results of the performance of TSA for the new benchmark of MCA instances

Instance	N*	IPOG-F		TSA	
		N	Time [sec]	N	Time [sec]
$MCA(N; 2, 6, 2^2 3^2 4^2)$	16	16	0.009	16	0.00202
$MCA(N; 3, 6, 2^2 3^2 4^2)$	48	51	0.002	48	0.01647
$MCA(N; 4, 6, 2^2 3^2 4^2)$	144	146	0.019	144	0.11819
$MCA(N; 5, 6, 2^2 3^2 4^2)$	288	295	0.014	288	0.17247
$MCA(N; 6, 6, 2^2 3^2 4^2)$	576	576	0.004	576	0.00162
$MCA(N; 2, 8, 2^2 3^2 4^2 5^2)$	25	25	0.003	25	0.00716
$MCA(N; 3, 8, 2^2 3^2 4^2 5^2)$	100	107	0.009	100	17.50079
$MCA(N; 4, 8, 2^2 3^2 4^2 5^2)$	400	433	0.035	400	94.88019
$MCA(N; 5, 8, 2^2 3^2 4^2 5^2)$	1200	1357	0.201	1200	11379.21255
$MCA(N; 6, 8, 2^2 3^2 4^2 5^2)$	3600	3743	0.995	3600	7765.91885
$MCA(N; 2, 10, 2^2 3^2 4^2 5^2 6^2)$	36	36	0.004	36	0.06124
$MCA(N; 3, 10, 2^2 3^2 4^2 5^2 6^2)$	180	207	0.034	185	991.70933
$MCA(N; 2, 12, 2^2 3^2 4^2 5^2 6^2 7^2)$	49	50	0.006	49	0.42382
$MCA(N; 3, 12, 2^2 3^2 4^2 5^2 6^2 7^2)$	294	356	0.061	330	528.76392
$MCA(N; 2, 14, 2^2 3^2 4^2 5^2 6^2 7^2 8^2)$	64	67	0.002	64	1.53441
$MCA(N; 2, 16, 2^2 3^2 4^2 5^2 6^2 7^2 8^2 9^2)$	81	86	0.012	81	26.93236
$MCA(N; 2, 18, 2^2 3^2 4^2 5^2 6^2 7^2 8^2 9^2 10^2)$	100	107	0.016	100	702.30086
$MCA(N; 2, 20, 2^2 3^2 4^2 5^2 6^2 7^2 8^2 9^2 10^2 11^2)$	121	131	0.017	122	3927.93448

5.3 Third Experiment: Comparison with State-of-the-Art Algorithms

In this experiment, the goal was applied TSA to construct different test sets with $t = \{2, \ldots, 5\}$ for a Traffic Collision Avoidance System (TCAS) module, which has been used in other evaluations of software testing [13,19]. TCAS module has 12 parameters, two parameters have 10 values, one has 4 values, two parameters have 3 values and seven have 2 values, this configuration can be represented like $MCA(N; t, 12, 10^2 4^1 3^2 2^7)$. The obtained results were compared with those of some tools that construct t-wise MCA [15]. Table 8 shows the size of the test sets of every tool and the spend time in seconds as the case, when the size of the test set of a tool is not available is labeled like NA.

Table 8. Results of the performance of TSA for TCAS module compared with other approaches

t-way	TSA		IPOG-F		ITCH		Jenny		TConfig		TVG		Best
	Size	Time	Size	Time	Size	Time	Size	Time	Size	Time	Size	Time	
2	100	0.03	100	0.8	120	0.73	108	0	108	1 hour	101	2.75	100
3	400	26.21	400	0.36	2388	1020	413	0.71	472	12 hour	9158	3.07	400
4	1200	10449.12	1361	3.05	1484	5400	1536	3.54	1476	21 hour	64696	127	1360
5	3600	627079.08	4219	18.41	NA	>1 day	4580	43.54	NA	1 day	313056	1549	4218

Figure 1 compares the results shown in Table 8 involving the TSA algorithm and IPOG-F. The results show that TSA provided a better quality of solution for all t-wise of TCAS, constructing the MCAs of optimal size in all the cases. The performance of the algorithm was better in both spent time and quality of solution for $t = 2$.

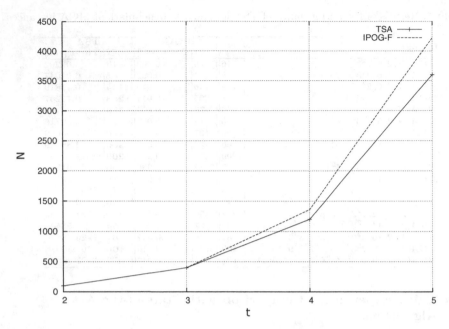

Fig. 1. Graphical comparison of the performance of TSA and IPOG-F when constructing $MCA(N; t, 12, 10^2 4^1 3^2 2^7)$. The graph shows how the size N of the MCA increases (y axis) when the strength increases (x axis). Note that the performance of TSA improves IPOG-F in every strength.

6   Conclusions

This paper presents a Tabu Search approach, referred as TSA, that deals with the problem of constructing Mixed Covering Arrays. The key features of TSA are: the use of mixture of three neighborhood functions to create neighbors, an efficient calculation of the objective function and a novel initialization function.

Each of the three different neighborhood functions had assigned a probability to be selected used by TSA to create a new neighbor. Given that the performance of TSA depends on the values of the probabilities assigned, this paper presented a fine tuning of the probabilities configurations based on a complete test set of discretized probabilities. The configuration used by TSA was the one that in this experiment produced the MCAs of smaller size and in less time, in average.

The TSA approach was compared against IPOG-F, one of the best state-of-art algorithms that has been used to construct MCAs of variable strength. Two benchmark were used in the comparison: a benchmark taken from the literature and a new benchmark proposed in this paper. With respect to the benchmark proposed in this paper, the TSA approach improved the size of the MCAs, in comparison with IPOG-F, and found the optimal solution in 15 instances of the 18 that constitutes the complete set; these instances ranges the number of cardinalities of the alphabets from 2 to 11, the number of columns from 6 to 20 and the strengths from 2 to 6. With respect to the benchmark taken from the literature, TSA improved the size of the matrices in comparison with the ones constructed by IPOG-F, finding the optimal solution in all the cases considered.

The empirical evidence presented in this paper showed that TSA improved the size of the MCAs in comparison with the tools that, to the best of our knowledge, are among the best found in the state-of-the-art of the construction of MCAs of variable strength.

Acknowledgments

This research was partially funded by the following projects: CONACyT 58554-Cálculo de Covering Arrays, 51623-Fondo Mixto CONACyT and Gobierno del Estado de Tamaulipas.

References

1. Bracho-Rios, J., Torres-Jimenez, J., Rodriguez-Tello, E.: A new backtracking algorithm for constructing binary covering arrays of variable strength. LNCS (LNAI), vol. 5845, pp. 397–407 (2009)
2. Bryce, R.C., Colbourn, C.J.: The density algorithm for pairwise interaction testing: Research articles. Softw. Test. Verif. Reliab. 17(3), 159–182 (2007)
3. Bryce, R.C., Colbourn, C.J.: One-test-at-a-time heuristic search for interaction test suites. In: GECCO 2007: Proceedings of the 9th Annual Conference on Genetic and Evolutionary Computation, pp. 1082–1089. ACM, New York (2007)
4. Bryce, R.C., Colbourn, C.J., Cohen, M.B.: A framework of greedy methods for constructing interaction test suites. In: Inverardi, P., Jazayeri, M. (eds.) ICSE 2005. LNCS, vol. 4309, pp. 146–155. Springer, Heidelberg (2006)
5. Cohen, D.M., Fredman, M.L., Patton, G.C.: The aetg system: An approach to testing based on combinatorial design. IEEE Transactions on Software Engineering 23, 437–444 (1997)
6. Cohen, M.B., Gibbons, P.B., Mugridge, W.B., Colbourn, C.J.: Constructing test suites for interaction testing. In: Proc. of the 25th Int. Conf. on Software Engeneering (ICSE 2003), pp. 38–48 (2003)
7. Colbourn, C.J., Cohen, M.B., Turban, R.C.: A deterministic density algorithm for pairwise interaction coverage. In: Proc. of the IASTED Intl. Conference on Software Engineering (2004)
8. Forbes, M., Lawrence, J., Lei, Y., Kacker, R.N., Kuhn, D.R.: Refining the in-parameter-order strategy for constructing covering arrays. Journal of Research of the National Institute of Standards and Technology 113(5), 287–297 (2008)
9. Glover, F., Laguna, M.: Tabu Search. Kluwer Academic Publishers, Dordrecht (1998)
10. Grindal, M., Offutt, J., Andler, S.F.: Combination testing strategies: a survey. Software testing verification & reliability 15, 167–199 (2005)
11. Hnich, B., Prestwich, S.D., Selensky, E., Smith, B.M.: Constraint models for the covering test problem. Constraints 11(2-3), 199–219 (2006)
12. Walker II, R.A., Colbourn, C.J.: Tabu search for covering arrays using permutation vectors. Journal of Statistical Planning and Inference 139(1), 69–80 (2009); Special Issue on Metaheuristics, Combinatorial Optimization and Design of Experiments

13. Kuhn, D.R., Okun, V.: Pseudo-exhaustive testing for software. In: Proceedings 30th Annual IEEE/NASA Software Engineering Workshop (SEW 2006), pp. 153–158 (April 2006)
14. Kuhn, D.R., Wallance, D.R., Gallo Jr., A.M.: Software fault interactions and implications for software testing. IEEE Transactions on Software Engineering 30, 418–421 (2004)
15. Lei, Y., Kacker, R., Kuhn, R.D., Okun, V., Lawrence, J.: Ipog: A general strategy for t-way software testing. In: 14th Annual IEEE International Conference and Workshops on the Engineering of Computer-Based Systems (ECBS 2007), pp. 549–556 (2007)
16. Lei, Y., Tai, K.C.: In-parameter-order: A test generation strategy for pairwise testing. In: Proceedings of Third IEEE International High-Assurance Systems Engineering Symposium, pp. 254–261 (1998)
17. Lopez-Escogido, D., Torres-Jimenez, J., Rodriguez-Tello, E., Rangel-Valdez, N.: Strength two covering arrays construction using a sat representation. In: Gelbukh, A., Morales, E.F. (eds.) MICAI 2008. LNCS (LNAI), vol. 5317, pp. 44–53. Springer, Heidelberg (2008)
18. Nurmela, K.J.: Upper bounds for covering arrays by tabu search. Discrete Applied Mathematics 138, 143–152 (2004)
19. Okun, V., Black, P.E., Yesha, Y.: Testing with model checker: Insuring fault visibility. WSEAS Transactions on Systems 2, 77–82 (2003)
20. Shiba, T., Kikuno, T., Tsuchiya, T.: Using artificial life techniques to generate test cases for combinatorial testing. In: Proceedings of the 28th Annual International Computer Software and Applications Conference (COMPSAC 2004), vol. 1, pp. 28–30 (2004)
21. Stardom, J.: Metaheuristics and the Search for Covering and Packing Arrays. PhD thesis, Simon Fraser University (2001)
22. Torres-Jimenez, J., de Alfonso, C., Hernández, V.: Computation of ternary covering arrays using a grid. In: Manandhar, S., Austin, J., Desai, U., Oyanagi, Y., Talukder, A.K. (eds.) AACC 2004. LNCS, vol. 3285, pp. 240–246. Springer, Heidelberg (2004)
23. Tung, Y.W., Aldiwan, W.S.: Automating test case generation for the new generation mission software system. In: Proceedings Aerospace Conference IEEE, vol. 1, pp. 431–437 (2000)
24. Williams, A.W.: Determination of test configurations for pair-wise interaction coverage. In: TestCom 2000: Proceedings of the IFIP TC6/WG6.1 13th International Conference on Testing Communicating Systems, Deventer, The Netherlands, pp. 59–74. Kluwer, B.V. (2000)
25. Williams, A.W., Probert, R.L.: A practical strategy for testing pair-wise coverage of network interfaces. In: Seventh Intl. Symp. on Software Reliability Engineering, pp. 246–254 (1996)
26. Yan, J., Zhang, J.: A backtraking search tool for constructing combinatorial test suites. The Journal of Systems and Software (2008), doi:10.1016/j.jss.2008.02.034

Feasibility-Based Bounds Tightening via Fixed Points*

Pietro Belotti[1], Sonia Cafieri[2], Jon Lee[3], and Leo Liberti[4]

[1] Dept. of Mathematical Sciences, Clemson University, Clemson SC 29634, USA
pbelott@clemson.edu
[2] Dept. Mathématiques et Informatique, ENAC, 7 av. E. Belin, 31055 Toulouse,
France
cafieri@recherche.enac.fr
[3] Dept. of Mathematical Sciences, IBM T.J. Watson Research Center, P.O. Box 218,
Yorktown Heights, NY 10598, USA
jonlee@us.ibm.com
[4] LIX, École Polytechnique, 91128 Palaiseau, France
liberti@lix.polytechnique.fr

Abstract. The search tree size of the spatial Branch-and-Bound algorithm for Mixed-Integer Nonlinear Programming depends on many factors, one of which is the width of the variable ranges at every tree node. A range reduction technique often employed is called Feasibility Based Bounds Tightening, which is known to be practically fast, and is thus deployed at every node of the search tree. From time to time, however, this technique fails to converge to its limit point in finite time, thereby slowing the whole Branch-and-Bound search considerably. In this paper we propose a polynomial time method, based on solving a linear program, for computing the limit point of the Feasibility Based Bounds Tightening algorithm applied to linear equality and inequality constraints.

Keywords: global optimization, MINLP, spatial Branch-and-Bound, range reduction, constraint programming.

1 Introduction

In this paper we discuss an important sub-step, called Feasibility Based Bounds Tightening (FBBT) of the spatial Branch-and-Bound (sBB) method for solving Mixed-Integer Nonlinear Programs (MINLP) of the form:

$$\left.\begin{array}{rl} \min & x_n \\ & g(x) \in G^0 \\ & x \in X^0 \\ \forall i \in Z & x_i \in \mathbb{Z}, \end{array}\right\} \tag{1}$$

where $x \in \mathbb{R}^n$ are decision variables, $\mathscr{I}$ is the set of all real intervals, $G^0 = [g^{0L}, g^{0U}] \in \mathscr{I}^m$ and $X^0 = [x^{0L}, x^{0U}] \in \mathscr{I}^n$ are vectors of real intervals, also

* Partially supported by grants ANR 07-JCJC-0151 "ARS", ANR 08-SEGI-023 "AsOpt", Digiteo Emergence "PASO", Digiteo Chair 2009-14D "RMNCCO", Digiteo Emergence 2009-55D "ARM", System@tic "EDONA".

W. Wu and O. Daescu (Eds.): COCOA 2010, Part I, LNCS 6508, pp. 65–76, 2010.
© Springer-Verlag Berlin Heidelberg 2010

called *boxes*, Z is a given subset of $\{1, \ldots, n\}$ encoding the integrality constraints on some of the variables, and $g : \mathbb{R}^n \to \mathbb{R}^m$ are continuous functions. Let $\mathscr{X}$ be the feasible region of (1). We remark that every MINLP involving a general objective function min $f(x)$ can be reformulated exactly to the formulation (1) at the cost of adjoining the constraint $x_n \geq f(x)$ to the constraints $g(x) \in G^0$.

The sBB is a ε-approximation algorithm (with a given $\varepsilon > 0$) for problems (1) which works by generating a sequence of upper bounds x'_n and lower bounds $\bar{x}_n$ to the optimal objective function value x^*_n. The upper bounding solution x' is found by solving (1) locally with MINLP heuristics [1,2], whilst the lower bound is computed by automatically constructing and solving a convex relaxation of (1). If $x'_n - \bar{x}_n \leq \varepsilon$ then x' is feasible and at most ε-suboptimal; if x' is the best optimum so far, it is stored as the *incumbent*. Otherwise X^0 is partitioned into two boxes X', X'' along a direction x_i at a branch point p_i, and the algorithm is called recursively on (1) with X^0 replaced by each of the two boxes X', X'' in turn. This generates a binary search tree: nodes can be pruned if the convex relaxation is infeasible or if $\bar{x}_n$ is greater than the objective function value at the incumbent. The sBB converges if the lower bound is guaranteed to increase strictly whenever the box X of ranges of x at the current node decreases strictly. In general, the sBB might fail to converge in finite time if $\varepsilon = 0$, although some exceptions exist [3,4].

An important step of the sBB algorithm is the reduction of the variable ranges X at each node. There are two commonly used range reduction techniques in Global Optimization (GO): Optimization Based Bounds Tightening (OBBT) and FBBT. The former is slower and more effective, involves the solution of $2n$ Linear Programs (LP) and is used either rarely or just at the root node of the sBB search tree [18]. The latter is an iterative procedure based on propagating the effect of the constraints $g(x) \in G^0$ on the variable ranges using interval arithmetic; FBBT is known to be practically efficient and is normally used at each sBB node. Practical efficiency notwithstanding, the FBBT sometimes converges to its limit point in infinite time, as the example of Eq. (3.11) in [5] shows. The same example also shows that an artificial termination condition enforced when the range reduction extent becomes smaller than a given tolerance might yield arbitrarily large execution times.

The FBBT was borrowed from Artificial Intelligence (AI) and Constraint Programming (CP) as a bounds filtering technique. Its origins can be traced to [6]; it is known *not* to achieve bound consistency [7] (apart from some special cases [8]), a desirable property for Constraint Satisfaction Problems (CSP): a CSP is bound consistent if every projection of its feasible region on each range X_i is X_i itself [5]. The FBBT was employed as a range reduction technique for Mixed Integer Linear Programs (MILP) in [9,10] and then for MINLPs in [11]. Within the context of GO, the FBBT was discussed in [12] and recently improved in [13] by considering its effect on common subexpressions of $g(x)$.

The main result of this paper is to show that if $g(x)$ are linear forms, then there exists an LP whose solution is exactly the limit point of the FBBT, which is therefore shown to be computable in polynomial time. If $g(x)$ are nonlinear functions (as is commonly the case for general MINLPs) we can replace them either by a linear relaxation $\bar{g}(x) \in \bar{G}^0$ of $\mathscr{X}$ or simply consider the largest linear subset $\hat{g}(x) \in \hat{G}^0$ of the constraints $g(x) \in G^0$, according to the usual trade-off between computational effort and result quality (we follow the latter approach in our computational results section).

The rest of this paper is organized as follows. In Sect. 2 we define an FBBT iteration formally as an operator in the interval lattice and show it has a fixed point. In Sect. 3 we show how to construct a linear relaxation of (1). In Sect. 4 we describe an LP the solution of which is the limit point of the FBBT algorithm. In Sect. 5 we discuss computational results on the MINLPLib showing the potential of the LP-based technique.

2   The FBBT Algorithm

The FBBT algorithm works by propagating the variable range vector $X^0 \in \mathscr{I}^n$ to the operators in $g(x)$ (using interval arithmetic) in order to derive the interval vector $G \in \mathscr{I}^m$ consisting of lower and upper bounds on $g(x)$ when $x \in X^0$; the vector $G \cap G^0$ is then propagated back to a variable range vector X using inverse interval arithmetic. The interval vector $X \cap X^0$ therefore contains valid and potentially tighter variable ranges for x. If $X \cap X^0 \subsetneq X^0$, then the procedure can be repeated with X^0 replaced by $X \cap X^0$ until no more change occurs. The FBBT algorithm can be shown to converge to a fixed point (fp), see Sect. 2.3. As mentioned in the introduction, the FBBT might fail to converge to its fp in finite time. With a slight abuse of terminology justified by Thm. 2.2 we shall refer to "nonconvergence" to mean "convergence to the fp in infinite time".

2.1   Expression Graphs

We make the assumption that the functions g appearing in (1) are represented by expressions built recursively as follows:

1. any element of $\{x_1, \ldots, x_n\} \cup \mathbb{R}$ is an expression (such primitive expressions are called *atoms*);
2. if e_1, e_2 are expressions, then $e = e_1 \otimes e_2$ is an expression for all operators $\otimes$ in a given set $\mathscr{O}$; e_1, e_2 are said to be *subexpressions* of e.

Let $\mathcal{E}$ be the set of all expressions built by the repeated applications of the two above rules, and for all $e \in \mathcal{E}$ let function(e) be the function $f : \mathbb{R}^n \to \mathbb{R}$ which e represents (this correspondence can be made precise using an evaluation function for expressions [19], p. 244); conversely, to all functions in $f \in$ function$(\mathcal{E})$ we let expression(f) be the expression e representing f. Each expression $e \in \mathcal{E}$ can also be associated to its recursion directed acyclic graph (DAG) dag(e) whose root node is e and whose other nodes are subexpressions of e, subexpressions of subexpressions of e and so on. An arc (u, v) in dag(e) implies a subexpression relation where v is a subexpression of u; with a slight abuse of notation we identify non-leaf nodes of dag(e) with operators (labelled $\otimes_v$ for an operator

$\otimes \in \mathcal{O}$ at a node v) and leaf nodes with atoms (labelled either x_i if the node is a variable, or the real constant that the node represents). For all nodes u of $\mathsf{dag}(e)$ we indicate by $\delta^+(u)$ the outgoing star of the node u, i.e. all vertices v such that (u, v) is an arc in $\mathsf{dag}(e)$.

We assume that $\mathcal{O}$ contains: the infix n-ary sum $+$, the infix binary difference $-$, left multiplication operators $a\times$ for each nonzero real constant $a \in \mathbb{R} \smallsetminus \{0\}$, the infix n-ary product $\times$, the infix binary division $\div$, right raising operators $(\cdot)^q$ for each rational constant $q \in \mathbb{Q}$, the left unary exponential operator exp, its inverse log and the left unary trigonometric operators sin, cos, tan. We also let $\mathcal{O}' = \{+, -, a\times\}$ be the set of linear operators. Endowed with the operator set $\mathcal{O}$, the MINLP formulation (1) can express all practically interesting MP problems from Linear Programming (LP), where $\mathcal{O}$ is replaced by $\mathcal{O}'$ and $Z = \emptyset$, Mixed-Integer Linear Programming (LP), where $\mathcal{O}$ is replaced by $\mathcal{O}'$ and $Z \neq \emptyset$, Nonlinear Programming (NLP), where $Z = \emptyset$, to MINLPs, where $Z \neq \emptyset$. By letting atoms range over matrices, formulation (1) can also encode Semidefinite Programming (SDP) [14]. Black-box optimization problems, however, cannot be described by formulation (1).

The expression representation for functions is well known in computer science [15], engineering [16,17] and GO [18,19,20] where it is used for two substeps of the sBB algorithm: lower bound computation and FBBT. More formal constructions of $\mathcal{E}$ can be found in [21], Sect. 3 and [22], Sect. 2.

2.2   The Problem DAG

We can also associate a DAG to the whole problem (1) to describe its symbolic nonlinear structure:

$$D' = \bigcup_{i \leq m} \mathsf{dag}(\mathsf{expression}(g_i)) \tag{2}$$

$$D = \mathsf{contract}(D', \{x_1, \ldots, x_n\}), \tag{3}$$

where $\mathsf{contract}(G, L)$ is the result of contracting all vertices of G labelled by ℓ for all $\ell \in L$, i.e. of replacing each subgraph of G induced by all vertices labelled by $\ell \in L$ by a single node labelled by ℓ. The difference between D' and D is that in D' some variable nodes are repeated; more precisely, if x_j occurs in both g_i and g_h, x_j will appear as a leaf node in both $\mathsf{dag}(\mathsf{expression}(g_i))$ and $\mathsf{dag}(\mathsf{expression}(g_h))$.

For all v in D (resp. D') we let $\mathsf{index}(v)$ be i if v is the root node for g_i, j if v is the node for x_j, and 0 otherwise. If v is a constant atom, we let c_v be the value of the constant.

2.3   Formal Definition of the FBBT Operator

An iteration of the FBBT consists of an upward phase, propagating variable ranges X to an interval vector G such that $g(x) \in G$, and of a downward phase, propagating $G \cap G^0$ down again to updated variable ranges X. Propagation occurs along the arcs of D. To each node v of D we associate an interval Y_v; initially we set $Y = Y^0$ where for all v representing variable nodes we have $Y_v^0 = X_{\text{index}(v)}^0$, for root nodes representing constraints we have $Y_v^0 = G_{\text{index}(v)}^0$, and $Y_v^0 = [-\infty, \infty]$ otherwise. Operators $\otimes_v$ act on intervals by means of interval arithmetic [23]. The upward propagation occurs along the arcs in the opposite direction, as shown in Alg. 1.

Algorithm 1. $\mathsf{up}(v, Y)$

Require: v (node of D), Y
Ensure: Y
1: **if** v is not an atom **then**
2:     **for all** $u \in \delta^+(v)$ **do**
3:        $Y = \mathsf{up}(u, Y)$
4:     **end for**
5:     Let $Y_v = Y_v \cap \otimes_v(Y_u \mid u \in \delta^+(v))$
6: **else if** v is a constant atom **then**
7:     $Y_v = [c_v, c_v]$
8: **end if**
9: **return** Y

The downward propagation is somewhat more involved and follows the arcs of D' in the natural direction. For each non-root node v in D' we define $\mathsf{parent}(v)$ as the unique node u such that (u, v) is an arc in D', the set $\mathsf{siblings}(v) = \delta^+(v) \setminus \{v\}$ and the set $\mathsf{family}(v) = \{\mathsf{parent}(v)\} \cup \mathsf{siblings}(v)$, i.e. the siblings and the parent of v. Let $z = \mathsf{parent}(v)$; then $\otimes_z$ induces a function $\mathbb{R}^{|\delta^+(z)|} \to \mathbb{R}$ such that:

$$w_z = \otimes_z(w_u \mid u \in \delta^+(z)). \tag{4}$$

We now define the operator $\otimes_v^{-1}$ as an "inverse" of $\otimes_z$ in the v coordinate. If there is an expression $e \in \mathcal{E}$ such that $w_v = \mathsf{function}(e)(w_u \mid u \in \mathsf{family}(v))$ if and only if (4) holds, let $\omega_v^{-1} = \mathsf{function}(e)$; otherwise, let $\otimes_v^{-1}$ map every argument tuple to the constant interval $[-\infty, \infty]$. The downward propagation, shown in Alg. 2, is based on applying $\otimes_v^{-1}$ recursively to D'.

We remark that down is defined on D' rather than D for a technical reason, i.e. family relies on nodes having a *unique* parent, which is the case for D' since it is the union of several DAGs; however, leaf nodes of D' representing the same variable $j \leq n$ are contracted to a single node in D, which therefore loses the parent uniqueness property at the leaf node level. Notwithstanding, Y is indexed on nodes of D rather than D', so that in Line 3 (Alg. 2), when u is a variable node we have $Y_u = X_{\text{index}(u)}$, so that the update on Y_u is carried out on the interval referring to the same variable independently of which DAG was used in the calling sequence.

Algorithm 2. down(v, Y)

Require: v (node of D'), Y
Ensure: Y
 1: **if** v is not an atom **then**
 2:     **for all** $u \in \delta^+(v)$ s.t. u is not a constant atom **do**
 3:       $Y_u = Y_u \cap \otimes_u^{-1}(\text{family}(u))$
 4:     **end for**
 5:     **for all** $u \in \delta^+(v)$ **do**
 6:       $Y = \text{down}(u, Y)$
 7:     **end for**
 8: **end if**
 9: **return** Y

If p is the number of vertices in D, up and down are operators $\mathscr{I}^p \to \mathscr{I}^p$. Since the up operator only changes the intervals in Y relating to a single expression DAG, we extend its action to the whole of D:

$$\mathcal{U}(Y) = \bigcap_{i \leq m} \text{up}(\bar{g}_i, Y), \tag{5}$$

where $\bar{g}_i$ denotes the root node of $\text{dag}(\text{expression}(g_i))$. Similarly, since down needs to be applied to each root node of D', we define:

$$\mathcal{D}(Y) = \bigcap_{i \leq m} \text{down}(\bar{g}_i, Y). \tag{6}$$

Finally, we define the FBBT operator:

$$\mathcal{F}(Y) = \mathcal{D}(\mathcal{U}(Y \cap Y^0) \cap Y^0). \tag{7}$$

Lemma 2.1. *The operators $\mathcal{U}, \mathcal{D}, \mathcal{F}$ are monotone and inflationary in the interval lattice $\mathscr{I}^p$ ordered by reverse inclusion $\supseteq$.*

Proof. Monotonicity follows because all the interval arithmetic operators in $\mathcal{O}$ are monotone [23], the composition of monotone operators is monotone, the functions represented by expressions in $\mathcal{E}$ are compositions of operators in $\mathcal{O}$. Inflationarity follows because of the intersection operators in Lines 5 of Alg. 1 and 3 of Alg. 2 and those in (5)-(7). $\square$

Theorem 2.2. *The operator $\mathcal{F}$ has a unique least fixed point.*

Proof. This follows by Lemma 2.1 and Thm. 12.9 in [24]. $\square$

We remark that since we ordered $\mathscr{I}^p$ by reverse inclusion, the least fixed point (lfp) of Thm. 2.2 is actually the greatest fixed point (gfp) with respect to standard interval inclusion.

3 Linear Relaxation of the MINLP

Different implementations of the sBB algorithm construct the lower bound $\bar{x}_n$ to the optimal objective function value x_n^* in different ways. Some are based on the factorability of the functions in $g(x)$ [25,26,27], whilst others are based on a symbolic reformulation of (1) based on the problem DAG D [18,19,20]. We employ the latter approach: each non-leaf node z in the vertex set of D is replaced by an added variable w_z and a corresponding constraint (4) is adjoined to to the formulation (usually, two variables w_v, w_u corresponding to identical defining constraints are replaced by one single added variable). The resulting reformulation, sometimes called *Smith standard form* [18], is exact [14]. A linear relaxation of (1) can automatically be obtained by the Smith standard form by replacing each nonlinear defining constraints by lower and upper linear approximations. In order for the sBB convergence property to hold, the coefficients of these linear approximations are functions of the variable ranges $X = [x^L, x^U]$, so that as the width of X_i decreases for some $i \leq n$, the optimal objective function value of the linear relaxation increases.

4 FBBT in the Linear Case

In this section we assume that the nonlinear part $g(x) \in G^0$ of (1) is either replaced by its linear relaxation $\bar{g}(x) \in \bar{G}^0$ as discussed in Sect. 3 or by the largest subset of linear constraints $\hat{g}(x) \in \hat{G}^0$ in (1), called the *linear part* of (1). If $\bar{\mathscr{X}} = \{x \in \mathbb{R}^n \mid \bar{g}(x) \in \bar{G}^0\}$ and $\hat{\mathscr{X}} = \{x \in \mathbb{R}^n \mid \hat{g}(x) \in \hat{G}^0\}$ are the feasible regions of the relaxation and of the linear part of (1), then $\bar{\mathscr{X}} \subseteq \hat{\mathscr{X}}$. Both $\bar{\mathscr{X}}$ and $\hat{\mathscr{X}}$ can be represented by systems of linear equations and inequalities:

$$Ax \in B^0, \tag{8}$$

where $A = (a_{ij})$ is a $m \times n$ real matrix. A system (8) encoding $\bar{\mathscr{X}}$ is a lifting in the w added variables and has in general many more constraints than a system (8) encoding $\hat{\mathscr{X}}$. Performing the FBBT on the relaxation will generally yield tighter ranges than on the linear part. For the purposes of this section, the construction of the LP yielding the gfp will be exactly the same in both cases.

The constraints in (8) are of the form $b^{0L} \leq a_i \cdot x \leq b^{0U}$, where $B^0 = [b^{0L}, b^{0U}]$ and a_i is the i-th row of A. Letting $J_i = \{j_1, \ldots, j_{k_i}\}$ be such that $a_{ij} \neq 0$ for all $j \in J_i$ and $a_{ij} = 0$ otherwise, we write $g_i(x) = \sum_{j \in J_i} a_{ij} x_j$, $e = \mathsf{expression}(g_i) = a_{ij_1} \times x_{j_1} + \cdots + a_{ij_{k_i}} \times x_{j_{k_i}}$ and hence $\mathsf{dag}(e)$ is the DAG shown in Fig. 1.

Because of their simple structure, the intervals Y_v where v represents operators such as $a_{ij} \times$ can be disposed of, and closed form interval expressions for $\mathcal{U}, \mathcal{D}$ can be derived that act on an interval vector (X, B) where $X \in \mathscr{I}^n$ and $B \in \mathscr{I}^m$. Specifically, $\mathcal{U}$ becomes:

$$\mathcal{U}(X, B) = (X, (B_i \cap \sum_{j \in J_i} a_{ij} X_j \mid i \leq m)) \tag{9}$$

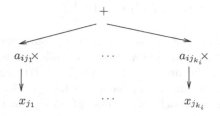

Fig. 1. The expression DAG of a row of (8)

and $\mathcal{D}$ becomes:

$$\mathsf{down}(X, B) = ((X_j \cap \bigcap_{\substack{i \leq m \\ a_{ij} \neq 0}} \frac{1}{a_{ij}}(B_i - \sum_{\ell \neq j} a_{i\ell} X_\ell) \mid j \leq n), B), \tag{10}$$

where the products and the sums are interval operations [23]. The equivalence of (9) with (5) in the linear case follows by simply replacing $a_{ij}X_j$ with an interval Y_v for a node v in D representing the operator $a_{ij}\times$, and similarly for the equivalence of (10) with (6), where we remark that the interval inverse $\otimes_v^{-1}$ of a linear form in each component is simply another linear form. We consequently re-define $\mathcal{F}$ for the linear case as follows:

$$\mathcal{F}(X, B) = \mathcal{D}(\mathcal{U}(X \cap X^0, B \cap B^0)). \tag{11}$$

4.1 Greatest Fixed Point via Linear Programming

The gfp of $\mathcal{F}$ with respect to interval inclusion is, by definition,

$$\mathrm{gfp}(\mathcal{F}) = \sup_{\subseteq}\{(X, B) \in \mathscr{I}^{n+m} \mid (X, B) = \mathcal{F}(X, B)\}. \tag{12}$$

Consider the interval vector width sum function $|\cdot| : \mathscr{I}^n \to \mathbb{R}_+$ given by $|X| = \sum_{j \leq n}(x_j^U - x_j^L)$. It is easy to see that it is monotonic with the lattice order $\subseteq$ on $\mathscr{I}^n$, in the sense that if $X \subseteq X'$ then $|X| \leq |X'|$ (the converse may not hold). Furthermore, for $(X, B) \in \mathscr{I}^{n+m}$ we have $|(X, B)| = |X| + |B|$.

4.1 Proposition

$|\mathrm{gfp}(\mathcal{F})| \geq |X| + |B|$ for all fixed points (X, B) of $\mathcal{F}$.

Proof. Since $(X^*, B^*) = \mathrm{gfp}(\mathcal{F})$ is the inclusion-wise greatest of all fixed points of $\mathcal{F}$, it is also maximal with respect to all the other fixed points that are included in it. Suppose, to get a contradiction, that there is a fixed point (X', B') of $\mathcal{F}$ with $|X'| + |B'| > |X^*| + |B^*|$; by assumption, (X', B') is not included in (X^*, B^*). Since the set of fixed points of $\mathcal{F}$ is a complete lattice by Tarski's Fixed Point theorem [28], there must be a fixed point (X'', B'') of $\mathcal{F}$ which includes both (X', B') and (X^*, B^*). By monotonicity of $|\cdot|$, $|X'| + |B'| \leq |X''| + |B''|$, and hence $|X^*| + |B^*| < |X''| + |B''|$, showing that $(X'', B'') \supsetneq (X^*, B^*)$ and hence that (X^*, B^*) is not the gfp of $\mathcal{F}$, against the hypothesis.   □

4.2 Theorem

The following interval programming problem:

$$\left.\begin{array}{l}\max |X| + |B| \\ (X, B) \subseteq (X^0, B^0) \\ (X, B) \subseteq \mathcal{U}(X, B) \\ (X, B) \subseteq \mathcal{D}(X, B)\end{array}\right\} \quad (13)$$

has a unique global optimum equal to $\mathrm{gfp}(\mathcal{F})$.

Proof. By Tarski's Fixed Point Theorem [28], the gfp of $\mathcal{F}$ is the join of all its pre-fixed points, so we can replace $=$ with $\subseteq$ in (12). The equivalence of (13) with (12) then follows by Prop. 4.1 and (11). Uniqueness of solution follows by uniqueness of the gfp.   □

We remark that (13) can be reformulated as an ordinary LP by simply replacing each interval $X_j = [x_j^L, x_j^U]$ by pairs of decision variables (x_j^L, x_j^U) with the constraint $x_j^L \le x_j^U$ (for all $j \le n$ and similarly for $B = [b^L, b^U]$). The constraints for $\mathcal{U}$ can be trivially reformulated from interval to scalar arithmetic; the equivalent reformulation for constraints in $\mathcal{D}$ is a little more involved but very well known [18,19,5]. The LP encoding the gfp of the FBBT is given in (14)-(34). The z variables have been added for clarity — they simply replace products $a_{ij}x_j$ appropriately, depending on the sign of a_{ij}. Let $S^+ = \{i \le m, j \le n \mid a_{ij} > 0\}$ and $S^-\{i \le m, j \le n \mid a_{ij} < 0\}$.

$$\max_{x,b,z} \sum_{j \le n} (x_j^U - x_j^L) + \sum_{i \le m} (b_i^U - b_i^L) \quad (14)$$

$$\forall j \le n \quad x_j^L \ge x_j^{0L} \quad (15)$$

$$\forall j \le n \quad x_j^U \le x_j^{0U} \quad (16)$$

$$\forall i \le m \quad b_j^L \ge b_i^{0L} \quad (17)$$

$$\forall i \le m \quad b_j^U \le b_i^{0U} \quad (18)$$

$$\forall (i,j) \in S^+ \quad z_{ij}^L = a_{ij}x_j^L \quad (19)$$

$$\forall (i,j) \in S^+ \quad z_{ij}^U = a_{ij}x_j^U \quad (20)$$

$$\forall (i,j) \in S^- \quad z_{ij}^L = a_{ij}x_j^U \quad (21)$$

$$\forall (i,j) \in S^- \quad z_{ij}^U = a_{ij}x_j^L \quad (22)$$

$$\forall i \le m \quad b_i^L \ge \sum_{j \le n} z_{ij}^L \quad (23)$$

$$\forall i \le m \quad b_i^U \le \sum_{j \le n} z_{ij}^U \quad (24)$$

$$\forall (i,j) \in S^+ \quad x_j^L \ge \frac{1}{a_{ij}}(b_i^L - \sum_{\ell \ne j} z_{i\ell}^U) \quad (25)$$

$$\forall (i,j) \in S^+ \quad x_j^U \le \frac{1}{a_{ij}}(b_i^U - \sum_{\ell \ne j} z_{i\ell}^L) \quad (26)$$

$$\forall (i,j) \in S^- \quad x_j^L \ge \frac{1}{a_{ij}}(b_i^U - \sum_{\ell \ne j} z_{i\ell}^L) \quad (27)$$

$$\forall (i,j) \in S^+ \quad x_j^U \le \frac{1}{a_{ij}}(b_i^U - \sum_{\ell \ne j} z_{i\ell}^U) \quad (28)$$

$$\forall j \le n \quad x_j^L \le x_j^U \quad (29)$$

$$\forall i \le m \quad b_i^L \le b_i^U \quad (30)$$

$$\forall i \le m, j \le n \quad z_{ij}^L \le z_{ij}^U \quad (31)$$

$$x^L, x^U \in \mathbb{R}^n \quad (32)$$

$$b^L, b^U \in \mathbb{R}^m \quad (33)$$

$$z^L, z^U \in \mathbb{R}^{mn}. \quad (34)$$

Constraints (15)-(18) encode $(X, B) \subseteq (X^0, B^0)$; constraints (19)-(24) encode $(X, B) \subseteq \mathcal{U}(X, B)$; constraints (25)-(28) encode $(X, B) \subseteq \mathcal{D}(X, B)$; and constraints (29)-(31) encode the fact that the decision variables x^L, x^U, b^L, b^U (as well as the auxiliary variables z^L, z^U) represent intervals.

If we formalize the problem of finding the gfp of the FBBT operator as a decision problem (for example deciding if the gfp width sum is smaller than the original bounds X^0 by at least a given constant $\gamma > 0$), then Thm. 4.2 shows that this problem is in **P**. It is interesting to remark that [29] proves that essentially the same problem (with a few more requirements on the type of allowed constraints in $Ax \in G^0$) is **NP**-complete as long as variable integrality constraints are enforced.

5   Computational Results

Our testbed consists of the MINLPLib [30] instance library. We first artificially restricted each instance to $X \in \{-10^4, 10^4\}$ in order for $|\cdot|$ to be bounded. Secondly, we ran the FBBT on the linear constraints $\hat{g}(x) \leq \hat{G}^0$, with a termination condition set at $|X^k \triangle X^{k-1}| \leq 10^{-6}$ on slow progress at iteration k. We then computed the fixed point by solving the LP (14)-(34) applied to $\hat{g}(x) \leq \hat{G}^0$. The FBBT, as well as the automatic construction of the LP (14)-(34), were implemented in ROSE [31,32]. LPs were solved using CPLEX 11 [33]. All results were obtained on one core of an Intel Core 2 Duo at 1.4GHz with 3GB RAM running Linux.

Out of 194 MINLPs in the MINLPLib we obtained results for 172 (the remaining ones failing on some AMPL [34] error). For each successful instance we computed the width sum $|X|$ of the obtained solution and the user CPU time taken by the traditional FBBT method and by the LP based one. As these methods would be typically used in a sBB algorithm, we ignored the LP construction time, since this would be performed just once at the root node and then simply updated with the current node interval bounds X^0. The full table can be accessed at http://www.lix.polytechnique.fr/~liberti/fbbtlp_table-1007.csv. Table 1 only reports the totals, averages and standard deviations of the sample.

Table 1. Totals, averages, standard deviations of the width sum and CPU times taken by FBBT and LP on 172 MINLPLib instances

Statistic	FBBT		LP-based	
	$\|X\|$	CPU	$\|X\|$	CPU
Total	9.38×10^7	394.37	9.16×10^7	9.07
Average	5.4×10^5	2.29	5.3×10^5	0.05
Std. dev.	1.365×10^6	17.18	1.362×10^6	0.18

Table 1 is consistent with what was empirically observed about the FBBT: it often works well but it occasionally takes a long time converging to the fixed point. The LP-based method addresses this weakness perfectly, as shown by the markedly better CPU time statistics. Since the LP finds the guaranteed gfp, it also produces somewhat tighter interval bounds, although the savings in terms of $|X|$ are not spectacular. The traditional FBBT was strictly faster than the LP-based method in 43% of the instances. The bulk of the CPU time savings of the LP-based method is due to twelve FBBT CPU time outliers taking $> 1s$ (the results are shown in Table 2). For nine of these the large CPU time is actually due to problem size, i.e. the number of FBBT iterations is low (< 5). Three of the instances displayed clear signs of nonconvergence.

Table 2. FBBT CPU time outlier instances

Name	FBBT CPU time	FBBT iterations
cecil_13	27.6	29
nuclear14b	16.25	2
nuclear14	5.96	1
nuclear24b	16.19	2
nuclear24	5.89	1
nuclear25b	18.97	2
nuclear25	6.78	1
product	11.21	4
risk2b	2.28	2
space960	218.41	2
super3t	39.96	15
util	12.78	321

References

1. D'Ambrosio, C., Frangioni, A., Liberti, L., Lodi, A.: Experiments with a feasibility pump approach for nonconvex MINLPs. In: Festa, P. (ed.) Experimental Algorithms. LNCS, vol. 6049, pp. 350–360. Springer, Heidelberg (2010)
2. Liberti, L., Mladenović, N., Nannicini, G.: A good recipe for solving MINLPs. In: Maniezzo, V., Stützle, T., Voß, S. (eds.) Hybridizing Metaheuristics and Mathematical Programming. Annals of Information Systems, vol. 10, pp. 231–244. Springer, New York (2009)
3. Al-Khayyal, F., Sherali, H.: On finitely terminating branch-and-bound algorithms for some global optimization problems. SIAM Journal of Optimization 10(4), 1049–1057 (2000)
4. Bruglieri, M., Liberti, L.: Optimal running and planning of a biomass-based energy production process. Energy Policy 36, 2430–2438 (2008)
5. Hooker, J.: Integrated methods for optimization. Springer, New York (2007)
6. Waltz, D.: Understanding the line drawings of scenes with shadows. In: Winston, P. (ed.) The Psychology of Computer Vision, pp. 19–91. McGraw-Hill, New York (1975)
7. Davis, E.: Constraint propagation with interval labels. Artificial Intelligence 32, 281–331 (1987)
8. Faltings, B.: Arc-consistency for continuous variables. Artificial Intelligence 65, 363–376 (1994)
9. Savelsbergh, M.: Preprocessing and probing techniques for mixed integer programming problems. INFORMS Journal on Computing 6(4), 445–454 (1994)
10. Andersen, D., Andersen, K.: Presolving in linear programming. Mathematical Programming 71, 221–245 (1995)
11. Shectman, J., Sahinidis, N.: A finite algorithm for global minimization of separable concave programs. Journal of Global Optimization 12, 1–36 (1998)
12. Schichl, H., Neumaier, A.: Interval analysis on directed acyclic graphs for global optimization. Journal of Global Optimization 33(4), 541–562 (2005)
13. Vu, X.H., Schichl, H., Sam-Haroud, D.: Interval propagation and search on directed acyclic graphs for numerical constraint solving. Journal of Global Optimization 45, 499–531 (2009)
14. Liberti, L.: Reformulations in mathematical programming: Definitions and systematics. RAIRO-RO 43(1), 55–86 (2009)
15. Knuth, D.: The Art of Computer Programming, Part II: Seminumerical Algorithms. Addison-Wesley, Reading (1981)

16. Bogle, I., Pantelides, C.: Sparse nonlinear systems in chemical process simulation. In: Osiadacz, A. (ed.) Simulation and Optimization of Large Systems. Clarendon Press, Oxford (1988)

17. Keeping, B., Pantelides, C.: Novel methods for the efficient evaluation of stored mathematical expressions on scalar and vector computers. AIChE Annual Meeting, Paper #204b (November 1997)

18. Smith, E., Pantelides, C.: A symbolic reformulation/spatial branch-and-bound algorithm for the global optimisation of nonconvex MINLPs. Computers & Chemical Engineering 23, 457–478 (1999)

19. Liberti, L.: Writing global optimization software. In: Liberti, L., Maculan, N. (eds.) Global Optimization: from Theory to Implementation, pp. 211–262. Springer, Berlin (2006)

20. Belotti, P., Lee, J., Liberti, L., Margot, F., Wächter, A.: Branching and bounds tightening techniques for non-convex MINLP. Optimization Methods and Software 24(4), 597–634 (2009)

21. Cafieri, S., Lee, J., Liberti, L.: On convex relaxations of quadrilinear terms. Journal of Global Optimization 47, 661–685 (2010)

22. Costa, A., Hansen, P., Liberti, L.: Formulation symmetries in circle packing. In: Mahjoub, R. (ed.) Proceedings of the International Symposium on Combinatorial Optimization. Electronic Notes in Discrete Mathematics. Elsevier, Amsterdam (accepted)

23. Moore, R., Kearfott, R., Cloud, M.: Introduction to Interval Analysis. SIAM, Philadelphia (2009)

24. Roman, S.: Lattices and Ordered Sets. Springer, New York (2008)

25. McCormick, G.: Computability of global solutions to factorable nonconvex programs: Part I — Convex underestimating problems. Mathematical Programming 10, 146–175 (1976)

26. Adjiman, C., Dallwig, S., Floudas, C., Neumaier, A.: A global optimization method, αBB, for general twice-differentiable constrained NLPs: I. Theoretical advances. Computers & Chemical Engineering 22(9), 1137–1158 (1998)

27. Tawarmalani, M., Sahinidis, N.: Global optimization of mixed integer nonlinear programs: A theoretical and computational study. Mathematical Programming 99, 563–591 (2004)

28. Tarski, A.: A lattice-theoretical fixpoint theorem and its applications. Pacific Journal of Mathematics 5(2), 285–309 (1955)

29. Bordeaux, L., Hamadi, Y., Vardi, M.: An analysis of slow convergence in interval propagation. In: Bessière, C. (ed.) CP 2007. LNCS, vol. 4741, pp. 790–797. Springer, Heidelberg (2007)

30. Bussieck, M., Drud, A., Meeraus, A.: MINLPLib — A collection of test models for mixed-integer nonlinear programming. INFORMS Journal on Computing 15(1) (2003)

31. Liberti, L., Cafieri, S., Savourey, D.: Reformulation-Optimization Software Engine. In: Fukuda, K., et al. (eds.): ICMS 2010. LNCS, vol. 6327, pp. 303–314. Springer, Heidelberg (2010)

32. Liberti, L., Cafieri, S., Tarissan, F.: Reformulations in mathematical programming: A computational approach. In: Abraham, A., Hassanien, A.E., Siarry, P., Engelbrecht, A. (eds.) Foundations of Computational Intelligence. Studies in Computational Intelligence, vol. 3(203), pp. 153–234. Springer, Berlin (2009)

33. ILOG: ILOG CPLEX 11.0 User's Manual. ILOG S.A., Gentilly, France (2008)

34. Fourer, R., Gay, D.: The AMPL Book. Duxbury Press, Pacific Grove (2002)

A Characterisation of Stable Sets in Games with Transitive Preference

Takashi Matsuhisa

Department of Natural Sciences, Ibaraki National College of Technology
Nakane 866, Hitachinaka-shi, Ibaraki 312-8508, Japan
mathisa@ge.ibaraki-ct.ac.jp

Abstract. This article characterises stable sets in an abstract game. We show that every stable subset of the pure strategies for the game is characterised as a fixed point of the mapping assigning to each upper boundedly preordered subset of the strategies the set of all its maximal elements.

Keywords: Abstract game, Jiang mapping, Stable sets, Transitive preference.

AMS 2010 Mathematics Subject Classification: Primary 91A10, Secondary 91A12.

Journal of Economic Literature Classification: C71, C72.

1 Introduction

This article investigates a stable property in strategies of abstract games. In cooperative game theory the central solution concept is stable sets, which are sets of outcomes on which a preference relation $\precsim$ satisfies the two property: Reflexivity ($x \precsim x$) and Transitivity (If $x \precsim y$ and $y \precsim z$ then $x \precsim z$.) A stable set consists of outcomes satisfying (i) the *internal stability* (for every outcomes being not stable, some coalition has an objection), and (ii) the *external stability* (no coalition has an objection to any stable outcome.)

This solution concept is introduced as a 'standard of behaviour' by von Neumann and Morgenstern [5]. The stable sets can be treated in the abstract game framework. Many mathematical difficulties still arise in the stable sets when the preference is not transitive.

Jiang [3] treats abstract games with *transitive* preferences, which arise from strategic games. He addresses the existence problem of the stable sets in the set of Nash equilibria of a strategic game. Regarding to the original intents of von Neumann and Morgenstern [5], it is unpleasant to restrict stable sets to subsets of the Nash equilibria set for the strategic game a priori.

This article aims to improve the point: Removing out the restriction of stable sets to subsets of Nash equilibria we treat stable sets in the framework of abstract games; we address a class of abstract games having stable sets, and characterise

W. Wu and O. Daescu (Eds.): COCOA 2010, Part I, LNCS 6508, pp. 77–84, 2010.
© Springer-Verlag Berlin Heidelberg 2010

the stable sets as the maximal sets in an upper bounded set of the outcomes. Our main result is as follows:

Characterisation theorem. *Every stable subset of the pure strategies for an abstract game is characterised as a fixed point of the mapping assigning to each upper boundedly preordered preference subset of the strategies the set of all its maximal elements.*

After reviewing basic notions and terminology, Section 2 presents the extended notion of stable sets (Definition 2) in an abstract game. The notion of upper bounded game (Definition 3) is also introduced, which plays crucial role in determining the existence of stable sets. Section 3 introduces the Jiang mapping for an abstract game and presents the main theorem (Theorem 1) and the characterisation theorem (Theorem 2). Section 4 establishes the main theorem, from which the theorem of Jiang [3] follows as a corollary. Finally I conclude with some remarks on the assumptions in the theorems.

2   Model

2.1   Preference

A binary relation R on a non-empty set X is a subset of $X \times X$ with (x, y) denoted by xRy. A relation R may satisfy one and more properties:

Ref (Reflexivity) For all $x \in X, xRx$;

Trn (Transitivity) For all $x, y, z \in X$, if xRy and yRz then xRz;

Sym (Symmetry) For all $x, y \in X, xRy$ implies yRx;

Asym (Antisymmetry) For all $x, y \in X$, if xRy and yRx then $x = y$;

Cmp (Completeness) For all $x, y \in X$, we have xRy or yRx (or both).

Definition 1. A *preference* relation on a non-empty X is a binary relation $\precsim$ on X. For $x, y \in X$ we will read '$x \precsim y$ as 'y is at least as preferable as x.' The *strict* preference relation $\prec$ is defined by

$$x \prec y \iff x \precsim y \text{ but not } y \precsim x.$$

The *indifference* relation $\sim$ is defined by

$$x \sim y \iff x \precsim y \text{ and } y \precsim x.$$

A set X together with a definite preference relation $\precsim$ will be called a *preference set* denoted by $(X, \precsim)$. A preference relation on X is called *rational* if it satisfies the properties **Trn** and **Cmp**. A *preorder* on X is a binary relation $\precsim$ on X satisfying the properties **Ref** and **Trn**. A preorder with **Asym** is called *partial order*.

Remark 1. Let $(X, \precsim)$ be a rational preference set. The following properties are true:

(i) $\precsim$ is a preorder.
(ii) $\prec$ is irreflexive ($x \prec x$ is never true) and transitive;
(iii) $\sim$ is an equivalence relation.

Let $(X, \precsim)$ be a preordered set and Y a subset of X. A *maximal* element of Y is an element a such that Y contains no element b with $a \prec b$. An element a is an *upper bound* of Y in case $x \precsim a$ for every x in Y. A subset Y of X is called *upper bounded* in X if it has at least one upper bound in it. A subset Y is called a *chain* if for every $x, y \in Y$, either $x \precsim y$ or $y \precsim y$ has to hold.

It is worthy noting that

Lemma 1. *Each of the following two statements is equivalent to the axiom of choice:*[1]

(i) (Zorn's Lemma) *Let $(X, \precsim)$ be a preordered set. If each chain in X has an upper bound then X has at least one maximal element.*
(ii) (Maximal Principle) *Let X be a partially ordered set. Each chain in X is contained in a maximal chain.*

2.2   Stable Sets and Bounded Games

An *n*-person *abstract game* is a tuple $\Gamma = \langle N, (A_i)_{i \in N}, (\precsim) \rangle$ consisting of

1. N is a set of *n players* with $n \geq 2$ and i denotes a player;
2. A_i is a non-empty set of player *i*'s *pure strategies* and $A = \prod_{i \in N} A_i$ is the set of *strategies*;
3. $\precsim$ is a binary relation on $A = \prod_{i \in N} A_i$, called *preference*.

Definition 2. Let $\Gamma = \langle N, (A_i)_{i \in N}, \precsim \rangle$ be an *n*-person abstract game and Y a non empty subset of $A = \prod_{i \in N} A_i$. A non-empty subset V of Y is said to be *von Neumann-Morgenstern stable in Y* or simply, *Y-stable* if the two conditions hold:

IS (Internal stability) For any $a, b \in V$, neither $a \prec b$ nor $b \prec a$ holds;
ES (External stability) For any $b \in Y \setminus V$ there exist an $a \in V$ such that $b \prec a$.

If a subset of A is an A-stable set then it will be simply called *N-M stable*.

We denote by $\mathrm{NMS}(\Gamma; Y)$ the set of all stable sets in Y of $A = \prod_{i \in N} A_i$, and denote by $\mathrm{NMS}(\Gamma)$ the set of all stable subsets in some non-empty set of $A = \prod_{i \in N} A_i$; i.e.,

$$\mathrm{NMS}(\Gamma) = \cup_{\emptyset \neq Y \subseteq A} \mathrm{NMS}(\Gamma; Y).$$

---

[1] See pp. 31-32 and p.58 in [1].

Definition 3. Let $\Gamma = \langle N, (A_i)_{i \in N}, \precsim \rangle$ be an n-person abstract game and Y a non empty subset of $A = \prod_{i \in N} A_i$. The game Γ is called Y-*upper bounded* if the following two conditions are true:

(i) $(Y, \precsim)$ is a preordred set, and
(ii) Each chain in Y has an upper bound in Y.

The game Γ will be called simply *upper bounded* if it is A-upper bounded.

Denote by $\mathrm{TUB}(\Gamma)$ the set of all non-empty upper bounded subsets of $A = \prod_{i \in N} A_i$. We will write by $\mathrm{NMS}^*(\Gamma)$ the set of all stable subsets in some upper bounded set of $A = \prod_{i \in N} A_i$; i.e.,

$$\mathrm{NMS}^*(\Gamma) = \cup_{Y \in \mathrm{TUB}(\Gamma)} \mathrm{NMS}(\Gamma; Y).$$

2.3   Classical Case

An n-person *strategic game* is a tuple $\Gamma = \langle N, (A_i)_{i \in N}, (\precsim_i)_{i \in N} \rangle$ consisting of

1. N and A_i are the same as above;
2. $\precsim_i$ is i's *rational* preference relation.

The *uniform preference* relation $\precsim$ on $A = \prod_{i=1}^n A_i$ is a binary relation on A defined by

$$a \precsim b \iff a \precsim_i b \text{ for any } i \in N.$$

The *strict* preference relation $\prec$ is defined by

$$a \prec b \iff a \precsim b \text{ but not } b \precsim a.$$

The *indifference* relation $\sim$ is defined by

$$x \sim y \iff x \precsim y \text{ and } y \precsim x.$$

Remark 2. The game $\Gamma = \langle N, (A_i)_{i \in N}, \precsim \rangle$ with the uniform preference is an abstract game. The preference $\precsim$ is a preorder on A, but it is not always rational; i.e., it satisfies **Ref** and **Trn**, but not **Cmp** in general.

A profile $a^* = (a_1^*, \cdots, a_i^*, \cdots, a_n^*)$ is a *pure* Nash equilibrium for a strategic game $\Gamma = \langle N, (A_i)_{i \in N}, (\precsim_i)_{i \in N} \rangle$ provided that for each $i \in N$ and for every $a_i \in A_i$, $(a_{-i}^*, a_i) \precsim_i a^*$. We denote by $\mathrm{PNE}(\Gamma)$ the set of all pure Nash equilibria for Γ.

Remark 3. In his paper [3], Jiang calls a strategic game Γ *regular* if Γ is $\mathrm{PNE}(G)$-upper bounded. By N-M stable set Jiang [3] means a stable set in $\mathrm{PNE}(\Gamma)$.

Example 1 (Jiang [3]). The *vagabonds game*, in which the preference relation is derived from individual utility functions, is the tuple $\langle \Gamma, (A_i), (\precsim_i) \rangle$ consisting of

1. $N = \{1, 2, \cdots, n\}$ is a set of n players called *vagabonds* ($n \in \mathbb{N}$) and i denotes a vagabond;
2. $A_i = \mathbb{R}_+$

3. i's utility function $u_i : \mathbb{R}^n_+ \to \mathbb{R}$ is given by

$$u_i(x_1, \cdots, x_i, \cdots, x_n) = \begin{cases} x_i & \text{if } \sum_{j \in N} x_j \in [0, 1] \\ 0 & \text{if } \sum_{j \in N} x_j \in (1, +\infty) \end{cases}$$

4. $\precsim_i$ is i's preference relation represented by the function u_i as follows: For any $x, x' \in A = \prod_{i=1}^{n} A_i$, $x \precsim_i x' \iff u_i(x) \le u_i(x')$.

Set

$$S_n = \{x = (x_1, \cdots, x_i, \cdots, x_n) \in \mathbb{R}^n_+ \mid \sum_{i \in N} x_i = 1\}$$

Then the game Γ is upper bounded with S_n a stable set in $\text{PNE}(\Gamma)$. Moreover, it is 'regular' in the sense of Jiang [3].

Remark 4. The game Γ actually contains the unique stable set S_n in $\text{PNE}(\Gamma)$. This can be verified by Corollary 1 that will be shown later in the next section.

3   Main Theorem

Let $\Gamma = \langle N, (A_i)_{i \in N}, \precsim \rangle$ be an n-person abstract game.

Definition 4. By the *Jiang mapping* for the abstract game Γ, we mean the mapping $J_\Gamma : \text{TUB}(\Gamma) \to \text{NMS}(\Gamma)$ which assigns to each Y of $\text{TUB}(\Gamma)$ the set V_Y of all maximal elements in Y: For each $Y \in \text{TUB}(\Gamma)$,

$$J_\Gamma(Y) = V_Y = \{y \in Y \mid y \text{ is maximal in } Y\} \quad \text{if } Y \neq \emptyset;$$
$$J_\Gamma(\emptyset) = \emptyset \quad \text{otherwise.}$$

We can now state our main result.

Theorem 1 (Main theorem). *Let* $\Gamma = \langle N, (A_i)_{i \in N}, \precsim \rangle$ *be an n-person abstract game. The Jiang mapping J_Γ is well-defined mapping with the property: $J_\Gamma \circ J_\Gamma = J_\Gamma$. Furthermore, it is a surjective map onto the set $NMS^*(\Gamma)$ of all stable sets in some upper bounded subset of strategies in the game Γ.*

Before proceeding with the proof we will establish the characterisation theorem for stable sets mentioned in Section 1, and we state the theorem explicitly: Let $\text{Fix}(J_\Gamma)$ denote the set of all fixed members of $\text{TUB}(\Gamma)$ for J_Γ:

$$\text{Fix}(J_\Gamma) = \{Y \in \text{TUB}(\Gamma) \mid J_\Gamma(Y) = Y \}.$$

Theorem 2 (Characterisation theorem). *Let J_Γ be the Jiang mapping for an n-person abstract game Γ with the preorder preference $\precsim$. Then the set $NMS^*(\Gamma)$ of all stable sets in some upper bounded subset of strategies in the game Γ coincides with the set of all fixed points of the Jiang mapping J_Γ in $TUB(\Gamma)$; i.e.,*

$$\text{NMS}^*(\Gamma) = \text{Fix}(J_\Gamma).$$

In particular, every $W \in NMS^*(\Gamma)$ can be uniquely expressed by the form $W = J_\Gamma(Y)$ for some $Y \in TUB(\Gamma)$.

Proof. For any $Y \in \text{Fix}(J_\Gamma)$, it immediately follows that $Y = J_\Gamma(Y) \in \text{TUB}(\Gamma) \cap \text{NMS}(\Gamma) \subseteq \text{NMS}^*(\Gamma)$, and hence $\text{Fix}(J_\Gamma) \subseteq \text{NMS}^*(\Gamma)$. The converse will be shown as follows: Let us take any $W \in \text{NMS}^*(\Gamma)$. By the surjectivity of J_Γ it follows that there is a $Y \in \text{TUB}(\Gamma)$ such that $W = J_\Gamma(Y)$, and thus it can be plainly seen by the property for J_Γ in Theorem 1 that $W \in \text{Fix}(J_\Gamma)$ because $J_\Gamma(W) = J_\Gamma(J_\Gamma(Y)) = J_\Gamma(W) = W$. Therefore we obtain that $\text{NMS}^*(\Gamma) \subseteq \text{Fix}(J_\Gamma)$, in completing the proof.   □

4   Proof of Theorem 1

We shall proceed with the proof by the following steps:

J_Γ is a well-defined mapping: This follows immediately from the below theorem:

Theorem 3. *Let $\Gamma = \langle N, (A_i)_{i \in N}, \precsim \rangle$ be an n-person abstract game with the preorder preference and Y a non-empty subset of $A = \prod_{i \in N} A_i$. If Γ is Y-upper bounded then it has the unique stable set in Y.*

Proof. **Existence**: Let V denote the set $J_\Gamma(Y)$ of all maximal elements in a preordered set $(Y, \precsim)$. By Lemma 1(i) we can observe that V is a non-empty set. We shall show that V is a stable set in Y.

For **IS**: On noting that each element in V is maximal, **IS** follows immediately.

For **ES**: Suppose to the contrary that there exists a $y_0 \in Y \setminus V$ such that for every $x \in V$, it is not true that $y_0 \prec x$. However, since y_0 is not maximal in Y, there is a $y_1 \in Y$ such that $y_0 \prec y_1$. Let $\mathcal{T}$ be the set of all the chains C satisfying the two conditions: (1) C consists of elements $x \in Y$ strictly prefereable than y_0 (i.e.; $x \succ y_0$), and (2) C contains the chain $T_0 = \{y_0 \prec y_1\}$. It is plainly seen that $T_0 \in \mathcal{T} \neq \emptyset$ and that $\mathcal{T}$ is a partially ordered set equipped with the set theoretical inclusion. Hence it follows by Lemma 1(ii) that the chain $T_0 \in \mathcal{T}$ is contained in a maximal chain T^* in Y. Since Γ is upper-bounded, the chain T^* has an upper bound $y^* \in Y$, and so it immediately follows that $T^* \cup \{y^*\} \in \mathcal{T}$ because $y_0 \prec y^*$ and $x \precsim y^*$ for all $x \in \mathcal{T}$ except y_0. This means that $T^* \cup \{y^*\} \in \mathcal{T}$ which properly contains T^*, in contradiction to the maximality of T^* in $\mathcal{T}$, as required.

Uniqueness: Suppose V and W are stable sets in Y with $V \neq W$. Without loss of generality we may assume $V \setminus W \neq \emptyset$. Take $a \in V \setminus W$. It follows from **ES** for W that there exists $b \in W$ such that $a \prec b$. By **IS** for V we obtain $b \in Y \setminus V$. From **ES** for V it follows that there exists $c \in V$ such that $b \prec c$, and thus $a \prec b \prec c$. By **Trn** on Y we obtain that $a \prec c$ for $a, c \in V$, in contradiction to **IS** for V, in completing the proof of Theorem 3.   □

$J_\Gamma \circ J_\Gamma = J_\Gamma$: It is easily seen that $\text{NMS}^*(\Gamma; Y) \subseteq \text{TUB}(\Gamma)$ for any $Y \in \text{TUB}(\Gamma)$, and so $\text{NMS}^*(\Gamma) \subseteq \text{TUB}(\Gamma)$. It follows that the composite mapping

$J_\Gamma \circ J_\Gamma$ is well-defined. For each $Y \in \mathrm{TUB}(\Gamma)$, in viewing of the definition of J_Γ it can be plainly observed that $J_\Gamma(Y)$ is the set of all maximal elements in Y, and so $J_\Gamma(Y) \subseteq Y$. Therefore it follows from **IS** that $Y \subseteq J_\Gamma(W)$, and $J_\Gamma \circ J_\Gamma(Y) = J_\Gamma(Y)$, as required. $\qquad\qquad\square$

J_Γ **is a surjection onto** $\mathbf{NMS}^*(\boldsymbol{\Gamma})$**:** For any $W \in \mathrm{NMS}^*(\Gamma)$, we can take $Y \in \mathrm{TUB}(\Gamma)$ such that $W \in \mathrm{NMS}^*(\Gamma; Y)$. By the same argument as above we can obtain that $W = J_\Gamma(W)$, as required. This completes the proof of Theorem 1.
$\qquad\qquad\square$

As consequence of Theorem 3 we obtain the Jiang' s theorem:

Corollary 1 (Jiang [3]). *Every PNE(Γ)-upper bounded strategic game Γ has the unique stable set in PNE(Γ).*

5   Concluding Remarks

It well ends this article by giving remarks on the assumptions on Theorem 3: Transitivity on preference and upper boundedness for a game. These assumptions play crucial role in the theorem.

Game with non-transitive preference having no stable set: We can easily construct such game: See the game in Figure 1 in Lucas [4] (p.545) has no stable set at all. $\qquad\qquad\square$

Non-upper bounded game having no stable set: Let $\langle \mathbb{R}_+, \leq \rangle$ be the real line equipped with the usual inequality, and we will consider it as one player strategic game. Then we can easily observe that the game is neither $\mathbb{R}_+$-upper bounded nor has stable set in $\mathbb{R}_+$. $\qquad\qquad\square$

Conclusion: This article treats the notion of stable sets in an abstract form game with transitive preference. We investigate conditions under which the stable sets are guaranteed. The main theorem shows that the stable sets for the abstract game is characterised as the fixed point of the mapping assigning to each upper bounded subset in the pure strategies the subset of the maximal elements of it. The key is to establish that the stable set uniquely exists in each inductive set of pure strategies for the abstract game. In the classical case of strategic game, we obtain Jiang's result as a consequence, which guarantees the unique stable set in the set of Nash equilibria for the game in the case the Nash equilibrium set is inductive.

The emphasis is that the continuity on the preferences is not assumed in the theorems as we can view in Example 1. However the two assumptions, transitivity on preference and upper boundedness for a game, play crucial role in guaranteeing existence of the stable set in the game. These comments show that the two assumptions are necessary to the theorems.

Acknowledgment. I thank the anonymous reviewers for their instructive comments to improve this article.

References

1. Dugunji, J. : Topology. xvi+447, pp. Wm. C. Brown Publishers, Iowa, USA (1966)
2. Jiang, D.: Realizability of of Expected Equilibria of N-Person Condition Game under Strong Knowledge System. International Journal of Innovative Computing, Information and Control 2(4), 761–770 (2006)
3. Jiang, D.: N-M Stable Set of a Regular Game and its Unique Existence Theorem in System ZFC. Journal of Systems Science and Mathematical Sciences (to appear, 2010)
4. Lucas, W.F.: Von Neumann-Morgenstern Stable Sets. In: Aumann, R.J., Hart, S. (eds.) Handbook of Game Theory, vol. I, ch. 17, pp. 543–590. Elsevier Science Publisher B.V., Amsterdam (1992)
5. von Neumann, J., Morgenstern, O.: Theory of Games and Economic Behavior. Princeton University Press, Princeton (1944); 2'nd edition (1947); 3'rd edition (1953)

Linear Coherent Bi-cluster Discovery via Beam Detection and Sample Set Clustering

Yi Shi*, Maryam Hasan, Zhipeng Cai, Guohui Lin, and Dale Schuurmans

Department of Computing Science, University of Alberta
Edmonton, Alberta T6G 2E8, Canada
{ys3,mhasan1,zhipeng,ghlin,dale}@cs.ualberta.ca

Abstract. We propose a new bi-clustering algorithm, LinCoh, for finding *linear coherent* bi-clusters in gene expression microarray data. Our method exploits a robust technique for identifying conditionally correlated genes, combined with an efficient density based search for clustering sample sets. Experimental results on both synthetic and real datasets demonstrated that LinCoh consistently finds more accurate and higher quality bi-clusters than existing bi-clustering algorithms.

Keywords: Bi-clustering, sample set clustering, gene expression microarray, gene ontology.

1 Introduction

Gene expression microarray data analysis interprets the expression levels of thousands of genes across multiple conditions (also called samples). Such a study enables the language of biology to be spoken in mathematical terms; however, it remains a challenge to extract useful information from the large volume of raw expression data.

One central problem in gene expression microarray data analysis is to identify groups of genes that have similar expression patterns in a common subset of conditions. Standard clustering methods, such as k-means clustering [4], hierarchical clustering [21] and self-organizing maps [20], are ill-suited to this purpose for two main reasons: that genes exhibit similar behaviors only under some, but not all conditions, and that genes may participate in more than one functional process and hence belong to multiple groups. Bi-clustering [9,16] is intended to overcome the limitations of standard clustering methods by identifying a group of genes that exhibit similar expression patterns in a subset of conditions. Bi-clustering was first applied to gene expression analysis a decade ago [3], subsequently leading to dozens of other bi-clustering algorithms. Nevertheless, the general bi-clustering problem is NP-hard [3]. Efforts were invested in designing bi-clustering algorithms, mostly heuristics, for finding postulated types of bi-clusters.

---

* Corresponding author. Tel.: +1 (780) 492 2285; Fax: +1 (780) 492 6393.

W. Wu and O. Daescu (Eds.): COCOA 2010, Part I, LNCS 6508, pp. 85–103, 2010.

There are several types of bi-clusters that have been sought previously, including (a) the constant value model, (b) the constant row model, (c) the constant column model, (d) the additive coherent model, where each row (or column) is obtained by adding a constant to another row (or column, respectively), and (e) the multiplicative coherent model, where each row (or column) is obtained by multiplying another row (or column, respectively) by a constant value. In this paper, we continue to exploit the most general type-(f) linear coherent model [7] (see Figure 1), in which each row is obtained by multiplying another row by a constant value and then adding a constant. We further assume that bi-clusters are arbitrarily positioned and may overlap each other [15]. The most biologically meaningful types of bi-clusters to be sought should map to the ultimate purpose of identifying groups of genes that co-participate in certain genetic regulatory process. For example, housekeeping genes are those that constitutively express in most conditions, and they could be identified in the first two bi-cluster models (a) and (b). Most of the existing bi-clustering algorithms seek either type-(d) additive bi-clusters or type-(e) multiplicative bi-clusters [7]. Mathematically, the type-(f) linear coherent model is strictly more general than all the other five models.

x	y	z	w
1.0	1.0	1.0	1.0
1.0	1.0	1.0	1.0
1.0	1.0	1.0	1.0
1.0	1.0	1.0	1.0

(a)

x	y	z	w
1.2	1.2	1.2	1.2
0.8	0.8	0.8	0.8
1.5	1.5	1.5	1.5
0.6	0.6	0.6	0.6

(b)

x	y	z	w
1.2	0.8	1.5	0.6
1.2	0.8	1.5	0.6
1.2	0.8	1.5	0.6
1.2	0.8	1.5	0.6

(c)

x	y	z	w
1.2	0.8	1.5	0.6
1.0	0.6	1.3	0.4
2.0	1.6	2.3	1.4
0.7	0.3	1.2	0.3

(d)

x	y	z	w
2.0	4.0	8.0	1.0
1.0	2.0	4.0	0.5
4.0	8.0	16.0	2.0
1.0	2.0	4.0	0.5

(e)

x	y	z	w
2.0	4.0	3.0	5.0
1.5	2.5	2.0	3.0
2.3	4.3	3.3	5.3
4.5	8.5	6.5	10.5

(f)

Fig. 1. The six different bi-cluster types

The key idea in our new algorithm, LinCoh, for finding type-(f) linear coherent bi-clusters is illustrated in Figure 2. Essentially, a pair of genes participates in a linear coherent bi-cluster must be evidenced by a non-trivial subset of samples in which these two genes are co-up-regulated (or co-down-regulated). Therefore, the scatter plot of their pairwise expression levels, see Figure 2, where every point (x, y) represents a sample in which the two genes have expression levels x and y respectively, must show a diagonal band with a sufficient number of sample points. The LinCoh algorithm starts with composing this non-trivial supporting sample set for each gene pair, then to cluster these so-called *outer* sample sets. Each outer sample set cluster, together with the associated genes and *inner* samples, is filtered to produce a final bi-cluster.

We compare our LinCoh algorithm to four most popular bi-clustering algorithms: Cheng and Church's algorithm named after CC [3]; the order preserving sub-matrix algorithm denoted as OPSM [2]; the iterative signature algorithm denoted as ISA [10]; and the maximum similarity bi-clustering algorithm denoted as MSBE [14]. The first three algorithms have been selected and implemented in a survey [17]. Cheng and Church defined a merit score called *mean squared*

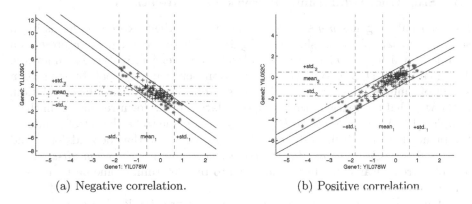

(a) Negative correlation.   (b) Positive correlation.

Fig. 2. (a) illustrates two yeast genes *YIL078W* and *YLL039C* that have negative expression correlation under a subset of conditions; the red conditions provide a stronger evidence than the blue conditions, whereas the green conditions do not suggest any correlation. Similarly in (b), genes *YIL078W* and *YIL052C* show a positive expression correlation.

residue to evaluate the quality of a bi-cluster, and CC is a greedy algorithm for finding bi-clusters of score no less than a given threshold [3]. OPSM is a heuristic algorithm attempting to find within a gene expression matrix the sub-matrices, *i.e.* bi-clusters, in each of which the genes have the same linear ordering of expression levels [2]. Another bi-cluster quality evaluation scheme is proposed in [10] using gene and condition signatures, and the ISA is proposed for finding the corresponding good quality bi-clusters. In particular, a randomized ISA is put in place when the prior information of the expression matrix is not available. The last algorithm, MSBE, is the first polynomial time bi-clustering algorithm that finds optimal solutions under certain constraints [14].

The rest of the paper is organized as follows: Section 2 presents the details of our LinCoh algorithm. In Section 3, we first introduce the quality measurements for bi-clustering results; then describe how synthetic datasets were generated, followed by the bi-clustering results and discussion; lastly we present the two real datasets on yeast and e.coli respectively, as well as the bi-clustering results and discussion. We conclude the paper in Section 4 with some remarks on the advantages and disadvantages of our LinCoh algorithm.

2   The LinCoh Algorithm

Let $E(G, S)$ be an $n \times m$ gene expression data matrix, where $G = \{1, 2, \ldots, n\}$ is the set of gene indices and $S = \{1, 2, \ldots, m\}$ is the set of sample (condition) indices. Its element e_{ij} is the expression level of gene i in sample j. Our LinCoh algorithm consists of two major steps, described in the next two subsections.

2.1   Step One: Establishing Pairwise Gene Relations

For each pair of genes $p, q \in \{1, 2, \ldots, n\}$, we plot their expression levels in all samples in a 2D plane, as shown in Figure 2, where a point (x, y) represents a sample in which gene p has expression level x and gene q has expression level y. The goal of this step is to detect a correlation between every pair of genes on a subset of samples, if any. Such a subset of samples must evidence the correlation, in the way that the two genes are co-up-regulated (or co-down-regulated) in these samples [13]. We define a *beam* $B_{\theta,\beta,\gamma}$ in the 2D plane to be the set of points on the plane that are within distance $\frac{1}{2}\beta$ to a straight line that depends on θ and γ. Here θ is the *beam angle*, β is the *beam width*, and γ is the *beam offset*. They are all search parameters, but we are able to pre-determine some best values or ranges of values for them.

Let μ_p and σ_p (μ_q and σ_q) denote the mean expression level of gene p (q, respectively) across all samples and the standard deviation. To identify the subset of supporting samples for this gene pair, $S_{\theta,\beta,\gamma} = S \cap B_{\theta,\beta,\gamma}$, we seek for a beam $B_{\theta,\beta,\gamma}$ in the 2D plane that aligns approximately the main diagonal (or the antidiagonal) of the rectangle defined by $\{(\mu_p - \sigma_p, \mu_q - \sigma_q), (\mu_p + \sigma_p, \mu_q - \sigma_q), (\mu_p + \sigma_p, \mu_q + \sigma_q), (\mu_p - \sigma_p, \mu_q + \sigma_q)\}$. Such an approximate alignment optimizes the following objective function, which robustly leads to good quality bi-clusters:

$$\max_{\theta,\beta,\gamma} \ W_{S_{\theta,\beta,\gamma}} \cdot D_{S_{\theta,\beta,\gamma}}$$

$$\text{subject to:} \ \left| corr\big(E(p, S_{\theta,\beta,\gamma}), E(q, S_{\theta,\beta,\gamma})\big) \right| \geq t_{cc}.$$

In the above maximization problem, $D_{S_{\theta,\beta,\gamma}}$ is the vector of the Euclidean distances of the samples inside the beam, *i.e.* $S_{\theta,\beta,\gamma}$, to the line passing through (μ_p, μ_q) and perpendicular to (called the *midsplit line* of) the beam center line; $W_{S_{\theta,\beta,\gamma}}$ is a weight vector over the samples in $S_{\theta,\beta,\gamma}$, and we use Shepard's function $w_j = d_j^r / \sum d_j^r$ with parameter $r \geq 0$ to weight sample $j \in S_{\theta,\beta,\gamma}$ (to weight more on distant samples but less on samples nearby the midsplit line); In the constraint, $|corr(\cdot, \cdot)|$ is the absolute correlation coefficient between the two genes p and q, calculated over only the samples in $S_{\theta,\beta,\gamma}$; t_{cc} is a pre-determined correlation threshold.

The output of the maximization problem is $S_{\theta,\beta,\gamma}$, which is either empty, indicating that no meaningful relationship between the two genes was found, or otherwise a subset of samples that evidence a meaningful correlation between genes p and q. According to our extensive preliminary experiments, the bi-clustering results are of high quality when the correlation threshold t_{cc} is larger than 0.75; and it is set to 0.90 and 0.75, respectively, on synthetic datasets and real datasets.

We implement a heuristic process to search for the beam, whose center line is initialized to be the line passing through the main diagonal (for positive correlation) or the antidiagonal (for negative correlation) of the rectangle in the expression plot. The beam width β is fixed at a certain portion of $4\sigma_p\sigma_q / \sqrt{\sigma_p{}^2 + \sigma_q{}^2}$; again supported by the preliminary experiments, a constant portion in between 0.8 and 1.0 is sufficient to capture most meaningful correlations; and we fix it at 0.8 for both synthetic datasets and real datasets. That is, $\beta = 0.8 \times 4\sigma_p\sigma_q/\sqrt{\sigma_p{}^2 + \sigma_q{}^2}$. To determine the beam angle θ in the positive correlation case, we define the search axis to be the midsplit line of the main diagonal; a small interval is placed on the search axis centering at (μ_p, μ_q), which is for a pivot point to float within; around each position of the pivot point, whose distance to the center point (μ_p, μ_q) is denoted as γ, different angles (the θ) are searched over to find a beam center line; each resultant beam is tested for the constraint satisfaction in the maximization problem, and discarded otherwise; among all those beams that satisfy the constraint, the one maximizing the objective function is returned as the target beam.

For evaluating the objective function, we have tested multiple values of r and found that 0 gives the most robust bi-clustering results. Therefore, r is set to 0 as default. For each sample j inside the beam, its distance d_j to the mid-split line of the beam center line is rounded to 0 or 1 using a threshold of $\sqrt{\sigma_p{}^2 + \sigma_q{}^2}$. When the target beam is identified, though might not be the true optimum to the objective function, the sample set $S_{\theta,\beta,\gamma}$ is further partitioned into *outer* sample set (containing the samples with distance d_j rounded to 1) and *inner* sample set (containing the rest). Gene pairs, together with non-empty outer and inner sample sets, are sent to Step two for clustering.

2.2 Step Two: Sample Set Clustering

Step one generates an outer sample set and an inner sample set for each gene pair. In this step, two $n \times n$ matrices are constructed: in the *outer matrix* M^o, the element m_{pq}^o is the outer sample set for gene pair p and q; likewise, in the *inner matrix* M^i, the element m_{pq}^i is the inner sample set for gene pair p and q. We next process these two matrices to robustly find bi-clusters.

First we want to filter out small outer sample sets that indicate insignificant correlations for gene pairs. To this purpose, we select roughly the largest 0.15% outer sample sets among all for clustering, which are of 99.7% confidence. This confidence level is set after testing on a randomly generated datasets, with 68%, 95%, 99.7% confidence levels according to the 68-95-99.7 rule. Observing that two disjoint gene pairs could have the same outer sample sets but very different expression patterns, simply using outer sample sets to construct bi-clusters might lead to meaningless bi-clusters. In our LinCoh algorithm, genes are used as bridges to group similar outer sample sets to form bi-clusters, since linear correlation is transitive. We first define the similarity between two outer sample sets m_{pq}^o and m_{rs}^o as $\mathrm{sim}(m_{pq}^o, m_{rs}^o) = |m_{pq}^o \cap m_{rs}^o|/|m_{pq}^o \cup m_{rs}^o|$, which is a most popular measure in the literature. Next, we rank genes in the descending order of the number of associated non-empty outer sample sets. Iteratively, the gene at the head of this list is used as the *seed* gene, to collect all its associated (non-empty) outer sample sets. These outer sample sets are partitioned into clusters using a density based clustering algorithm similar to DBSCAN [5], and the densest cluster is returned, which is defined as a cluster whose central point has the most close neighbors (see the pseudocode in the Appendix). An initial bi-cluster is formed on the union S_1 of the outer sample sets in the densest cluster, and the set G_1 of the involved genes.

The quality of the bi-cluster (G_1, S_1) is evaluated by its *average absolute correlation coefficient*,

$$\text{aacc}(G_1, S_1) = \frac{\sum_{p,q \in G_1} |corr(E(p, S_1), E(q, S_1))|}{(|G_1|^2 - |G_1|)}. \tag{1}$$

The initial bi-cluster (G_1, S_1) is refined in three steps to locally improve its quality. In the first step, all samples in S_1 are sorted in decreasing frequency of occurrence in all the outer sample sets of the seed gene; the lowest ranked sample is removed if this removal improves the quality of the bi-cluster, or otherwise the first step is done. Secondly, every gene in G_1 is checked to see whether its removal improves the quality of the bi-cluster, and if so it is removed from G_1. At the end, if the minimum gene pairwise absolute correlation coefficient of the bi-cluster is smaller than a threshold, the bi-cluster is considered as of low quality and discarded. By examining values from 0.50 to 0.99, our preliminary experiments showed that a high threshold in between 0.7 and 0.9 is able to ensure good quality bi-clusters, and it is set to 0.8 in all our final experiments. In the last step of bi-cluster (G_1, S_1) refinement, the inner sample sets of the gene pairs from G_1 are collected; their samples are sorted in decreasing frequencies in these inner sample sets; using this order, samples are added to S_1 as long as their addition passes the 0.8 minimum pairwise absolute correlation coefficient. A final bi-cluster (G_1, S_1) is thus produced.

Subsequently, all genes from G_1 are removed from the gene list, and the next gene is used as the seed gene for finding the next bi-cluster. The process iterates till the gene list becomes empty. We remark that a gene can participate in multiple bi-clusters, but it serves as a seed gene at most once. At the end, when two bi-clusters overlap more than 60% area, the one of smaller size is treated as redundant and discarded [13]. A pseudocode of our LinCoh algorithm 1 is provided in the Appendix for the interested readers.

3   Results and Discussion

We examine the LinCoh algorithm, and make comparisons with four other existing bi-clustering algorithms, CC, OPSM, ISA, and MSBE (their parameter settings follow the previous works [17,14]), on many synthetic datasets and two real gene expression microarray datasets on *Saccharomyces cerevisiae* (yeast) and *Escherichia coli* (e.coli) respectively. Essentially, synthetic datasets are used for evaluating absolute performance, since we know the ground truth; while the real datasets are mainly used for evaluating relative performance. Consequently we have different sets of performance measurements on synthetic and real datasets.

On synthetic datasets, bi-clustering algorithms are evaluated on their ability to recover the implanted (true) bi-clusters. Prelić's gene match score and overall match score [17] are adopted. Let $\mathcal{C}$ and $\mathcal{C}^*$ denote the set of output bi-clusters from an algorithm and the set of true bi-clusters for a dataset. The gene match score of $\mathcal{C}$ with respect to the target $\mathcal{C}^*$ is defined as $\text{score}_G(\mathcal{C}, \mathcal{C}^*) = \frac{1}{|\mathcal{C}|} \sum_{(G_1, S_1) \in \mathcal{C}} \max_{(G_1^*, S_1^*) \in \mathcal{C}^*} \frac{|G_1 \cap G_1^*|}{|G_1 \cup G_1^*|}$, which is essentially the average of the maximum gene match scores of bi-clusters in $\mathcal{C}$ with respect to the target bi-clusters. Similarly, the sample match score $\text{score}_S(\mathcal{C}, \mathcal{C}^*)$ can be defined by replacing gene sets with the corresponding sample sets in the above. The overall match score is then defined as their geometric mean, $i.e.$

$$\text{score}(\mathcal{C}, \mathcal{C}^*) = \sqrt{\text{score}_G(\mathcal{C}, \mathcal{C}^*) \times \text{score}_S(\mathcal{C}, \mathcal{C}^*)}.$$

On real datasets, the bi-clusters discovered by an algorithm are mapped to known biological pathways, defined in the GO functional classification scheme [1], the KEGG pathways [11], the MIPS yeast functional categories [18] (for yeast dataset), and the EcoCyc database [12] (for e.coli dataset), to obtain their *gene functional enrichment score* as implemented in [13]. The average absolute correlation coefficients (aacc's) of the discovered bi-clusters are also used to compare different algorithms.

3.1   Synthetic Datasets

Noise resistance test: This experiment examines how well a bi-clustering algorithm can recover implanted bi-clusters. We follow Prelić's testing strategy to first generate a 100×50 background matrix (*i.e.*, 100 genes and 50 samples), using a standard normal distribution for the matrix elements; we then embed ten 10×5 non-overlapping linear coherent bi-clusters along the diagonal; later for each vector of the five expression values, the first two of them are set to down-regulated, the last two are set to up-regulated, and the middle one is non-regulated; lastly, we add noise to the embedded bi-clusters at six different noise levels ($\ell = 0.00, 0.05, 0.10, 0.15, 0.20, 0.25$) by perturbing the entry values so that the resultant values are ℓ away from the original values. The generation is repeated ten times to give ten matrices.

The same simulation process is done to generate synthetic datasets containing additive bi-clusters, when we compare the bi-clustering algorithms on their performance to recover additive bi-clusters only (which is a special case of linear coherent bi-clusters).

Figure 3 shows the gene match scores of all five bi-clustering algorithms at six different noise levels, on their performance of recovering linear coherent bi-clusters and additive bi-clusters, respectively. Their overall match scores and gene discovery rates (defined as the percentage of genes in the output bi-clusters over all the genes in the true bi-clusters) can be found in Figures 8 and 9 in the Appendix. In terms of match scores, Figures 3 and 8 clearly show that our LinCoh outperformed all the other four algorithms, ISA ranked the second, and the other three performed quite poorly. In terms of gene discovery rate, again LinCoh outperformed all the other four algorithms. We remark that gene discovery rate can be trivially lifted up by simply output more bi-clusters. It is not a main measure used in this work, but a useful measure in conjunction with match scores.

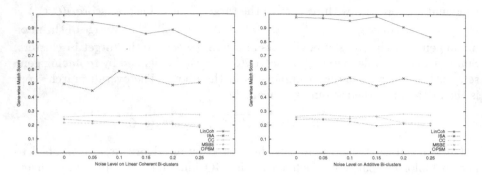

Fig. 3. The gene match scores of the five algorithms on recovering linear coherent bi-clusters and additive bi-clusters at six different noise levels

Overlapping test: Individual genes can participate in multiple biological processes, yielding bi-clusters that overlap with common genes in an expression matrix. Bi-clusters might also overlap with a subset of samples. This experiment is designed to examine the ability of different bi-clustering algorithms to recover overlapping bi-clusters. As before, we consider type-(f) linear coherent bi-clusters and type-(d) additive bi-clusters, at a fixed noise level of $\ell = 0.1$.

Again, ten 100×50 background matrices are generated using a standard normal distribution for the matrix elements; into each of them, two 10×10 bi-clusters are embedded, overlapping with each other by one of the following six cases: $0 \times 0, 1 \times 1, 2 \times 2, 3 \times 3, 4 \times 4$, and 5×5. Previous simulation studies suggested to replace the matrix elements in the overlapped area with a random value; we expect, however, these overlapping genes to obey a reasonable logic such as the AND gate and the OR gate leading to a *union* and an *additive* behavior. Therefore, in the union overlap model, the matrix entries in the overlapped area preserve linear coherency in both bi-clusters (consequently, the overlapped area extends its linear coherency into both bi-clusters on those samples in the overlapped area); and in the additive overlap model, these entries take the sum of the gene expression levels from both bi-clusters.

Figure 4 shows the gene match scores of the five bi-clustering algorithms in this experiment. Their overall match scores and gene discovery rates under the union overlap model are plotted in Figures 10 and 11 in the Appendix. The results of the additive overlap model are in Figures 12, 13, and 14 in the Appendix. From all these results, one can see that our LinCoh outperformed the other four algorithms; OPSM and MSBE performed worse, but similarly to each other; CC performed the worst; and ISA demonstrated varying performance.

3.2   Real Datasets

The yeast dataset is obtained from [8], containing 2993 genes on 173 samples; the e.coli dataset (version 4 built 3) is from [6], which contains initially 4217 genes on 264 samples. Genes with small expression deviations were removed from the

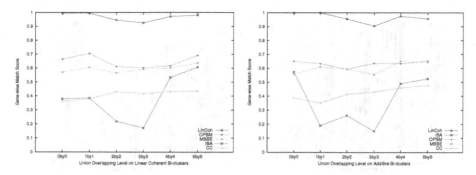

Fig. 4. The gene match scores of the five algorithms for recovering the overlapping linear coherent and additive bi-clusters, under the union overlap model

second dataset, giving rise to 3016 genes. Such a process ensures that all five bi-clustering algorithms can run on the dataset. In particular, it took two weeks for LinCoh to run on each dataset using a 2.2GHz CPU node of 2.5GB memory. The performance of an algorithm on these two real datasets is measured in gene functional enrichment score [13]. First, the P-value of each output bi-cluster is defined using its most enriched functional class (biological process).

The probability of having r genes of the same functional class in a bi-cluster of size n from a genome with a total of N genes can be computed using the hypergeometric function, where p is the percentage of that functional class of genes over all functional classes of genes encoded in the whole genome. Numerically [13],

$$Pr(r|N,p,n) = \binom{pN}{r} \cdot \binom{(1-p)N}{n-r} / \binom{N}{n}.$$

Such a probability is taken as the P-value of the output bi-cluster enriched with genes from that functional class [13]. The smallest P-value over all functional classes is defined as the P-value of the output bi-cluster — the smaller the P-value of a bi-cluster the more likely its genes come from the same biological process. For each algorithm, we calculate the fraction of its output bi-clusters whose P-values are smaller than a significance cutoff α.

Figure 5 compares the five algorithms using six different P-value cutoffs, evaluated on the GO database. Results on the KEGG, MIPS, and Regulon databases are in Figures 15 and 16 in the Appendix. All these results show that our LinCoh consistently performed well; OPSM and ISA did not perform consistently on the two datasets across databases; and that MSBE and CC did not perform as well as the other three algorithms.

One potential issue with the P-value based performance measurement is that P-values are sensitive to the bi-cluster size [13]; in general, larger bi-clusters tend to produce more significant P-values. Table 1 in the Appendix summarizes the statistics of the bi-clusters produced by the five algorithms. The last column in the table records the numbers of unique functional terms enriched by the produced bi-clusters. On yeast dataset, when measured by the gene enrichment significance test, OPSM performed very well (Figure 5, left); yet its

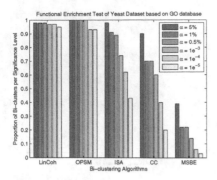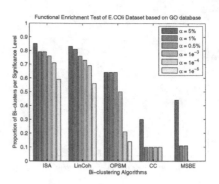

Fig. 5. Portions of discovered bi-clusters by the five algorithms on the two real datasets that are significantly enriched the GO biological process, using six different P-value cutoffs

bi-clusters only cover one functional term on the GO and KEGG databases and two terms on MIPS database. Such a phenomenon suggests that its bi-clustering result is biased to a group of correlated genes, missed by the P-value based significance test. Furthermore, we generated all the gene pairs with absolute correlation coefficient greater than or equal to 0.8 over all the samples for both the yeast and e.coli datasets. Table 2 in the Appendix shows the numbers of common GO terms and their counts. Among these strongly correlated gene pairs, many do not even have one common GO term. Table 3 in the Appendix shows the top 10 counted common GO terms (full table can be found at 'http://www.cs.ualberta.ca/~ys3/LinCoh').

The above two potential issues hint that the P-value based evaluation is meaningful but has limitations. We propose to use the average absolute correlation coefficient over all gene pairs in a bi-cluster defined in Eq. (1) as an alternative assessment of the quality of a linear coherent bi-cluster. Figure 6 shows the box plot of these correlation values for the bi-clusters produced by the five algorithms on the two real datasets. From the figure, one can see that our LinCoh and OPSM significantly outperformed the other three algorithms. Additionally, the minimum absolute correlation coefficient over all gene pairs in a bi-cluster can also be adopted as a quality measurement. Figure 17 in the Appendix shows these results.

Figure 6 shows that OPSM produced bi-clusters with very high linear coherence. But the numbers of samples in its bi-clusters are much smaller than those in LinCoh's bi-clusters, as shown in Table 1 in the Appendix (tens versus hundreds). This suggests that very closely interacting gene pairs can have small empirical correlation coefficients on a subset of samples, largely due to noise and measurement errors. In fact, there is always a trade-off between bi-cluster coherence and its size. Thus, to compare algorithms in a less sample-size biased way, we replaced for each bi-cluster its average absolute correlation coefficient by the 99% confidence threshold using the number of samples in the bi-cluster [19], and

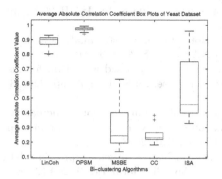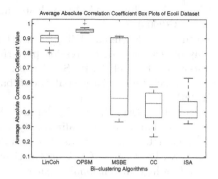

Fig. 6. Box plots of the average absolute correlation coefficients obtained by the five bi-clustering algorithms on yeast and e.coli datasets, respectively

box plotted these values in Figure 18 in the Appendix. They show much more comparable performance between LinCoh and OPSM.

4   Conclusions

In this paper, we proposed a new bi-clustering algorithm, LinCoh, for finding linear coherent bi-clusters in a gene expression matrix. The algorithm has two important steps, beam detection for pairwise gene correlations and density based outer sample set clustering. Our experiments on synthetic and real datasets demonstrate that LinCoh consistently performed well. Using real datasets, we also showed some limitations of the widely adopted functional enrichment measurement, and proposed to use average absolute correlation coefficient as an alternative measure for bi-clustering quality. With its outperformance over the compared four popular algorithms, LinCoh can serve as another useful tool for microarray data analysis, including bi-clustering and genetic regulatory network inference.

One disadvantage of LinCoh is its large memory and extensive computing time requirement, due to constructing the outer and inner sample set matrices. It takes $O(n^2 m_p)$ to compute the sample set matrices where n is the number of genes and m_p is the number of parameters θ, β and γ. The memory required for storing the matrices is $O(n^2 m_s)$ where n is the number of genes and m_s is the average size of the sample set elements. It takes weeks and up to 1 Gigabyte memory to run experiments on the e.coli dataset. Improvements in beam detection and sample set clustering can also achieve significant speed-ups.

References

1. Ashburner, M., Ball, C.A., Blake, J.A., et al.: Gene ontology: tool for the unification of biology. Nature Genetics 25, 25–29 (2000)
2. Ben-Dor, A., Chor, B., Karp, R., Yakhini, Z.: Discovering local structure in gene expression data: The order-preserving sub-matrix problem. In: RECOMB 2002, pp. 49–57 (2002)

3. Cheng, Y., Church, G.M.: Biclustering of expression data. In: ISMB 2000, pp. 93–103 (2000)
4. Eisen, M.B., Spellman, P.T., Brown, P.O., Botstein, D.: Cluster analysis and display of genome-wide expression patterns. PNAS 95, 14863–14868 (1998)
5. Ester, M., Kriegel, H.-P., Sander, J., Xu, X.: A density-based algorithm for discovering clusters in large spatial databases with noise. In: KDD 1996, pp. 226–231 (1996)
6. Faith, J.J., Driscoll, M.E., Fusaro, V.A., et al.: Many microbe microarrays database: uniformly normalized Affymetrix compendia with structured experimental metadata. Nucleic Acids Research 36, D866–D870 (2008)
7. Gan, X., Liew, A.W.-C., Yan, H.: Discovering biclusters in gene expression data based on high-dimensional linear geometries. BMC Bioinformatics 9, 209 (2008)
8. Gasch, A.P., Spellman, P.T., Kao, C.M., et al.: Genomic expression programs in the response of yeast cells to environmental changes. Nucleic Acids Research 11, 4241–4257 (2000)
9. Hartigan, J.A.: Direct clustering of a data matrix. Journal of the American Statistical Association 67, 123–129 (1972)
10. Ihmels, J., Bergmann, S., Barkai, N.: Defining transcription modules using large scale gene expression data. Bioinformatics 20, 1993–2003 (2004)
11. Kanehisa, M.: The KEGG database. In: Novartis Foundation Symposium, vol. 247, pp. 91–101 (2002)
12. Keseler, I.M., Collado-Vides, J., Gama-Castro, S., et al.: EcoCyc: a comprehensive database resource for escherichia coli. Nucleic Acids Research 33, D334–D337 (2005)
13. Li, G., Ma, Q., Tang, H., Paterson, A.H., Xu, Y.: QUBIC: A qualitative biclustering algorithm for analyses of gene expression data. Nucleic Acids Research 37, e101 (2009)
14. Liu, X., Wang, L.: Computing the maximum similarity bi-clusters of gene expression data. Bioinformatics 23, 50–56 (2006)
15. Madeira, S.C., Oliveira, A.L.: Biclustering algorithms for biological data analysis: A survey. Journal of Computational Biology and Bioinformatics 1, 24–45 (2004)
16. Mirkin, B.: Mathematical classification and clustering. Kluwer Academic Publishers, Dordrecht (1996)
17. Prelić, A., Bleuler, S., Zimmermann, P., Wille, A.: A systematic comparison and evaluation of biclustering methods for gene expression data. Bioinformatics 22, 1122–1129 (2006)
18. Ruepp, A., Zollner, A., Maier, D., et al.: The FunCat, a functional annotation scheme for systematic classification of proteins from whole genomes. Nucleic Acids Research 32, 5539–5545 (2004)
19. Shen, D., Lu, Z.: Computation of correlation coefficient and its confidence interval in SAS, http://www2.sas.com/proceedings/sugi31/170-31.pdf
20. Tamayo, P., Slonim, D., Mesirov, J., et al.: Interpreting patterns of gene expression with self-organizing maps: Methods and application to hematopoietic differentiation. PNAS 96, 2907–2912 (1999)
21. Tavazoie, S., Hughes, J.D., Campbell, M.J., Cho, R.J., Church, G.M.: Systematic determination of genetic network architecture. Nature Genetics 22, 281–285 (1999)

Appendix

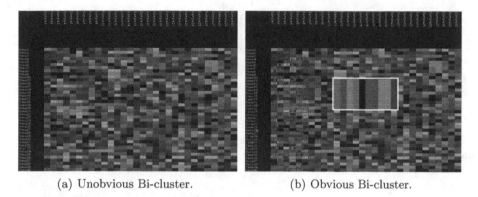

(a) Unobvious Bi-cluster.   (b) Obvious Bi-cluster.

Fig. 7. An example of a constant row bi-cluster in the gene expression matrix. (a) shows a gene expression matrix without any obvious bi-clusters; (b) shows after swapping rows and columns, a constant row bi-cluster becomes salient.

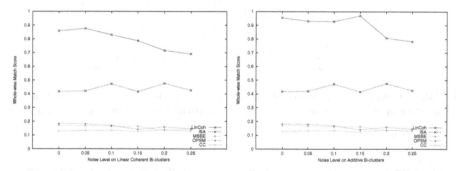

Fig. 8. The overall match scores of the five algorithms for recovering linear coherent and additive bi-clusters, at six different noise levels

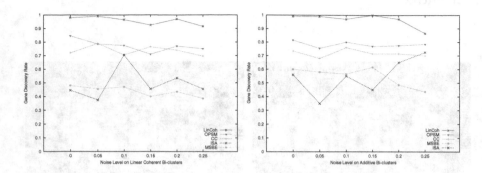

Fig. 9. The gene discovery rates of the five algorithms for recovering linear coherent and additive bi-clusters, at six different noise levels

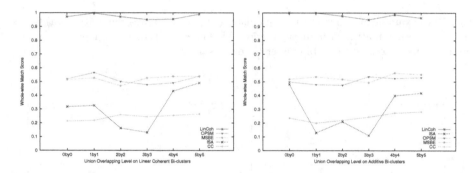

Fig. 10. The overall match scores of the five algorithms for recovering linear coherent and additive bi-clusters, under six different amounts of overlap using the union overlap model

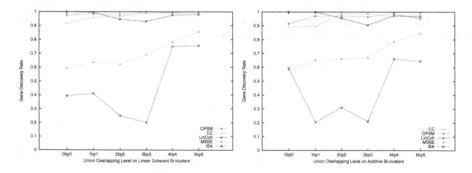

Fig. 11. The gene discovery rates of the five algorithms for recovering linear coherent and additive bi-clusters, under six different amounts of overlap using the union overlap model

Algorithm 1. The LinCoh Algorithm

Input An $n \times m$ real value matrix $A(I, J)$, T_{close}, T_{minCC}
Output A set of bi-clusters $A(g_i, s_i)$, where $g_i \subseteq I$ and $s_i \subseteq J$.

for $i = 1$ to n **do**
  for $j = i + 1$ to n **do**
    $M_O(i, j) = NULL$, $M_I(i, j) = NULL$;
    $\theta_{rec} = NULL$, $\beta_{rec} = NULL$, $\gamma_{rec} = NULL$;
    for A set of beam parameters (θ, β, γ) **do**
      if $W_{S_{outer(\theta,\beta,\gamma)}} \cdot D^T_{S_{outer(\theta,\beta,\gamma)}} > |M_O(i, j)|$ **then**
        $M_O(i, j) = S_{outer(\theta,\beta,\gamma)}$;
        $\theta_{rec} = \theta$, $\beta_{rec} = \beta$, $\gamma_{rec} = \gamma$;
      end if
    end for
    $M_I(i, j) = S_{inner(\theta_{rec}, \beta_{rec}, \gamma_{rec})}$;
  end for
end for

for $i = 1$ to n **do**
  for $j = i + 1$ to n **do**
    if $M_O(i, j) < \mu_{ss} + \alpha \cdot \sigma_{ss}$ **then**
      $M_O(i, j) = NULL$, $M_I(i, j) = NULL$;
    end if
  end for
end for

for $i = 1$ to n **do**
  $SS_i = \bigcup_{j \in J}(M_O(i, j))$;
end for
$GeneList_{ss} = DescendSort(Genes)$ based on $|SS_i| \neq NULL$ of each $i \in I$;
$BiclusterPool = NULL$;
while $GeneList_{ss} \neq EMPTY$ **do**
  $SeedGene = Pop(GeneList_{ss})$;
  Construct similarity matrix $Matrix_{ss}$ for $SS_{seedGene}$ elements based on
  $M_S(SS_i, SS_j) = \frac{|SS_i \cap SS_j|}{|SS_i \cup SS_j|}$;
  Find the centroid sample set $SS_{centroid}$ that has the most $close$ $(M_S(S_i, S_j) \geq T_{close})$ neighbors, $SS_{neighbors}$;
  $GenePool = \bigcup(G_i \in G_{SS_{centroid}} \bigcup G_{SS_{neighbors}})$;
  $SamplePool = \bigcup(S_j \in SS_{centroid} \bigcup SS_{neighbors})$;
  $BiCluster_{initial} = A(GenePool, SamplePool)$;
  $BiCluster_{refined} = RefineBicluster(BiCluster_{initial})$
  if $MinAbsCC(BiCluster_{refined}) \geq T_{minCC}$ **then**
    $BiclusterPool.add(BiCluster_{refined})$;
  end if
end while
$Biclusters_{final} = RedundantRemoval(BiclusterPool)$

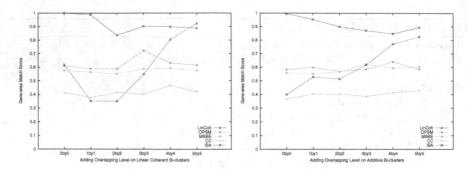

Fig. 12. The gene match scores of the five algorithms for recovering linear coherent and additive bi-clusters, under six different amounts of overlap using the additive overlap model

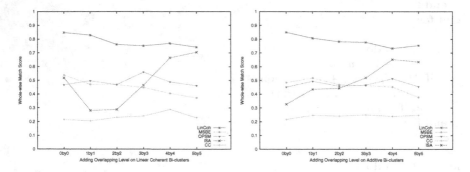

Fig. 13. The overall match scores of the five algorithms for recovering linear coherent and additive bi-clusters, under six different amounts of overlap using the additive overlap model

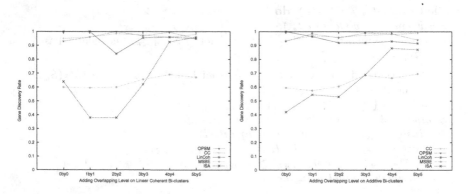

Fig. 14. The gene discovery rates of the five algorithms for recovering linear coherent and additive bi-clusters, under six different amounts of overlap using the additive overlap model

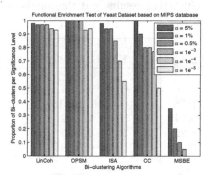
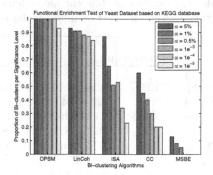

Fig. 15. Portions of yeast bi-clusters that are significantly enriched over different P-values in the MIPS pathway and KEGG pathway

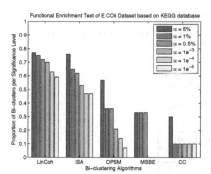
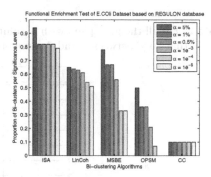

Fig. 16. Portions of e.coli bi-clusters that are significantly enriched over different P-values in the KEGG pathway and experimentally verified regulons

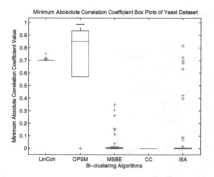
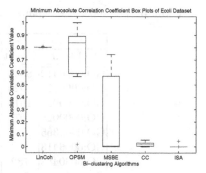

Fig. 17. The box plots of minimum absolute correlation coefficients of the bi-clusters produced by the five algorithms on the yeast and e.coli datasets

Table 1. Statistics of different algorithms' bi-clustering results and the numbers of functional terms enriched on different databases

	#Bi-clusters	$\mu_{|gene|}$	$\sigma_{|gene|}$	$\mu_{|sample|}$	$\sigma_{|sample|}$	#Unique terms enriched (GO, KEGG, MIPS/regulons)
yeast:						
LinCoh	100	61.84	38.43	133.09	18.09	5, 7, 5
ISA	47	67	34.54	8.4	1.78	15, 13, 18
OPSM	14	423.29	728.95	9.07	5.14	1, 1, 2
MSBE	40	19.25	8.32	18.68	8.22	8, 4, 6
CC	10	297.7	304.18	60.8	23.46	6, 4, 8
e.coli:						
LinCoh	100	9.63	7.66	141.43	34.04	24, 24, 22
ISA	34	124.21	42.18	13.88	6.11	11, 10, 13
OPSM	14	419.29	744.35	8.93	4.8	8, 4, 5
MSBE	9	82.67	18.1	80.22	19.18	1, 3, 4
CC	10	309.9	950.15	31.4	81.74	2, 2, 2

Table 2. For all the gene pairs with absolute correlation coefficient ≥ 0.8, the number of pairs that have between 0 and 7 common GO terms

Term count	yeast	e.coli
0	909	2680
1	18860	3485
2	7898	1533
3	1839	490
4	165	239
5	30	52
6	6	18
7	4	0
Overall	28802	5817

Table 3. The top 10 gene pairs' common GO terms and their counts

yeast		e.coli	
GO term	Count	GO term	Count
GO:0006412	8353	GO:0006412	811
GO:0000723	1920	GO:0008652	680
GO:0000027	1615	GO:0001539	388
GO:0000028	1070	GO:0006810	317
GO:0006365	969	GO:0006355	234
GO:0006413	893	GO:0006811	195
GO:0030488	782	GO:0006865	183
GO:0006364	720	GO:0006260	127
GO:0030490	683	GO:0046677	115
GO:0006360	424	GO:0008152	111

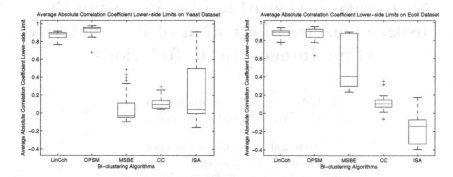

Fig. 18. The box plots of the 99% confidence thresholds of the average absolute correlation coefficients of the bi-clusters, using the number of samples in each bi-cluster, produced by the five algorithms on the yeast and e.coli datasets

An Iterative Algorithm of Computing the Transitive Closure of a Union of Parameterized Affine Integer Tuple Relations

Bielecki Wlodzimierz[1], Klimek Tomasz[1],
Palkowski Marek[1], and Anna Beletska[2]

[1] Technical University of Szczecin,
ul. Zolnierska 49, Szczecin, Poland
{bielecki,tklimek,mpalkowski}@wi.ps.pl
[2] INRIA Saclay, Parc Club Universite
3, rue J. Rostand, 91893 Orsay Cedex, France
anna.beletska@inria.fr

Abstract. A novel iterative algorithm of calculating the exact transitive closure of a parameterized graph being represented by a union of simple affine integer tuple relations is presented. When it is not possible to calculate exact transitive closure, the algorithm produces its upper bound. To calculate the transitive closure of the union of all simple relations, the algorithm recognizes the class of each simple relations, calculates its exact transitive closure, forms the union of calculated transitive closures, and applies this union in an iterative procedure. Results of experiments aimed at the comparison of the effectiveness of the presented algorithm with those of related ones are outlined and discussed.

Keywords: parameterized graph, tuple relation, transitive closure, iterative algorithm.

1 Introduction

The computation of the transitive closure of a directed graph is a necessary operation in many algorithms in software engineering, real-time process control, data bases, optimizing compilers, etc. For example, in the domain of programming languages and compilation transitive closure is used for redundant synchronization removal [1], testing the legality of iteration reordering transformations [1], computing closed form expressions for induction variables [1], iteration space slicing and code generation [3,9,10]. Graphs can be represented in different ways. In this paper, we consider a class of parameterized affine integer tuple relations whose constraints are represented with Presburger formulas and which describe graphs with the parameterized number of vertices (the number of vertices is represented by an expression including symbolic parameters). To our best knowledge, techniques for computing the transitive closure of a parameterized affine integer tuple relation describing such graphs were the subject of the investigation of a few papers only [1,4,5,6,8]. This paper presents an iterative algorithm to calculate the exact transitive closure of relation R being a union of n simple relations R_i, $i=1,2,\ldots,n$ as well as results of experiments with NAS 3.2 benchmarks to

W. Wu and O. Daescu (Eds.): COCOA 2010, Part I, LNCS 6508, pp. 104–113, 2010.

compare the presented algorithm effectiveness with the semi-naive iterative algorithm [11,12] and that presented in paper [1].

2 Background

The transitive closure of a directed graph $G=(V,E)$ is a graph $H=(V,F)$ with edge (v,w) in H if and only if there is a path from v to w in G. A graph can be represented by an integer tuple relation whose domain consists of integer k-tuples and whose range consists of integer k'-tuples, for some fixed k and k'. The general form of a parameterized affine integer tuple relation is as follows [1]

$$\{[s_1, s_2, \ldots, s_k] \rightarrow [t_1, t_2, \ldots, t_{k'}] \mid \bigvee_{i=1}^{n} \exists \alpha_{i1}, \alpha_{i2} \ldots, \alpha_{im_i} \; s.t. \, F_i \, \} \qquad (1)$$

where the F_i are conjunctions of affine equalities and inequalities on the input variables $s_1, s_2, \ldots, s_k$, the output variables $t_1, t_2, \ldots, t_{k'}$, the existentially quantified variables $\alpha_{i1}, \alpha_{i2} \ldots, \alpha_{im_i}$, and symbolic constants. The Omega library is used for computations over such relations [2]. Different operations on relations are permitted, such as intersection ($\cap$), union ($\cup$), difference (-), domain of relation ($\text{domain}(R)$), range of relation ($\text{range}(R)$), relation application ($R(S)$), positive transitive closure R^+, transitive closure ($R^* = R^+ \cup I$, where I is the identity relation). These operations are described in detail in [2].

We say a relation is simple if it is represented neither as a union of other relations nor its constraints include multiple F_i (only one conjunct F is allowed).

An integer tuple relation describes the corresponding graph $G = V, E$, where the input and output tuples of the relation represent a set of vertices V of the graph G while affine equalities and inequalities describe the existence of edges in the graph, i.e, define set E of edges (an edge exists if corresponding affine equalities and inequalities are honored for a given pair of vertices represented with input and output tuples).

The transitive closure of a relation with arbitrary Presburger constraints are not computable in general [1].

Kelly et al. [1] define a class of relations, called d-form (simple conjunct relations with constraints only on the difference of the output and input tuples) for which the transitive closure is easily computable.

According to Kelly *et al.* [1], a relation R is said to be in d-form if it can be represented as follows

$$R=\{[i_1, i_2, ..., i_n] \rightarrow [j_1, j_2, ..., j_n] : \qquad \forall p, 1 \leq p \leq n,$$
$$\exists \alpha_p s.t. (L_p \leq j_p - i_p \leq U_p \wedge j_p - i_p = M_p \alpha_p\},$$

where L_p and U_p are constants and M_p is an integer. If $L_p = -\infty$ or $U_p = \infty$, the corresponding constraints are omitted.

Because a d-form relation has constraints only on the difference between the corresponding elements of the input and output tuples, it is described on the unbounded region.

For multiple conjunct relations (or a union of simple conjunct relations), the computation of transitive closure is more complex. To our knowledge, it is not yet known whether the exact transitive closure of a union of d-form relations is computable. The solution presented by Kelly et al. [1] is based on heuristics that guarantee neither calculating the exact result nor its conservative approximation. Approaches proposed by Boigelot in [4] or Kelly et al. in [1] compute approximations of transitive closures for such relations.

Despite the limited ability to compute relation R^+ exactly, some of useful algorithms have been proposed. They can be divided into iterative, matrix-based, graph-based, and hybrid algorithms [5]. In this paper, we concentrate on iterative algorithms that compute the transitive closure of a relation R, R^+, by evaluating the least fixed point of the following equation [5]:

$$R^+ = R^+ \circ R \cup R \tag{2}$$

A simple iterative algorithm for computing the least fixed point is presented below, where R is the source relation. Algorithm 1 is known to be semi-naive [11,12], it is clear that after executing each iteration, it produces a more accurate lower bound of R^+ until the result becomes exact.

Algorithm 1. Semi-naive iterative algorithm

$R^+ \leftarrow R$
$\Delta \leftarrow R$
repeat
$\quad \Delta \leftarrow \Delta \circ R - R^+$
$\quad R^+ \leftarrow R^+ \cup \Delta$
until $\Delta = \varnothing$

Although this technique works in some cases, there is no guarantee of the termination of the algorithm. For example, the exact transitive closure of the relation $R = \{[i] \rightarrow [i+1]\}$ cannot be computed using this approach. Computing the transitive closure of a relation using Algorithm 1 is prohibitively expensive due to the possible exponential growth in the number of conjuncts. Thus more sophisticated techniques are required. In the following section, we present an improvement of Algorithm 1 permitting for increasing its effectiveness.

3   Iterative Algorithm

We suppose that the input of the algorithm presented below is a union of simple parameterized affine integer tuple relations. To improve Algorithm 1, we suggest to firstly recognize the class of each simple relation.

In paper [7], we consider the following classes of relations describing dependences in program loops: d-form relations, uniform relations, relations describing chains only, relations with coupled index variables, relations with non-coupled index variables, and finally relations with different numbers of index variables of input and output relation tuples. We propose techniques and implementation permitting for recognizing these types of relations. When the class of a relation is recognized, we proceed to the calculation of the transitive closure of this relation using techniques presented in papers [1,6,7].

Given a union of simple relations, Algorithm 2 presented below recognizes the class of each relation, computes its transitive closure and produces the union of transitive closures of all relations.

Algorithm 2. Recognizing classes of simple relations and calculating the union of transitive closures of these relations

---

Input : $R_{in} = \bigcup\limits_{i=1}^{m} R_i$, where m is the number of simple relations

/* input is represented by the union of simple relations */

Output : $R_{out} = \bigcup\limits_{i=1}^{m} R_i^*$

$R_{out} \leftarrow \varnothing$
for all relations $R_i \in R_{in}$ **do**
  1   recognize the class of relation R_i using techniques presented in [7] and calculate R_i^* using the appropriate algorithm of those described in [1,6,7]
  2    $R_{out} \leftarrow R_{out} \cup R_i^*$
end for
return R_{out}

Because an iterative algorithm does not always guarantee the convergence to exact transitive closure, in some cases we may prefer to calculate an upper bound of transitive closure(its over-approximation) that describes all direct and transitive connections in a graph and some additional connections not existing in the graph. For this purpose, we need to convert input relations to d-form relations and calculate the union of transitive closures of these relations. Algorithm 3 realizes this task.

Algorithm 3. Converting input relations to d-form relations and calculating the union of transitive closures of these relations

---

Input : $R_{in} = \bigcup\limits_{i=1}^{m} R_i$, where m is the number of simple relations

/* input is represented by the union of simple relations */

Output : $R_{out} = \bigcup\limits_{i=1}^{m} d_i^*$, where d_i is the approximation of input relation R_i to the d-form relation described on the unbounded region

$R_{out} \leftarrow I$ /* variable R_{out} is initialized to the identity relation */
for all relations $R_i \in R_{in}$ **do**
  1   for relation R_i, create d-form relation d_i of the form
      $d_i \leftarrow \{[s_1, s_2, \ldots, s_m] \rightarrow [t_1, t_2, \ldots, t_m] \mid \forall p, 1 \leq p \leq m, \exists \alpha_p\, s.t.\, L_p \leq t_p - s_p \leq U_p \wedge t_p - s_p = M_p \alpha_p\}$,where d_i is described on the unbounded region, L_p and U_p are constants, and M_p is an integer. If L_p is $-\infty$ or U_p is $+\infty$, the corresponding constraints are not included in the above equation.
  2   calculate d_i^* using the algorithm described in [1] as follows
      $d_i^+ \leftarrow \{[s_1, s_2, \ldots, s_m] \rightarrow [t_1, t_2, \ldots, t_m] \mid \exists k > 0\, s.t.\, \forall p, 1 \leq p \leq m, \exists \alpha_p\, s.t.\, L_p k \leq t_p - s_p \leq U_p k \wedge t_p - s_p = M_p \alpha_p\}$
  3    $R_{out} \leftarrow R_{out} \cup d_i^+$
end for
return R_{out}

Algorithm 4. Calculating transitive closure of the union of simple relations

Input : $R_{in} = \bigcup\limits_{i=1}^{m} R_i$, where m is the number of simple relations
/* input is represented by the union of simple relations */

N /* the maximal number of iterations to be run */

mode [*exact, upper bound*]
/* *exact* asks for trying to calculate the exact transitive closure of relation R_{in} while
upper bound asks for calculating an upper bound of transitive closure */

Output : exact R_{in}^* or its lower or upper bound

if mode is *exact* **then**
    $R_{old} \leftarrow R' \leftarrow \Delta \leftarrow$ Algorithm 2(R_{in})
    /* variables R_{old}, R', and Δ are initialized by calling Algorithm 2. */
    $i \leftarrow 1$
    loop
        $R_{new} \leftarrow R' \circ \Delta$
        /* variable R_{new} is the composition of relation R' and relation Δ */
        $\Delta \leftarrow R_{new} - R_{old}$
        /* the variable Δ is the difference between relations R_{new} and R_{old} */
        $i \leftarrow i + 1$
        if $\Delta = \varnothing$ OR $i > N$ **then**
            break /* the algorithm terminates when relation Δ is empty or when the
            algorithm meets the maximal number of iterations to be run */
        end if
        $\Delta \leftarrow$ Algorithm 2(Δ)
        /* Algorithm 2 is called with parameter Δ */
        $R_{old} \leftarrow R_{old} \cup \Delta$
    end loop
    if $i \leq N$ **then**
        $R^* \leftarrow R_{old}$
        print ”R^* is exact transitive closure”
    else
        $R^* \leftarrow R_{new}$
        print ”R^* is a lower bound of transitive closure”
    end if
    return R^*
else
    $d^* \leftarrow$ Algorithm 3(R_{in})
    $R^* \leftarrow \prod\limits_{i=1}^{m} d_i^*$
    /* the variable R^* is an upper bound of the transitive closure of relation R_{in}. */
    $R^* \leftarrow R^* \setminus domain(R_{in})$
    /* Restrict the domain of relation R^* to the domain of relation R_{in} */
    $R^* \leftarrow R^* / range(R_{in})$
    /* Restrict the range of relation R^* to the range of relation R_{in} */
    print ”R^* is an upper bound of transitive closure”
    return R^*
end if

Algorithm 4 is the basic iterative algorithm permitting for calculating either the exact transitive closure of a union of simple relations or its lower or upper bound.

To prove that $(\prod_{i=1}^{m} d_i^* \setminus domain(R_{in}))\,/\,range(R_{in})$ (produced by Algorithm 4) represents an upper bound of the transitive closure of a union of simple relations, we take into account the following. Since $R_i \in d_i$, $d = \bigcup_{i=1}^{m} d_i$, where m is the number of simple relations in the union of relations, represents an upper bound for $R_{in} = \bigcup_{i=1}^{m} R_i$. Hence $(d^* \setminus domain(R_{in}))\,/\,range(R_{in})$ is an upper bound of R_{in}^*. It is well-known that simple uniform relations and relations being formed as a union of simple uniform relations are commutative [8]. A d-form relation can be represented as a union of uniform relations on the unbounded space, hence d-form relations are commutative. Taking into consideration that for commutative relations the following is true:

i) the operator $(\cup)$ is commutative and associative $(R_1 \cup R_2 = R_2 \cup R_1,\ R_1 \cup (R_2 \cup R_3) = (R_1 \cup R_2) \cup R_3)$
ii) the operator $(\circ)$ is associative $(R_1 \circ (R_2 \circ R_3) = (R_1 \circ R_2) \circ R_3)$
iii) the operator $(\circ)$ distributes over $(\cup)$ $(R_1 \circ (R_2 \cup R_3) = R_1 \circ R_2 \cup R_1 \circ R_3)$
iv) $R^n \cup R^n = R^n$, where n is any integer, we can rewrite

$$d^* = \bigcup_{i=0}^{\infty} R^i = I \cup (d_1 \cup d_2 \cup \ldots \cup d_m) \cup (d_1 \cup d_2 \cup \ldots \cup d_m)^2 \cup$$
$$(d_1 \cup d_2 \cup \ldots \cup d_m)^3 \cup \ldots$$

as

$$d^* = \prod_{i=0}^{m} d_i^*.$$

To illustrate Algorithm 4, let us consider the following example.

$$R = \{[i,5] \rightarrow [i, i+7] \mid 1 \le i \le n-7\} \cup$$
$$\{[i,5] \rightarrow [i', i+7] \mid 1 \le i < i' \le n \wedge i \le n-7\} \cup$$
$$\{[i,j] \rightarrow [j-7,5] \mid 1 \le i \le j-8 \wedge j \le n\} \cup$$
$$\{[i,j] \rightarrow [i',j] \mid 1 \le i < i' \le n \wedge 1 \le j \le n\}$$

The values of differences Δ calculated according to Algorithm 4 are presented in Table 1.

After three iterations, the algorithm produces the exact transitive closure of the form:

$$R^* = \{[i,5] \rightarrow [i', i'+7] \mid 1 \le i < i' \le n-7\} \cup$$
$$\{[i,5] \rightarrow [i', j'] \mid i+8 \le j' \le i'+6, n \wedge i' \le n \wedge 1 \le i\} \cup$$
$$\{[i,j] \rightarrow [i',5] \mid i+8 \le j \le j \le i'+6, n \wedge i' \le n \wedge 1 \le i\} \cup$$
$$\{[i,5] \rightarrow [i, i+7] \mid 1 \le i \le n-7\} \cup$$
$$\{[i,5] \rightarrow [i', i+7] \mid 1 \le i < i' \le n \wedge i \le n-7\} \cup$$
$$\{[i,j] \rightarrow [j-7,5] \mid 1 \le i \le j-8 \wedge j \le n\} \cup$$
$$\{[i,j] \rightarrow [i',j] \mid 1 \le i < i' \le n \wedge 1 \le j \le n\} \cup$$

Table 1. Values of differences Δ

i	$\Delta = R_{new} - R_{old}$
1	$\{[i,5] \to [i',i'+7] \mid 1 \le i < i' \le n-7\} \cup$ $\{[i,5] \to [i',j'] \mid i+8 \le j' \le i'+6, n \wedge i' \le n \wedge i\} \cup$ $\{[i,j] \to [i',5] \mid i+8 \le j \le i'+6, n \wedge i' \le n \wedge 1 \le i\}$
2	$\{[i,j] \to [i',i'+7] \mid 1 \le i \le j-8 \wedge j-6 \le i' \le n-7\} \cup$ $\{[i,j] \to [i',j'] \mid i+8 \le j < j' \le i'+6, n \wedge i' \le n \wedge 1 \le i\}$
3	$\{[i,j] \to [i',j'] \mid FALSE\}$

$$\{[i,j] \to [i',i'+7] \mid 1 \le i \le j-8 \wedge j-6 \le i' \le n-7\} \cup$$
$$\{[i,j] \to [i',j'] \mid i+8 \le j < j' \le i'+6, n \wedge i' \le n \wedge 1 \le i\} \cup$$
$$\{[i,j] \to [i,j]\}$$

The above result contains ten "union" operators, while calculating R^* in Omega according to the algorithm presented in [1] results in an approximation of R^* that contains the twelve "union" operators.

4   Experiments

Algorithms 2, 3, and 4 were implemented using the Omega library [2]. The version of the Omega library permitting for calculating transitive closure using Algorithms 2, 3, and 4 can be download from: `http://www.sfs.zut.edu.pl/files/omega3.tar.gz`.

The following functions were added to the Omega library to produce the exact transitive closure of an input relation (being represented by a union of simple relations) or its upper bound:

```
Relation IterateClosure(Relation &R)
```

and

```
Relation ApproxClosure(Relation &R),
```

where the dimensions of input and output tuples of R are the same. The first function calculates the exact transitive closure of input relation R or its lower bound. The second one produces an upper bound of the transitive closure of input relation R. Because the Omega library does not permit for non-linear constraints of relations, the current implementation does not permit for maintaining such constraints in relations.

Transitive closure is a basic operation underlying Iteration Space Slicing techniques to extract coarse-grained parallelism in program loops [9,10]. That is why we have studied the effectiveness of the presented iterative algorithms by experimenting with program loops of NAS 3.2 [13] benchmarks. We had two goals. The first one was to recognize the number and percentage of loops for which the presented algorithms are able to calculate the transitive closure of a union of dependence relations. The second goal was to compare obtained results with whose yielded by Algorithm 1 and the algorithm presented in [1] and being implemented in the Omega library and calculator [2]. Results of experiments are presented in Table 2.

A dependence relation is a parameterized affine integer tuple relation that describes all dependences yielded by instances of a single loop statement or a pair of statements [1]. A set of all dependence relations (being extracted by a dependence analysis) describes all dependences in a program loop. We used Petit [2] to extract dependence relations in a program loop. We have considered only such loops for which Petit was able to carry out a dependence analysis.

Table 2. Results of experiments

All loops	Transitive closure algorithm, R^*	Result	Loops	%
133	Semi-naive iterative algorithm, (Algorithm 1)	exact	92	69%
		lower bound	8	6%
		upper bound	-	-
		no result	33	25%
	Algorithm presented in [1]	exact	64	48%
		lower bound	-	-
		upper bound	7	6%
		no result	62	46%
	Presented algorithm, (Algorithm 4)	exact	96	72%
		lower bound	-	-
		upper bound	4	3%
		no result	33	25%

In Table 2, "no result" means that the Omega library (used for implementation of R^* in all algorithms) fails to produce any result due to the termination of calculation (exceeding the number of equations permitted, exceeding size memory permitted, or other exceptions). The number of dependence relations extracted by Petit for some loops exceeds 500, and the Omega library fails to produce any result in such a case.

The proposed iterative algorithm allowed us to calculate the exact transitive closure of dependence relations for 72% loops while that described in [1] only for 48% loops. The semi-naive iterative algorithm does not terminate for 6% loops. It fails to produce transitive closure, because there exists a simple dependence relation describing dependence graphs of the chain topology, so we can conclude that the semi-naive iterative algorithm never stops if there exists a simple relation in a union of relations that describes a graph of the chain topology.

Let us now consider the following example of R:

$$R = \{[i,j] \to [i+1,j] \mid 1 \le i < n \land 1 \le j \le m\} \cup$$
$$\{[i,j] \to [i+1,j+1] \mid 1 \le i < n \land 1 \le j < m\}$$

For the above relation, the semi-naive iterative algorithm fails to calculate transitive closure, while Algorithm 4 does calculate exact R^*.

We have compared the number of conjuncts in the resulting transitive closure produced by the algorithm described in paper [1] and in that computed according to Algorithm 4. For 15 loops from NAS 3.2 benchmarks, the number of conjuncts in transitive closure obtained applying Algorithm 4 is fewer than that produced by the algorithm presented in [1]. This permits for generating more efficient parallel code [10]. For 49 loops, the number of conjuncts in the resulting transitive closure is the same. Below, we present the transitive closure calculated for the *LU_HP_pintgr.f2p_3* loop from NAS 3.2 benchmarks. The union of dependence relations for the *LU_HP_pintgr.f2p_3* loop is as follows

$$R = \{[j,i] \rightarrow [j^{'},i^{'}] \mid j = j^{'} \wedge N3 \le i < i^{'} \le N4 \wedge N1 \le j^{'} \le N2 \wedge N3 \le$$
$$i^{'} \wedge i \le N4 \quad OR \quad N1 \le j < j^{'} \le N2 \wedge N3 \le i^{'}, i \le N4 \wedge N1 \le j^{'} \wedge j \le N2\}$$

Algorithm proposed in [1] and implemented in the Omega library outputs the following result for transitive closure

$$R^* = \{[j,i] \rightarrow [j^{'},i^{'}] \mid j = N2 \wedge N1 = N2 \wedge j^{'} = N2 \wedge N3 \le i < i^{'} \le$$
$$N4 \quad OR \quad i^{'} = i \wedge N3 = i \wedge N4 = i \wedge N1 \le j < j^{'} \le N2 \quad OR \quad N1 \le j <$$
$$j^{'} \le N2 \wedge N3 \le N4 - 1, i^{'}, i \wedge i \le N4 \wedge i^{'} \le N4 \quad OR \quad j^{'} = j \wedge N3 \le i < i^{'} \le$$
$$N4 \wedge N1 \le j \le N2 \wedge N1 < N2 \quad OR \quad j^{'} = j \wedge i^{'} = i\}$$

while Algorithm 4 yields

$$R^* = \{[j,i] \rightarrow [j^{'},i^{'}] \mid j = j^{'} \wedge N3 \le i < i^{'} \le N4 \wedge N1 \le j^{'} \le N2 \quad OR \quad N1 \le$$
$$j < j^{'} \le N2 \wedge N3 \le i^{'}, i \le N4 \quad OR \quad j^{'} = j \wedge i^{'} = i\}$$

Both the results represent the exact transitive closure of the above relation R, but that yielded by Algorithm 4 has fewer conjuncts.

5   Conclusion

The presented iterative algorithm for calculating either exact transitive closure of a union of simple integer tuple relations or its upper bound demonstrates better effectiveness than that of the algorithm presented in [1] and that yielded by Algorithm 1. The experimental study carried out with NAS 3.2 benchmarks showed that it allows us to calculate exact transitive closure for most graphs representing dependences in program loops. This permits us to extract all coarse-grained parallelism in those program loops by means of well-known techniques based of applying exact transitive closure. In our future research, we intend to advance the presented algorithm in order to improve its convergence.

References

1. Kelly, W., Pugh, W., Rosser, E., Shpeisman, T.: Transitive clousure of infinite graphs and its applications, Languages and Compilers for Parallel Computing (1995)
2. Kelly, W., Maslov, V., Pugh, W., Rosser, E., Shpeisman, T., Wonnacott, D.: The Omega library interface guide, Technical Report CS-TR-3445, Dept. of Computer Science, University of Maryland, College Park (March 1995)
3. Pugh, W., Rosser, E.: Iteration Space Slicing and its Application to Communication Optimization. In: Proceedings of International Conference on Supercomputing (1997)
4. Boigelot, B.: Symbolic Methods for Exploring Infnite State Spaces. PhD thesis, Universite de Liµege (1998)
5. Nuutila, E.: Efficient Transitive Closure Computation in Large Digraphs, Mathematics and Computing in Engineering Series No. 74 PhD thesis Helsinki University of Technology. Helsinki (1995)
6. Bielecki, W., Klimek, T., Trifunovi, K.: Calculating Exact Transitive Closure for a Normalized Affine Integer Tuple Relation. Electronic Notes in Discrete Mathematics 33, 7–14 (2009)
7. Bielecki, W., Klimek, T., Pietrasik, M.: An experimental study on recognizing classes of dependence relations. Measurement Automation and Monitoring 55(10) (2009)
8. Beletska, A., Barthou, D., Bielecki, W., Cohen, A.: Computing the Transitive Closure of a Union of Affine Integer Tuple Relations. In: COCOA 2009. LNCS, vol. 5573, pp. 98–109. Springer, Heidelberg (2009)
9. Beletska, A., Bielecki, W., Siedlecki, K., San Pietro, P.: Finding synchronization-free slices of operations in arbitrarily nested loops. In: Gervasi, O., Murgante, B., Laganà, A., Taniar, D., Mun, Y., Gavrilova, M.L. (eds.) ICCSA 2008, Part II. LNCS, vol. 5073, pp. 871–886. Springer, Heidelberg (2008)
10. Bielecki, W., Beletska, A., Palkowski, M., San Pietro, P.: Extracting synchronization-free chains of dependent iterations in non-uniform loops. In: ACS 2007: Proceedings of International Conference on Advanced Computer Systems (2007)
11. Bancilhon, F., Ramakrishnan, R.: An amateur's introduction to recursive query processing strategies. In: ACM-SIGMOD 1986 Conference on Management of Data, pp. 16–52 (1986)
12. Ioannidis, Y.E.: On the computation of the transitive closure of relational operators. In: Proceedings of the 12th International VLDB Conference, pp. 403–411 (1986)
13. NAS benchmarks suite, http://www.nas.nasa.gov
14. The Omega project, http://www.cs.umd.edu/projects/omega

Bases of Primitive Nonpowerful Sign Patterns

Guanglong Yu[1,2], Zhengke Miao[2,*], and Jinlong Shu[1,**]

[1] Department of Mathematics, East China Normal University,
Shanghai, 200241, China
[2] Department of Mathematics, Xuzhou Normal University, Xuzhou, 221116, China
yglong01@163.com, zkmiao@xznu.edu.cn, jlshu@math.ecnu.edu.cn

Abstract. For a square primitive nonpowerful sign pattern A, the base of A, denoted by $l(A)$, is the least positive integer l such that every entry of A^l is #. In this paper, we consider the base set of the primitive nonpowerful sign pattern matrices. Some bounds on the bases for the sign pattern matrices with base at least $\frac{3}{2}n^2 - 2n + 4$ is given. Some sign pattern matrices with given bases is characterized and some *"gaps"* in the base set are shown.

AMS Classification: 05C50

Keywords: Sign pattern; Primitive; Nonpowerful; Base.

1 Introduction

We adopt the standard conventions, notations and definitions for sign patterns and generalized sign patterns, their entries, arithmetics and powers. The reader who is not familiar with these matters is referred to [5], [11].

The sign pattern of a real matrix A, denoted by $\text{sgn}(A)$, is the $(0, 1, -1)$-matrix obtained from A by replacing each entry by its sign. Notice that in the computations of the entries of the power A^k, an *"ambiguous sign"* may arise when we add a positive sign to a negative sign. So a new symbol "#" has been introduced to denote the ambiguous sign.

For convenience, we call the set $\Gamma = \{0, 1, -1, \#\}$ the generalized sign set and define the addition and multiplication involving the symbol # as follows (the addition and multiplication which do not involve # are obvious):

$$(-1) + 1 = 1 + (-1) = \#, \ a + \# = \# + a = \# \ (for \ all \ a \in \Gamma),$$

$$0 \cdot \# = \# \cdot 0 = 0, \ b \cdot \# = \# \cdot b = \# \ (for \ all \ b \in \Gamma \backslash \{0\}).$$

It is straightforward to check that the addition and multiplication in Γ defined in this way are commutative and associative, and the multiplication is distributive with respect to addition. It is easy to see that a $(0, 1)$-Boolean matrix is a non-negative sign pattern matrix.

---

* Supported by NSFC(No. 10871166).
** Corresponding author. Supported by the NSFC (No. 10671074, No. 11075057, No. 11071078 and No. 60673048).

W. Wu and O. Daescu (Eds.): COCOA 2010, Part I, LNCS 6508, pp. 114–127, 2010.

Definition 1.1. *Let A be a square sign pattern matrix of order n with powers sequence $A, A^2, \cdots$. Because there are only 4^{n^2} different generalized sign pattern matrices of order n, there must be repetitions in the powers sequence of A. Suppose $A^l = A^{l+p}$ is the first pair of powers that are repeated in the sequence. Then l is called the generalized base (or simply base) of A, and is denoted by $l(A)$. The least positive integer p such that $A^l = A^{l+p}$ holds for $l = l(A)$ is called the generalized period (or simply period) of A, and is denoted by $p(A)$. For a square $(0, 1)$-Boolean matrix A, $l(A)$ is also known as the convergence index of A, denoted by $k(A)$.*

In 1994, Z. Li, F. Hall and C. Eschenbach [5] extended the concept of the base (or convergence index) and period from nonnegative matrices to sign pattern matrices. They defined powerful and nonpowerful for sign pattern matrices, gave a sufficient and necessary condition that an irreducible sign pattern matrix is powerful and also gave a condition for the nonpowerful case.

Definition 1.2. *A square sign pattern matrix A is powerful if all the powers A^1, A^2, A^3, $\cdots$ are unambiguously defined, namely there is no $\#$ in A^k ($k = 1, 2, \cdots$). Otherwise, A is called nonpowerful.*

If A is a sign pattern matrix, then $|A|$ is the nonnegative matrix obtained from A by replacing a_{ij} with $|a_{ij}|$.

Definition 1.3. *An irreducible $(0, 1)$-Boolean matrix A is primitive if there exists a positive integer k such that all the entries of A^k are non-zero, such least k is called the primitive index of A, denoted by $exp(A) = k$. A square sign pattern matrix A is called primitive if $|A|$ is primitive. The primitive index of A is equal to $exp(|A|)$, denoted by $exp(A)$.*

It is well known that graph theoretical methods are often useful in the study of the powers of square matrices, so we now introduce some graph theoretical concepts.

Definition 1.4. *Let A be a square sign pattern matrix of order n. The associated digraph of A, denoted by $D(A)$, has vertex set $V = \{1, 2, \cdots, n\}$ and arc set $E = \{(i, j)|a_{ij} \neq 0\}$. The associated signed digraph of A, denoted by $S(A)$, is obtained from $D(A)$ by assigning sign of a_{ij} to arc (i, j) for all i and j. Let S be a signed digraph of order n and A be a square sign pattern matrix of order n; A is called associated sign pattern matrix of S if $S(A)=S$. The associated sign pattern matrix of a signed digraph S is always denoted by $A(S)$. Note that $D(A) = D(|A|)$, so $D(A)$ is also called the underlying digraph of the associated signed digraph of A or is called the underlying digraph of A simply. We always denote by $D(A(S))$ or $|S|$ simply for the underlying digraph of a signed digraph S. Sometimes, $|A(S)|$ is called the associated or underlying matrix of signed digraph S.*

In this paper, we permit loops but no multiple arcs in a signed digraph. Denote by $V(S)$ the vertex set and denoted by $E(S)$ the arc set for a signed digraph S.

Let $W = v_0 e_1 v_1 e_2 \cdots e_k v_k$ ($e_i = (v_{i-1}, v_i)$, $1 \leq i \leq k$) be a directed walk of signed digraph S. The sign of W, denoted by sgn(W), is $\prod_{i=1}^{k}$ sgn(e_i). Sometimes a directed walk can be denoted simply by $W = v_0 v_1 \cdots v_k$, $W = (v_0, v_1, \cdots, v_k)$ or $W = e_1 e_2 \cdots e_k$ if there is no ambiguity. The positive integer k is called the length of the directed walk W, denoted by $L(W)$. The definitions of directed cycle and directed path are given in [1]. The length of the shortest directed path from v_i to v_j is called the distance from v_i to v_j in signed digraph S, denoted by $d(v_i, v_j)$. A cycle with length k is always called a k-cycle, a cycle with even (odd) length is called an *even cycle* (*odd cycle*). The length of the shortest directed cycle in digraph S is called the *girth* of S usually. When there is no ambiguity, a directed walk, a directed path or a directed cycle will be called a walk, a path or a cycle. A walk is called a *positive walk* if its sign is positive, and a walk is called a *negative walk* if its sign is negative. If p is a positive integer and if C is a cycle, then pC denotes the walk obtained by traversing through C p times. If a cycle C passes through the end vertex of W, $W \bigcup pC$ denotes the the walk obtained by going along W and then going around the cycle C p times; $pC \bigcup W$ is similarly defined. We use the notation $v \xrightarrow{k} u$ ($v \xrightarrow{k} u$) to denote that there exists (exists no) a directed walk with length k from vertex v to u. For a digraph S, let $R_k(v) = \{u | v \xrightarrow{k} u, u \in V(S)\}$ and $R_t(v) \xrightarrow{k} u$ mean that there exists a $s \in R_t(v)$ such that $s \xrightarrow{k} u$.

Definition 1.5. *Assume that W_1, W_2 are two directed walks in signed digraph S, they are called a pair of $SSSD$ walks if they have the same initial vertex, the same terminal vertex and the same length, but they have different signs.*

From [5] or [11], we know that a signed digraph S is powerful if and only if there is no pair of $SSSD$ walks in S.

Definition 1.6. *A strongly connected digraph G is primitive if there exists a positive integer k such that for all vertices $v_i, v_j \in V(G)$ (not necessarily distinct), there exists a directed walk of length k from v_i to v_j. The least such k is called the primitive index of G, and is denoted by $\exp(G)$. Let G be a primitive digraph. The least l such that there is a directed walk of length t from v_i to v_j for any integer $t \geq l$ is called the local primitive index from v_i to v_j, denoted by $\exp_G(v_i, v_j) = l$. Similarly, $\exp_G(v_i) = \max_{v_j \in V(G)} \{\exp_G(v_i, v_j)\}$ is called the local primitive index at v_i, so $\exp(G) = \max_{v_i \in V(G)} \{\exp_G(v_i)\}$.*

For a square sign pattern A, let $W_k(i, j)$ denote the set of walks of length k from vertex i to vertex j in $S(A)$; notice that the entry $(A^k)_{ij}$ of A^k satisfies
$$(A^k)_{ij} = \sum_{W \in W_k(i,j)} \text{sgn}(W).$$
Then we have

(1) $(A^k)_{ij} = 0$ if and only if there is no walk of length k from i to j in $S(A)$ (i.e., $W_k(i, j) = \phi$);

(2) $(A^k)_{ij} = 1$ (or -1) if and only if $W_k(i, j) \neq \phi$ and all walks in $W_k(i, j)$ have the same sign 1 (or -1);

(3) $(A^k)_{ij} = \#$ if and only if there is a pair of $SSSD$ walks of length k from i to j.

So the associated signed digraph can be used to study the properties of the powers sequence of a sign pattern matrix, and the signed digraph is taken as the tool in this paper. From the relation between sign pattern matrices and signed digraphs, we know that it is logical to define a sign pattern A to be primitive and to define $\exp(A) = \exp(D(A)) = \exp(|A|)$ if A is primitive.

Definition 1.7. *A signed digraph S is primitive and nonpowerful if there exists a positive integer l such that for any integer $t \geq l$, there is a pair of SSSD walks of length t from any vertex v_i to any vertex v_j $(v_i, v_j \in V(S))$. Such least integer l is called the base of S, denoted by $l(S)$. Let S be a primitive nonpowerful signed digraph of order n. Let $u, v \in V(S)$. The local base from u to v, denoted by $l_S(u, v)$, is defined to be the least integer k such that there is a pair of SSSD walks of length t from u to v for any integer $t \geq k$. The local base at a vertex $u \in V(S)$ is defined to be $l_S(u) = \max\limits_{v \in V(S)} \{l_S(u, v)\}$. So*

$$l(S) = \max_{u \in V(S)} l_S(u) = \max_{u,v \in V(S)} l_S(u, v).$$

Therefore, a sign pattern A is primitive nonpowerful if and only if $S(A)$ is primitive nonpowerful, and the base $l(A) = l(S(A))$ is the least positive integer l such that every entry of A^l is $\#$.

From [5], we know that $l(A) = l(|A|)$ for a powerful sign pattern A. So $l(A) = \exp(A)$ if A is a primitive powerful sign pattern. Moreover, if A is a powerful sign pattern, then A is primitive if and only if every real matrix B in $Q(A)$ $(Q(A) = \{B| \text{ real matrix } B \text{ with pattern } A\})$ is primitive. Thus, when A is a primitive, powerful sign pattern, every real matrix B with pattern A is primitive, has $D(B) = D(A)$, and has $\exp(B) = \exp(D(|A|))$. But we say that the result about the base of a powerful sign pattern fails to hold for a nonpowerful sign patterns, see an example as follow:

$$A = \begin{pmatrix} 1 & 1 \\ -1 & 1 \end{pmatrix}.$$

Note that A is trivially primitive since $D(A)$ has all possible arcs, that $l(|A|) = 1$, that A^2 contains no 0, but $l(A) = 3$. In particular, a real matrix B with sign pattern A can behave very differently from A:

$$B = \begin{pmatrix} 1 & 1 \\ -1 & 1 \end{pmatrix}$$

gives $B^4 = -4I$, which means B is not primitive in the usual sense. So the treatments of the bases about the nonpowerful sign patterns require greater care than the treatments for the powerful sign patterns.

Let S be a primitive nonpowerful signed digraph of order n and $V(S) = \{1, 2, \cdots, n\}$; for convenience, the vertices can be ordered so that $l_S(1) \leq l_S(2) \leq \cdots \leq l_S(n)$. We call $l_S(k)$ the kth local base of S. Thus $l(S) = l_S(n)$ and it is easy to see that $l_S(k)$ is the smallest integer l such that there are k all $\sharp$ rows in $[A(S)]^l$. Similarly, $\exp_G(k)$ is defined to be the smallest integer l such that there are l all "1" rows in $|A(G)|^l$ for a primitive digraph G.

Primitivity, base, local base, extremal patterns and other properties of powers sequence of a square sign pattern matrix are of great significance. The bases of sign patterns are closely related to many other problems in various areas of pure and applied mathematics (see [3], [4], [6], [7], [9], [12]). In practice, we consider the *memoryless communication system* [6] in communication field, which is depicted as a digraph D of order n. If D is primitive, the least time t such that each vertex in D receive the n pieces of different information from any vertex is equal to the index of D. If D is a primitive non-powerful signed digraph, the least time t such that each vertex in D receive the n pieces of ambiguous information from any vertex is just equal to the base of D. So studying the bases of the primitive non-powerful signed digraphs is very useful in information communication field, and hence studying the bases of the primitive non-powerful signed digraphs is also very useful for net works and theoretical computer science.

This paper is organized as follows: Section 1 introduces the basic ideas of patterns and their supports. Section 2 introduces series of working lemmas. Section 3 and Section 4 characterize the cycle properties in the associated signed digraphs and some bounds about the bases for the sign pattern matrices with base at least $\frac{3}{2}n^2 - 2n + 4$. Section 5 characterizes some sign pattern matrices with given bases and shows that there are some "*gaps*" in the base set.

2   Preliminaries

Let S be a strongly connecte digraph of order n and $C(S)$ denote the set of all cycle lengths in S.

Definition 2.1. *Let $\{s_1, s_2, \cdots, s_\lambda\}$ be a set of distinct positive integers, $gcd(s_1, s_2, \cdots, s_\lambda) = 1$. The Frobenius number of $s_1, s_2, \cdots, s_\lambda$, denoted by $\phi(s_1, s_2, \cdots, s_\lambda)$, is the smallest positive integer m such that $k = \sum_{i=1}^{\lambda} a_i s_i$ for any positive integer $k \geq m$ where a_i $(i = 1, 2, \cdots, \lambda)$ is non-negative integer.*

Lemma 2.2. *([6]) If $gcd(s_1, s_2) = 1$, then $\phi(s_1, s_2) = (s_1 - 1)(s_2 - 1)$.*

From Definition 2.1, it is easy to know that $\phi(s_1, s_2, \cdots, s_\lambda) \leq \phi(s_i, s_j)$ if there exist $s_i, s_j \in \{s_1, s_2, \cdots, s_\lambda\}$ such that $gcd(s_i, s_j) = 1$. So, if $\min\{s_i : 1 \leq i \leq \lambda\} = 1$, then $\phi(s_1, s_2, \cdots, s_\lambda) = 0$.

Lemma 2.3. *([4]) A Boolean matrix A is primitive if and only if $D(A)$ is strongly connected and $gcd(p_1, p_2, \cdots, p_t) = 1$ where $C(D(A)) = \{p_1, p_2, \cdots, p_t\}$.*

Definition 2.4. *For a primitive digraph S, suppose $C(S) = \{p_1, p_2, \ldots, p_u\}$. Let $d_{C(S)}(v_i, v_j)$ denote the length of the shortest walk from v_i to v_j which meets at least one p_i-cycle for each i $(i = 1, 2, \cdots, u)$. Such shortest directed walk is called a $C(S)$-walk from v_i to v_j. Further, $d_{C(S)}(v_i)$, $d_1(C(S))$ and $d(C(S))$ are defined as follows: $d_{C(S)}(v_i) = \max\{d_{C(S)}(v_i, v_j): v_j \in V(S)\}$, $d(C(S)) = \max\{d_{C(S)}(v_i, v_j): v_i, v_j \in V(S)\}$, $d_i(C(S))$ $(1 \leq i \leq n)$ is the ith smallest one in $\{d_{C(S)}(v_i) | 1 \leq i \leq n\}$, $d_n(C(S)) = d(C(S))$. In particular, if $C(S) = \{p, q\}$, $d(C(S))$ can be simply denoted by $d\{p, q\}$.*

Lemma 2.5. ([2]) *Let S be a primitive digraph of order n and $C(S) = \{p_1, p_2, \ldots, p_u\}$. Then $exp(v_i, v_j) \leq d_{C(S)}(v_i, v_j) + \phi(p_1, p_2, \ldots, p_u)$ for $v_i, v_j \in V(S)$. We have $\exp(S) \leq d(C(S)) + \phi(p_1, p_2, \ldots, p_u)$ furthermore.*

Lemma 2.6. ([2]) *Let S be a primitive digraph of order n whose girth is s. Then $exp_S(k) \leq s(n-2) + k$ for $1 \leq k \leq n$.*

Lemma 2.7. ([8]) *Let S be a primitive digraph of order n and $|C(S)| \geq 3$. Then $exp_S(k) \leq \lfloor \frac{1}{2}(n-2)^2 \rfloor + k$ for $1 \leq k \leq n$.*

Lemma 2.8. *Let D be a primitive digraph of order n which has a $s-$cycle C, $v \in V(C)$, and $|R_1(v)| \geq 2$. Then $exp(1) \leq exp(v) \leq 1 + s(n-2)$.*

Proof. We can take $w, z \in R_1(v)$ such that $(v, w) \in E(C)$ and $(v, z) \notin E(C)$ because of $v \in V(C)$ and $|R_1(v)| \geq 2$. We consider strongly connected digraph D^s (where $A(D^s) = [A(D)]^s$) in which the arc corresponds to the walk of length s in S. In D^s, w has a loop and there is arc (w, z). Thus $R_1(v) \xrightarrow{n-2} u$ for any vertex u in D^s. So there exists a walk of length $1 + s(n-2)$ from vertex v to any vertex u in D. $\qquad\qquad \square$

Lemma 2.9. ([11]) *Let S be a primitive nonpowerful signed digraph. Then S must contain a p_1-cycle C_1 and a p_2-cycle C_2 satisfying one of the following two conditions:*

(1) p_i is odd, p_j is even and $sgnC_j = -1$ $(i, j = 1, 2; i \neq j)$.
(2) p_1 and p_2 are both odd and $sgnC_1 = -sgnC_2$.

C_1, C_2 satisfying condition (1) or (2) are always called *a distinguished cycle pair.* It is easy to prove that $W_1 = p_2C_1$ and $W_2 = p_1C_2$ have the same length p_1p_2 but different signs if p_1-cycle C_1 and p_2-cycle C_2 are a distinguished cycle pair, namely $(sgnC_1)^{p_2} = -((sgnC_2)^{p_1})$.

Lemma 2.10. ([12]) *Let S be a primitive signed digraph. Then S is nonpowerful if and only if S contains a distinguished cycle pair.*

Lemma 2.11. ([12]) *Let S be a primitive nonpowerful signed digraph of order n and $C(S) = \{p_1, p_2, \ldots, p_m\}$. If the cycles in S with the same length have the same sign, p_1-cycle C_1 and p_2-cycle C_2 form a distinguished cycle pair, then*

(i) $l_S(v_i, v_j) \leq d_{C(S)}(v_i, v_j) + \phi(p_1, p_2, \ldots, p_m) + p_1p_2$, $v_i, v_j \in V(S)$.
(ii) $l_S(v_i) \leq d_{C(S)}(v_i) + \phi(p_1, p_2, \ldots, p_m) + p_1p_2$.
(iii) $l(S) \leq d(C(S)) + \phi(p_1, p_2, \ldots, p_m) + p_1p_2$.

Lemma 2.12. ([10]) *Let S be a primitive nonpowerful signed digraph of order n and $u \in V(S)$. If there exists a pair of SSSD walks with length r from u to u, then $l_S(u) \leq exp_S(u) + r$.*

Lemma 2.13. ([10]) *Let S be a primitive nonpowerful signed digraph of order n. Then we have $l_S(k) \leq l_S(k-1) + 1$ for $2 \leq k \leq n$.*

Let D_1 consist of cycle $(v_n, v_{n-1}, \cdots, v_2, v_1, v_n)$ and arc (v_1, v_{n-1}) and $D_2 = D_1 \bigcup \{(v_2, v_n)\}$. Then we have the following lemmas 2.14, 2.15.

Lemma 2.14. ([11]) *Let S_2 be a nonpowerful signed digraph of order $n \geq 3$ with D_2 as its underlying digraph. Then we have*

(1) if the (only) two cycles of length $n-1$ of S_2 have different signs, then $l(S_2) \leq n^2 - n + 2$;
(2) if the two cycles of length $n-1$ of S_2 have the same sign, then $l(S_2) = 2(n-1)^2 + (n-1)$.

Lemma 2.15. ([11]) *Let A be an irreducible generalized sign pattern matrix of order $n \geq 3$. Then*

(i) $l(A) \leq 2(n-1)^2 + n$; (1)
(ii) equality holds in (1) if and only if A is a nonpowerful sign pattern matrix and the associated digraph $D(A)$ of A is isomorphic to D_1;
(iii) for each integer k with $2n^2 - 4n + 5 < k < 2n^2 - 3n + 1$, there is no irreducible generalized sign pattern matrix A of order n with $l(A) = k$.

3   Cycle Properties

Theorem 3.1. *Let S be a primitive nonpowerful signed digraph of order $n \geq 6$ whose underlying digraph is $|S|$. If $|C(S)| \geq 3$, then $l(S) \leq \dfrac{3}{2}n^2 - 2n + 3$.*

Proof. By Lemma 2.9, there exists a distinguished cycle pair p_1-cycle C_1 and p_2-cycle C_2 in S, $p_1 C_2$ and $p_2 C_1$ have different signs.

Case 1. C_1, C_2 have no common vertex.

Then $p_1 + p_2 \leq n$. Suppose $p_1 \leq \dfrac{n}{2}$ for convenience, Q_1 is one of the shortest walks with length q_1 from C_1 to C_2, $\{v_1\} = V(Q_1) \bigcap V(C_1)$, $\{v_2\} = V(Q_1) \bigcap V(C_2)$, and Q_2 is one of the shortest walks with length q_2 from v_2 to v_1, then $q_1 \leq n - p_1 - p_2 + 1, q_2 \leq n - 1$, $p_2 C_1 \bigcup Q_1 \bigcup Q_2$ and $Q_1 \bigcup p_1 C_2 \bigcup Q_2$ are a pair of $SSSD$ walks with length $p_1 p_2 + q_1 + q_2$ from v_1 to v_1.

Note that

$$p_1 p_2 + q_1 + q_2 \leq p_1 p_2 + 2n - p_1 - p_2 = (p_1 - 1)(p_2 - 1) + 2n - 1$$

$$\leq [\frac{1}{2}(p_1 + p_2 - 2)]^2 + 2n - 1 \leq [\frac{1}{2}(n-2)]^2 + 2n - 1 = \frac{n^2}{4} + n,$$

and $\exp(v_1) \le p_1(n-2)+1$ by Lemma 2.8, then

$$l_S(1) \le l_S(v_1) \le \exp_S(v_1) + p_1 p_2 + q_1 + q_2 \le \frac{n}{2}(n-2) + 1 + \frac{n^2}{4} + n = \frac{3n^2}{4} + 1$$

and $l_S(n) \le l_S(1) + n - 1 \le \dfrac{3n^2}{4} + n$ by Lemmas 2.12, 2.13.

Case 2. C_1, C_2 have common vertices.

Subcase 2.1. $p_1 = p_2$. It is easy to see that p_1 is odd.

$1°$ $p_1 = n$, let $v_1 \in V(S)$, $\exp_S(v_1) = \exp_S(1)$. The underlying digraph $|S|$ is not isomorphic to D_1 or D_2 because $|C(S)| \ge 3$. Thus the girth s of S is at most $n-2$. By Lemma 2.8, then $\exp_S(1) = \exp(v_1) \le s(n-2)+1 \le (n-2)^2+1$. Note that C_1 and C_2 form a pair of $SSSD$ walks from v_1 to itself now, by Lemmas 2.12, 2.13, then

$$l_S(1) \le l_S(v_1) \le n+(n-2)^2+1 = n^2-3n+5, \ l_S(n) \le l_S(1)+n-1 \le n^2-2n+4.$$

$2°$ $p_1 \le n-1$, suppose $v_1 \in V(C_1) \bigcap V(C_2)$ and $|R_1(v_1)| \ge 2$. By Lemmas 2.8, 2.12, 2.13, then

$$\exp_S(v_1) \le p_1(n-2)+1 \le n^2-3n+3, \ l_S(1) \le l_S(v_1) \le p_1+\exp_S(v_1) < n^2-2n+2$$

and $l_S(n) \le l_S(1) + n - 1 \le n^2 - n + 1$.

Subcase 2.2. $\text{Min}(p_1, p_2) = p_1 \le n-2$.

Suppose $V(C_1) \bigcap V(C_2) = \{v_1, v_2, \cdots, v_t\}$ and $\exp_D(u) = \exp_D(1)$. Because of $|C(S)| \ge 3$, so $\exp_S(u) \le \lfloor \frac{1}{2}(n-2)^2 \rfloor + 1$ by Lemma 2.7. Let $q_i = d(u, v_i)$, $1 \le i \le t$ and suppose $q_1 = \min\limits_{1 \le i \le t} \{q_i\}$ for convenience, then $q_1 \le n - (p_1 + p_2 - t) + p_2 - t = n - p_1$. So there exists a pair of $SSSD$ walks with length $q_1 + d(v_1, u) + p_1 p_2$ from u to u. Because of

$$d(v_1, u) \le n-1, \ q_1 + d(v_1, u) + p_1 p_2 \le 2n-1+p_1(p_2-1) \le 2n-1+(n-2)(n-1) \le n^2-n+1,$$

so

$$l_S(1) \le l_S(u) \le q_1+d(v_1,u)+p_1p_2+\exp_S(u) \le \lfloor \tfrac{1}{2}(n-2)^2 \rfloor+1+n^2-n+1 \le \frac{3n^2}{2}-3n+4$$

by Lemma 2.12 and $l_S(n) \le l_S(1) + n - 1 \le \dfrac{3n^2}{2} - 2n + 3$ by Lemma 2.13.

Subcase 2.3. $\{p_1, p_2\} = \{n-1, n\}$.

Let $C_1 = C_{n-1}$, $C_2 = C_n$. Suppose $\exp_S(u) = \exp_D(1)$ for convenience, then $\exp_S(u) \le \frac{1}{2}(n-2)^2 + 1$ by Lemma 2.7. Because there exists a pair of $SSSD$ walks with length $n(n-1)$ from u to u if $u \in V(C_{n-1})$, by Lemmas 2.12 and 2.13, so

$$l_S(1) \le l_S(u) \le \frac{(n-2)^2}{2} + 1 + n(n-1) = \frac{3n^2-6n}{2} + 3, \ l_S(n) \le \frac{3n^2}{2} - 2n + 2.$$

If $u \notin V(C_{n-1})$, there exists a vertex $v \in V(S)$ towards u such that (v, u) is an arc in S and $v \in V(C_{n-1})$. Then $\exp_S(v) \leq \exp_S(u) + 1 \leq \frac{1}{2}(n-2)^2 + 2$. Because there exists a pair of $SSSD$ walks of length $n(n-1)$ from v to v, by Lemmas 2.12 and 2.13, so

$$l_S(1) \leq l_S(v) \leq \frac{(n-2)^2}{2} + 2 + n(n-1) = \frac{3n^2 - 6n}{2} + 4, \ l_S(n) \leq \frac{3n^2}{2} - 2n + 3. \quad \square$$

Corollary 3.2. *Let S be a primitive nonpowerful signed digraph of order $n \geq 6$. Then $|C(S)| = 2$ if $l(S) \geq \frac{3}{2}n^2 - 2n + 4$.*

Theorem 3.3. *Let S be a primitive nonpowerful signed digraph of order $n \geq 6$. Cycle C_1 with length p_1 and cycle C_2 with length p_2 form a distinguished cycle pair $(p_1 \leq p_2)$. If $p_1 + p_2 \leq n$, then $l(S) \leq \frac{3}{4}n^2 + n$.*

Proof. **Case 1.** C_1 and C_2 have no common vertex.

As proved in case 1 of Theorem 3.1, $l_S(n) \leq \frac{3}{4}n^2 + n$ can be obtained.

Case 2. C_1 and C_2 have at least one common vertex.

Subcase 2.1. If $p_1 = p_2$, then $p_1 \leq \frac{1}{2}n$. Let $v_1 \in V(C_1) \bigcap V(C_2)$ and $|R_1(v_1)| \geq 2$. By Lemmas 2.8, 2.12, 2.13, then $\exp_S(v_1) \leq p_1(n-2) + 1 \leq \frac{1}{2}n^2 - n + 1$, $l_S(1) \leq l_S(v_1) \leq p_1 + \exp_S(v_1) \leq \frac{1}{2}n^2 - \frac{1}{2}n + 1$ and $l_S(n) \leq l_S(1) + n - 1 \leq \frac{1}{2}n^2 + \frac{1}{2}n$.

Subcase 2.2. If $p_1 < p_2$, then $p_1 < \frac{1}{2}n$. Let $v_1 \in V(C_1) \bigcap V(C_2)$ and $|R_1(v_1)| \geq 2$. Similar to Subcase 2.1, note that there is a pair of $SSSD$ walks with length $p_1 p_2$ from v_1 to itself and $p_1 p_2 \leq (\frac{p_1 + p_2}{2})^2$, we get $l_S(v_1) < \frac{3}{4}n^2 - n + 1$ and $l_S(n) < \frac{3}{4}n^2$. $\quad \square$

Corollary 3.4. *Let S be a primitive nonpowerful signed digraph of order $n \geq 6$. Cycle C_1 with length p_1 and cycle C_2 with length p_2 form a distinguished cycle pair $(p_1 \leq p_2)$. Then $p_1 + p_2 > n$ if $l(S) \geq \frac{3}{4}n^2 + n + 1$.*

Theorem 3.5. *Let S be a primitive nonpowerful signed digraph with order $n \geq 3$. If there exist two cycles with the same length but different signs, then we have $l(S) \leq n^2$.*

Proof. Let C_1 and C_2 be two cycles such that $L(C_1) = p = L(C_2)$ but $\mathrm{sgn}(C_1) = -\mathrm{sgn}(C_2)$.

Case 1. C_1 and C_2 have at least one common vertex.

Let $v_1 \in V(C_1) \bigcap V(C_2)$ and $|R_1(v_1)| \geq 2$. Note that $p \leq n$ and there is a pair of $SSSD$ walks with length p from v_1 to itself, similar to the proof in Subcase 2.1 of Theorem 3.3, we get $l_S(v_1) \leq n^2 - n + 1$ and $l_S(n) \leq n^2$.

Case 2. C_1 and C_2 have no common vertex. Then $p \leq \dfrac{n}{2}$.

Let Q_1 be one of the shortest walks with length q_1 from C_1 to C_2, $V(Q_1) \bigcap V(C_1) = \{v_1\}$, $V(Q_1) \bigcap V(C_2) = \{v_2\}$ and Q_2 is one of the shortest walks with length q_2 from v_2 to v_1. Then $q_1 \leq n - 2p + 1, q_2 \leq n - 1$. $C_1 + Q_1 + Q_2$ and $C_2 + Q_1 + Q_2$ are a pair of $SSSD$ walks with length $p + q_1 + q_2$ from v_1 to v_1. Similar to Subcase 2.1 of Theorem 3.3, we get $\exp_S(v_1) \leq p(n-2) + 1, l_S(v_1) \leq \dfrac{n^2}{2} + \dfrac{n}{2} + 1$ and $l_S(n) \leq \dfrac{n^2}{2} + \dfrac{3n}{2}$.

To sum up, the theorem is proved. □

Corollary 3.6. *Let S be a primitive nonpowerful signed digraph of order $n \geq 6$. Then any two cycles with the same length have the same sign if $l(S) \geq n^2 + 1$.*

Theorem 3.7. *Let A be a primitive nonpowerful square sign pattern matrix with order $n \geq 6$. If $l(A) \geq \dfrac{3}{2}n^2 - 2n + 4$, then we have the results as follows:*

(i)$|C(S(A))| = 2$. Suppose $C(S(A)) = \{p_1, p_2\}$ $(p_1 < p_2)$, then $\gcd(p_1, p_2) = 1, p_1 + p_2 > n$;

(ii) In $S(A)$, all p_1-cycles have the same sign, all p_2-cycles have the same sign, and every pair of p_1-cycle and p_2-cycle form a distinguished cycle pair.

Proof. The theorem follows from Corollaries 3.2, 3.4, 3.6. □

4   Bounds of the Bases

Lemma 4.1. *Let A be a primitive nonpowerful square sign pattern matrix with order $n \geq 6$. If $C(S(A)) = \{p_1, p_2\}$ $(p_1 < p_2, p_1 + p_2 > n)$, all p_1-cycles have the same sign, all p_2-cycles have the same sign in $S(A)$, then $l(A) \leq 2n - 1 + (2p_1 - 1)(p_2 - 1)$.*

Especially, if $p_1 = n - 1$, $p_2 = n$, then $l(A) \leq 2n^2 - 3n + 2$.

Proof. Let C_1, C_2 form a distinguished cycle pair and $L(C_1) = p_1$, $L(C_2) = p_2$ in $S(A)$. Suppose $V(C_1) \bigcap V(C_2) = \{v_1, v_2, \cdots, v_t\}$. Let $d_0 = \min\limits_{1 \leq i \leq t} \{d(x, v_i)\}$ for $x \in V(S(A))$. Thus $d_0 \leq n - (p_1 + p_2 - t) + p_2 - t = n - p_1$. Let $d_0 = d(x, v_k)$ $(1 \leq k \leq t)$, then $d(v_k, y) \leq n - 1$ for $y \in V(S(A))$. So there exist a pair of $SSSD$ walks of length $d_0 + \phi(p_1, p_2) + p_1 p_2 + d(v_k, y)$ $(d_0 + \phi(p_1, p_2) + p_1 p_2 + d(v_k, y) \leq 2n - 1 + (2p_1 - 1)(p_2 - 1))$ from x to y. Because x, y are arbitrary, thus $l(S(A)) \leq 2n - 1 + (2p_1 - 1)(p_2 - 1)$. Therefore, the lemma is proved. □

Theorem 4.2. *Let A be a primitive nonpowerful square signed pattern matrix with order $n \geq 10$. If $l(A) \geq \dfrac{3}{2}n^2 - 2n + 4$ and the girth of $S(A)$ is at most $n - 3$, then $l(A) \leq 2n^2 - 7n + 6$.*

Proof. Note that $\min C(S(A)) \leq n - 3$, then the theorem follows from Theorem 3.7 and Lemma 4.1. □

5   Gaps and Some Digraphs with Given Bases

Theorem 5.1. *Let $D_{k,i}$ consist of cycle $C_n = (v_1, v_n, v_{n-1}, v_{n-2}, \ldots, v_2, v_1)$ and arcs $(v_1, v_{n-k}), (v_2, v_{n-k+1}), \ldots, (v_i, v_{n-k+i-1})$ $(1 \le i \le \min\{k+1, n-k-1\})$ (see Fig. 1) where $\gcd(n, n-k) = 1$. Let $S_{k,i}$ be a primitive nonpowerful signed digraph with underlying digraph $D_{k,i}$ $(1 \le i \le \min\{k+1, n-k-1\})$. If all $(n-k)$-cycles have the same sign in $S_{k,i}$, then $l(S_{k,i}) = (2n-2)(n-k)+1-i+n$.*

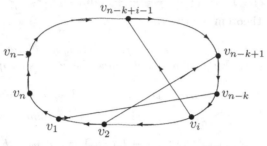

Fig. 1. $D_{k,i}$

Proof. Every pair of $(n-k)$-cycle and n-cycle form a distinguished cycle pair because $S_{k,i}$ is a primitive nonpowerful signed digraph.

Case 1. $i - 1 < k$. Then $n - k + i - 1 < n$.

Now $d(C(S_{k,i})) = d_{C(S_{k,i})}(v_n, v_{n-k+i}) = n + k - i$, by Lemma 2.11, then

$$l(S_{k,i}) \le d(C(S_{k,i})) + \phi(n, n-k) + n(n-k) = (2n-2)(n-k) + 1 - i + n.$$

We assert $l(S_{k,i}) = (2n-2)(n-k)+1-i+n$. Now we prove that there is no pair of $SSSD$ walks of length $(2n-2)(n-k) - i + n$ from v_n to v_{n-k+i}.

Otherwise, suppose W_1, W_2 are a pair of $SSSD$ walks with length $(2n-2)(n-k) - i + n$ from v_n to v_{n-k+i}. Let P be the unique path from v_n to v_{n-k+i} on cycle C_n and $W = P \bigcup C_n$. Then W_j $(j = 1, 2)$ must consist of W, some n-cycles and some $(n-k)$-cycles, namely, $|W_j| = (2n-2)(n-k) - i + n = n + k - i + a_j n + b_i(n-k)$ $(a_j, b_j \ge 0, \ j = 1, 2)$. Because of $\gcd(n, n-k) = 1$, so $(a_1 - a_2)n = (b_2 - b_1)(n-k)$, $n|(b_2 - b_1)$, $(n-k)|(a_1 - a_2)$, and then $b_2 - b_1 = nx$, $a_1 - a_2 = (n-k)x$ for some integer x.

We assert $x = 0$. If $x \ge 1$, then $b_2 \ge n$. Thus we have

$$(2n-2)(n-k) - i + n = n + k - i + a_2 n + (b_2 - n)(n-k) + n(n-k)$$

and $\phi(n, n-k) - 1 = a_2 n + (b_2 - n)(n-k)$, which contradicts the definition of $\phi(n, n-k)$. In a same way, we can get analogous contradiction when $x \le -1$. The assertion $x = 0$ is proved. So W_1, W_2 have the same sign because $b_2 = b_1$, $a_1 = a_2$ and all $(n-k)$-cycles have the same sign. This contradicts that W_1, W_2 are a pair of $SSSD$ walks. Thus there is no pair of $SSSD$ walks of length $(2n-2)(n-k) - i + n$ from v_n to v_{n-k+i}, and so

$$l(S_{k,i}) = l_{S_{k,i}}(v_n, v_{n-k+i}) = (2n-2)(n-k) + 1 + n - i.$$

Case 2. $k = i - 1$. Then $n - k + i - 1 = n$.

Now $d(C(S_{k,i})) = d_{C(S_{k,i})}(v_n, v_1) = n - 1$. As the proof of case 1, we can prove

$$l(S_{k,i}) = l_{S_{k,i}}(v_n, v_1) = (2n - 2)(n - k) + 1 + n - i. \qquad \square$$

Suppose that n is odd, let $\mathscr{L}$ consist of cycle $C_n = (v_1, v_n, v_{n-1}, v_{n-2}, v_{n-3}, \cdots, v_2, v_1)$ $(n \geq 6)$ and arcs (v_1, v_{n-2}), (v_3, v_n). Let F consist of cycle $C_{n-1} = (v_1, v_n, v_{n-1}, v_{n-3}, v_{n-4}, \ldots, v_2, v_1)$ $(n \geq 6)$ and arcs (v_1, v_{n-2}), (v_{n-2}, v_{n-3}). Let F_1 consist of cycle $(v_1, v_{n-1}, v_{n-2}, \ldots, v_2, v_1)$ and arcs (v_1, v_{n-2}), (v_2, v_n), (v_n, v_{n-1}). Let F_2 consist of cycle $(v_1, v_n, v_{n-2}, v_{n-3}, v_{n-4}, \ldots, v_2, v_1)$ and arcs (v_1, v_{n-2}), (v_n, v_{n-1}), (v_{n-1}, v_{n-3}). Let F_3 consist of cycle $(v_1, v_{n-2}, v_{n-3}, v_{n-4}, \ldots, v_2, v_1)$ and arcs (v_1, v_{n-1}), (v_{n-1}, v_{n-2}), (v_1, v_n), (v_n, v_{n-2}). Let F_i' $(2 \leq i \leq n - 3)$ consist of cycle $(v_1, v_{n-1}, v_{n-2}, \ldots, v_2, v_1)$ and arcs (v_1, v_{n-2}), (v_{i+1}, v_n), (v_n, v_{i-1}). Let F_4 consist of cycle $(v_1, v_{n-1}, v_{n-2}, \ldots, v_2, v_1)$ and arcs (v_1, v_{n-2}), (v_1, v_n), (v_n, v_{n-3}). Let F_5 consist of cycle $(v_1, v_{n-1}, v_{n-2}, \cdots, v_2, v_1)$ and arcs (v_1, v_{n-2}), (v_2, v_n), (v_n, v_{n-2}). Let F_6 consist of cycle $(v_1, v_{n-1}, v_{n-2}, \ldots, v_2, v_1)$ and arcs (v_1, v_n), (v_n, v_{n-3}), (v_2, v_{n-1}). Let F_7 consist of cycle $(v_1, v_{n-1}, v_{n-2}, \ldots, v_2, v_1)$ and arcs (v_1, v_{n-2}), (v_3, v_n), (v_n, v_{n-1}). Let $\mathscr{B}_1$ consist of cycle $C_n = (v_1, v_n, v_{n-1}, \ldots, v_2, v_1)$ and arcs (v_1, v_{n-3}), (v_3, v_{n-1}). Let $\mathscr{B}_2$ consist of cycle $C_n = (v_1, v_n, v_{n-1}, \ldots, v_2, v_1)$ and arcs (v_1, v_{n-3}), (v_4, v_n). Let $\mathscr{B}_3$ consist of cycle $C_n = (v_1, v_n, v_{n-1}, \ldots, v_2, v_1)$ and arcs (v_1, v_{n-3}), (v_2, v_{n-2}), (v_4, v_n). Let $\mathscr{B}_3$ consist of cycle $C_n = (v_1, v_n, v_{n-1}, \ldots, v_2, v_1)$ and arcs (v_1, v_{n-3}), (v_2, v_{n-2}), (v_4, v_n). Let $\mathscr{B}_4$ consist of cycle $C_n = (v_1, v_n, v_{n-1}, \ldots, v_2, v_1)$ and arcs (v_1, v_{n-3}), (v_3, v_{n-1}), (v_4, v_n).

Suppose that n is odd, let $\mathscr{T}$ be a primitive nonpowerful signed digraph with underlying digraph $\mathscr{L}$, all $(n - 2)$-cycles have the same sign in $\mathscr{T}$. Let $\mathscr{S}_0$ be a primitive nonpowerful signed digraph with underlying digraph F, all $(n - 1)$-cycles have the same sign, all $(n - 2)$-cycles have the same sign in $\mathscr{S}_0$. Let $\mathscr{S}_1$ be a primitive nonpowerful signed digraph with underlying digraph F_1, all $(n - 1)$-cycles have the same sign, all $(n - 2)$-cycles have the same sign in $\mathscr{S}_1$. Let $\mathscr{S}_2$ be a primitive nonpowerful signed digraph with underlying digraph F_2, all $(n - 1)$-cycles have the same sign, all $(n - 2)$-cycles have the same sign in $\mathscr{S}_2$. Let $\mathscr{S}_3$ be a primitive nonpowerful signed digraph with underlying digraph F_3, all $(n - 1)$-cycles have the same sign, all $(n - 2)$-cycles have the same sign in $\mathscr{S}_3$. Let $\mathscr{S}_4$ be a primitive nonpowerful signed digraph with underlying digraph F_4, all $(n - 1)$-cycles have the same sign, all $(n - 2)$-cycles have the same sign in $\mathscr{S}_4$. Let $\mathscr{S}_5$ be a primitive nonpowerful signed digraph with underlying digraph F_5, all $(n - 1)$-cycles have the same sign, all $(n - 2)$-cycles have the same sign in $\mathscr{S}_5$. Let $\mathscr{S}_6$ be a primitive nonpowerful signed digraph with underlying digraph F_6, all $(n - 1)$-cycles have the same sign, all $(n - 2)$-cycles have the same sign in $\mathscr{S}_6$. Let $\mathscr{S}_7$ be a primitive nonpowerful signed digraph with underlying digraph F_7, all $(n - 1)$-cycles have the same sign, all $(n - 2)$-cycles have the same sign in $\mathscr{S}_7$. Let $\mathscr{S}_i$ be a primitive nonpowerful signed digraph with underlying digraph F_i', all $(n - 1)$-cycles have the same sign, all $(n - 2)$-cycles have the same sign in $\mathscr{S}_i$. Let $\mathscr{Q}_1$ be a primitive nonpowerful signed digraph with underlying digraph $\mathscr{B}_1$, all $(n-3)$-cycles have the same sign in $\mathcal{Q}_1$. Let $\mathcal{Q}_2$ be a primitive nonpowerful signed digraph with underlying digraph $\mathcal{B}_2$, all $(n-3)$-cycles have the same sign in $\mathcal{Q}_2$. Let $\mathcal{Q}_3$ be a primitive nonpowerful signed digraph with underlying digraph $\mathcal{B}_3$, all $(n-3)$-cycles have the same sign in $\mathcal{Q}_3$. Let $\mathcal{Q}_4$ be a primitive nonpowerful signed digraph with underlying digraph $\mathcal{B}_4$, all $(n-3)$-cycles have the same sign in $\mathcal{Q}_4$.

Similar to the proof of Theorem 5.1, we can prove the following Theorem 5.2:

Theorem 5.2. *Let S be a primitive nonpowerful signed digraph with order n. Then*

$$
l(S) = \begin{cases}
2n^2 - 5n + 2, & S = \mathcal{T}, \ n \text{ is odd}; \\
2n^2 - 7n + 8, & S \in \{\mathcal{S}_0, \mathcal{S}_1, \mathcal{S}_2\}; \\
2n^2 - 7n + 7, & S \in \{\mathcal{S}_3, \cdots, \mathcal{S}_7\} \bigcup \{\mathcal{S}_i | 2 \leq i \leq n-3\}; \\
2n^2 - 7n + 4, & S = \mathcal{Q}_1; \\
2n^2 - 7n + 3, & S \in \{\mathcal{Q}_2, \mathcal{Q}_3, \mathcal{Q}_4\}.
\end{cases}
$$

Theorem 5.3. *Let A be a primitive nonpowerful square sign pattern matrix with order $n(n \geq 10)$. Then we have:*

(1) There is no A such that $l(A) \in [2n^2 - 7n + 9, 2n^2 - 3n]$ if n is a positive even integer.

(2) If n is a positive odd integer, then
(i) There is no A such that $l(A) \in ([2n^2 - 7n + 9, 2n^2 - 5n + 1] \bigcup [2n^2 - 5n + 5, 2n^2 - 3n])$;
(ii) $l(A) = 2n^2 - 5n + 4$ if and only if $D(A) \cong D_{2,1}$;
$l(A) = 2n^2 - 5n + 3$ if and only if $D(A) \cong D_{2,2}$, the cycles with the same length have the same sign in $S(A)$;
$l(A) = 2n^2 - 5n + 2$ if and only if $D(A) \cong D_{2,3}$ or $D(A) \cong \mathcal{L}$, the cycles with the same length have the same sign in $S(A)$;

(3) For any integer $n \geq 10$, $l(A) = 2n^2 - 7n + 8$ if and only if $D(A)$ is isomorphic to one in $\{F, F_1, F_2\}$, the cycles with the same length have the same sign in $S(A)$;
$l(A) = 2n^2 - 7n + 7$ if and only if $D(A)$ is isomorphic to one in $\{F_3, F_4, F_5, F_6, F_7\} \bigcup \{F_i' \mid 2 \leq i \leq n-3\}$, the cycles with the same length have the same sign in $S(A)$.

(4) $\{2n^2 - 7n + m| \ 3 \leq m \leq 6\} \subset E_n$ if $\gcd(n, n-3) = 1$ (namely $3 \nmid n$), where $E_n = \{l(A)| \ A \text{ is a primitive nonpowerful square sign pattern matrix with order } n \ (n \geq 10)\}$.

Proof. Note that $n \geq 10$, then $2n^2 - 7n + 7 \geq \dfrac{3}{2}n^2 - 2n + 4$. By Theorem 3.7, then $C(S(A)) = \{p_1, p_2\}, p_1 < p_2, p_1 + p_2 > n$, all p_1-cycles have the same sign, all p_2-cycles have the same sign in $S(A)$. By Theorem 4.2, we know that $l(A) \leq 2n^2 - 7n + 6$ if $p_1 \leq n - 3$. So, if $l(A) \geq 2n^2 - 7n + 7$, there are just the following possible cases:

(1) $p_2 = n, p_1 = n - 1$;
(2) $p_2 = n, p_1 = n - 2$;
(3) $p_2 = n - 1, p_1 = n - 2$.

Note that $l(S_{3,i}) = 2n^2 - 7n + 7 - i$ ($1 \leq i \leq 4$), then the theorem follows from the Lemmas 2.14, 2.15 and Theorems 5.1, 5.2. □

References

1. Bondy, J.A., Murty, U.S.R.: Graph Theory. Springer, Heidelberg (2008)
2. Brualdi, R.A., Liu, B.: Generalized Exponents of primitive directed graphs. J. of Graph Theory 14, 483–499 (1994)
3. Dulmage, A.L., Mendelsohn, N.S.: Graphs and matrices. In: Harary, F. (ed.) Graph Theory and Theoretical Physics, ch. 6, pp. 167–227
4. Kim, K.H.: Boolean Matrix Theory and Applications. Marcel Dekkez, New York (1982)
5. Li, Z., Hall, F., Eschenbach, C.: On the period and base of a sign pattern matrix. Linear Algebra Appl. 212(213), 101–120 (1994)
6. Liu, B.: Combinatorical matrix theory, second published. Science Press, Beijing (2005)
7. Ljubic, J.I.: Estimates of the number of states that arise in the determinization of a nondeterministic autonomous automaton. Dokl. Akad. Nauk SSSR 155, 41–43 (1964); Soviet Math. Dokl., 5, 345–348 (1964)
8. Miao, Z., Zhang, K.: The local exponent sets of primitive digraphs. Linear Algebra Appl. 307, 15–33 (2000)
9. Schwarz, S.: On the semigroup of binary relations on a finite set. Czech. Math. J. 20(95), 632–679 (1970)
10. Wang, L., Miao, Z., Yan, C.: Local bases of primitive nonpowerful signed digraphs. Discrete Mathematics 309, 748–754 (2009)
11. You, L., Shao, J., Shan, H.: Bounds on the basis of irreducible generalized sign pattern matrices. Linear Algebra Appl. 427, 285–300 (2007)
12. Yu, G.L., Miao, Z.K., Shu, J.L.: The bases of the primitive, nonpowerful sign patterns with exactly d nonzero diagonal entries (revision is submitted)

Extended Dynamic Subgraph Statistics
Using h-Index Parameterized Data Structures

David Eppstein, Michael T. Goodrich, Darren Strash, and Lowell Trott

Computer Science Department, University of California, Irvine, USA

Abstract. We present techniques for maintaining subgraph frequencies in a dynamic graph, using data structures that are parameterized in terms of h, the h-*index* of the graph. Our methods extend previous results of Eppstein and Spiro for maintaining statistics for undirected subgraphs of size three to directed subgraphs and to subgraphs of size four. For the directed case, we provide a data structure to maintain counts for all 3-vertex induced subgraphs in $O(h)$ amortized time per update. For the undirected case, we maintain the counts of size-four subgraphs in $O(h^2)$ amortized time per update. These extensions enable a number of new applications in Bioinformatics and Social Networking research.

1 Introduction

Deriving inspiration from work done on fixed-parameter tractable algorithms for NP-hard problems (e.g., see [5,6,18]), the area of *parameterized algorithm design* involves defining numerical parameters for input instances, other than just the input size, and designing data structures and algorithms whose performance can be characterized in terms of those parameters. The goal, of course, is to find useful parameters and then design data structures and algorithms that are efficient for typical values of those parameters (e.g., see [9,10]). In this paper, we are interested in extending previous applications of this approach in the context of dynamic subgraph statistics—where one maintains the counts of all (induced and non-induced) subgraphs of certain types—from undirected size-three subgraphs [10] to applications involving directed size-three subgraphs and undirected subgraphs of size four.

Upon cursory examination this contribution may seem incremental, but these extensions allow for the possibility of significant computational improvement in several important applications. For instance, in bioinformatics, statistics involving the frequencies of certain small subgraphs, called *graphlets*, have been applied to protein-protein interaction networks [16,21] and cellular networks [20]. In these applications, the frequency statistics for the subgraphs of interest have direct bearing on biological network structure and function. In particular, in these graphlets applications, the undirected subgraphs of interest include one size-two subgraph (the 1-path), two size-three subgraphs (the 3-cycle and 2-path), and six size-four subgraphs (the 3-star, 3-path, triangle-plus-edge, 4-cycle, K_4 minus an edge, and K_4), which we respectively illustrate later in Fig. 7 as Q_4, Q_6, Q_7, Q_8, Q_9, and Q_{10}.

In addition, maintaining subgraph counts in a dynamic graph is of crucial importance to statisticians and social-networking researchers using the *exponential random graph model* (ERGM) [12,22,24] to generate random graphs. ERGMs can be tailored to generate random graphs that possess specific properties, which makes ERGMs an

W. Wu and O. Daescu (Eds.): COCOA 2010, Part I, LNCS 6508, pp. 128–141, 2010.

ideal tool for Social Networking research [24,22]. This tailoring is accomplished by a Markov Chain Monte Carlo (MCMC) method [22], which generates random graphs via a sequence of incremental changes. These incremental changes are accepted or rejected based on the values of subgraph statistics, which must be computed exactly for each incremental change in order to facilitate acceptance or rejection. Thus, there is a need for dynamic graph statistics in ERGM applications.

Typical graph attributes of interest in ERGM applications include the frequencies of undirected stars and triangles, which are used in the triad model [13] to study friends-of-friends relationships, as well as other more-complex subgraphs [23], including undirected 4-cycles and two-triangles (K_4 minus an edge), and directed transitive triangles, which we illustrate as graph T_9 in Fig. 3. Therefore, there is a salient need for algorithms to maintain subgraph statistics in a dynamic graph involving directed subgraphs of size three and undirected subgraphs of size four.

Interestingly, extending the previous approach, of Eppstein and Spiro [10], for maintaining undirected size-three subgraphs to these new contexts involves overcoming some algorithmic challenges. The previous approach uses a parameterized algorithm design framework for counting three-vertex induced subgraphs in a dynamic undirected graph. Their data structure has running time $O(h)$ amortized time per graph update (assuming constant-time hash table lookups), where h is the largest integer such that there exists h vertices of degree at least h, which is a parameter known as the h-index of the graph. This parameter was introduced by Hirsch [14] as a combined way of measuring productivity and impact in the academic achievements of researchers. As we will show, extending the approach of Eppstein and Spiro to directed subgraphs of size three and undirected subgraphs of size four involves more than doubling the complexity of the algebraic expressions and supporting data structures needed. Ensuring the directed size-three procedure maintains the complexity bounds of previous work required extensive understanding of dynamic graph composition. Developing the approach for size-four subgraphs that would allow only the addition of a single factor of h required innovative work with the structure of stored graph elements.

1.1 Other Related Work

Although subgraph isomorphism is known to be NP-complete, it is solvable in polynomial time for small subgraphs. For example, all triangles and four-cycles can be found in an n-vertex graph with m edges in $O(m^{3/2})$ time [15,3]. All cycles up to length seven can be counted (but not listed) in $O(n^\omega)$ time [2], where $\omega \approx 2.376$ is the exponent for the asymptotically fastest known matrix multiplication algorithm [4]. Also, in planar graphs, the number of copies of any fixed subgraph may be found in linear time [7,8]. These previous approaches run too slowly for the iterative nature of ERGM Markov Chain Monte Carlo simulations, however.

1.2 Our Results

In this paper, we present an extension of the h-index parameterized data structure design from statistics for undirected subgraphs of size three to directed subgraphs of size three and undirected subgraphs of size four. We show that in a dynamic directed graph one can maintain the counts of all directed three-vertex subgraphs in $O(h)$ amortized time per update, and in a dynamic undirected graph one can maintain the four-vertex sub-graph counts in $O(h^2)$ amortized time per update, assuming constant-time hash-table lookups (or worst-case amortized times that are a logarithmic factor larger). These results therefore provide techniques for application domains, in Bioinformatics and Social Networking, that can take advantage of these extended types of statistics. In addition, our data structures are based a number of novel insights into the combinatorial structure of these different types of subgraphs.

2   Preliminaries

As mentioned above, we define the *h-index* of a graph to be the largest h such that the graph contains h vertices of degree at least h. We define the h-partition of a graph to be the sets $(H, V \setminus H)$, where H is the set of vertices that form the h-index.

2.1   The H-Index

It is easy to see that the h-index of a graph with m edges is $O(\sqrt{m})$; hence it is $O(\sqrt{n})$ for sparse graphs with a linear number of edges, where n is the number of vertices. Moreover, this bound is optimal in the worst-case, e.g., for a graph consisting of $\sqrt{n}$ stars of size $\sqrt{n}$ each. As can be seen in Fig. 1 Eppstein and Spiro [10] show experimentally that real-world social networks often have h-indices much lower than the indicated worst-case bound. These indices, perhaps more easily viewed in log-log scale in Fig 2, were calculated on networks with a range of ten to just over ten-thousand nodes. The h-index of these networks were consistently below forty with only a few exceptions, none greater than slightly above one-hundred. Moreover, many large real-world networks possess *power laws*, so that their number of vertices with degree d is proportional to $nd^{-\lambda}$, for some constant $\lambda > 1$. Such networks are said to be *scale-free* [1,17,19], and it is often the case that the parameter λ is between 2 and 3 in real-world networks. Note that the h-index of a scale-free graph is $h = \Theta(n^{1/(1+\lambda)})$, since it must satisfy the

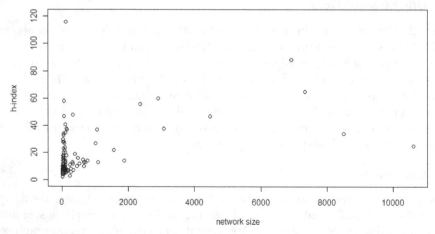

Fig. 1. Scatter plot of h-index and network size from Eppstein and Spiro [11]

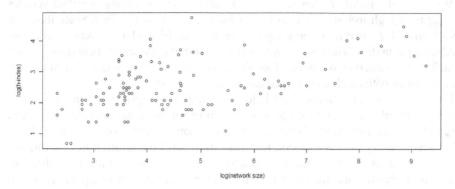

Fig. 2. Scatter plot of h-index and network size, on log-log scale from Eppstein and Spiro [11]

equation $h = nh^{-\lambda}$. Thus, for instances of scale-free graphs with λ between 2 and 3, an algorithmic performance of $O(h)$ is much better than the worst-case $O(\sqrt{n})$ bound for graphs without power-law degree distributions. For example, an $O(h)$ time bound for a scale-free graph with $\lambda = 2$ would give a bound of $O(n^{1/3})$ while for $\lambda = 3$ it would give an $O(n^{1/4})$ bound. Likewise, an algorithmic performance of $O(h^2)$ is much better than a worst-case performance of $O(n)$ for these instances, for $\lambda = 2$ would give a bound of $O(n^{2/3})$ while for $\lambda = 3$ it would give an $O(n^{1/2})$ bound. Thus, by taking a parametric algorithm design approach, we can, in these cases, achieve running times better than worst-case bounds characterized strictly in terms of the input size, n.

2.2 Maintaining Undirected Size-3 Subgraph Statistics

As mentioned above, Eppstein and Spiro [10] develop an algorithm for maintaining the h-index and the h-partition of a graph among edge insertions, edge deletions, and insertions/deletions of isolated vertices in constant time plus a constant number of dictionary operations per update. Observing that the h-index doubles after $\Omega(h^2)$ updates, Eppstein and Spiro further show a partitioning scheme requiring amortized $O(1/h)$ partition changes per graph update. This partitions the graph into sets of *low-* and *high*-degree vertices, which we summarize in Theorem 1.

Theorem 1 ([10]). *For a dynamic graph $G = (V, E)$, we can maintain a partition $(H, V \setminus H)$ such that for $v \in H$, degree$(v) = \Omega(h)$ and $|H| = O(h)$; and for $u \in V \setminus H$, degree$(u) = O(h)$ in constant time per update, with amortized $O(1/h)$ changes to the partition per update.*

Using this partitioning scheme, one can develop a triangle-counting algorithm as follows. For each pair of vertices i and j, store the number of length-two paths $P[i, j]$ that have an intermediate low-degree vertex. Whenever an edge (u, v) is added to the graph, increase the number of triangles by $P[u, v]$, and update the number of length-two paths containing (u, v) in $O(h)$ time. One can then iterate over all the high-degree vertices, adding to a triangle count when a high-degree vertex is adjacent to both u and v. Since there are $O(h)$ high-degree vertices, this method takes $O(h)$ time. These same steps can be done in reverse for an edge removal.

Whenever the partition changes, one must update $P[\cdot,\cdot]$ values to reflect vertices moving from high to low, or low to high, which requires $O(h^2)$ time. Since there are amortized $O(1/h)$ partition changes per graph update, this updating takes $O(h)$ amortized time per update. The randomization comes from the choice of dictionary scheme used. The data structure as described requires $O(mh)$ space, which is sufficient to store the length-two paths with an intermediate low-degree vertex.

Finally, to maintain counts of all induced undirected subgraphs on three vertices, it suffices to solve a simple four-by-four system of linear equations relating induced subgraphs and non-induced subgraphs. This allows one to keep counts of the induced subgraphs of every type with a constant amount of work in addition to counting triangles. Extending this to directed subgraphs of size three and undirected subgraphs of size four requires that we come up with a much larger system of equations, which characterize the combinatorial relationships between such types of subgraphs.

3   Directed Three-Vertex Induced Subgraphs

Using the partitioning scheme detailed in Theorem 1, we can maintain counts for the all possible induced subgraphs on three vertices (see Fig. 3) in $O(h)$ amortized time per update for a dynamic directed graph. We begin by maintaining counts for induced subgraphs that are a directed triangle, we then show how to maintain counts of all induced subgraphs on three vertices.

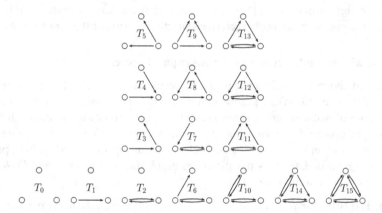

Fig. 3. The 16 possible directed graphs on three vertices, excluding isomorphisms, organized in left-to-right order by number of edges in the graph. We label these graphs T_0 to T_{15}.

3.1   Counting Directed Triangles

Let a *directed triangle* be a three-vertex directed graph with at least one directed edge between each pair of vertices. There are seven possible directed triangles, labeled D_0 to D_6 in Fig. 4. We let d_k denote the count of induced directed triangles of type D_k in the dynamic graph. We now show how to maintain each count d_i by extending Eppstein and Spiro's technique.

For a pair of vertices i and j, we define a *joint* to be a third vertex l that is adjacent to both i and j. Vertices i, l and j are said to form an *elbow*. Fixing a vertex to be a joint,

Fig. 4. The 7 directed triangles, labeled D_0 to D_6

there are nine unique elbows which we label E_0 to E_8 (see Fig. 5). We store a dictionary mapping pairs of vertices i and j to the number of elbows of type E_k formed by i and j and a low-degree joint, denoted $e_k[i, j]$.

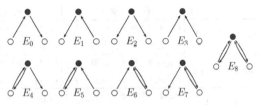

Fig. 5. The nine elbows with a fixed joint

We now discuss how the directed triangle counts change when adding an edge (u, v). We do not discuss edge removal since its effects are symmetric to edge insertion.

For directed triangles with a third low-degree vertex, we update our counts using the dictionary of elbow counts. If edge (v, u) is not in the graph, directed triangle counts increase as follows.

$$d_0 = d_0 + e_1[u, v]$$
$$d_1 = d_1 + e_0[u, v] + e_2[u, v] + e_3[u, v]$$
$$d_2 = d_2 + e_5[u, v] + e_7[u, v]$$
$$d_3 = d_3 + e_4[u, v]$$
$$d_4 = d_4 + e_6[u, v]$$
$$d_5 = d_5 + e_8[u, v]$$

If edge (v, u) is present in the graph, adding (u, v) destroys some directed triangles containing (v, u). Therefore, the directed triangle counts change as follows.

$$d_0 = d_0 - e_1[v, u]$$
$$d_1 = d_1 - (e_0[v, u] + e_2[v, u] + e_3[v, u])$$
$$d_2 = d_2 + (e_0[u, v] + e_1[u, v]) - (e_5[v, u] + e_7[v, u])$$
$$d_3 = d_3 + e_3[u, v] - e_4[v, u]$$
$$d_4 = d_4 + e_2[u, v] - e_6[v, u]$$
$$d_5 = d_5 + (e_4[u, v] + e_5[u, v] + e_6[u, v] + e_7[u, v]) - e_8[v, u]$$
$$d_6 = d_6 + e_8[u, v]$$

To complete the directed triangle counting step, we iterate over the $O(h)$ high-degree vertices to account for directed triangles formed with u and v and a high-degree vertex, taking $O(h)$ time.

If either u or v is a low-degree vertex, we must also update the elbow counts involving the added edge (u, v). We consider, without loss of generality, the updates when u is considered the low-degree elbow joint. For ease of notation, we categorize the different relationships between adjacent vertices as follows:

$$\text{inneighbor}(u) = \{w \in V : (w, u) \in E \land (u, w) \notin E\}$$
$$\text{outneighbor}(u) = \{w \in V : (u, w) \in E \land (w, u) \notin E\}$$
$$\text{neighbor}(u) = \{w \in V : (u, w) \in E \land (w, u) \in E\}.$$

We summarize the elbow count updates in Table 1.

Table 1. Summary of updating elbow counts when u is considered a low-degree joint

	$(v, u) \notin E$	$(v, u) \in E$
$w \in \text{inneighbor}(u) \setminus \{v\}$	$e_0[w, v] = e_0[w, v] + 1$	$e_6[w, v] = e_6[w, v] + 1$
	$e_1[v, w] = e_1[v, w] + 1$	$e_5[v, w] = e_5[v, w] + 1$
$w \in \text{outneighbor}(u) \setminus \{v\}$	$e_0[v, w] = e_0[v, w] + 1$	$e_4[v, w] = e_4[v, w] + 1$
	$e_1[w, v] = e_1[w, v] + 1$	$e_7[w, v] = e_7[w, v] + 1$
$w \in \text{neighbor}(u) \setminus \{v\}$	$e_4[w, v] = e_4[w, v] + 1$	$e_8[w, v] = e_8[w, v] + 1$
	$e_7[v, w] = e_7[v, w] + 1$	$e_8[v, w] = e_8[v, w] + 1$

Finally, when there is a partition change, we must update the elbow counts. If node w moves across the partition, then we consider all pairs of neighbors of w and update their elbow counts appropriately. Since there are $O(h^2)$ pairs of neighbors, and a constant number of elbows, this step takes $O(h^2)$ time. Since $O(1/h)$ amortized partition changes occur with each graph update, this step requires $O(h)$ amortized time.

3.2 Subgraph Multiplicity

Let the count for induced subgraph T_i be called t_i. Furthermore, for a vertex v, let $i(v) = |\text{inneighbor}(v)|$, $o(v) = |\text{outneighbor}(v)|$ and $r(v) = |\text{neighbor}(v)|$. We can represent the relationship between the number of induced and non-induced subgraphs using the matrix equation

$$
\begin{bmatrix}
1 1 1 1 1 1 1 1 1 1 1 1 1 1 1 1 \\
0 1 2 2 2 2 3 3 3 3 4 4 4 4 5 6 \\
0 0 1 0 0 0 1 1 0 0 2 1 1 1 2 3 \\
0 0 0 1 0 0 1 1 3 1 2 2 2 3 4 6 \\
0 0 0 0 1 0 0 1 0 1 1 1 2 1 2 3 \\
0 0 0 0 0 1 1 0 0 1 1 2 1 1 2 3 \\
0 0 0 0 0 0 1 0 0 0 2 2 0 1 3 6 \\
0 0 0 0 0 0 0 1 0 0 2 0 2 1 3 6 \\
0 0 0 0 0 0 0 0 1 0 0 0 0 1 1 2 \\
0 0 0 0 0 0 0 0 0 1 0 2 2 1 3 6 \\
0 0 0 0 0 0 0 0 0 0 1 0 0 0 1 3 \\
0 0 0 0 0 0 0 0 0 0 0 1 0 0 1 3 \\
0 0 0 0 0 0 0 0 0 0 0 0 1 0 1 3 \\
0 0 0 0 0 0 0 0 0 0 0 0 0 1 2 6 \\
0 0 0 0 0 0 0 0 0 0 0 0 0 0 1 6 \\
0 0 0 0 0 0 0 0 0 0 0 0 0 0 0 1
\end{bmatrix}
\begin{bmatrix}
t_0 \\ t_1 \\ t_2 \\ t_3 \\ t_4 \\ t_5 \\ t_6 \\ t_7 \\ t_8 \\ t_9 \\ t_{10} \\ t_{11} \\ t_{12} \\ t_{13} \\ t_{14} \\ t_{15}
\end{bmatrix}
=
\begin{bmatrix}
n_0 = \binom{n}{3} \\
n_1 = m(n-2) \\
n_2 = \frac{1}{2}(n-2) \sum_{v \in V} r(v) \\
n_3 = \sum_{(u,v) \in E} \sum_{(v,w) \in E, w \neq u} 1 \\
n_4 = \sum_{v \in V} \binom{\text{indegree}(v)}{2} \\
n_5 = \sum_{v \in V} \binom{\text{outdegree}(v)}{2} \\
n_6 = \sum_{v \in V} (\binom{r(v)}{2} + o(v) \cdot r(v)) \\
n_7 = \sum_{v \in V} (\binom{r(v)}{2} + i(v) \cdot r(v)) \\
n_8 = d_0 + d_2 + d_5 + 2d_6 \\
n_9 = d_1 + d_2 + 2d_3 + 2d_4 + 3d_5 + 6d_6 \\
n_{10} = \sum_{v \in V} \binom{r(v)}{2} \\
n_{11} = d_3 + d_5 + 3d_6 \\
n_{12} = d_4 + d_5 + 3d_6 \\
n_{13} = d_2 + 2d_5 + 6d_6 \\
n_{14} = d_5 + 6d_6 \\
n_{15} = d_6
\end{bmatrix}.
$$

On the right hand side, each n_i is the count of the number of non-induced T_i subgraphs in the dynamic graph. Each n_i (excluding directed triangle counts) is maintained in constant time per update by storing a constant amount of structural information at each node, such as indegree, outdegree, and reciprocity of neighbors. On the left hand side, position i, j in the matrix counts how many non-induced subgraphs of type T_i appear in T_j. We are counting non-induced subgraphs in two ways: (1) by counting the number of appearances within induced subgraphs and (2) by using the structure of the graph. Since the multiplicand is an upper (unit) triangular matrix, this matrix equation is easily solved, yielding the induced subgraph counts. Thus, we can maintain the counts for three-vertex induced subgraphs in a directed dynamic graph in $O(h)$ amortized time per update, and $O(mh)$ space, plus the additional overhead for the choice of dictionary.

4   Four-Vertex Subgraphs

We begin by describing the data structure for our algorithm. It will be necessary to maintain the counts of various subgraph structures. The data structure in whole consists of the following information:

- Counts of the non-induced subgraph structures, m_3 through m_{10}.
- A set E of the edges in the graph, indexed such that given a pair of endpoints there is a constant-time lookup to determine if they are linked by an edge.
- A partition of the vertices of the graph into two sets H and $V \setminus H$.
- A dictionary P_1 mapping each vertex u to a pair $P_1[u] = (s_0[u], s_1[u])$. This pair contains the counts for the structures S_0 and S_1 that involve vertex u (see Fig. 6). That is, the count of the number of two-edge paths that begin at u and pass through two vertices in $V \setminus H$ and the number of these paths that connect back to u forming a triangle. We only maintain nonzero values for these numbers in P_1; if there is no entry in $P_1[u]$ for the vertex u then there exist no such path from u.
- A dictionary P_2 mapping each pair of vertices u, v to a tuple $P_2[u, v] = (s_2[u, v], s_3[u, v], s_4[u, v], s_5[u, v], s_6[u, v])$. This tuple contains the counts for the structures S_2 through S_6 that involve vertices u and v (see Fig. 6). That is, the number of two-edge paths from u to v via a vertex of $V \setminus H$, the number of three-edge paths from u to v via two vertices of $V \setminus H$, the number of structures in which both u and v connect to the same vertex in $V \setminus H$ which connects to another vertex in $V \setminus H$, the number of structures similar to the last in which the final vertex in $V \setminus H$ shares an edge connection with u or v, and the number of structures where between u and v there are two two-edge paths through vertices of $V \setminus H$ in which the two vertices in $V \setminus H$ share an edge connection. Again, we only maintain nonzero values.
- A dictionary P_3 mapping each triple of vertices u, v, w to a number $P_2[u, v, w] = (s_7[u, v, w])$. This value is the count for the structure S_7 that involves vertices u, v, and w (see Fig. 6). This is, the number of vertices in $V \setminus H$ that share edge connections with all three vertices. As before, we only maintain nonzero values for these numbers.

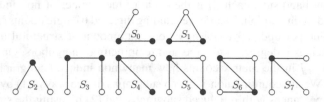

Fig. 6. We store counts of these eight non-induced subgraphs to maintain counts of four-vertex non-induced subgraphs Q_3 to Q_{10}. The counts are indexed by the labels of the white vertices, and the blue vertices indicate a vertex has low-degree.

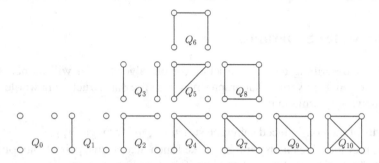

Fig. 7. The 11 possible graphs on four vertices, excluding isomorphisms, organized in left-to-right order by number of edges in the graph

Upon insertion of an edge between vertices v_1 and v_2 we will need to update the dictionaries P_1, P_2, and P_3. If both v_1 and v_2 are in H, no update is necessary.

If v_1 and v_2 are both in $V \setminus H$ then we will need to update the counts s_0 through s_6. First find which vertices in H connect to v_1 or to v_2. Increment s_0 for these vertices. If both vertices in $V \setminus H$ connect to the same vertex in H then increment s_1 for this vertex. Increment s_2 for v_1 and the vertices that connect to v_2, and for v_2 and the vertices that connect to v_1. Then increment s_3 based on pairs of neighbors of v_1 and v_2 and neighbors of neighbors in $V \setminus H$. If either v_1 or v_2 connect to two vertices in H increment s_4 for the vertices in H. Considering v_1 to be the vertex with edge connections to two vertices in H, for each vertex in H that connects to v_2 increment s_5. For two vertices in H such that v_1 and v_2 each connect to both, increment s_6 for the vertices in H.

If v_1 and v_2 are such that one is in $V \setminus H$ and the other in H we will proceed as follows. Consider v_1 to be the vertex in $V \setminus H$. First, determine the number of vertices in $V \setminus H$ connected to v_1 and increase s_0 for v_2 by that amount. Upon discovering these adjacent vertices in $V \setminus H$ test their connection to v_2. For each of those connected, increment s_1 for v_2. It is necessary to determine which vertices in H share an edge with v_1. After these connections have been determined increment the appropriate dictionary entries. Form pairs with v_2 and the connected vertices in H and update the s_2 counts. Form triples with v_2 and two other connected vertices in H and update the counts in s_7. The s_5 update comes from determining the triangles formed by the additional edge and using the degree of the vertices in H, and the count of the connected triangles, which can be calculated by searching for attached vertex pairs in H and using s_2. In order to update the count for s_6 begin with location of vertex pairs as with the elbow update. For each of the H vertex pairs increase the stored value by the number of vertices in $V \setminus H$ that share an edge with v_1 and with both of the vertices in H, which can be retrieved from s_2.

Examining the time complexity we can see that in order to generate the dictionary updates the most complex operation involves examination of two sets of connected vertices consecutively that are $O(h)$ in size each. This results in $O(h^2)$ operations to determine which updates are necessary. Since it is possible to see from the structure of the stored items that no single edge insertion can result in more than $O(h^2)$ new structures, this will be the upper bound on dictionary updates, and make $O(h^2)$ the time complexity bound.

These maintained counts will have to be modified when the vertex partition is updated. If a vertex is moved from H to $V \setminus H$ then it is necessary to count the connected structures it now forms. This can be done by examining all edges formed by this vertex, and following the procedure for edge additions. When a vertex is moved into H it is necessary to count the structures it had been forming as a vertex in $V \setminus H$ and decrement the appropriate counts. This can be done similarly to the method for generating new structures. In analysis of the partition updates we see that since we are working with a single vertex with $O(h)$ degree the complexity has an additional $O(h)$ factor to use the edge-based dictionary update scheme. This results in $O(h^3)$ time per update. Since this partition update is done an average of $O(1/h)$ times per operation, the amortized time for updates, per change to the input graph, is $O(h^2)$.

4.1 Subgraph Structure Counts

The following section covers the update of the subgraph structure counts after an edge between vertices v_1 and v_2 has been inserted. Let these vertices have degree count d_1 and d_2 respectively. Recall that m_i refers to the count of the non-induced subgraph of the structure Q_i (see Fig. 7).

The m_3 count will be increased by $(m - (d_1 + d_2 - 2))$, where m is the number of edges in the graph. Since this structure consists of two edges that do not share vertices, the increase of the count comes from a selection of a second edge to be paired with the inserted edge. The second term in the update value reflects the number of edges that connected to the inserted edge.

The m_4 count will be updated as follows. Each of the two vertices can be the end of a claw structure. From each end two edges in addition to the newly inserted edge must be selected. Thus the value to update the count is $\binom{d_1-1}{2} + \binom{d_2-1}{2}$.

The m_5 count is updated by calculating the number of additional triangles the edge addition would add, which can be done with the Eppstein-Spiro [10] method, and multiplying that by a factor of $(n-3)$ to reflect the selection of the additional vertex, where n is the number of vertices in the graph.

The update for m_6 is done in parts based on which position in the structure the edge is forming. The increase to the count for the new structures in which the additional edge is the center in the three-edge path is $((d_1 - 1)(d_2 - 1))$.

This value will be increased by the count when the new edge is not the center of the structure. The process to calculate the count increase will assume that v_1 connects to the rest of this structure. The same process can be done without loss of generality with the assumption v_2 connects to the rest of the structure. These values will then both be added to form the final part of the count update. If v_1 is an element of H then we will sum the results from the following subcases. First we consider the case where the vertex adjacent to v_1 is in H. The number of these paths of length two originating at v_1 can be counted by summing the degree of these vertices minus 1. We must also subtract one for each of the adjacent vertices in H that are adjacent to v_2. If v_1 is not an element of H, then it has h or less neighbors. Sum over all neighbors the following value. If the vertex does not have an edge connecting it to v_2 then the degree of the vertex; if it does the degree minus one.

The m_7 count is updated as follows. An inserted edge can form the structure in three positions, so our final update will be the sum of those three counts. For the first case let the inserted edge be the additional edge connected to the triangle. For this case, we must do all of the following for both vertices and sum the result. If the vertex is in H retrieve s_1. This gives us the connected triangles through vertices in $V \setminus H$. Then determine which vertices in H connect to the vertex. Form the triangle counts with all vertices in H. Form those with one additional vertex in H using s_2. If the vertex is in $V \setminus H$, then determine its neighbors connections and form a connected triangle count.

In the second case the edge is in the triangle and shares a vertex with the additional edge. The count can be determined in two parts. First the triangles. If either v_1 or v_2 are in $V \setminus H$ then the triangle count can be calculated. If both are in H then a lookup to s_2 will determine the number of triangles. The number of additional edges can then be calculated using the degrees of the vertices of the inserted edge, with care to not count the edges used to form the triangle. The product of the triangle and additional edge will form the increase for this case.

The final case occurs when the inserted edge is part of the triangle, but does not share a vertex with the additional edge. If either v_1 or v_2 are in $V \setminus H$ then the triangle count can be calculated, and the degree of the vertices used to form these triangles can be used to calculate the count increase. If both v_1 or v_2 are in H then there are three remaining subcases. The count if all vertices are in H can be determined. If the vertex on the additional edge that is not in the triangle is in H, then using the three known vertices in H and a lookup from P_2 can yield the counts. If both remaining vertices are in $V \setminus H$ this is the structure stored in s_4, and counts can be retrieved. Sum the counts for these subcases to calculate the total increase for this case.

The count for m_8 is increased upon edge update by a sum of the following. The count of the length three path through vertices in $V \setminus H$ can be looked up in s_3. There are two possible types of length three paths remaining. In the first, both vertices are in H. These paths can be counted be examining the connections between v_1, v_2, and all vertices in H. The second contains one vertex in H and one in $V \setminus H$. These paths can be counted by establishing which vertices in H connect to either v_1 or v_2, and then using the count in s_2 of the length two paths from the vertices in H to v_2 or v_1 respectively.

The m_9 count can be increased by an edge insert in two positions. The first is between the opposite ends of the cycle. If either v_1 or v_2 is in $V \setminus H$ then the edge connections can be determined and the count calculated. If both v_1 and v_2 are in H then the count of the two two-edge paths that form the cycle must be determined. These paths will either pass through a vertex in H or a vertex in $V \setminus H$. The former can be counted by examining the vertices in H, and the latter by a lookup to s_2.

The second possible position for an edge insert is on the outer path of a cycle that already has an edge through it. If either v_1 or v_2 are in $V \setminus H$ calculate the count as follows, summing with an additional calculation considering the vertices reversed. If the vertex connected to the triangle is in $V \setminus H$ then the count can be determined by examining neighbors and their edge connections. If the vertex not connected to the triangle is in $V \setminus H$ then examine the neighbors. For those neighbors that are in $V \setminus H$ the count can be determined by examining additional edge connections of neighbors. For the neighbors in H a lookup s_2 is required to completely determine the counts. If both v_1 and v_2 are in H then the count is calculated as follows. If all vertices of the structure are in H, determine the count by examining edge connections. If both remaining vertices are in $V \setminus H$ the count can be determined by lookup to s_5. Otherwise, one of the two remaining vertices is in H. This will leave a structure that can be completed and provide a count by using a lookup to s_2, or s_7.

The m_{10} count update is separated by the membership of v_1 and v_2. If either vertex is contained in $V \setminus H$, consider v_1, then it is possible to determine which vertices connect to v_1 and which of these share edges with v_2 and each other. This count can be calculated and the total count can be updated. If both v_1 and v_2 are in H then we will sum the values determined in the following three subcases. First, all four vertices are in H. This count can be determined by examining the edge connections of the vertices in H. If three vertices in H form the correct structure, the count of cliques formed with one vertex in $V \setminus H$ can be determined by a look up to s_7. These counts should be summed for all vertices in H that form the correct structure with v_1 and v_2. The final count, with both of the remaining vertices in $V \setminus H$ can be determined by an s_6 lookup.

The time complexity for the updates of the stored subgraphs is $O(h^2)$. Calculations and lookups can be performed in constant time, and subcase calculations can be done independently. The most complicated subcase count computations involve examination of two sets of connected vertices consecutively that are $O(h)$ in size each. This results in $O(h^2)$ operations. The space complexity for our data structure is $O(1)$ for the maintained subgraph counts, $O(m)$ for E, $O(n)$ for the partition to maintain H, and $O(mh^2)$ for the dictionaries, because each edge belongs to at most $O(h^2)$ subgraph structures.

4.2 Subgraph Multiplicity

The data structure in the previous section only maintains counts of certain subgraph structures. With the addition of m, n, and the count of length two paths, where m is the number of edges and n the number of vertices, it is possible to use these counts to determine the counts of all subgraphs on four vertices. The additional values m, n, and the count of length two paths can be maintained in constant time per update. Values for m and n are modified incrementally. Adding an edge uv will increase the count of length two paths by $d_u + d_v$, the degrees of u and v respectively. Removing the edge will decrease the value by $d_u + d_v - 2$.

$$
\begin{bmatrix}
1 & 1 & 1 & 1 & 1 & 1 & 1 & 1 & 1 & 1 & 1 \\
0 & 1 & 2 & 2 & 3 & 3 & 3 & 4 & 4 & 5 & 6 \\
0 & 0 & 1 & 0 & 3 & 3 & 2 & 5 & 4 & 8 & 12 \\
0 & 0 & 0 & 1 & 0 & 0 & 1 & 1 & 2 & 2 & 3 \\
0 & 0 & 0 & 0 & 1 & 0 & 0 & 1 & 0 & 2 & 4 \\
0 & 0 & 0 & 0 & 0 & 1 & 0 & 1 & 0 & 2 & 4 \\
0 & 0 & 0 & 0 & 0 & 0 & 1 & 2 & 4 & 6 & 12 \\
0 & 0 & 0 & 0 & 0 & 0 & 0 & 1 & 0 & 4 & 12 \\
0 & 0 & 0 & 0 & 0 & 0 & 0 & 0 & 1 & 1 & 3 \\
0 & 0 & 0 & 0 & 0 & 0 & 0 & 0 & 0 & 1 & 6 \\
0 & 0 & 0 & 0 & 0 & 0 & 0 & 0 & 0 & 0 & 1
\end{bmatrix}
\begin{bmatrix}
q_0 \\ q_1 \\ q_2 \\ q_3 \\ q_4 \\ q_5 \\ q_6 \\ q_7 \\ q_8 \\ q_9 \\ q_{10}
\end{bmatrix}
=
\begin{bmatrix}
m_0 = \binom{n}{4} \\
m_1 = m\binom{n-2}{2} \\
m_2 = (n-3)\sum_{v \in V} \binom{\deg(v)}{2} \\
m_3 \\
m_4 \\
m_5 \\
m_6 \\
m_7 \\
m_8 \\
m_9 \\
m_{10}
\end{bmatrix}
$$

Similar to the matrix for size three subgraphs, we can use the counts of the non-induced subgraphs on the right and the composition of the induced subgraphs to determine the counts of any desired subgraph.

5 Conclusion

The work we present here can maintain counts for all 3-vertex directed subgraphs $O(h)$ amortized time per update. This can be done in $O(mh)$ space. For the undirected case, we maintain counts of size-four subgraphs in $O(h^2)$ amortized time per update and $O(mh^2)$ space. Although we do not discuss the specifics in this paper, the methodology presented can be used to count directed size-four subgraphs with similar complexity. These developments open significant possibility for improvement in calculating *graphlet* frequencies within Bioinformatics and in ERGM applications for social network analysis.

References

1. Albert, R., Jeong, H., Barabasi, A.-L.: The diameter of the world wide web. Nature 401, 130–131 (1999)
2. Alon, N., Yuster, R., Zwick, U.: Finding and counting given length cycles. Algorithmica 17(3), 209–223 (1997)
3. Chiba, N., Nishizeki, T.: Arboricity and subgraph listing algorithms. SIAM J. Comput. 14(1), 210–223 (1985)
4. Coppersmith, D., Winograd, S.: Matrix multiplication via arithmetic progressions. Journal of Symbolic Computation 9(3), 251–280 (1990)
5. Demaine, E.D., Fomin, F.V., Hajiaghayi, M., Thilikos, D.M.: Fixed-parameter algorithms for (k, r)-center in planar graphs and map graphs. ACM Trans. Algorithms 1(1), 33–47 (2005)
6. Downey, R.G., Fellows, M.R.: Fixed-parameter tractability and completeness i: Basic results. SIAM J. Comput. 24(4), 873–921 (1995)
7. Eppstein, D.: Subgraph isomorphism in planar graphs and related problems. Journal of Graph Algorithms & Applications 3(3), 1–27 (1999)
8. Eppstein, D.: Diameter and treewidth in minor-closed graph families. Algorithmica 27, 275–291 (2000)
9. Eppstein, D., Goodrich, M.T.: Studying (non-planar) road networks through an algorithmic lens. In: GIS 2008: Proceedings of the 16th ACM SIGSPATIAL International Conference on Advances in Geographic Information Systems, pp. 1–10. ACM, New York (2008)

10. Eppstein, D., Spiro, E.S.: The h-index of a graph and its application to dynamic subgraph statistics. In: Dehne, F., Gavrilova, M., Sack, J.-R., Tóth, C.D. (eds.) WADS 2009. LNCS, vol. 5664, pp. 278–289. Springer, Heidelberg (2009)
11. Eppstein, D., Spiro, E.S.: The h-index of a graph and its application to dynamic subgraph statistics. arXiv:0904.3741 (2009)
12. Frank, O.: Statistical analysis of change in networks. Statistica Neerlandica 45, 283–293 (1991)
13. Frank, O., Strauss, D.: Markov graphs. J. Amer. Statistical Assoc. 81, 832–842 (1986)
14. Hirsch, J.E.: An index to quantify an individual's scientific research output. Proc. National Academy of Sciences 102(46), 16569–16572 (2005)
15. Itai, A., Rodeh, M.: Finding a minimum circuit in a graph. SIAM J. Comput. 7(4), 413–423 (1978)
16. Milenković, T., Pržulj, N.: Uncovering biological network function via graphlet degree signatures. Cancer Informatics 6, 257–273 (2008)
17. Newman, M.E.J.: The structure and function of complex networks. SIAM Review 45, 167–256 (2003)
18. Niedermeier, R., Rossmanith, P.: On efficient fixed-parameter algorithms for weighted vertex cover. J. Algorithms 47(2), 63–77 (2003)
19. Price, D.J.d.S.: Networks of scientific papers. Science 149(3683), 510–515 (1965)
20. Pržulj, N.: Biological network comparison using graphlet degree distribution. Bioinformatics 23(2), e177–e183 (2007)
21. Pržulj, N., Corneil, D.G., Jurisica, I.: Efficient estimation of graphlet frequency distributions in protein–protein interaction networks. Bioinformatics 22(8), 974–980 (2006)
22. Snijders, T.A.B.: Markov chain Monte Carlo estimation of exponential random graph models. Journal of Social Structure 3(2), 1–40 (2002)
23. Snijders, T.A.B., Pattison, P.E., Robins, G., Handcock, M.S.: New specifications for exponential random graph models. Sociological Methodology 36(1), 99–153 (2006)
24. Wasserman, S., Pattison, P.E.: Logit models and logistic regression for social networks, I: an introduction to Markov graphs and p^*. Psychometrika 61, 401–425 (1996)

Discrete Optimization with Polynomially Detectable Boundaries and Restricted Level Sets

Yakov Zinder[1], Julia Memar[1], and Gaurav Singh[2]

[1] University of Technology, Sydney, P.O. Box 123, Broadway, NSW 2007, Australia
yakov.zinder@uts.edu.au, julia.memar@uts.edu.au
[2] CSIRO, Private Bag 33, South Clayton, VIC 3169 Australia
Gaurav.Singh@csiro.au

Abstract. The paper describes an optimization procedure for a class of discrete optimization problems which is defined by certain properties of the boundary of the feasible region and level sets of the objective function. It is shown that these properties are possessed, for example, by various scheduling problems, including a number of well-known NP-hard problems which play an important role in scheduling theory. For an important particular case the presented optimization procedure is compared with a version of the branch-and-bound algorithm by means of computational experiments.

Keywords: discrete optimization, scheduling theory, parallel machines, unit execution times.

1 Introduction

We consider the discrete optimization problem

$$\min_{(x_1, x_2, ..., x_n) \in X} F(x_1, x_2, ..., x_n), \tag{1}$$

where $F(x_1, x_2, ..., x_n)$ is a nondecreasing function defined on the n-dimensional hypercube of points with integer coordinates satisfying the inequalities $0 \le x_i \le p(n)$, $1 \le i \le n$, where p is a polynomial. Without loss of generality it will be assumed that $p(n)$ is integer. The feasible region X is a subset of this hypercube.

The above description is too general for any specific optimization procedure. The following three additional properties narrow the considered class of discrete optimization problems but are not very restrictive - the resultant class, for example, contains various well-known NP-hard problems of scheduling theory. In what follows, the expression "in polynomial time" has the standard meaning that, for all instances of (1), the number of operations is bounded above by the same polynomial in n. Similarly, the expression "cardinality is bounded above by some polynomial in n" means that, for all instances of (1), the number of elements in the considered set is bounded above by the value of the same polynomial in n.

W. Wu and O. Daescu (Eds.): COCOA 2010, Part I, LNCS 6508, pp. 142–156, 2010.

The first property is concerned with the boundary of X which definition is based on the notion of dominance: point $a = (a_1, a_2, ..., a_n)$ dominates point $b = (b_1, b_2, ..., b_n)$ if $b_i \leq a_i$ for all $1 \leq i \leq n$, and a strictly dominates b if at least one of these inequalities is strict. The boundary of X is the set of all points in X which do not strictly dominate any point in X.

Property 1. There is an algorithm which for any point in X in polynomial time determines whether or not this point is on the boundary of X.

The second property pertains to the notion of a level set defined as follows. For any value $\bar{F}$ of F, a set D is $\bar{F}$-dominant if $F(x_1, x_2, ..., x_n) = \bar{F}$ for all $(x_1, x_2, ..., x_n) \in D$ and $F(y_1, y_2, ..., y_n) = \bar{F}$ implies that $(y_1, y_2, ..., y_n)$ is dominated by some point in D. For any value $\bar{F}$ of F, a level set, denoted by $A(\bar{F}, F)$, is an F-dominant set with the smallest cardinality among all $\bar{F}$-dominant sets. The next section justifies this definition by showing that for any value $\bar{F}$ of F the corresponding level set is unique.

Property 2. For any value F' of F, the corresponding level set can be found in polynomial time.

According to *Property 2* the cardinalities of all level sets are bounded above by some polynomial in n. Observe that if $\bar{F} = F(p(n), ..., p(n))$, then $A(\bar{F}, F)$ is comprised of only one point $(p(n), ..., p(n))$.

The third property is the existence of an algorithm that for any value F' of F such that $F' < F(p(n), ..., p(n))$ finds the smallest value of F greater than F' with a number of operations bounded above by some polynomial in n.

Property 3. For any value F' of F such that $F' < F(p(n), ..., p(n))$ the value

$$F'' = \min_{\{(x_1, x_2, ..., x_n): F(x_1, x_2, ..., x_n) > F')\}} F(x_1, x_2, ..., x_n).$$

can be found in polynomial time.

As has been mentioned above, Section 2 justifies the definition of level sets by showing that, for any value of F, the corresponding level set is unique. The next section, Section 3, presents some examples. In particular, Section 3 shows that, even when the cardinality of the range of F is bounded above by some polynomial in n, *Property 2* and *Property 3* do not imply the existence of a polynomial-time optimization procedure. Indeed, in general, the problem remains NP-hard in the strong sense because for some (and even for all) values $\bar{F}$ of the objective function F, it is quite possible to have $A(\bar{F}, F) \cap X = \emptyset$. The proposed optimization procedure, which is based on *Property 1*, *Property 2* and *Property 3*, is described in Section 4. Section 5 illustrates the optimization procedure presented in Section 4 by considering its application to one of the scheduling problems which plays an important role in scheduling theory. The results of computational experiments aimed at comparing this application with a version of the branch-and-bound method are reported in Section 6.

2 Level Sets

Let K be any set of points with integer coordinates. Denote K^c the set of all points $x \in K$ such that there is no point in K which strictly dominates x.

Lemma 1. *For any value $\bar{F}$ of F and any $\bar{F}$-dominant set D, the set D^c is a level set.*

Proof. Consider an arbitrary $x = (x_1, x_2, ..., x_n)$ such that $F(x_1, x_2, ..., x_n) = \bar{F}$. Since D is an $\bar{F}$-dominant set, there exists $y \in D$ that dominates x. Since the domain of F is a hypercube of points with integer coordinates, there exists $z \in D^c$ that dominates y and therefore x. (Observe that $z = y$ if z does not dominate y strictly.) Hence, D^c is an $\bar{F}$-dominant set.

Suppose that D^c is not a level set. Then, its cardinality $|D^c| > |A(\bar{F}, F)|$, and therefore, there exists $x \in D^c$ such that $x \notin A(\bar{F}, F)$. By the definition of a level set, there exists $y \in A(\bar{F}, F)$ which dominates x, and by the definition of an $\bar{F}$-dominant set, there exists $z \in D^c$ which dominates y. Since $x \notin A(\bar{F}, F)$, y strictly dominates x. Hence, z strictly dominates x which contradicts the definition of D^c. □

The following theorem justifies the definition of a level set by establishing its uniqueness. The theorem is a straightforward consequence of Lemma 1.

Theorem 1. *For any value $\bar{F}$ of F, $A(\bar{F}, F)$ is unique.*

Proof. Suppose that for some value $\bar{F}$ of F where exist two different level sets A and D. Then by Lemma 1, $A \subseteq D$ and $D \subseteq A$. Hence, $A = D$. □

3 Particular Cases of the Considered Problem

3.1 Scheduling on Parallel Machines: The Boundary of the Feasible Region

An important source of NP-hard problems with *Property 1* and *Property 2* is scheduling theory. As an example, consider the following classical scheduling problem. A set $N = \{1, \ldots, n\}$ of n tasks is to be processed on $m > 1$ identical machines subject to precedence constraints in the form of an anti-reflexive, anti-symmetric and transitive relation on N. If in this relation task i precedes task j, denoted $i \rightarrow j$, then task i must be completed before task j can be processed. If $i \rightarrow j$, then i is called a predecessor of j and j is called a successor of i. Each task can be processed on any machine, and each machine can process at most one task at a time. The processing time of each task is one unit of time. For each $j \in N$, the processing of task j can commence only after its release time r_j, where r_j is a nonnegative integer. Without loss of generality it is assumed that the smallest release time is zero. If a machine starts processing a task, it continues until completion, i.e. no preemptions are allowed. Since no preemptions are allowed, a schedule is specified by tasks' completion times. In the scheduling literature the completion time of task j is normally denoted by C_j, but for the purpose of our discussion it is convenient to denote the completion time of task j by x_j. The goal is to minimize $F(x_1, ..., x_n)$, where F is a nondecreasing function. In the scheduling literature (see for example [3]) this problem is denoted by $P|prec, r_j, p_j = 1|F$. Here P signifies parallel identical machines, $prec$ indicates presence of precedence constraints, and $p_j = 1$ shows that all processing times are equal to one unit of time. Correspondingly, $P|prec, p_j = 1|F$ denotes the same problem under the assumption that all release times are zero.

Since $i \to j$ implies $x_i + 1 \leq x_j$, without loss of generality we assume that $i \to j$ implies $r_i + 1 \leq r_j$. Given this assumption, we can assume that for any task i with $r_i > 0$, the number of tasks j with $r_j < r_i$ is greater than r_i. Indeed, suppose that this does not hold, and among all i, violating this assumption, g is a task with the smallest release time. Then, even processing only one task at a time, one can complete all tasks j with $r_j < r_g$ before time r_g, and therefore the problem can be split into two separate problems: one with all tasks j satisfying $r_j < r_g$ and another with all remaining tasks. The above assumption implies that there exists an optimal schedule with all completion times less than or equal to n. Consequently, we can consider the objective function F only on the n-dimensional hypercube defined by the inequalities $0 \leq x_j \leq n$ for all $1 \leq j \leq n$. Then, the feasible region X can be viewed as the set of points $(x_1, x_2, ..., x_n)$ specifying all feasible schedules with completion times less than or equal to n. In other words, X is the set of all points $(x_1, ..., x_n)$ with integer coordinates satisfying the following three conditions

(a) $r_j + 1 \leq x_j \leq n$ for all $1 \leq j \leq n$;
(b) $|\{i : x_i = t\}| \leq m$ for all $1 \leq t \leq n$;
(c) $x_i \leq x_j + 1$ for all i and j such that $i \to j$.

The property specified in the following lemma can be checked in $O(n^2)$ operations. Therefore, this lemma shows that the $P|prec, r_j, p_j = 1|F$ scheduling problem has *Property 1.*

Lemma 2. *A point $(x_1, ..., x_n) \in X$ is on the boundary of X if and only if, for each integer $t \geq 1$ such that $|\{g : x_g = t\}| < m$ and each $x_j > t$, either $r_j \geq t$ or there exists i such that $x_i = t$ and $i \to j$.*

Proof. Suppose that $x = (x_1, ..., x_n) \in X$ is on the boundary of X, and let $t \geq 1$ be any integer such that $|\{g : x_g = t\}| < m$ and j be any task such that $x_j > t$. Consider the point $x' = (x'_1, ..., x'_n)$, where $x'_j = t$ and $x'_g = x_g$ for all $g \neq j$. Since x is on the boundary of X and x strictly dominates x', $x' \notin X$. Hence, either $r_j \geq t$ or there exists g such that $g \to j$ and $x_g \geq t$. If $r_j \geq t$, then the desired property holds. Suppose that $r_j < t$ and among all g such that $g \to j$ and $x_g \geq t$ select one with the smallest x_g. Let it be task i. If $x_i = t$, then the desired property holds. Suppose that $x_i > t$. The relation $i \to j$ implies $r_i < r_j < t$. On the other hand, consider the point $x'' = (x''_1, ..., x''_n)$, where $x''_i = t$ and $x''_g = x_g$ for all $g \neq i$. Since x strictly dominates x'' and x is on the boundary of X, $x'' \notin X$. Hence, there exists q such that $q \to i$ and $x_q \geq t$, which by the transitivity of precedence constraints gives $q \to j$ and therefore contradicts the selection of i.

Conversely, consider $x = (x_1, ..., x_n) \in X$ and suppose that, for each integer $t \geq 1$ such that $|\{g : x_g = t\}| < m$ and each j such that $x_j > t$, either $r_j \geq t$ or there exists i such that $x_i = t$ and $i \to j$. Suppose that x is not on the boundary of X, i.e. x strictly dominates some $x' = (x_1', ..., x_n') \in X$. Then, among all g such that $x_g' < x_g$ select one with the smallest x_g'. Let it be task j. Then, $x_g = x_j'$ implies $x_g' = x_g$. Hence $|\{g : x_g = x_j'\}| < m$, which together with $x_j' < x_j$ implies that either $r_j \geq x_j'$ or there exists i such that $x_i = x_j'$ and $i \to j$. The inequality $r_j \geq x_j'$ contradicts $(x_1', ..., x_n') \in X$. On the other hand, the existence of i such that $x_i = x_j'$ and $i \to j$ also contradicts $x' \in X$ because $x_i' = x_i$. $\square$

3.2   The Level Sets of $\max\limits_{1 \leq j \leq n} \varphi_j(x_j)$

In order to give an example of a problem with *Property 2* and *Property 3*, consider (1) with the objective function

$$F(x_1, x_2, ..., x_n) = \max_{1 \leq j \leq n} \varphi_j(x_j), \tag{2}$$

where each $\varphi_j(x_j)$ is a nondecreasing function defined for all integer $0 \leq x_j \leq p(n)$. Let $\bar{F}$ be an arbitrary value of F, and let a_j be the largest among all integer x_j satisfying the inequalities $\varphi_j(x_j) \leq \bar{F}$ and $x_j \leq p(n)$. It is easy to see that $F(a_1, ..., a_n) = \bar{F}$. Moreover, if $F(x_1, x_2, ..., x_n) = \bar{F}$, then $(a_1, ..., a_n)$ dominates $(x_1, x_2, ..., x_n)$. Hence, for each $\bar{F}$ the level set $A(\bar{F}, F)$ is comprised of only one point. This point can be found in polynomial time, for example by using the binary search on the interval $[0, p(n)]$ separately for each φ_j. Hence, the objective function (2) has *Property 2*.

Let $F' < F''$ be two consecutive values of F, i.e. there is no value of F between these two values. Let $(a_1', ..., a_n')$ and $(a_1'', ..., a_n'')$ be the points constituting $A(F', F)$ and $A(F'', F)$, respectively. Let J be the set of all j satisfying $a_j' < p(n)$. Observe that $(a_1'', ..., a_n'')$ strictly dominates $(a_1', ..., a_n')$ and $a_j' < a_j''$ implies $j \in J$. Moreover, for all $j \in J$, $\varphi_j(a_j' + 1) > F'$ and therefore $\varphi_j(a_j' + 1) \geq F''$. The above observations lead to the following inequalities

$$F'' \leq \min_{j \in J} \varphi_j(a_j' + 1) \leq \max_{1 \leq j \leq n} \varphi_j(a_j'') = F''.$$

Hence,

$$F'' = \min_{j \in J} \varphi_j(a_j' + 1). \tag{3}$$

As has been shown above, for a given F', the corresponding point $(a_1', ..., a_n')$, constituting $A(F', F)$, can be found in polynomial time. Then, F'' can be obtained using (3). Therefore, the objective function (2) has *Property 3*.

Objective functions of the form (2) are common, for example, in scheduling theory. Thus, one of the most frequently used objective functions in scheduling is the function with all $\varphi_j(t) = t - d_j$, where d_j is interpreted as a due date of task j. In this case, the objective function is referred to as the maximum lateness and is denoted by L_{max}. If all due dates are zero, the maximum lateness problem converts into the so-called makespan problem and the corresponding objective function is denoted by C_{max}.

Since the domain of each φ_j is the set of all integer points in the interval $[0, p(n)]$, φ_j takes on at most $p(n) + 1$ different values. Consequently, the cardinality of the range of (2) cannot exceed $n(p(n) + 1)$. So, the union of all level sets cannot contain more than $n(p(n) + 1)$ points. Starting with value $F(0, ..., 0)$ and with the corresponding level set (which is comprised of only one point), one can enumerate the union of all level sets in polynomial time. Nevertheless, in general, the problem remains NP-hard because the elements of level sets do not necessarily belong to the feasible region X. Thus, it is well known that the $P|prec, p_j = 1|C_{max}$ problem is NP-hard in the strong sense [4,2].

3.3 *Property 2* and *Property 3* in Multi-objective Optimization

Consider an optimization problem with k nondecreasing objective functions $F_1, ..., F_k$, each defined on the same n-dimensional hypercube of points with integer coordinates satisfying the inequalities $0 \le x_i \le p(n)$, $1 \le i \le n$, where $p(n)$ is a polynomial in n. A common approach in multi-objective optimization is the replacement of several objective functions by a single function

$$F(x_1, ..., x_n) = \psi(F_1(x_1, ..., x_n), ..., F_k(x_1, ..., x_n)), \tag{4}$$

where ψ is a nondecreasing function.

Lemma 3. *For any value $\bar{F}$ of (4) and for any $(a_1, ..., a_n) \in A(\bar{F}, F)$, there are points $(a_1^{(i)}, ..., a_n^{(i)}) \in A(F_i(a_1, ..., a_n), F_i)$, $1 \le i \le k$, such that $a_j = \min\limits_{1 \le i \le k} a_j^{(i)}$ for all $1 \le j \le n$.*

Proof. Consider an arbitrary point $(a_1, ..., a_n) \in A(\bar{F}, F)$, and for each $1 \le i \le k$ denote $\bar{F}_i = F_i(a_1, ..., a_n)$. By the definition of $A(\bar{F}_i, F_i)$, there exists $(a_1^{(i)}, ..., a_n^{(i)}) \in A(\bar{F}_i, F_i)$ such that $a_j \le a_j^{(i)}$ for all $1 \le j \le n$. Consider the point $(\tilde{a}_1, ..., \tilde{a}_n)$, where $\tilde{a}_j = \min\limits_{1 \le i \le k} a_j^{(i)}$ for all $1 \le j \le n$. Since ψ is a nondecreasing function, each F_i is a nondecreasing function, each $(a_1^{(i)}, ..., a_n^{(i)})$ dominates $(\tilde{a}_1, ..., \tilde{a}_n)$, and $(\tilde{a}_1, ..., \tilde{a}_n)$ dominates $(a_1, ..., a_n)$,

$$F(a_1, ..., a_n) \le F(\tilde{a}_1, ..., \tilde{a}_n) \le \psi(F_1(a_1^{(1)}, ..., a_n^{(1)}), ..., F_k(a_1^{(k)}, ..., a_n^{(k)}))$$

$$= \psi(\bar{F}_1, ..., \bar{F}_k) = F(a_1, ..., a_n).$$

Hence, $F(\tilde{a}_1, ..., \tilde{a}_n) = F(a_1, ..., a_n)$. On the other hand, by the definition of $A(\bar{F}, F)$, $F(a_1, ..., a_n) = \bar{F}$, and therefore $F(\tilde{a}_1, ..., \tilde{a}_n) = \bar{F}$. Moreover, by the same definition, there exists a point $(a_1', ..., a_n') \in A(\bar{F}, F)$ which dominates $(\tilde{a}_1, ..., \tilde{a}_n)$. Consequently, $a_j \le \tilde{a}_j \le a_j'$ for all $1 \le j \le n$. If at least one of these inequalities is strict, then $(a_1', ..., a_n')$ strictly dominates $(a_1, ..., a_n)$ which contradicts Lemma 1 and Theorem 1. So, $a_j = \min\limits_{1 \le i \le k} a_j^{(i)}$ for all $1 \le j \le n$. $\square$

According to Lemma 3 all level sets of F can be obtained from the level sets of $F_1,..., F_k$. Therefore, in some cases, the fact that each of $F_1,..., F_k$ has *Property 2* or *Property 3* or both may imply that (4) also has these properties. One such case is considered in Theorem 2 and Theorem 3. Both theorems consider the case when the cardinality of the range of each $F_1,..., F_k$ is bounded above by a polynomial in n, which is typical for example for scheduling theory.

Theorem 2. *If each of $F_1,..., F_k$ has Property 3 and the cardinality of the range of each $F_1,..., F_k$ is bounded above by a polynomial in n, then (4) also has Property 3.*

Proof. Each F_i is a nondecreasing function defined on the n-dimensional hypercube of points with integer coordinates satisfying the inequalities $0 \le x_i \le p(n)$, $1 \le i \le n$. Therefore, its smallest value is $F_i(0, ..., 0)$. Since F_i has *Property 3* and the cardinality of the range of F_i is bounded above by some polynomial in n, it is possible to enumerate all values of F_i in polynomial time by starting with $F_i(0, ..., 0)$ and using *Property 3* of F_i. Consequently, it is possible in polynomial time to generate all combinations $(\bar{F}_1, ..., \bar{F}_k)$, where each $\bar{F}_i$ is some value of the corresponding F_i. Therefore, (4) has *Property 3*.   □

Theorem 3. *If each of $F_1,..., F_k$ has Property 2 and Property 3 and the cardinality of the range of each $F_1,..., F_k$ is bounded above by a polynomial in n, then (4) also has Property 2.*

Proof. Let $\bar{F}$ be an arbitrary value of F, and let $(\bar{F}_1, ..., \bar{F}_k)$ be an arbitrary combination of values of $F_1, ..., F_k$ such that $\bar{F} = \psi(\bar{F}_1, ..., \bar{F}_k)$. Since each F_i has *Property 2*, there exists an algorithm which in polynomial time finds all elements of $A(\bar{F}_i, F_i)$. Hence, it is possible to find in polynomial time all combinations $(a^{(1)}, ..., a^{(k)})$, where each $a^{(i)} = (a_1^{(i)}, ..., a_n^{(i)})$ is an element of the corresponding $A(\bar{F}_i, F_i)$. Each combination $(a^{(1)}, ..., a^{(k)})$ gives the point $\left(\min_{1 \le i \le k} a_1^{(i)}, ..., \min_{1 \le i \le k} a_n^{(i)} \right)$. So, the cardinality of the set of all such points is bounded above by some polynomial in n. Therefore, it is possible to find in polynomial time the subset $D(\bar{F}_1, ..., \bar{F}_k)$ of this set comprised of all points $(a_1, ..., a_n)$ satisfying $F_i(a_1, ..., a_n) = \bar{F}_i$ for all $1 \le i \le n$.

Since the cardinality of the range of each F_i is bounded above by some polynomial in n and since each F_i has *Property 3*, it is possible to find in polynomial time all combinations $(\bar{F}_1, ..., \bar{F}_k)$ of values of $F_1,..., F_k$ satisfying the condition $\bar{F} = \psi(\bar{F}_1, ..., \bar{F}_k)$. Furthermore, as has been shown above, there exists a polynomial-time algorithm which for each such combination finds all elements of the corresponding set $D(\bar{F}_1, ..., \bar{F}_k)$. Therefore, the union D of $D(\bar{F}_1, ..., \bar{F}_k)$ for all combinations $(\bar{F}_1, ..., \bar{F}_k)$, satisfying $\bar{F} = \psi(\bar{F}_1, ..., \bar{F}_k)$, can be found in polynomial time. According to Lemma 3, $A(\bar{F}, F) \subseteq D$, and therefore D is an $\bar{F}$-dominant set. Then, by Lemma 1, D^c is the level set. Since the cardinality of D is bounded above by a polynomial in n, D^c can be found in polynomial time which implies *Property 2*.   □

4   The Optimization Procedure

The approach outlined below assumes that the problem (1) has *Property 1*, *Property 2* and *Property 3*. The considered iterative optimization procedure at each iteration uses some lower bound on the optimal value of F. All these lower bounds belong to the range of F. Although $F(0, 0, ..., 0)$ is always available, a better choice of the initial lower bound possibly can be made from the analysis of the actual problem.

For each lower bound $\bar{F}$, the optimization procedure strives to find a feasible point with this value of the objective function (in such case the procedure terminates with this point as an optimal solution) or to detect that there is no feasible point that corresponds to $\bar{F}$ (after that a larger lower bound is calculated). This is accomplished by considering in succession all elements of $A(\bar{F}, F)$. Each element of $A(\bar{F}, F)$ initiates a search tree. The root of the search tree corresponds to the considered element of $A(\bar{F}, F)$, say point $(a_1, ..., a_n)$. All nodes in this search tree correspond to points in the domain of F, i.e. points with integer coordinates satisfying the inequalities $0 \leq x_i \leq p(n)$, $1 \leq i \leq n$. At each stage of constructing the search tree, the procedure chooses a node which does not have a successor in the already constructed fragment of this tree and connects this node to one or several new nodes (branching). Branching is conducted in such a way that the point associated with the chosen node strictly dominates the points that correspond to the new nodes introduced by branching. Hence, all points associated with nodes of the search tree cannot have value of F greater than $F(a_1, ..., a_n) = \bar{F}$. Furthermore, since $\bar{F}$ is a lower bound on the optimal value of F, branching adds only new points $(y_1, ..., y_n)$ satisfying the condition $F(y_1, ..., y_n) = \bar{F}$. If one of the new points introduced by branching is in X, the procedure terminates with this point as an optimal solution.

Since each variable is a nonnegative integer and is bounded above by $p(n)$, the number of nodes in any path of any search tree cannot exceed $np(n)$ which guarantees convergence. Furthermore, branching at any node of the search tree is conducted in such a way that the point which corresponds to this node strictly dominates some point in X if and only if at least one of the new points introduced by branching also dominates some point in X. This guarantees that eventually the procedure terminates with an optimal solution.

As in the case of the branch-and-bound method, the implementation of the approach outlined above varies from problem to problem. Nevertheless, *Property 1*, *Property 2* and *Property 3* allow some general techniques, including the idea of projection described below and the idea of function ϱ and the corresponding lower bounds $\underline{\varrho}$ also described below. Again, the calculation of each lower bound $\underline{\varrho}$ depends on the problem in hand.

Let $b = (b_1, ..., b_n) \notin X$ be an arbitrary point associated with a node of the already constructed fragment of the search tree, and assume that b does not have any successor in this fragment. Denote

$$\varrho(b_1, ..., b_n) = \min_{(x_1, ..., x_n) \in X} \max_{1 \leq j \leq n} [x_j - b_j].$$

Then, b dominates a feasible point if and only if $\varrho(b_1, ..., b_n) \leq 0$. In general, the question whether or not $\varrho(b_1, ..., b_n) \leq 0$ is an NP-complete problem. Thus, the NP-completeness in the strong sense of this question for the $P|prec, p_j = 1|C_{max}$ scheduling problem, which is a particular case of (1), follows from [4] and [2]. Given the above observation, it is a good idea to calculate instead of $\varrho(b_1, ..., b_n)$ its lower bound $\underline{\varrho}$. The actual method of calculating $\underline{\varrho}$ depends on a specific problem. If $\underline{\varrho} > 0$, then b is fathomed, i.e. no branching at b is required. If $\underline{\varrho} \leq 0$ or $\underline{\varrho}$ is even not calculated at all, then b can be projected onto X, where the projection of b onto X is a point with the smallest t among all points $(b_1 + t, ..., b_n + t)$ satisfying $(b_1 + t, ..., b_n + t) \in X$. The point $(b_1, ..., b_n)$ can be projected onto X in polynomial time, since this requires to consider only points $(b_1 + t, ..., b_n + t)$ with integer t satisfying the inequality $|t| \leq p(n)$. Of course, the projection may not exist.

Let $(b_1 + \tau, ..., b_n + \tau)$ be the projection of b onto X. Recall that $b \notin X$. On the other hand, by the definition $(b_1 + \tau, ..., b_n + \tau) \in X$. Hence, $\tau \neq 0$.

Theorem 4. *If $(b_1 + \tau, ..., b_n + \tau)$ is on the boundary of X and $\tau > 0$, then b does not dominate any feasible point.*

Proof. Since $(b_1 + \tau, ..., b_n + \tau)$ is on the boundary of X, by the definition of the boundary of X, for any $(x_1, ..., x_n) \in X$, there exists j such that $x_j \geq b_j + \tau$, and therefore $\max_{1 \leq i \leq n} (x_i - b_i) \geq \tau$. Hence, $\varrho(b_1, ..., b_n) \geq \tau > 0$ and b does not dominate any feasible point. $\qquad\square$

Observe that if $\tau < 0$, then b strictly dominates $(b_1 + \tau, ..., b_n + \tau)$. Since b is a point associated with a node of the search tree and since $(b_1 + \tau, ..., b_n + \tau)$ is a feasible point, $(b_1 + \tau, ..., b_n + \tau)$ is an optimal solution. The two remaining cases that have not been covered in Theorem 4 and in the observation above are the case when $\tau > 0$ but the projection does not belong to the boundary of X and the case when the projection does not exist. These two cases are addressed by Theorem 5. Let $X(b_1, ..., b_n)$ be the set of all $(x_1, ..., x_n) \in X$ such that

$$\max_{1 \leq j \leq n} [x_j - b_j] = \varrho(b_1, ..., b_n).$$

Theorem 5. *If $\tau > 0$ and $(b_1 + \tau, ..., b_n + \tau)$ does not belong to the boundary of X or if the projection does not exist, then there exists $(x_1, ..., x_n) \in X(b_1, ..., b_n)$ and i such that*

$$x_i - b_i < \varrho(b_1, ..., b_n). \tag{5}$$

Proof. Observe that the statement of this theorem does not hold if and only if $X(b_1, ..., b_n)$ is comprised of only point $(b_1 + \varrho(b_1, ..., b_n), ..., b_n + \varrho(b_1, ..., b_n))$. If the projection does not exist, then $(b_1 + t, ..., b_n + t) \notin X$ for all integer t. In particular, $(b_1 + \varrho(b_1, ..., b_n), ..., b_n + \varrho(b_1, ..., b_n))$ is not in X and therefore is not in $X(b_1, ..., b_n)$, because $X(b_1, ..., b_n)$ is a subset of X. Hence, for any $(x_1, ..., x_n) \in X(b_1, ..., b_n)$, there exists i satisfying (5).

Suppose that the projection exists but $(b_1 + \tau, ..., b_n + \tau) \notin X(b_1, ..., b_n)$. Assume that $(b_1 + \varrho(b_1, ..., b_n), ..., b_n + \varrho(b_1, ..., b_n)) \in X(b_1, ..., b_n)$. Then, by the definition of projection, $\tau < \varrho(b_1, ..., b_n)$, which by virtue of $(b_1+\tau, ..., b_n+\tau) \in X$ leads to the following contradiction:

$$\varrho(b_1, ..., b_n) > \tau \geq \min_{(x_1, ..., x_n) \in X} \max_{1 \leq j \leq n} [x_j - b_j] = \varrho(b_1, ..., b_n).$$

So, $(b_1 + \varrho(b_1, ..., b_n), ..., b_n + \varrho(b_1, ..., b_n)) \notin X(b_1, ..., b_n)$, and therefore for any $(x_1, ..., x_n) \in X(b_1, ..., b_n)$ there exists i satisfying (5).

Finally, assume that $(b_1 + \tau, ..., b_n + \tau) \in X(b_1, ..., b_n)$. Then, $\tau = \varrho(b_1, ..., b_n)$. Furthermore, since $(b_1+\tau, ..., b_n+\tau)$ is not on the boundary of X, $(b_1+\tau, ..., b_n+\tau)$ strictly dominates some $(x_1, ..., x_n) \in X$, i.e. $x_j \leq b_j +\tau$ for all $1 \leq j \leq n$ and at least one of these inequalities is strict. Then, taking into account the definition of $\varrho(b_1, ..., b_n)$,

$$\varrho(b_1, ..., b_n) = \min_{(y_1, ..., y_n) \in X} \max_{1 \leq j \leq n} [y_j - b_j] \leq \max_{1 \leq j \leq n} [x_j - b_j] \leq \tau = \varrho(b_1, ..., b_n).$$

Hence, $(x_1, ..., x_n) \in X(b_1, ..., b_n)$, and (5) holds for this $(x_1, ..., x_n)$.   □

Let b_i be a coordinate satisfying (5), and let $b' = (b'_1, ..., b'_n)$ be the point obtained from b by replacing b_i by $b'_i = x_i - \varrho(b_1, ..., b_n)$. Hence, b and b' differ only by one coordinate. Point b strictly dominates b' and $\varrho(b'_1, ..., b'_n) = \varrho(b_1, ..., b_n)$. Hence, b strictly dominates some point in X if and only if b' also dominates some point in X. Therefore, if b_i and $x_i - \varrho(b_1, ..., b_n)$ are known, then it is possible to branch at the node which corresponds to b with only one new node - the node which corresponds to b'. Even if $x_i - \varrho(b_1, ..., b_n)$ is not known, branching with only one new point is still possible by replacing b_i by $b_i - \delta_i$, where δ_i is a lower bound on $b_i - x_i + \varrho(b_1, ..., b_n)$. Since $b_i - x_i + \varrho(b_1, ..., b_n)$ is integer and is greater than zero, it is always possible to choose $\delta_i = 1$, although a better lower bound improves convergence.

Not surprisingly, in general a coordinate b_i satisfying (5) is not known, and the optimization procedure instead of finding a desired coordinate b_i, finds some subset $B \subseteq \{1, ..., n\}$ such that the set $\{b_i : i \in B\}$ contains the desired coordinate. Different instances of (1) may have different methods of finding B. An example can be found in the next section. In the unlikely absence of a better idea, the entire set $\{1, ..., n\}$ can be taken as a subset containing a coordinate with the desired property. Once the set B is found, the optimization procedure connects the node associated with $(b_1, ..., b_n)$ with several new nodes (branching) each corresponding to a point obtained from $(b_1, ..., b_n)$ by reducing one coordinate b_i with $i \in B$ by $b_i - \delta_i$.

5   Minimization of the Maximum Weighted Lateness

Consider the $P|prec, p_j = 1|F$ scheduling problem, introduced in Subsection 3.1, with the objective function

$$F(x_1, ..., x_n) = \max_{1 \leq j \leq n} w_j(x_j - d_j), \tag{6}$$

where d_j is a due date for completion of task j (the desired time by which task j needs to be completed) and w_j is a positive weight. Given this interpretation, the considered problem requires to minimize the maximum weighted lateness. An approach similar to one discussed below was briefly outlined in [5].

Observe that (6) is a particular case of the objective function (2) considered in Subsection 3.2. Hence, for any lower bound $\bar{F}$ on the optimal value of F, the corresponding level set $A(\bar{F}, F)$ is comprised of only one point, say $a = (a_1, ..., a_n)$. According to Subsection 3.1 and Subsection 3.2, for each $1 \le i \le n$,

$$a_i = \min \left\{ \left\lfloor \frac{\bar{F}}{w_i} \right\rfloor + d_i, n \right\}. \tag{7}$$

According to Section 4, in the search tree induced by a, each node represents a point in the domain of F. This set of candidates for being a point associated with a node of the search tree can be reduced to the set of so called consistent points. For any point $(b_1, ..., b_n)$ in the domain of F, the corresponding consistent point $(b'_1, ..., b'_n)$ is computed as follows. For each task i, let $K(i)$ be the set of all tasks j such that $i \to j$, i.e. $K(i)$ is the set of all successors of i. The coordinates of $(b'_1, ..., b'_n)$ are computed iteratively. At each iteration, b'_i is computed for i satisfying the condition that all b'_j, $j \in K(i)$, have been already computed, including the case $K(i) = \emptyset$. If $K(i) = \emptyset$, then $b'_i = b_i$. Otherwise,

$$b'_i = \min \left\{ b_i, \min_{d \ge h} \left(d - \left\lceil \frac{|\{j : j \in K(i) \text{ and } b'_j \le d\}|}{m} \right\rceil \right) \right\}, \tag{8}$$

where $h = \min_{j \in K(i)} b'_j$ and d is integer.

The notion of consistency was originally introduced in [1] for the $P2|prec, p_j = 1|L_{max}$ scheduling problem, where $P2$ indicates that the set of tasks is to be processed on two machines. Similar to [1], we have the following lemma.

Lemma 4. *Let $b = (b_1, ..., b_n)$ be an arbitrary point in the domain of F, $b' = (b'_1, ..., b'_n)$ be the corresponding consistent point, and $x = (x_1, ..., x_n) \in X$ be an arbitrary feasible point dominated by b. Then b' also dominates x.*

Proof. Suppose that this statement is not true, and consider the first iteration of the procedure computing b' that produces some b'_i such that $b'_i < x_i$. Since b dominates x, $x_i \le b_i$. Hence, according to (8), $K(i) \ne \emptyset$ and there exists integer $d' \ge \min_{j \in K(i)} b'_j$ such that

$$b'_i = d' - \left\lceil \frac{|\{j : j \in K(i) \text{ and } b'_j \le d'\}|}{m} \right\rceil.$$

By the selection of b'_i, for all $j \in K(i)$, $x_j \le b'_j$ and therefore

$$\{j : j \in K(i) \text{ and } b'_j \le d'\} \subseteq \{j : j \in K(i) \text{ and } x_j \le d'\}.$$

Then, by the feasibility conditions (b) and (c) in Subsection 3.1,

$$x_i \le d' - \left\lceil \frac{|\{j : j \in K(i) \text{ and } x_j \le d'\}|}{m} \right\rceil \le d' - \left\lceil \frac{|\{j : j \in K(i) \text{ and } b'_j \le d'\}|}{m} \right\rceil = b'_i$$

which contradicts $b'_i < x_i$. $\qquad\qquad\qquad\qquad\qquad\qquad\qquad\qquad\qquad\qquad\quad\square$

Given Lemma 4, in order to narrow the search for a feasible point, each point associated with a node in the search tree should be replaced by the corresponding consistent point. This is equally applicable to the root of the search tree, i.e. the point computed according to (7) should be replaced by the corresponding consistent point.

The initial lower bound on the optimal value of F can be calculated as

$$\bar{F} = \max_{1 \leq j \leq n} w_j(c_j - d_j), \tag{9}$$

where the point $(c_1, ..., c_n)$ is computed according to an iterative procedure which is a mirror reflection of the one used for computing consistent points. For each task i, let $Q(i)$ be the set of all tasks j such that $j \to i$, i.e. $Q(i)$ is the set of all predecessors of i. The coordinates of $(c_1, ..., c_n)$ are computed iteratively. At each iteration, c_i is computed for i satisfying the condition that all c_j, $j \in Q(i)$, have been already computed, including the case $Q(i) = \emptyset$. If $Q(i) = \emptyset$, then $c_i = 1$. Otherwise, $l(i) = \max_{j \in Q(i)} c_j$ and

$$c_i = \max_{1 \leq l \leq l(i)} \left\{ l + \left\lceil \frac{|\{j : j \in Q(i) \text{ and } c_j \geq l\}|}{m} \right\rceil \right\}, \tag{10}$$

where l is integer. Similar to Lemma 4, it is easy to show that, for any $(x_1, ..., x_n) \in X$ and for all $1 \leq j \leq n$, $c_j \leq x_j$ which justifies that (9) is a lower bound on the optimal value of F.

Let $b = (b_1, ..., b_n)$ be a consistent point associated with some node of the search tree. If $b \in X$, then the optimization procedure terminates with b as an optimal solution. Assume that $b \notin X$. Then, the optimization procedure calculates a lower bound $\underline{\varrho}$ on $\varrho(b_1, ..., b_n)$ as follows. Consider an arbitrary $(x_1, ..., x_n) \in X$ and arbitrary $\bar{S} \subseteq \{1, ..., n\}$. Then, according to the feasibility condition (b),

$$\min_{j \in S} x_j + \left\lceil \frac{|S|}{m} \right\rceil - 1 \leq \max_{j \in S} x_j.$$

Hence, $\underline{\varrho}$ can be calculated as

$$\underline{\varrho} = \max_{1 \leq l \leq \bar{l}} \left\{ \max_{d \geq b(l)} \left(l + \left\lceil \frac{|\{j : c_j \geq l \text{ and } b_j \leq d\}|}{m} \right\rceil - 1 - d \right) \right\}, \tag{11}$$

where l and d are integer, $\bar{l} = \max_{1 \leq i \leq n} c_i$ and $b(l) = \min_{\{i:\ c_i \geq l\}} b_i$.

If $\underline{\varrho} > 0$, then the considered node is fathomed. The minimum of all $\underline{\varrho}$ for fathomed nodes will be used for calculating a larger lower bound on the optimal value of F. Assume that $\underline{\varrho} \leq 0$. Then, by (11), $c_i \leq b_i$ for all $1 \leq i \leq n$. Since each c_i is greater than or equal to 1, b satisfies the first of the three feasibility conditions stated in Subsection 3.1, i.e. b satisfies condition (a) and does not satisfy either condition (b), or condition (c), or both. Then, for any integer t, the point $(b_1 + t, ..., b_n + t)$ does not satisfy the same condition either. Hence, the projection of b onto X does not exist, and according to Theorem 5, there exists b_i satisfying (5). As has been discussed in Section 4, instead of finding b_i, the optimization procedure finds a subset $B \subseteq \{1, ..., n\}$ such that for at least one $i \in B$ the coordinate b_i satisfies (5). In order to find B, a schedule $(z_1, ..., z_n)$ is constructed using the following iterative algorithm.

According to this algorithm, each iteration corresponds to some point in time t. At the first iteration $t = 1$. Each iteration deals with the set of all tasks i such that either $Q(i) = \emptyset$ or $z_j < t$ for all $j \in Q(i)$. If the cardinality of this set is less than or equal to m (recall that m is the number of machines), then the algorithm sets $z_i = t$ for all i in this set. If the cardinality is greater than m, then the algorithm sets $z_i = t$ only for m tasks by considering tasks in a nondecreasing order of b_i. Then, t is replaced by $t + 1$ and, if there are any unscheduled tasks, the next iteration starts. This cycle repeats until all tasks have been scheduled.

Denote $\bar{\varrho} = \max_{1 \leq j \leq n}(z_j - b_j)$. If $\bar{\varrho} \leq 0$, then $(z_1, ..., z_n)$ is an optimal solution and the optimization procedure terminates. Assume that $\bar{\varrho} > 0$. Let z_g be the smallest among all z_j such that $z_j - b_j = \bar{\varrho}$. Then, the inequality $\underline{\varrho} \leq 0$ guarantees the existence of an integer t such that

$$1 \leq t \leq z_g - 1 \qquad \text{and} \qquad |\{j: z_j = t \text{ and } b_j \leq b_g\}| < m. \qquad (12)$$

Indeed, if $z_g = 1$ or if $z_g > 1$ but $|\{j: z_j = t \text{ and } b_j \leq b_g\}| = m$ for all $1 \leq t \leq z_g - 1$, then

$$\underline{\varrho} \geq \left\lceil \frac{|\{j: c_j \geq 1 \text{ and } b_j \leq b_g\}|}{m} \right\rceil - b_g \geq z_g - b_g = \bar{\varrho} > 0,$$

which contradicts $\underline{\varrho} \leq 0$. Let τ be the largest among all integer t satisfying (12). Denote $U = \{j: \tau < z_j < z_g\} \cup \{g\}$.

Lemma 5. *For any $j \in U$, there exists i such that $i \to j$ and $z_i = \tau$.*

Proof. Consider an arbitrary $j \in U$. Since $z_j > \tau$ and $b_j \leq b_g$, according to the algorithm, which was used for constructing $(z_1, ..., z_n)$, there exists v such that $z_v \geq \tau$ and $v \to j$. Among all such v select one with the smallest z_v. Let it be i. If $z_i = \tau$, then the lemma holds. Suppose that $z_i > \tau$. Then, since $i \to j$, $z_i < z_j$ and consequently $i \in U$. Since $i \in U$, there exists u such that $z_u \geq \tau$ and $u \to i$. This implies that $z_u < z_i$ and, by the transitivity of precedence constraints, $u \to j$, which contradicts the selection of i. $\square$

Let B be the set of all j such that $z_j = \tau$ and $K(j) \cap U \neq \emptyset$. Then, by virtue of Lemma 5,

$$U \subseteq \cup_{j \in B} K(j). \qquad (13)$$

Consider an arbitrary $(x_1, ..., x_n) \in X(b_1, ..., b_n)$, and let $x_i = \min_{j \in B} x_j$. Then, taking into account (13), the fact that $b_j \leq b_g$ for all $j \in U$, and the definition of $X(b_1, ..., b_n)$,

$$x_i - b_i \leq \max_{j \in U} x_j - \left\lceil \frac{|U|}{m} \right\rceil - b_i = \max_{j \in U} x_j - (z_g - z_i) - b_i$$

$$= \max_{j \in U} x_j - b_g - (z_g - b_g) + z_i - b_i \leq \varrho(b_1, ..., b_n) - (z_g - b_g) + z_i - b_i.$$

By the choice of g and i, $z_g - b_g > z_i - b_i$, and therefore x_i and b_i satisfy (5). Furthermore,

$$b_i - x_i + \varrho(b_1, ..., b_n) \geq z_g - b_g - (z_i - b_i).$$

So, the right-hand side of this inequality can be used in branching as a lower bound δ_i on $b_i - x_i + \varrho(b_1, ..., b_n)$. That is, branching at the node associated with b introduces $|B|$ new nodes corresponding to the points each of which is obtained form b by replacing one b_i by $b_i - \delta_i$ where $i \in B$ and $\delta_i = z_g - b_g - (z_i - b_i)$.

Suppose that the search tree does not give a feasible point. Then, each fathomed node $(b_1, ..., b_n)$ in this tree has an associated lower bound $\varrho > 0$ on $\varrho(b_1, ..., b_n)$. Then, the smallest among all these lower bounds, say ϱ^*, is a lower bound on $\varrho(a_1, ..., a_n)$, where $(a_1, ..., a_n)$ is the point computed according to (7). This leads to a new lower bound

$$\bar{F} = \min_{\{j: \, a_j < n\}} w_j(a_j + \varrho^* - d_j).$$

6   Computational Experiments

The algorithm described in Section 5 was compared by means of computational experiments with an implementation of the branch-and-bound method. These computational experiments were conducted in CSIRO by the third author on a 64-bit 28x dual 3.2 GHz Xeon machines with 2Gb of virtual memory. The partially ordered sets for these experiments were provided by Dr Tatjana Davidović.

The computational experiments used an implementation of the branch-and-bound method which constructs a search tree with nodes corresponding to partial schedules. A partial schedule is defined by the set $S = \{x_{j_1}, ..., x_{j_k}\}$ of completion times that have been already determined. The root of the search tree corresponds to the empty partial schedule with $S = \emptyset$. Consider a node defined by $S = \{x_{j_1}, ..., x_{j_k}\}$, and let $t = \max_{i \in S} x_i$. Branching at this node is based on the set $R \subseteq (\{1, ..., n\} - \{j_1, ..., j_k\})$ of all j such that either $x_i \leq t$, for all $i \in Q(j)$, or $Q(j) = \emptyset$. If $|R| \leq m$, then the branching introduces only one new node which is obtained by expanding the current S by setting $x_j = t + 1$ for all $j \in R$. If $|R| > m$, then all combinations of m elements of R are considered and each combination gives a new node by expanding the current S by setting $x_j = t + 1$ for all j in this combination.

Consider a partial schedule $S = \{x_{j_1}, ..., x_{j_k}\}$ corresponding to a node of the search tree. Then, lower and upper bounds on the best value of the objective function which can be obtained from this node are computed as follows. The set S induces the scheduling problem $P|prec, p_j = 1, r_j| \max_{j \notin \{j_1, ..., j_k\}} w_j(x_j - d_j)$, where $i \rightarrow j$ if and only if this relation exists in the original problem, and each release time r_j is $r = \max_{x_j \in S} x_j$. Let $\underline{f}$ and $\bar{f}$ be lower and upper bounds on the optimal value of the objective function of the induced problem, and let $\bar{F}$ be the value of the objective function for the partial schedule. Then, the lower and upper bounds for the considered node are $\min\{\underline{f}, \bar{F}\}$ and $\max\{\bar{f}, \bar{F}\}$, respectively. The upper bound $\bar{f}$ is a value of the objective function for a schedule constructed in the same way as $(z_1, ..., z_n)$ in Section 5, i.e. the original due dates d_j are replaced by consistent due dates d'_j and then the tasks are scheduled according to these new due dates. In order to compute $\underline{f}$, a lower bound c_j on the completion time of each j is computed similarly to Section 5. The only difference is the smallest value of such lower bounds which instead of 1 is now r. Then, similar to (11),

$$\underline{f} = \max_{r \leq l \leq \bar{l}} \left\{ \max_{d \geq b(l)} W_{ld} \left(l + \left\lceil \frac{|\{j : c_j \geq l \text{ and } d'_j \leq d\}|}{m} \right\rceil - 1 - d \right) \right\},$$

where l and d are integer; $\bar{l}$ is the largest c_i; $b(l) = \min_{\{i: \, c_i \geq l\}} d'_i$; and

$$W_{ld} = \min_{\{j: \, c_j \geq l \text{ and } d'_j \leq d\}} w_j.$$

Four groups, each containing 30 partially ordered sets, were used. Each partially ordered set in the first group contained 50 tasks, in the second group - 100 tasks, in the third group - 150 tasks and in the fourth group - 200 tasks. Both algorithms were terminated after the first 15 minutes. The table below gives the percentage of all problems solved by the branch-and-bound method (BB) and by the method described in this paper (A).

Machines	3		4		5		6	
Tasks	A	BB	A	BB	A	BB	A	BB
50	100	60	100	73.33	100	90	100	86.67
100	86.67	53.33	96.67	60	100	43.33	100	63.33
150	60	50	86.67	53.33	90	43.33	93.33	56.67
200	70	56.67	66.67	46.67	70	50	66.67	46.67

The next table gives the percentage of all problems where one method gives a better solution than the other.

Machines	3		4		5		6	
Tasks	A	BB	A	BB	A	BB	A	BB
50	13.33	0.00	20.00	0.00	13.33	0.00	13.33	0.00
100	13.33	0.00	16.67	0.00	26.67	0.00	10.00	0.00
150	16.67	0.00	23.33	3.33	33.33	3.33	20.00	0.00
200	26.67	3.33	46.67	0.00	40.00	0.00	33.33	0.00

References

1. Garey, M.R., Johnson, D.S.: Scheduling tasks with nonuniform deadlines on two processors. J. of ACM 23, 461–467 (1976)
2. Lenstra, J.K., Rinnooy Kan, A.H.G.: Complexity of scheduling under precedence constraints. Oper. Res. 26, 22–35 (1978)
3. Pinedo, M.: Scheduling: theory, algorithms, and systems, 3rd edn. Springer, Heidelberg (2008)
4. Ullman, J.D.: NP-complete scheduling problems. J. Comp. Syst. Sci. 10, 384–393 (1975)
5. Zinder, Y.: The strength of priority algorithms. In: Proceedings, MISTA, pp. 531–537 (2007)

Finding Strong Bridges and
Strong Articulation Points in Linear Time[*]

Giuseppe F. Italiano[1], Luigi Laura[2], and Federico Santaroni[3]

[1] Dipartimento di Informatica, Sistemi e Produzione
Università di Roma "Tor Vergata", via del Politecnico 1, 00133 Roma, Italy
italiano@disp.uniroma2.it
[2] Dipartimento di Informatica e Sistemistica
"Sapienza" Università di Roma, via Ariosto 25, 00185, Roma, Italy
laura@dis.uniroma1.it
[3] Università di Roma "Tor Vergata", 00133 Roma, Italy
federico.santaroni@gmail.com

Abstract. Given a directed graph G, an edge is a strong bridge if its removal increases the number of strongly connected components of G. Similarly, we say that a vertex is a strong articulation point if its removal increases the number of strongly connected components of G. In this paper, we present linear-time algorithms for computing all the strong bridges and all the strong articulation points of directed graphs, solving an open problem posed in [2].

1 Introduction

We assume that the reader is familiar with the standard graph terminology, as contained for instance in [5]. Let $G = (V, E)$ be a directed graph, with m edges and n vertices. A *directed path* in G is a sequence of vertices v_1, v_2, ..., v_k, such that edge $(v_i, v_{i+1}) \in E$ for $i = 1, 2, \ldots, k - 1$. A directed graph G is *strongly connected* if there is a directed path from each vertex in the graph to every other vertex. The *strongly connected components* of G are its maximal strongly connected subgraphs. Given a directed graph G, a vertex $v \in V$ is a *strong articulation point* if its removal increases the number of strongly connected components of G. Similarly, we say that an edge $e \in E$ is a *strong bridge* if its removal increases the number of strongly connected components of G. Figure 1 illustrates the notion of strong articulation points and strong bridges of a directed graph.

The notions of strong articolation points and strong bridges are related to the notion of 2-vertex and 2-edge connectivity of directed graphs. We recall that a strongly connected graph G is said to be 2-vertex-connected if the removal of

---

[*] This work has been partially supported by the 7th Framework Programme of the EU (Network of Excellence "EuroNF: Anticipating the Network of the Future - From Theory to Design") and by MIUR, the Italian Ministry of Education, University and Research, under Project AlgoDEEP.

W. Wu and O. Daescu (Eds.): COCOA 2010, Part I, LNCS 6508, pp. 157–169, 2010.

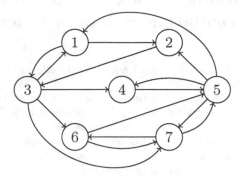

Fig. 1. A strongly connected graph G. Vertices 3 and 5 are strong articulation points in G, while edges $(2,3)$ and $(4,5)$ are strong bridges in G

any vertex leaves G strongly connected; similarly, a strongly connected graph G is said to be 2-edge-connected if the removal of any edge leaves G strongly connected. It is not difficult to see that the strong articulation points are exactly the vertex cuts for 2-vertex connectivity, while the strong bridges are exactly the edge cuts for 2-edge connectivity: G is 2-vertex-connected (respectively 2-edge-connected) if and only if G does not contain any strong articulation point (respectively strong bridge). The 2-vertex and 2-edge connectivity of a directed graph can be tested in linear time. For 2-vertex connectivity, there is a very recent algorithm of Georgiadis [11]. Although there is no specific linear-time algorithm in the literature, the 2-edge connectivity of a directed graph can be tested in $O(m+n)$ as observed somewhat indirectly in [9]: one can test whether a directed graph is 2-edge-connected by using Tarjan's algorithm [14] to compute two edge-disjoint spanning trees in combination with the disjoint set-union algorithm of Gabow and Tarjan [10]. However, both the algorithm of Georgiadis and the combination of Tarjan's algorithm with the set-union data structure of Gabow and Tarjan do not find *all* the strong articulation points or *all* the strong bridges of a directed graph.

In this paper, we show how to find *all* the strong bridges and *all* the strong articulation points of a directed graph in $O(m + n)$ worst-case time. Besides being natural concepts in graph theory, strong articulation points and strong bridges find applications in other areas. For instance, the computation of strong articulation points arise in constraint programming, and in particular in the tree constraint defined by Beldiceanu et al. [2]. The detection of the strong articulation points in a directed graph is indeed a crucial step in their filtering algorithm for the tree constraint, and Beldiceanu et al. [2] leaves as an open problem the design of a linear-time algorithm for computing all the strong articulation points of a directed graph. The algorithm presented in this paper solves exactly this problem.

Besides being interesting on their own, the algoritms presented in this paper have further applications. A first application is verifying the restricted edge connectivity of strongly connected graphs, as defined by Volkman in [15]: our algorithm is able to improve the bounds in [15] from $O(m(m+n))$ to $O(n(m+n))$. As a further application, we cite that our method is able to simplify substantially the approach of Georgiadis [11] to test the 2-vertex connectivity of a strongly connected graph and it also provides a more practical linear-time algorithm for this problem.

The remainder of this paper is organized as follows. Section 2 introduces few preliminary definitions and proves some basic properties on strong articulation points and strong bridges. In Section 3 we analyze the relationship between strong articulation points, strong bridges and dominators in flowgraphs. In Section 4 and Section 5 we show how to exploit effectively this relationship and present, respectively, our linear-time algorithms for computing the strong bridges and the strong articulation points of a directed graph, whilst in Section 6 we consider some applications. Finally, Section 7 contains some concluding remarks. Some details are omitted from this extended abstract for lack of space.

2 Preliminaries

In this section we prove some properties about strong bridges and strong articulation points. Throughout the paper, we assume that we need to compute strong bridges and strong articulation points only in the case of strongly connected graphs. This is without loss of generality, since the strong bridges (respectively strong articulation points) of a directed graph G are given by the union of the strong bridges (respectively strong articulation points) of the strongly connected components of G.

We start with some technical lemmas. Let $G = (V, E)$ be a directed graph. Let $S \subseteq E$ be a subset of edges of G: we denote by $G \setminus S$ the graph obtained after deleting from G all the edges in S. Let $X \subseteq V$ be a subset of vertices of G: we denote by $G \setminus X$ the graph obtained after deleting from G all the vertices in X together with their incident edges.

Lemma 1. Let $G = (V, E)$ be a strongly connected graph, and let $v \in V$ be a vertex of G. Then v is a strong articulation point in G if and only if there exist vertices x and y in G, $x \neq v$, $y \neq v$, such that all the paths from x to y in G contain vertex v.

Proof. Assume that v is a strong articulation point: then $G \setminus \{v\}$ is not strongly connected. This implies that there must be two vertices x and y, $x \neq v$, $y \neq v$, such that there is no path from x to y in $G \setminus \{v\}$, which is equivalent to saying that all the paths from x to y in G must contain vertex v. Conversely, assume that there exist vertices x and y in G, $x \neq v$, $y \neq v$, such that all the paths from x to y in G contain vertex v. Removing v from G leaves no path from x to y, and thus $G \setminus \{v\}$ is no longer strongly connected. This implies that v must be an articulation point of G. $\qquad \square$

Lemma 2. *Let $G = (V, E)$ be a strongly connected graph, and let $(u, v) \in E$ be an edge of G. Then (u, v) is a strong bridge in G if and only if there exist vertices x and y in G such that all the paths from x to y in G contain edge (u, v).*

Proof. Similar to the proof of Lemma 1.   □

Note that as a special case of Lemma 2 we can have $x = u$ and $y = v$. We now need the following definition.

Definition 1. *Given a directed graph $G = (V, E)$, we say that an edge (u, v) is redundant if there is an alternative path from vertex u to vertex v avoiding edge (u, v). Otherwise, we say that (u, v) is non-redundant.*

The following lemma is a consequence of Lemma 2.

Lemma 3. *Let $G = (V, E)$ be a strongly connected graph. Then the edge $(u, v) \in E$ is a strong bridge if and only if (u, v) is non-redundant in G.*

By Lemma 3, in a strongly connected graph the problem of finding strong bridges is equivalent to the problem of finding redundant edges. We remark that finding all the redundant edges in a directed acyclic graph is essentially the transitive reduction problem. This is equivalent to the problem of computing the transitive closure [1], which is known to be equivalent to Boolean matrix multiplication [7,8,12]. Thus, while for directed acyclic graphs the best known bound for computing redundant edges is $O(n^\omega)$, where ω is the exponent of the fastest matrix multiplication algorithm (currently $\omega < 2.376$), we show in this paper that for strongly connected graphs all the redundant edges can be computed faster, in optimal $O(m + n)$ time.

A directed graph can have at most n strong articulation points. This bound is realized by the graph consisting of a simple cycle: indeed in this graph each vertex is a strong articulation point. To bound the number of strong bridges in a directed graph, we need a different argument. Let $G = (V, E)$ be a strongly connected graph, and fix any vertex v of G. Let $T^+(v)$ to be an out-branching rooted at v, i.e., a directed spanning tree rooted at v with all edges directed away from v. Similarly, let $T^-(v)$ to be an in-branching rooted at v, i.e., a directed spanning tree rooted at v with all edges directed towards v.

Lemma 4. *The graph $G' = T^+(v) \cup T^-(v)$ contains at most $(2n - 2)$ edges, can be computed in $O(m + n)$ time and includes all the strong bridges of G.*

Proof. G' is the union of two branchings, each having at most $(n - 1)$ edges: this gives immediately the bound on the size of G'. G' can be computed in linear time using either depth-first or breadth-first search. We now show that all the strong bridges of G must be contained in G'. For any given pair of vertices x, y in V, there is a directed path from x to y in G': the in-branching $T^-(v)$ contains a path from x to v, and the out-branching $T^+(v)$ contains a path from v to y. Thus, G' is a strongly connected subgraph of G, and therefore all the edges not in G' are redundant and cannot be strong bridges.   □

The following corollary is an immediate consequence of Lemma 4:

Corollary 1. *A directed graph G can have at most $(2n - 2)$ strong bridges.*

The bound given in Corollary 1 is tight, as it is easy to construct graphs having exactly $(2n-2)$ strong bridges. Lemma 4 gives immediately a simple $O(n(m+n))$ algorithm to compute strong bridges: first compute the subgraph $G' = T^+(v) \cup T^-(v)$ in $O(m + n)$ time; since all the strong bridges of G are contained in G', for each edge (u, v) in G' test whether (u, v) is a strong bridge by computing the strongly connected components of $G \setminus \{(u,v)\}$.

3 Strong Articulation Points, Strong Bridges and Dominators

Our linear-time algorithms for computing strong articulation points and strong bridges exploits a connection between strong articulation points, strong bridges and dominators in flowgraphs. We start with few definitions, and next we show how those notions are related.

A flowgraph $G(s) = (V, E, s)$ is a directed graph with a *start vertex* $s \in V$ such that every vertex in V is reachable from s. The *dominance relation* in $G(s)$ is defined as follows: a vertex u is a *dominator* of vertex v if every path from vertex s to vertex v contains vertex u. Let $dom(v)$ be the set of dominators of v. Clearly, $dom(s) = \{s\}$ and for any $v \neq s$ we have that $\{s, v\} \subseteq dom(v)$: we say that s and v are the *trivial dominators* of v in the flowgraph $G(s)$. The dominance relation is transitive and its transitive reduction is referred to as the *dominator tree $DT(s)$*. Note that the dominator tree $DT(s)$ is rooted at vertex s. Furthermore, vertex u dominates vertex v if and only if u is an ancestor of v in $DT(s)$. If u is a dominator of v, and every other dominator of u also dominates v, we say that u is an *immediate dominator* of v. It is known that if a vertex v has any dominators, then v has a unique immediate dominator: the immediate dominator of v is the parent of v in the dominator tree $DT(s)$.

Let $G = (V, E)$ be a strongly connected graph, and let s be any vertex in G. Since G is strongly connected, every vertex of V is reachable from s: thus for every vertex $s \in V$, $G(s) = (V, E, s)$ is a flowgraph. Note that there are n flowgraphs for each strongly connected graph. As an example, Figure 2 shows the dominator trees of the flowgraphs relative to the graph of Figure 1. The following lemmas show a close relationship between strong articulation points in strongly connected graphs and non-trivial dominators in flow graphs.

Lemma 5. *Let $G = (V, E)$ be a strongly connected graph, and let s be any vertex in G. Let $G(s) = (V, E, s)$ be the flowgraph with start vertex s. If u is a non-trivial dominator of a vertex v in $G(s)$, then u is a strong articulation point in G.*

Proof. If u is a non-trivial dominator of v in the flowgraph $G(s) = (V, E, s)$, then $u \neq s$, $u \neq v$ and all the paths in G from s to v must include u. Consequently, $G \setminus \{u\}$ is not strongly connected and thus u must be a strong articulation point in G. $\qquad\square$

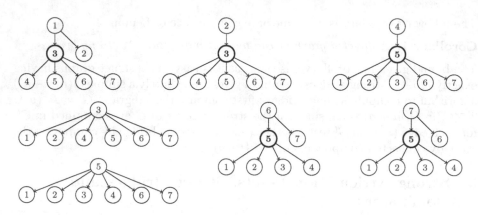

Fig. 2. Dominator trees of the flowgraphs relative to the graph of Figure 1. Non-trivial dominators are shown in bold.

Lemma 6. *Let $G = (V, E)$ be a strongly connected graph. If u is a strong artic-ulation point in G, then there must be a vertex $s \in V$ such that u is a non-trivial dominator of a vertex v in the flowgraph $G(s) = (V, E, s)$.*

Proof. If u is a strong articulation point of G, then by Lemma 1 there must be two vertices s and v in G, $s \neq u$, $v \neq u$, such that every path from s to v contains vertex u. This implies that u must be a non-trivial dominator of vertex v in the flowgraph $G(s)$. □

We observe that Lemmas 5 and 6 are still not sufficient to achieve a linear-time algorithm for our problem: indeed, to compute all the strong articulation points of a strongly connected graph G, we need to compute all the non-trivial dominators in the flowgraphs $G(s)$, for each vertex s in V. Since the dominators of a flowgraph can be computed in $O(m + n)$ time [4] and there are exactly n flowgraphs to be considered, the running time of this algorithm is $O(n(m+n))$. In the next sections, we will show how a more careful exploitation of the relationship between strong articulation points and dominators yields a linear-time algorithm for computing the strong articulation points of a directed graph.

We now show how to exploit similar properties for strong bridges. We say that an edge (u, v) is a *dominator edge* of vertex w if all every path from vertex s to vertex w contains edge (u, v). Furthermore, if (u, v) is an edge dominator of w, and every other edge dominator of u also dominates w, we say that (u, v) is an *immediate edge dominator* of w. Similar to the notion of dominators, if a vertex has any edge dominators, then it has a unique immediate edge dominator. As an example, Figure 3 shows the out-branching trees of the flowgraphs relative to the graph of Figure 1. The following lemmas are completely analogous to Lemmas 5 and 6.

Lemma 7. *Let $G = (V, E)$ be a strongly connected graph, and let s be any vertex in G. Let $G(s) = (V, E, s)$ be the flowgraph with start vertex s. If (u, v) is a dominator edge in $G(s)$, then (u, v) is a strong bridge in G.*

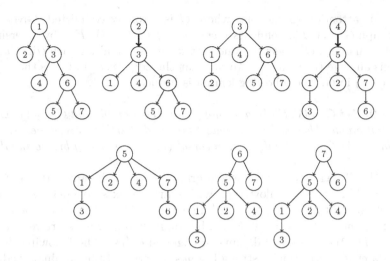

Fig. 3. Out-branchings of the flowgraphs relative to the graph of Figure 1. Dominator edges are shown in bold.

Lemma 8. *Let $G = (V, E)$ be a strongly connected graph. If (u, v) is a strong bridge in G, then there must be a vertex $s \in V$ such that (u, v) is a dominator edge in the flowgraph $G(s) = (V, E, s)$.*

Similar to dominators, also the dominator edges of a flow graph can be computed in linear time. To do this, we compute two edge-disjoint out-branchings rooted at a fixed vertex s with the algorithm of Tarjan [14]: by definition, the dominator edges of the flowgraph $G(s)$ are all the edges that appear in both out-branchings[1]. Tarjan's algorithm is able to identify the dominator edges before building the out-branchings, and can be made to run in time $O(m + n)$ with the help of the linear time disjoint-set union algorithm of Gabow and Tarjan [10].

As previously explained for the case of dominators and strong articulation points, Lemmas 7 and 8 are still not sufficient to obtain a linear-time algorithm for computing all the strong bridges of a strongly connected graph G, since we would need to compute the edge dominators for all n flowgraphs of a given strongly connected graph. We will show in the next section how to obtain this goal by exploiting more carefully the relationship between strong bridges and dominator edges.

4   Finding Strong Bridges

In this section we present a linear-time algorithm for computing the strong bridges of a directed graph G. We recall from Section 2 that it is enough to

---

[1] This result can be seen also as a consequence of Edmond's Theorem [6], that states that a k-connected directed graph admits k edge-disjoint out-branchings rooted at a fixed vertex r.

restrict our attention to the case where G is strongly connected. Given a directed graph $G = (V, E)$, define its *reversal graph* $G^R = (V, E^R)$ by reversing all edges of G: namely, G^R has the same vertex set as G and for each edge (u, v) in G there is an edge (v, u) in G^R. We say that the edge (v, u) in G^R is the *reversal* of edge (u, v) in G. The following lemma is immediate:

Lemma 9. *Let $G = (V, E)$ be a strongly connected graph and $G^R = (V, E^R)$ be its reversal graph. Then G^R is strongly connected. Furthermore, edge (u, v) is a strong bridge in G if and only if its reversal (v, u) is a strong bridge in G^R.*

Let $G = (V, E)$ be a strongly connected graph, let $s \in V$ be any vertex in G, and let $G(s) = (V, E, s)$ be the flowgraph with start vertex s. We denote by $DE(s)$ the set of dominator edges in $G(s)$. Similarly, let $G^R = (V, E^R)$ be the reversal graph of G, and let $G^R(s) = (V, E^R, s)$ be the flowgraph with start vertex s. We denote by $DE^R(s)$ the set of dominator edges in $G^R(s)$. The following theorem provides a characterization of strong bridges in terms of the dominator edges in the two flowgraphs $G(s) = (V, E, s)$ and $G^R(s) = (V, E^R, s)$.

Theorem 1. *Let $G = (V, E)$ be a strongly connected graph, and let $s \in V$ be any vertex in G. Then edge (u, v) is a strong bridge in G if and only if $(u, v) \in DE(s)$ or $(v, u) \in DE^R(s)$.*

Proof. We first prove that if (u, v) is a strong bridge in G, $v \neq s$, then we must have $(u, v) \in DE(s)$ or $(v, u) \in DE^R(s)$. Assume not: namely, assume that (u, v) is a strong bridge in G, $v \neq s$, but $(u, v) \notin DE(s)$ and $(v, u) \notin DE^R(s)$. Since v is a strong bridge in G, then $G \setminus \{(u, v)\}$ is not strongly connected. As a consequence, there must be a vertex w in G, $w \neq s$, such that the following is true: w is in the same strongly connected component as s in G, but w is not the same strongly connected component as s in $G \setminus \{(u, v)\}$. Namely, either (a) there is a path from s to w in G, but there is no path from s to w in $G \setminus \{(u, v)\}$, or (b) there is a path from w to s in G, but there is no path from w to s in $G \setminus \{(u, v)\}$. If we are in case (a), then all the paths from s to w in G must contain edge (u, v). This is equivalent to saying that (u, v) is a dominator edge of w in the flowgraph $G(s) = (V, E, s)$, i.e., $(u, v) \in DE(s)$, which clearly contradicts the assumption $(u, v) \in DE(s)$ or $(v, u) \in DE^R(s)$ If we are in case (b), then all the paths from w to s in G must contain edge (u, v). This is equivalent to saying that (v, u) is a dominator edge of w in the flowgraph $G^R(s) = (V, E^R, s)$, i.e., $(v, u) \in DE^R(s)$, which again contradicts the assumption $(u, v) \in DE(s)$ or $(v, u) \in DE^R(s)$. This shows that if (u, v) is a strong bridge in G, then (u, v) must be in $DE(s)$ or (v, u) must be in $DE^R(s)$.

To prove the converse, let (u, v) be any edge such that $(u, v) \in DE(s)$ or $(v, u) \in DE^R(s)$. If $(u, v) \in DE(s)$, (u, v) is a dominator edge in $G(s)$, and thus (u, v) must be a strong bridge in G by Lemma 7. Analogously, if $(v, u) \in DE^R(s)$, again by Lemma 7 (v, u) must be a strong bridge in G^R: Lemma 9 now ensures that (u, v) must be a strong bridge in G as well. This completes the proof of the theorem. □

We now present our algorithm for computing the strong bridges of a strongly connected graph.

Algorithm StrongBridges(G)

Input : A strongly connected graph $G = (V, E)$, with n vertices and m edges.

Output : The strong bridges of G.

1. Choose arbitrarily a vertex $s \in V$ in G.
2. Compute $DE(s)$, the set of dominator edges in the flowgraph $G(s) = (V, E, s)$.
3. Compute the reversal graph $G^R = (V, E^R)$.
4. Compute $DE^R(s)$, the set of dominator edges in the flowgraph $G^R(s) = (V, E^R, s)$.
5. Output the union of the edges in $DE(s)$ and the reversal of the edges in $DE^R(s)$.

Theorem 2. *Algorithm* StrongBridges *computes the strong bridges of a strongly connected graph G in time $O(m + n)$.*

Proof. The correctness of the algorithm hinges on Theorem 1. We now analyze its running time. Since the edge dominators of a flowgraph can be computed in linear time, Steps 2 and 4 can be implemented in time $O(m + n)$. Finally, the reversal graph G^R in Step 3 can be computed in linear time, and thus the theorem follows. $\square$

Theorem 2 can be generalized to general directed graphs:

Theorem 3. *The strong bridges of a directed graph G can be computed in time $O(m + n)$.*

Proof. To compute the strong bridges of G it is enough to find the strongly connected components of G in $O(m + n)$ time [13], and then to return the strong bridges of each connected component of G. By Theorem 2, this require overall $O(m + n)$ time. $\square$

5   Finding Strong Articulation Points

In the previous section we have seen how to compute the strong bridges of a directed graph. Now we show that to adapt the same approach to compute the strong articulation points. The following are the analogs of Lemma 9 and of Theorem 1.

Lemma 10. *Let $G = (V, E)$ be a strongly connected graph and $G^R = (V, E^R)$ be its reversal graph. Vertex v is a strong articulation point in G if and only if v is a strong articulation point in G^R.*

Theorem 4. *Let $G = (V, E)$ be a strongly connected graph, and let $s \in V$ be any vertex in G. Then vertex $v \neq s$ is a strong articulation point in G if and only if $v \in D(s) \cup D^R(s)$.*

Note that Theorem 4 provides no information on whether vertex s is a strong articulation point. However, this can be easily checked as shown in the following algorithm.

Algorithm StrongArticulationPoints(G)

Input : A strongly connected graph $G = (V, E)$, with n vertices and m edges.

Output : The strong articulation points of G.

1. Choose arbitrarily a vertex $s \in V$ in G, and test whether s is a strong articulation point in G. If s is an articulation point, output s.
2. Compute $D(s)$, the set of non-trivial dominators in the flowgraph $G(s) = (V, E, s)$.
3. Compute the reversal graph $G^R = (V, E^R)$.
4. Compute $D^R(s)$, the set of non-trivial dominators in the flowgraph $G^R(s) = (V, E^R, s)$.
5. Output $D(s) \cup D^R(s)$.

Theorem 5. *Algorithm* StrongArticulationPoints *computes the strong articulation points of a strongly connected graph G in time $O(m + n)$.*

Proof. The correctness of the algorithm hinges on Theorem 4. We now analyze its running time. Step 1 can be implemented in time $O(m + n)$ by simply computing the strongly connected components of $G \setminus \{s\}$ [13]. Since computing the dominators of a flowgraph requires linear time [3], also Steps 2 and 4 can be implemented in time $O(m + n)$. Finally, the reversal graph G^R in Step 3 can be computed in linear time, and thus the theorem follows.   □

Theorem 5 can be generalized to general directed graphs:

Theorem 6. *The strong articulation points of a directed graph G can be computed in time $O(m + n)$.*

Proof. To compute the strong articulation points of G it is enough to find the strongly connected components of G in $O(m + n)$ time [13], and then to return the strong articulation points of each connected component of G. By Theorem 5, this require overall $O(m + n)$ time.   □

We conclude this section by observing that is possible to reduce the problem of computing strong bridges to the problem of computing strong articulation points:

Lemma 11. *If there is an algorithm to compute the strong articulation points of a strongly connected graph in time $T(m, n)$, then there is algorithm to compute the strong bridges of a strongly connected graph in time $O(m+n+T(2m, n+m))$.*

Proof. Let $G = (V, E)$ be a strongly connected graph with m edges and n vertices. We define a new graph G' as follows. G' contains all the vertices of G; furthermore, for each edge $e = (u, v)$ in G, we introduce a new vertex $\varphi(e)$ in G'. The edge set of G' is defined as follows: for each edge $e = (u, v)$ in G, there are edges $(u, \varphi(e))$ and $(\varphi(e), v)$ in G'. Note that G' has exactly $n + m$ vertices and $2m$ edges, it is still strongly connected and can be computed in time $O(m + n)$. The lemma now follows from the observation that an edge e is a strong bridge in G if and only if the corresponding vertex $\varphi(e)$ is a strong articulation point in G'. $\square$

By Lemma 11, a linear-time algorithm for computing the strong articulation points implies immediately a linear-time algorithm for computing the strong bridges. This provides an alternative algorithm to the one described in Section 4.

6   Applications of Strong Articulation Points and Strong Bridges

Strong articulation points and strong bridges are natural concepts in graph theory, and appear to be interesting on their own. In this section, we list a few other applications of our linear-time algorithms for computing strong articulation points and strong bridges.

Verifying restricted edge connectivity. A notion related to edge connectivity is *restricted edge connectivity* [15]. Let G be a strongly connected graph: an edge set S is a restricted edge-cut of G if $G \setminus S$ has a non-trivial strongly connected component D such that $G \setminus V(D)$ contains (at least) one edge. The restricted edge connectivity $\lambda'(G)$ is the minimum cardinality over all restricted edge-cuts S. A strongly connected graph G is called λ'-connected if $\lambda'(G)$ exists. In [15] Volkmann presented an $O(m(m + n))$ algorithm to verify the λ'-connectedness of a strongly connected graph G. The algorithm by Volkmann needs to carry out m distinct computations of strongly connected components, one for each edge of the graph. It is possible to reduce to $O(n)$ the total number of computations of strongly connected components required, i.e., one for each strong bridge of G (recall that the strong bridges of a directed graph are $O(n)$ by Corollary 1). This improves to $O(n(m + n))$ the overall time needed to verify the λ'-connectedness of a strongly connected graph.

Testing 2-vertex connectivity. Very recently Georgiadis [11] gave a linear-time algorithm for testing the 2-vertex connectivity of a strongly connected graph G. The algorithm in [11] work as follows: first, choose two distinct vertices a and b in G, and next check whether the four flowgraphs $G(a)$, $G(b)$, $G^R(a)$, and $G^R(b)$ have only trivial dominators. The proof of correctness given in [11] is rather involved and lengthy. Theorem 4 in this paper provides a much simpler proof: if $G(a)$ and $G^R(a)$ have only trivial dominators, then only vertex a could be an articulation point in G; if also the flowgraphs $G(b)$ and $G^R(b)$ have only trivial dominators, then a cannot be a strong articulation point, and hence G must be 2-vertex connected. Conversely, if G is 2-vertex connected, then G has no strong articulation points, and therefore, for any choice of the start vertex s, the flowgraphs $G(s)$ and $G^R(s)$ have only trivial dominators. We also observe that the algorithm presented in this paper is likely to be faster in practice than the algorithm in [11]. Indeed, while the algorithm in [11] needs to work with four different flowgraphs, the algorithm presented here needs only two different flowgraphs, plus one simple computation of the strongly connected components. We recall that, besides simplifying the correctness proofs and providing algorithms that are likely to be faster in practice, our algorithm is able to report all the cut vertices, i.e., all the strong articulation points in G.

7   Conclusions

Strong articulation points and strong bridges in directed graphs are a natural generalization of the notions of articulation points and bridges in undirected graphs. Surprisingly, they have not received much attention in the literature, despite the fact that they seem to arise in several applications (see, e.g., [2,15]). In this paper, we have presented linear-time algorithms for computing all the strong articulation points and all the strong bridges of a directed graph. Our algorithms are simple, and thus appear to be amenable to practical implementations.

Acknowledgments

We are indebted to Nicolas Beldiceanu for suggesting this problem, and to Fabrizio Grandoni for initial discussions on this topic.

References

1. Aho, A.V., Garey, M.R., Ullman, J.D.: The transitive reduction of a directed graph. SIAM J. Comput. 1(2), 131–137 (1972)
2. Beldiceanu, N., Flener, P., Lorca, X.: The tree constraint. In: Barták, R., Milano, M. (eds.) CPAIOR 2005. LNCS, vol. 3524, pp. 64–78. Springer, Heidelberg (2005)
3. Buchsbaum, A.L., Georgiadis, L., Kaplan, H., Rogers, A., Tarjan, R.E., Westbrook, J.R.: Linear-time algorithms for dominators and other path-evaluation problems. SIAM Journal on Computing 38(4), 1533–1573 (2008)
4. Buchsbaum, A.L., Kaplan, H., Rogers, A., Westbrook, J.R.: A new, simpler linear-time dominators algorithm. ACM Trans. Program. Lang. Syst. 20(6), 1265–1296 (1998)
5. Cormen, T.H., Leiserson, C.E., Rivest, R.L., Stein, C.: Introduction to Algorithms, 3rd edn. MIT Press, Cambridge (2009)
6. Edmonds, J.: Edge-disjoint branchings. In: Proceedings of the 9th Courant Computer Science Symposium, Combinatorial Algorithms, pp. 91–96. Algorithmics Press (1972)
7. Fischer, M.J., Meyer, A.R.: Boolean matrix multiplication and transitive closure. In: Proceedings of 12th FOCS, pp. 129–131. IEEE, Los Alamitos (1971)

8. Furman, M.E.: Applications of a method of fast multiplication of matrices in the problem of finding the transitive closure of a graph. Soviet Math. Dokl. 11(5), 1252 (1970)
9. Gabow, H.N.: A matroid approach to finding edge connectivity and packing arborescences. J. Comput. Syst. Sci. 50(2), 259–273 (1995)
10. Gabow, H.N., Tarjan, R.E.: A linear-time algorithm for a special case of disjoint set union. Journal of Computer and System Sciences 30(2), 209–221 (1985)
11. Georgiadis, L.: Testing 2-vertex connectivity and computing pairs of vertex-disjoint s-t paths in digraphs. In: ICALP 2010: Proceedings of the 37th International Colloquium on Automata, Languages and Programming, pp. 433–442 (2010)
12. Munro, I.: Efficient determination of the transitive closure of a directed graph. Inform. Process. Lett. 1(2), 56–58 (1971)
13. Tarjan, R.E.: Depth-first search and linear graph algorithms. SIAM Journal on Computing 1(2), 146 160 (1972)
14. Tarjan, R.E.: Edge-disjoint spanning trees and depth-first search. Acta Inf. 6, 171–185 (1976)
15. Volkmann, L.: Restricted arc-connectivity of digraphs. Inf. Process. Lett. 103(6), 234–239 (2007)

Robust Optimization of Graph Partitioning and Critical Node Detection in Analyzing Networks

Neng Fan and Panos M. Pardalos

Center for Applied Optimization, Department of Industrial and Systems Engineering,
University of Florida, Gainesville, FL, USA
{andynfan,pardalos}@ufl.edu

Abstract. The graph partitioning problem (GPP) consists of partitioning the vertex set of a graph into several disjoint subsets so that the sum of weights of the edges between the disjoint subsets is minimized. The critical node problem (CNP) is to detect a set of vertices in a graph whose deletion results in the graph having the minimum pairwise connectivity between the remaining vertices. Both GPP and CNP find many applications in identification of community structures or influential individuals in social networks, telecommunication networks, and supply chain networks. In this paper, we use integer programming to formulate GPP and CNP. In several practice cases, we have networks with uncertain weights of links. Some times, these uncertainties have no information of probability distribution. We use robust optimization models of GPP and CNP to formulate the community structures or influential individuals in such networks.

1 Introduction

The graph partitioning problem (GPP) consists of partitioning the vertex set of a graph into several disjoint subsets so that the sum of weights of the edges between the disjoint subsets is minimized. The critical node problem (CNP) is to detect a set of vertices in a graph whose deletion results in the graph having the minimum pairwise connectivity between the remaining vertices. Both GPP and CNP are NP-complete [8,1].

Let $G = (V, E)$ be an undirected graph with a set of vertices $V = \{v_1, v_2, \cdots, v_N\}$ and a set of edges $E = \{(v_i, v_j) : \text{edge between vertices } v_i \text{ and } v_j, 1 \leq i, j \leq N\}$, where N is the number of vertices. The weights of the edges are given by a matrix $W = (w_{ij})_{N \times N}$, where $w_{ij} (> 0)$ denotes the weight of edge (v_i, v_j) and $w_{ij} = 0$ if no edge (v_i, v_j) exists between vertices v_i and v_j. This matrix is symmetric for undirected graphs G and is the adjacency matrix of G if $w_{ij} \in \{0, 1\}$.

For the graph partitioning problem, we are given the cardinalities $n_1, \cdots, n_K$ of subsets that we want to partition V, and K is the number of subsets. Let x_{ik} be the indicator that vertex v_i belongs to the kth subset if $x_{ik} = 1$ or not if $x_{ik} = 0$, and y_{ij} be the indicator that the edge (v_i, v_j) with vertices v_i, v_j are in different subsets if $y_{ij} = 1$ and v_i, v_j in the same subset if $y_{ij} = 0$. Thus, the sum of weights of the edges between the disjoint subsets can be expressed as $\frac{1}{2} \sum_{i=1}^{N} \sum_{j=1}^{N} w_{ij} y_{ij}$ or $\sum_{i=1}^{N} \sum_{j=i+1}^{N} w_{ij} y_{ij}$ because of $w_{ij} = w_{ji}$ and $w_{ii} = 0$ for non-existence of loops. Each vertex v_i has to be partitioned into one and only one subset, i.e., $\sum_{k=1}^{K} x_{ik} = 1$, and the kth subset has the number n_k of vertices, i.e., $\sum_{i=1}^{N} x_{ik} = n_k$. The relation between x_{ik} and y_{ij} can be expressed as $y_{ij} = 1 - \sum_{k=1}^{K} x_{ik} x_{jk}$

W. Wu and O. Daescu (Eds.): COCOA 2010, Part I, LNCS 6508, pp. 170–183, 2010.

and this can be linearized as $-y_{ij} - x_{ik} + x_{jk} \leq 0, -y_{ij} + x_{ik} - x_{jk} \leq 0$ for $k = 1, \cdots, K$ under the objective of minimization. Therefore, the feasible set of deterministic formulation of graph partitioning problem for a graph $G = (V, E)$ with weight matrix W is

$$
X = \left\{ (x_{ik}, y_{ij}) : \begin{array}{l} \sum_{k=1}^{K} x_{ik} = 1, \sum_{i=1}^{N} x_{ik} = n_k, \\ -y_{ij} - x_{ik} + x_{jk} \leq 0, \\ -y_{ij} + x_{ik} - x_{jk} \leq 0, \\ x_{ik} \in \{0,1\}, y_{ij} \in \{0,1\}, \\ i = 1, \cdots, N, j = i+1, \cdots, N, k = 1, \cdots, K \end{array} \right\}, \tag{1}
$$

and the objective function is

$$
\min_{(x_{ik}, y_{ij}) \in X} \sum_{i=1}^{N} \sum_{j=i+1}^{N} w_{ij} y_{ij} \tag{2}
$$

where we minimize the total weight of edges connecting distinct subsets. The GPP is to solve the program with the objective (2) and the constraints in (1) of X. This is a binary integer linear programming problem.

For the critical node problem, we are given the number K as the number of vertices we want to delete in V. Let v_i be the indicator that vertex v_i belongs to the deleted subset V_d of V if $v_i = 1$ and otherwise $v_i = 0$, and let u_{ij} be the indicator that the edge (v_i, v_j) with two ends v_i, v_j are in the resulted graph after deletion of the subset V_d of V if $u_{ij} = 1$ and otherwise $u_{ij} = 0$. Here, we use the notation v_i as one vertex of V and the indicator of this vertex to be deleted or not. Thus, the pairwise connectivity between the remaining vertices in $V \setminus V_d$ can be expressed as $\sum_{i=1}^{N} \sum_{j=i+1}^{N} w_{ij} u_{ij}$ because of the symmetric $w_{ij} = w_{ji}$ and no loop of G with $w_{ii} = 0$. The constraints of CNP include that the number of vertices in the deleted subset is K, i.e., $\sum_{i=1}^{N} v_i = K$, and all three edges in E have the relation that if two edges are in the resulted graph, another edge is also in the resulted graph, i.e., $\max\{u_{ij} + u_{jk} - u_{ik}, u_{ij} - u_{jk} + u_{ik}, -u_{ij} + u_{jk} + u_{ik}\} \leq 1$. The relation between u_{ij} and v_i, v_j can be expressed as $u_{ij} + v_i + v_j \geq 1$ under the objective of minimization, which implies that the edge (v_i, v_j) between the subsets V_d and $V \setminus V_d$ should be deleted if one or two of the vertices are deleted. Therefore, the feasible set of deterministic formulation of critical node problem for a graph $G = (V, E)$ with weight matrix W is

$$
Y = \left\{ (v_i, u_{ij}) : \begin{array}{l} u_{ij} + v_i + v_j \geq 1, \\ u_{ij} + u_{jk} - u_{ik} \leq 1, \\ u_{ij} - u_{jk} + u_{ik} \leq 1, \\ -u_{ij} + u_{jk} + u_{ik} \leq 1, \\ \sum_{i=1}^{N} v_i = K, \\ v_i \in \{0,1\}, u_{ij} \in \{0,1\}, \\ i = 1, \cdots, N, j = i+1, \cdots, N, k = j+1, \cdots, N \end{array} \right\}, \tag{3}
$$

and the objective function is

$$
\min_{(v_i, u_{ij}) \in Y} \sum_{i=1}^{N} \sum_{j=i+1}^{N} w_{ij} u_{ij} \tag{4}
$$

where we minimize the pairwise connectivity between the remaining vertices. The CNP is to solve the program with the objective (4) and the constraints in (3) of Y. This is also a binary integer linear programming problem.

The graph partitioning problem has been studied for a long time and recently studied by linear and quadratic programming approaches [6]. The critical node problem is proposed recently by [1], and solved with several heuristic methods and exact integer programming methods. In this paper, we minimize the pairwise connectivity between the remaining vertices after deleting some vertices instead of another form named the cardinality constrained critical node problem, which is to minimize the number of deleted nodes to limit the connectivity.

Both GPP and CNP have been applied in analyzing networks [5,2,4]. Two important elements or structures in networks are communities and influential individuals, where a community is a dense group of nodes with high connectivity and the influential individual is a node which plays a leader role in the network. The GPP is to partition the nodes into several subsets which are quite related to communities, while the CNP is trying to find the critical nodes or influential individuals in the network. However, the influential individual or the critical node always have the dense part with itself as the center of this community. This leads us to use both GPP and CNP to analyze the networks in the meantime.

In practice, the links between nodes always change their connectivity in a network. That is, the weights are uncertain along the time. In this paper, we use robust optimization models to formulate both GPP and CNP to deal with these problems with uncertain weights. In addition, we use our models to analyze several networks arising from social networks or some artificial networks.

The rest of this paper is organized as follows: section 2 discusses the robust optimization models and algorithms based on a decomposition method, for GPP and CNP with uncertain weights; In section 3, we discuss the application of GPP and CNP in networks with detailed explanations; In section 4, several numerical experiments are performed to analyze some social networks; Section 5 concludes the paper.

2   Robust Models for GPP and CNP

2.1   Uncertainty Assumptions

In this paper, we consider the uncertainty for the weight matrix $W = (w_{ij})_{N \times N}$. Assume that each entry w_{ij} is modeled as independent, symmetric and bounded random but unknown distribution variable $\tilde{w}_{ij}$, with values in $[w_{ij} - \hat{w}_{ij}, w_{ij} + \hat{w}_{ij}]$. Note that we require $w_{ij} = w_{ji}$ for undirected graph G and thus $\hat{w}_{ij} = \hat{w}_{ji}$ for $i, j = 1, \cdots, N$. Assume $\hat{w}_{ij} \geq 0$, $w_{ij} \geq \hat{w}_{ij}$ and $w_{ii}, \hat{w}_{ii} = 0$ for all $i, j = 1, \cdots, N$.

2.2   Robust Optimization Models for GPP and CNP

For the graph $G = (V, E)$ with the weighted matrix $W = (w_{ij})_{N \times N}$, the uncertainties satisfy $\tilde{w}_{ij} \in [w_{ij} - \hat{w}_{ij}, w_{ij} + \hat{w}_{ij}]$. For the positive integer K, the GPP requires the given cardinalities $n_k(k = 1, \cdots, K)$. For general graph partitioning [6], n_k is not necessarily to be given and only required to satisfy $n_k > 1$; For equal partitioning, $n_k \in \{\lfloor N/K \rfloor, \lfloor N/K \rfloor + 1\}$ with total $\sum_k n_k = N$. In CNP, we also delete exactly K vertices to minimize the pairwise connectivity between the remaining vertices.

Because of the existence of uncertain weights of edges in G, we use robust optimization models to formulate the GPP and CNP in order to optimize against the worst cases by min-max objective functions. These models find the best partitioning in GPP and best deletion of CNP in the worst cases of the uncertainties $\tilde{w}_{ij}$.

Let J be the index set of W with uncertain changes, i.e., $J = \{(i, j) : \hat{w}_{ij} > 0, i = 1, \cdots, N, j = i + 1, \cdots, N\}$, where we assume that $j > i$ since W is symmetric. Let Γ be a parameter, not necessarily integer, that takes values in the interval $[0, |J|]$. This parameter Γ, which is introduced in [3], is to adjust the robustness of the proposed method against the level of conservatism of the solution. The number of coefficients w_{ij} is allowed to change up to $\lfloor \Gamma \rfloor$ and another w_{i_t, j_t} changes by $(\Gamma - \lfloor \Gamma \rfloor)$.

Thus, the robust optimization model of GPP (RGPP) with given cardinalities n_k can be formulated as follows,

$$\min_{(x_{ik}, y_{ij}) \in X} z \tag{5}$$

$$s.t. \quad \sum_{i=1}^{N} \sum_{j=i+1}^{N} w_{ij} y_{ij} + \max_{\left\{ \begin{array}{c} S : S \subseteq J, |S| \leq \Gamma \\ (i_t, j_t) \in J \setminus S \end{array} \right\}} \left(\sum_{(i,j) \in S} \hat{w}_{ij} y_{ij} + (\Gamma - \lfloor \Gamma \rfloor) \hat{w}_{i_t, j_t} y_{i_t, j_t} \right) - z \leq 0.$$

and as shown in the following theorem, it can be reformulated as an equivalent binary integer linear programming. The method used in this proof was first proposed in [3].

Theorem 1. *The formulation (5) is equivalent to the following linear programming formulation:*

$$\min \quad \sum_{i=1}^{N} \sum_{j=i+1}^{N} w_{ij} y_{ij} + \Gamma p_0 + \sum_{(i,j) \in J} p_{ij} \tag{6}$$

$$s.t. \quad p_0 + p_{ij} - \hat{w}_{ij} y_{ij} \geq 0, \quad (i, j) \in J$$

$$p_{ij} \geq 0, \quad (i, j) \in J$$

$$p_0 \geq 0,$$

$$(x_{ik}, y_{ij}) \in X.$$

Proof. For a given matrix $(y_{ij})_{i=1, \cdots, N, j=i+1, \cdots, N}$, the part

$$\max_{\left\{ \begin{array}{c} S : S \subseteq J, |S| \leq \Gamma \\ (i_t, j_t) \in J \setminus S \end{array} \right\}} \left(\sum_{(i,j) \in S} \hat{w}_{ij} y_{ij} + (\Gamma - \lfloor \Gamma \rfloor) \hat{w}_{i_t, j_t} y_{i_t, j_t} \right),$$

in (5) can be linearized by introducing z_{ij} for all $(i,j) \in J$ with the constraints $\sum_{(i,j)\in J} z_{ij} \leq \Gamma, 0 \leq z_{ij} \leq 1$, or equivalently, by the following formulation

$$\max \quad \sum_{(i,j)\in J} \hat{w}_{ij} y_{ij} z_{ij} \tag{7}$$

$$s.t. \quad \sum_{(i,j)\in J} z_{ij} \leq \Gamma,$$

$$0 \leq z_{ij} \leq 1, \quad (i,j) \in J$$

The optimal solution of this formulation should have $\lfloor \Gamma \rfloor$ variables $z_{ij} = 1$ and one $z_{ij} = \Gamma - \lfloor \Gamma \rfloor$, which is equivalent to the optimal solution in the maximizing part in (5).

By strong duality, for a given matrix $(y_{ij})_{i=1,\cdots,N, j=i+1,\cdots,N}$, the problem (7) is linear and can be formulated as

$$\min \quad \Gamma p_0 + \sum_{(i,j)\in J} p_{ij}$$

$$s.t. \quad p_0 + p_{ij} - \hat{w}_{ij} y_{ij} \geq 0, \quad (i,j) \in J$$

$$p_{ij} \geq 0, \quad (i,j) \in J$$

$$p_0 \geq 0.$$

Combing this formulation with (5), we obtain the equivalent formulation (6), which finishes the proof.   □

Similarly, the robust optimization model for CNP (RCNP) to delete K vertices is as follows,

$$\min_{(v_i, u_{ij})\in Y} \quad z' \tag{8}$$

$$s.t. \quad \sum_{i=1}^{N} \sum_{j=i+1}^{N} w_{ij} u_{ij} + \max_{\left\{ \begin{array}{c} S: S \subseteq J, |S| \leq \Gamma \\ (i_t, j_t) \in J \setminus S \end{array} \right\}} \left(\sum_{(i,j)\in S} \hat{w}_{ij} u_{ij} + (\Gamma - \lfloor \Gamma \rfloor) \hat{w}_{i_t, j_t} u_{i_t, j_t} \right) - z' \leq 0.$$

and its equivalent binary integer linear programming formulation is

$$\min \quad \sum_{i=1}^{N} \sum_{j=i+1}^{N} w_{ij} u_{ij} + \Gamma p_0' + \sum_{(i,j)\in J} p_{ij}' \tag{9}$$

$$s.t. \quad p_0' + p_{ij}' - \hat{w}_{ij} u_{ij} \geq 0, \quad (i,j) \in J,$$

$$p_{ij}' \geq 0, \quad (i,j) \in J,$$

$$p_0' \geq 0,$$

$$(v_i, u_{ij}) \in Y.$$

By comparing the robust optimization models of GPP ((5),(6)) and CNP ((8),(9)), they are quite similar except that the feasible sets for x_{ik}, y_{ij} in GPP and v_i, u_{ij} in CNP. Both programs in (6) and (9) are formulated as binary integer linear programs, which can be solved through commercial software, such as CPLEX. In the following section, we use a decomposition method on a variable to reformulate robust optimization models of GPP and CNP. In addition, the probability distribution of the gap, between the objective value of the robust optimization model RGPP (5) and the objective value of the robust optimization model with respect to uncertainty $\tilde{w}_{ij}$ by choosing different Γ, is studied in [3]. Since the robust optimization model for RCNP (8) is quite similar to RGPP (5), the probability distribution of the gap for RCNP can be studied similarly.

2.3 Algorithm Based on a Decomposition on One Variable

Next, we will construct an algorithm based on a decomposition method on a variable to solve the programs (6) and (9). The numerical experiments in Section 4 show that this algorithm is much more efficient than the direct branch and bound method by CPLEX. For all $(i,j) \in J$, let e_l $(l = 1, \cdots, |J|)$ be the corresponding value of $\hat{w}_{ij}$ in the increasing order. For example, $e_1 = \min_{(i,j)\in J} \hat{w}_{ij}$ and $e_{|J|} = \max_{(i,j)\in J} \hat{w}_{ij}$. Let $(i^l, j^l) \in J$ be the corresponding index of l, i.e., $\hat{w}_{(i^l,j^l)} = e_l$. In addition, we define $e_0 = 0$. Thus, $[0, e_1], [e_1, e_2], \cdots, [e_{|J|}, \infty)$ is a decomposition of $[0, \infty)$.

For $l = 0, 1, \cdots, |J|$, we define the program G^l as follows:

$$G^l = \Gamma e_l + \min_{(x_{ik}, y_{ij}) \in X} \left\{ \sum_{i=1}^{N} \sum_{j=i+1}^{N} w_{ij} y_{ij} + \sum_{(i,j):\hat{w}_{ij} \geq e_{l+1}} (\hat{w}_{ij} - e_l) y_{ij} \right\}. \quad (10)$$

Totally, there are $|J| + 1$ of G^ls. In the following theorem, we prove that the decomposition method based on p_0 can solve the program (6). The method in the proof was first proposed in [3].

Theorem 2. *Solving robust graph partitioning problem (6) is equivalent to solve the* $|J| + 1$ *problems* G^ls *in (10) for* $l = 0, 1, \cdots, |J|$.

Proof. From (6), the optimal solution $(x_{ik}^*, y_{ij}^*, p_0^*, p_{ij}^*)$ satisfies

$$p_{ij}^* = \max\{\hat{w}_{ij} y_{ij}^* - p_0^*, 0\},$$

and therefore, the objective function of (6) can be expressed as

$$\min_{\{p_0 \geq 0, (x_{ik}, y_{ij}) \in X\}} \Gamma p_0 + \sum_{i=1}^{N} \sum_{j=i+1}^{N} w_{ij} y_{ij} + \sum_{(i,j)\in J} \max\{\hat{w}_{ij} y_{ij} - p_0, 0\}$$

$$= \min_{\{p_0 \geq 0, (x_{ik}, y_{ij}) \in X\}} \Gamma p_0 + \sum_{i=1}^{N} \sum_{j=i+1}^{N} w_{ij} y_{ij} + \sum_{(i,j)\in J} y_{ij} \max\{\hat{w}_{ij} - p_0, 0\}, \quad (11)$$

where the equality is obtained by the fact y_{ij} is binary in the feasible set X.

By the composition $[0, e_1], [e_1, e_2], \cdots, [e_{|J|}, \infty)$ of $[0, \infty)$ for p_0, we have

$$\sum_{(i,j)\in J} y_{ij} \max\{\hat{w}_{ij} - p_0, 0\} = \begin{cases} \sum_{(i,j):\hat{w}_{ij} \geq \hat{w}_{i^l,j^l}} (\hat{w}_{ij} - p_0) y_{ij}, & \text{if } p_0 \in [e_{l-1}, e_l], l = 1, \cdots, |J|; \\ 0, & \text{if } p_0 \in [e_{|J|}, \infty). \end{cases}$$

Thus, the optimal objective value of (6) is $\min_{l=1,\cdots,|J|,|J|+1}\{Z^l\}$, where

$$Z^l = \min_{\{p_0\in[e_{l-1},e_l],(x_{ik},y_{ij})\in X\}}\left(\Gamma p_0 + \sum_{i=1}^{N}\sum_{j=i+1}^{N} w_{ij}y_{ij} + \sum_{(i,j):\hat{w}_{ij}\geq\hat{w}_{il,jl}} (\hat{w}_{ij}-p_0)y_{ij}\right), \quad (12)$$

for $l = 1,\cdots,|J|$, and

$$Z^{|J|+1} = \min_{\{p_0\geq e_{|J|},(x_{ik},y_{ij})\in X\}} \Gamma p_0 + \sum_{i=1}^{N}\sum_{j=i+1}^{N} w_{ij}y_{ij}.$$

For $l = 1,\cdots,|J|$, since the objective function (12) is linear over the interval $p_0 \in [e_{l-1},e_l]$, the optimal is either at the point $p_0 = e_{l-1}$ or $\dot{p}_0 = e_l$. For $l = |J|+1$, Z^l is obtained at the point $e_{|J|}$ since $\Gamma \geq 0$.

Thus, the optimal value $\min_{l=1,\cdots,|J|,|J|+1}\{Z^l\}$ with respect to p_0 is obtained among the points $p_0 = e_l$ for $l = 0,1,\cdots,|J|$. Let G^l be the value at point $p_0 = e_l$ in (12), i.e.,

$$G^l = \Gamma e_l + \min_{(x_{ik},y_{ij})\in X}\left\{\sum_{i=1}^{N}\sum_{j=i+1}^{N} w_{ij}y_{ij} + \sum_{(i,j):\hat{w}_{ij}\geq e_{l+1}} (\hat{w}_{ij}-e_l)y_{ij}\right\}.$$

We finish the proof.   □

As shown in Theorem 2, $G^{|J|} = \Gamma e_{|J|} + \sum_{i=1}^{N}\sum_{j=i+1}^{N} w_{ij}y_{ij}$ is the original nominal problem. Our Algorithm is based on this theorem.

Algorithm
Step 1: For $l = 0,1,\cdots,|J|$, solving G^l in (10);
Step 2: Let $l^* = \arg\min_{l=0,1,\cdots,|J|} G^l$;
Step 3: Then $\{x_{ik}^*,y_{ij}^*\} = \{x_{ik},y_{ij}\}^{l^*}$.

Similarly, for the robust critical node problem, we have the following theorem, and we omit the proof here.

Theorem 3. *Solving robust critical node problem (9) is equivalent to solve the $|J|+1$ problems H^ls for $l = 0,1,\cdots,|J|$, where the problem H^l is formulated as follows:*

$$H^l = \Gamma e_l + \min_{(v_i,u_{ij})\in Y}\left\{\sum_{i=1}^{N}\sum_{j=i+1}^{N} w_{ij}u_{ij} + \sum_{(i,j):\hat{w}_{ij}\geq e_{l+1}} (\hat{w}_{ij}-e_l)u_{ij}\right\}. \quad (13)$$

In the case of $\Gamma = 0$, which means none of w_{ij}s is allowed to change, both RGPP and RCNP become nominal problems (2) and (4), respectively.

3   Networks Analysis by GPP and CNP

Before presenting the relations of GPP, CNP and networks, we first introduce the two important properties in networks: community structure and influential individuals. As mentioned in [9], many systems take the form of networks, such as co-author networks, telecommunication networks, supply chain networks, Internet and Worldwide Web, power grids, networks in social society, as well as many biological networks, including neural networks, food webs and metabolic networks. Two well-known networks that have been studied much are scale-free networks, which mean the degree distribution of them follows a power law, and small-world networks, or known as six degrees of separation.

If a network can be divided into several groups such that the nodes within a group have denser connections than nodes from different groups, this network is said to have **community structure**. A **community** is the group of nodes in such division. For example, in a co-author network, the groups may mean different research scientists under different research topics. Mathematically, in a graph $G = (V, E)$ for a network, the vertex set can be divided according to the weights between them into several subsets such that the vertices within a subset have heavier sum weights of edges than that among distinct subsets. The partitioning of graph is exactly the process to detect community structure in a network while the subsets are corresponding to communities. As we have mentioned above, the graph partitioning is NP-complete and is hard to solve. Thus, community detection in a network, especially for arising complex networks for some real world problems, is a difficult task.

Despite of these difficulties, several methods for community detection have been developed and employed with application in different areas successfully, such as minimum-cut method [11], hierarchical clustering, Girvan-Newman algorithm [9], modularity maximization [12], and clique percolation method [13]. Our proposed method graph partitioning can be considered as a generalization of minimum-cut method with more than two subsets. It is different from clique percolation method where each subset is a complete graph. Although these methods work in different situations, we still have the question such that how many possible groups we have, and what the sizes of groups are in a network. These are the parameters K and $n_1, \cdots, n_K$ required in graph partitioning. In graph partitioning, the paper [6] has discussed how to obtain K and presented that the cardinalities $n_1, \cdots, n_K$ can be relaxed to an interval region $[C_{min}, C_{max}]$.

In fact, the constraint of the sum $\sum_{k=1}^{K} x_{ik} = 1$ requires each vertex is partitioned into one and only one subset. The $\sum_{i=1}^{N} x_{ik} = n_k$ defines the size of the kth subset. Since $x_{ik} \in \{0, 1\}$, the sum $\sum_{i=1}^{N} x_{ik}$ takes integer values between the lower size bound C_{min} and the upper bound C_{max}. These two bounds are known parameters and can be chosen roughly from $\{1, \cdots, N\}$. The two kinds of constraints ensure that each vertex belongs to exactly one subset and all vertices have corresponding subsets. The later one is guaranteed by the fact that $\sum_{k=1}^{K} n_k = \sum_{k=1}^{K} \sum_{i=1}^{N} x_{ik} = \sum_{i=1}^{N} \sum_{k=1}^{K} x_{ik} = \sum_{i=1}^{N} 1 = N$, which means that n_k can take any integer values in $[C_{min}, C_{max}]$ but their sum is fixed as N. Thus, in analyzing networks, we only need the information about the rough region of the sizes for communities.

The weight of an edge in a graph describes the closeness of the connections between two nodes in a network. However, the weight is not always certain. It has dynamic changes along the time. The proposed robust optimization model of graph partitioning is to deal with such uncertainty.

Another property arising in networks is the **influential individuals**, which represent the most important nodes in networks. For example, in a co-author network, some scientists may have bigger contributions to or influences in the research society, and they can be considered as influential nodes in this network. In a supply chain network, some logistics centers have the most important positions for satisfying the supply and demand in the supply chain and these centers are the influential nodes. Obviously, detecting these nodes are important for logistics companies so that they can design emergency plans early before having possible destroy or other problems arising in these centers. Different from wide studies in community structures in networks, the research in influence individuals is rare.

In graph theory, the centrality is used to determine the relative importance of a vertex within the graph [7,14]. Four measures of centrality include degree centrality, betweenness, closeness and eigenvector centrality. Different from the methods used in [10,17], we use the critical node detection methods in graphs [1] and the importance of vertices is measured based on the connectivity. A node is said to have influence in the network if deletion of it results in the maximum number of disconnect components in the network. Similarly, the weights between vertices are always uncertain, and we also construct the robust optimization model of critical node detection.

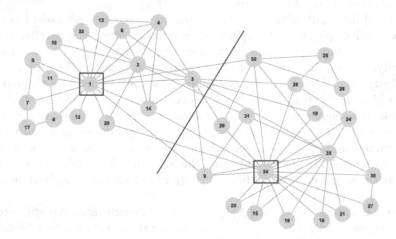

Fig. 1. Zachary's karate club with two communities and two influential nodes: 1, 34

When concentrating on some networks, we always find that several dense groups appear and these groups always have centers within them. Using the concepts mentioned above, we say that this is the phenomenon of community with influence individual center. This is the reason why we use both GPP and CNP to analyze the networks in the meantime.

Given a network with the graph notation $G = (V, E)$ with the weight matrix W to measure the closeness of connectivity between nodes, we are deciding to detect K communities and also K influential individuals in the network.

In addition, we can study the influential individuals within each community. After obtaining the community structure of the network, we use the the CNP model on each community and the two or three influential individuals with each community can be found. These parameters $K(K \in \{2, \cdots, N-1\}), n_1, \cdots, n_K$ for analyzing are obtained from experiences or direct observations. For robust models, we assume that every weight w_{ij} of edge (v_i, v_j) has the uncertainty which is a more close description of real networks. In next section, we discuss several numerical experiments on networks.

4 Numerical Experiments

In this section, we consider a well-known social networks: Zachary's karate club [15] and a generated artificial network with uncertainties on edges. For the network of Zachary's karate club, we use GPP and CNP models to identify the two groups in this club and the leaders, respectively. For the artificial network, we use robust models of GPP and CNP to study its community structures and influential individuals. The algorithms based on our models are implemented using CPLEX 11.0 via ILOG Concert Technology 2.5, and all computations are performed on a SUN UltraSpace-III with a 900 MHz processor and 2.0 GB RAM. Computational times are reported in CPU seconds.

The Zachary's karate club [15] is a network consists of 34 nodes and 78 links representing friendships between members of the club over two years' period. In the study of this network [15], a disagreement, between administrator of the club and the club's instructor, resulted two groups. One group is the members leaving with the instructor to start a new club. The corresponding parameters are $K = 2, n_1 = 16, n_2 = 18$ in our models of GPP and CNP. It shows that our algorithm based on GPP model can find the two groups correctly as shown in Fig. 1, which is the same as that in the study of [9]. In addition, by our model of CNP, we correctly find the two influential individuals with in two groups: instructor (node 1) and the administrator (node 34). If assume $K = 5$ in our CNP model, we can find influential nodes 1, 3, 2 in the first group and nodes 34, 33 in the second group, which are named as core vertices in [9].

In the following, we use Matlab to generate an artificial network with 22 nodes and 48 links with weight in interval ranges (see Table 1). The values of w_{ij} and $\hat{w}_{ij}$ in

Table 1. Uncertainties of 48 links

Edge	$[l_{ij}, u_{ij}]$	Edge	$[l_{ij}, u_{ij}]$	Edge	$[l_{ij}, u_{ij}]$	Edge	$[l_{ij}, u_{ij}]$
(1, 2)	[0.11, 0.67]	(4, 10)	[0.62, 1.54]	(6, 17)	[1.49, 2.97]	(9, 20)	[0.21, 3.01]
(1, 3)	[0.07, 0.49]	(4, 14)	[0.78, 0.98]	(6, 19)	[0.79, 0.95]	(10, 20)	[0.69, 0.71]
(1, 6)	[0.80, 3.50]	(4, 20)	[0.34, 1.78]	(6, 21)	[0.43, 0.93]	(11, 20)	[0.36, 1.48]
(1, 7)	[0.50, 1.28]	(5, 13)	[0.29, 1.15]	(7, 10)	[0.35, 0.79]	(12, 13)	[0.35, 2.01]
(1, 17)	[0.65, 2.17]	(5, 14)	[1.35, 0.53]	(7, 18)	[0.01, 1.03]	(12, 19)	[0.78, 3.70]
(1, 18)	[0.29, 0.22]	(5, 15)	[0.45, 1.73]	(7, 20)	[0.30, 0.66]	(12, 22)	[1.04, 2.12]
(1, 19)	[0.08, 0.22]	(5, 16)	[0.86, 3.32]	(8, 11)	[1.37, 0.63]	(13, 14)	[0.40, 0.90]
(2, 11)	[0.71, 1.41]	(5, 19)	[0.94, 1.18]	(8, 13)	[0.22, 0.44]	(13, 21)	[0.39, 0.59]
(2, 20)	[1.21, 1.63]	(5, 21)	[0.94, 2.20]	(8, 20)	[0.14, 2.20]	(14, 15)	[0.79, 2.63]
(3, 8)	[0.27, 1.69]	(5, 22)	[0.70, 1.16]	(9, 11)	[0.67, 3.41]	(15, 16)	[0.50, 0.66]
(3, 9)	[0.42, 1.40]	(6, 7)	[0.51, 0.77]	(9, 13)	[0.07, 0.17]	(17, 18)	[0.56, 1.26]
(4, 5)	[0.76, 0.86]	(6, 12)	[0.48, 0.80]	(9, 15)	[0.51, 0.71]	(21, 22)	[0.84, 4.24]

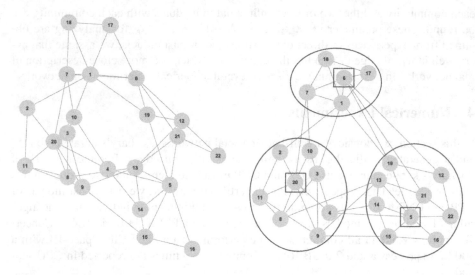

Fig. 2. An artificial network with 22 nodes and 48 links

problems (5) and (8) can be easily computed from the interval values $[l_{ij}, u_{ij}]$. We use robust optimization models of GPP and CNP to analyze this network (see Fig. 2). In this network, 3 communities are found with 5 nodes, 8 nodes and 9 nodes respectively. Assume $K = 3$ and $\Gamma = 48$ in RCNP, three influential nodes (nodes 5, 6, and 20) are found, and each of them is in a community. Observing the node 1 and node 6 in the same group, there are 6 edges incident with node 6 and 7 edges incident with node 1. The weighted degree for node 6 is in $[4.50, 9.92]$, while the weighted degree for node 1 is in $[2.50, 8.55]$. Node 6 is chosen as the influential node in this group in the worst cases $(4.50 > 2.50$ and $9.92 > 8.55)$.

Next, we compare the methods proposed in Section 2.3 with the direct method in CPLEX to solve the equivalent formulations (6) and (9). In Table 2, we present the computational seconds and computational results. The methods in Theorem 2 and

Table 2. Comparisons of two computational methods

Graphs and Parameters						RGPP				RCNP			
						CPLEX (6)		Method Thm 2		CPLEX (9)		Method Thm 3	
N	r	$[l,u]$	$\|J\|$	Γ K	$n_1, \cdots, n_K$	Seconds	Results	Seconds	Results	Seconds	Results	Seconds	Results
10	0.1	[0,1]	4	2 3	3,3,4	0	0	0	0	0.04	0	0.08	0
	0.2		9	5 3	3,3,4	0.08	**2.55**	0.03	**2.55**	0.03	**2.46**	0.02	**2.46**
	0.3		13	7 3	3,3,4	0.13	**6.29**	0.10	6.55	0.03	**4.22**	0.12	**4.22**
20	0.1	[0,1]	19	10 4	4,5,5,6	0.36	**6.39**	0.28	7.07	0.65	**8.66**	0.41	9.46
	0.2		37	19 4	4,5,5,6	9.07	**22.10**	5.05	**22.10**	1.58	**26.30**	1.66	28.63
	0.3		57	29 4	4,5,5,6	35.93	**43.25**	14.37	43.77	1.31	**41.16**	1.35	41.16
30	0.1	[0,1]	42	22 4	5,7,8,10	10.35	**16.38**	2.02	16.58	18.23	**31.52**	6.87	33.27
40	0.1	[0,1]	78	39 4	8,9,10,13	153.67	**33.85**	194.44	**33.85**	187.42	78.27	49.22	**78.02**
50	0.1	[0,1]	122	61 4	9,12,13,16	>3000	**65.46**	2167.83	67.43	668.66	**118.98**	398.00	120.26

Table 3. Subsets and critical nodes

Graphs		RGPP	RCNP
10	0.1	(2,3,5) (4,7,9) (1,6,8,10)	9 6,10
10	0.2	(1,2,9) (4,7,10) (3,5,6,8)	9 5 10
10	0.3	(2,3,6) (1,5,10) (4,7,8,9)	5 7,9
20	0.1	(3,4,6,12) (2,8,9,13,17) (5,11,14,18,19) (1,7,10,15,16,20)	8 14,18 1
20	0.2	(10,16,17,19) (4,8,11,14,18) (1,3,7,13,15) (2,5,6,9,12,20)	4,18 1 12
20	0.3	(5,6,13,19) (2,3,7,10,16) (8,9,12,14,18) (1,4,11,15,17,20)	13 2 4,11
30	0.1	(11,17,20,21,29) (4,10,13,14,16,19,23) (1,6,7,9,12,22,27,28) (2,3,5,8,15,18,23,24,26,30)	13 12 8,18
40	0.1	(4,7,11,15,18,28,29,40) (1,3,6,8,12,2,4,30,33,39) (2,5,9,10,14,16,20,21,25,38) (13,17,19,22,23,26,27,31,32,34,35,36,37)	39 5 19,23
50	0.1	(4,8,18,20,21,28,30,33,40) (1,9,12,17,22,24,25,26,37,41,42,43) (2,6,11,13,15,16,19,23,31,45,46,48,49) (3,5,7,10,14,27,29,32,34,35,36,38,39,44,47,50)	21 35,38,50

Theorem 3 are also implemented in ILOG Concert Technology 2.5. The gaps for all these methods in CPLEX are set as 0.1. The parameter r in a graph is the density, which is the ratio of the number of edges and the number of possible edges.

From Table 2, for robust graph partitioning problems, the method by Theorem 2 is more efficient than default CPLEX method (6) in most cases; for robust critical node problem, the method by Theorem 3 is also more efficient than default CPLEX method (9) in most cases. The better results are in bold. In Table 3, we present the subsets and critical nodes from the better results. For K critical nodes and K subsets, in many cases, the K critical nodes are distributed in $K - 1$ subsets.

5   Conclusions

In this paper, for a given network modeled by a graph model with certain and uncertain weights of links, we have presented the optimization models for both graph partitioning problem and critical node problem. These models are formulated as binary integer linear programs. An algorithm based on a decomposition method on one variable is presented. After that, we introduce two important structures: community and influential individuals in networks. We have established the relationship between these two structures with GPP and CNP in graph theory.

Because of the uncertainties arising in real social or biological networks, the robust optimization models of GPP and CNP are quite useful to analyze these complex networks. We also present several numerical experiments to analyze the networks by RGPP and RCNP. It shows our models are quite useful. However, because of the NP-completeness of the nominal problems GPP and CNP, the RGPP and RCNP are quite complex in computation. Designing efficient algorithms for such problems is still under discussion. On the other hand, since the real networks in practice are always random and have some properties, such as scale-free and small-world, the further research can concentrate on combining such problems with these propositions.

Moreover, a set of critical nodes has some specific functions in some networks with dynamic situations. For example, the network modeling for epileptic brain is constructed by nonlinear dynamic measurements [16]. Part of the brain has special functions to control the movement of body. This network can be analyzed by our proposed method to find the critical sites of such functional nodes.

References

1. Arulselvan, A., Commander, C.W., Elefteriadou, L., Pardalos, P.M.: Detecting critical nodes in sparse graphs. Computers and Operations Research 36(7), 2193–2200 (2009)
2. Arulselvan, A., Commander, C.W., Pardalos, P.M., Shylo, O.: Managing network risk via critical node identification. In: Gulpinar, N., Rustem, B. (eds.) Risk Management in Telecommunication Networks. Springer, Heidelberg (2010)
3. Bertsimas, D., Sim, M.: The price of robustness. Operations Research 52(1), 35–53 (2004)
4. Boginski, V., Commander, C.W.: Identifying critical nodes in protein-protein interaction networks. In: Butenko, S.I., Chaovilitwongse, W.A., Pardalos, P.M. (eds.) Clustering Challenges in Biological Networks, pp. 153–167. World Scientific, Singapore (2008)
5. Fan, N., Pardalos, P.M., Chinchuluun, A., Pistikopoulos, E.N.: Graph partitioning approaches for analyzing biological networks. In: Mondaini, R.P. (ed.) BIOMAT 2009 International Symposium on Mathematical and Computational Biology, pp. 250–262. World Scientific, Singapore (2010)
6. Fan, N., Pardalos, P.M.: Linear and quadratic programming approaches for the general graph partitioning problem. Journal of Global Optimization 48(1), 57–71 (2010)
7. Freeman, L.C.: A set of measures of centrality based on betweenness. Sociometry 40, 35–41 (1977)
8. Garey, M.R., Johnson, D.S., Stockmeyer, L.: Some simplified NP-complete graph problems. Theor. Comput. Sci. 1, 237–267 (1976)
9. Girvan, M., Newman, M.E.J.: Community structure in social and biological networks. Proc. Natl. Acad. Sci. 99, 7821–7826 (2002)

10. Kempe, D., Kleinberg, J., Tardos, E.: Influential nodes in a diffusion model for social networks. In: Caires, L., Italiano, G.F., Monteiro, L., Palamidessi, C., Yung, M. (eds.) ICALP 2005. LNCS, vol. 3580, pp. 1127–1138. Springer, Heidelberg (2005)
11. Newman, M.E.J.: Detecting community structure in networks. Eur. Phys. J. B 38, 321–330 (2004)
12. Newman, M.E.J.: Fast algorithm for detecting community structure in networks. Phys. Rev. E 69, 066133 (2004)
13. Palla, G., Derenyi, I., Farkas, I., Vicsek, T.: Uncovering the overlapping community structure of complex networks in nature and society. Nature 435, 814–818 (2005)
14. Sabidussi, G.: The centrality index of a graph. Psychometrika 31(4), 581–603 (1966)
15. Zachary, W.W.: An information flow model for conflict and fission in small groups. Journal of Anthropological Research 33, 452–473 (1977)
16. Zhang, J., et al.: Real-time differentiation of nonconvulsive status epilepticus from other encephalopathies using quantitative EEG analysis. a pilot study. Epilepsia 51(2), 243–250 (2010)
17. Zou, F., Zhang, Z., Wu, W.: Latency-bounded minimum influential node selection in social networks. In: Liu, B., Bestavros, A., Du, D.-Z., Wang, J. (eds.) Wireless Algorithms, Systems, and Applications. LNCS, vol. 5682, pp. 519–526. Springer, Heidelberg (2009)

An Efficient Algorithm for Chinese Postman Walk on Bi-directed de Bruijn Graphs

Vamsi Kundeti, Sanguthevar Rajasekaran, and Heiu Dinh

Department of Computer Science and Engineering
University of Connecticut
Storrs, CT 06269, USA
{vamsik,rajasek,hieu}@engr.uconn.edu

Abstract. Sequence assembly from short reads is an important problem in biology. It is known that solving the sequence assembly problem exactly on a bi-directed de Bruijn graph or a string graph is intractable. However finding a Shortest Double stranded DNA string (SDDNA) containing all the k-long words in the reads seems to be a good heuristic to get close to the original genome. This problem is equivalent to finding a cyclic Chinese Postman (CP) walk on the underlying un-weighted bi-directed de Bruijn graph built from the reads. The Chinese Postman walk Problem (CPP) is solved by reducing it to a general bi-directed flow on this graph which runs in $O(|E|^2 \log^2(|V|))$ time.

In this paper we show that the cyclic CPP on bi-directed graphs can be solved without reducing it to bi-directed flow. We present a $\Theta(p(|V|+|E|)\log(|V|)+(d_{max}p)^3)$ time algorithm to solve the cyclic CPP on a weighted bi-directed de Bruijn graph, where $p = \max\{|\{v|d_{in}(v) - d_{out}(v) > 0\}|, |\{v|d_{in}(v) - d_{out}(v) < 0\}|\}$ and $d_{max} = \max\{|d_{in}(v) - d_{out}(v)|\}$. Our algorithm performs asymptotically better than the bi-directed flow algorithm when the number of *imbalanced* nodes p is much less than the nodes in the bi-directed graph. From our experimental results on various datasets, we have noticed that the value of $p/|V|$ lies between 0.08% and 0.13% with 95% probability.

Many practical bi-directed de Bruijn graphs do not have cyclic CP walks. In such cases it is not clear how the bi-directed flow can be useful in identifying contigs. Our algorithm can handle such situations and identify maximal bi-directed sub-graphs that have CP walks. We also present a $\Theta((|V| + |E|)\log(V))$ time algorithm for the single source shortest path problem on bi-directed de Bruijn graphs, which may be of independent interest.

1 Introduction

Sequencing the human genome was one of the major scientific breakthroughs in the last seven years. Analysis of the sequenced genome can give us vital information about the expression of genes, which in turn can help scientists to develop drugs for diseases. Thus sequencing the genome of an organism is of fundamental importance in both medicine and biology. Unfortunately the

W. Wu and O. Daescu (Eds.): COCOA 2010, Part I, LNCS 6508, pp. 184–196, 2010.
© Springer-Verlag Berlin Heidelberg 2010

technology used in major human genome sequencing projects – Human Genome Project (HGP) [1] and Celera [2], was too expensive to be adopted in a large scale. This led to the research on *next-generation sequencing* methods. *Pyrosequencing* technologies such as SOLiD, 454 and Solexa generate a large number of short reads which have acceptable accuracy but are several times cheaper compareted to the Sanger technology adopted in the HGP project.

Directed de Bruijn graph based sequence assembly algorithms such as [3] and [4] seem to handle these short read data efficiently compared to the string graph based algorithms (see e.g., [5]). Unfortunately solving the sequence assembly problem exactly on both these graph models seems intractable [6]. However heuristics such as finding a shortest string which includes all the k-mers (sub strings of length k) seem to yield results close to the original genome. In the case of directed de Bruijn graphs finding an Eulerian tour seems to yield good results. If the graph is not Eulerian then a Chinese Postman (CP) tour has been suggested in [4]. To account for the double strandedness of the DNA molecule we need to simultaneously search for two complimentary CP tours. In [6] the directed de Bruijn graphs are replaced with bi-directed de Bruijn graphs to find two complimentary CP tours simultaneously. A CP tour on the *un-weighted* bi-directed graph constructed from the reads serves as a solution to the *Shortest Double Stranded DNA* string (SDDNA) problem. The solution presented in [6] solves the SDDNA problem by reducing it to a general weighted bi-directed flow problem. This algorithm runs in $O(|E|^2 \log^2(V))$ time.

In this paper we present algorithms for SDDNA/CPP on bi-directed de Bruijn graphs without using a bi-directed flow algorithm. Our algorithms are based on identifying shortest bi-directed paths and use of weighted bi-partite matching. Our algorithms perform asymptotically better than the bi-directed flow algorithm when the *imbalanced* nodes in the bi-directed graphs are much smaller in number than $|V|$. This restriction seems to be true in practice from what we have observed in our experiments. On the other hand it turns out that in many practical situations these bi-directed de Bruijn graphs fail to have *cyclic* CP tours. In these cases it is not clear how the bi-directed flow algorithm [6] can help us in identifying a set of *contigs* covering every k-long word at least once. In contrast to this flow algorithm, our algorithm can be useful in obtaining a *minimal* set of contigs when a *cyclic* CP tour does no exist. We now summarize our results as follows. Firstly our deterministic algorithm to solve the *cyclic* CPP on a general bi-directed graph takes $\Theta(p(|V| + |E|) \log(|V|) + (d_{max} p)^3)$ time, where $d_{max} = \max\{|d_{in}(v) - d_{out}(v)|, v \in V\}$, $p = \max\{|V^+|, |V^-|\}$, $V^+ = \{v|v \in V, d_{in}(v) - d_{out} > 0\}$ and $V^- = \{v|v \in V, d_{in}(v) - d_{out} < 0\}$. Secondly we solve the SDDNA problem on an un-weighted bi-directed de Bruijn graph deterministically in $\Theta(p(|V|+|E|) + (d_{max} p)^3)$ time. As a consequence we also present a $\Theta((|V| + |E|) \log(V))$ time single source shortest bi-directed path algorithm, which may be of independent interest to some assembly algorithms such as Velvet [3] – TourBus heuristic.

The organization of the paper is as follows. In Section 2 we provide some preliminaries. Section 3 defines the CPP and SDDNA problems. In Section 4 we introduce our algorithm for single source shortest bi-directed paths, which is used as a component in our main algorithm. The main algorithm is introduced in Section 7 along with algorithms for several sub-problems. Section 8 briefly explains how we can handle situations when the bi-directed graphs do not have cyclic CP tours. Finally experimental studies are reported in Section 9.

2   Preliminaries

Let $s \in \Sigma^n$ be a string of length n. Any substring s_j (i.e., $s[j, \ldots j + k - 1], n - k + 1 \geq j \geq 1$) of length k is called a $k-$mer of s. The set of all $k-$mer's of a given string s is called the $k-$spectrum of s and is denoted by $\mathbb{S}(s, k)$. Given a $k-$mer s_j, $\bar{s}_j$ denotes the *reverse compliment* of s_j (e.g., if $s_j = AAGTA$ then $\bar{s}_j = TACTT$). Let $\leq$ be the partial ordering among the strings of equal length, then $s_i \leq s_j$ indicates that string s_i is lexicographically smaller than s_j. Given any $k-$mer s_i, let $\hat{s}_i$ be the lexicographically smaller string between s_i and $\bar{s}_i$. We call $\hat{s}_i$ the *canonical $k-$mer* of s_i. More formally, if $s_i \leq \bar{s}_i$ then $\hat{s}_i = s_i$ else $\hat{s}_i = \bar{s}_i$. A $k-$molecule of a given $k-$mer s_i is a tuple $(s_i, \bar{s}_i)$ consisting of s_i and its reverse compliment $\bar{s}_i$, the first entry in this tuple is called the positive strand and the second entry is called the negative strand.

A *bi-directed* graph is a generalized version of a standard directed graph. In a directed graph every edge ($-\triangleright$ or $\triangleleft-$) has only one arrow head. On the other hand, in a bi-directed graph every edge ($\triangleleft-\triangleright$, $\triangleleft-\triangleleft$, $\triangleright-\triangleleft$ or $\triangleright-\triangleright$) has two arrow heads attached to it. Formally, let V be the set of vertices of a bi-directed graph, $E = \{(v_i, v_j, o_1, o_2) | v_i, v_j \in V \wedge o_1, o_2 \in \{\triangleleft, \triangleright\}\}$ is the set of bi-directed edges in a bi-directed graph $G(V, E)$. A *walk* $w(v_i, v_j)$ between two nodes $v_i, v_j \in V$ of a bi-directed graph $G(V, E)$ is a sequence $v_i, e_{i_1}, v_{i_1}, e_{i_2}, v_{i_2} \ldots v_{i_m}, e_{i_{m+1}}, v_j$, such that for every intermediate vertex $v_{i_l}, 1 \leq l \leq m$, the orientation of the arrow heads on either side is opposite. To make this more clear let $e_{i_l}, v_{i_l}, e_{i_{l+1}}$ be the subsequence in the walk $w(v_i, v_j)$, $e_{i_l} = (v_{i_{l-1}}, v_{i_l}, o_1, o_2), e_{i_{l+1}} = (v_{i_l}, v_{i_{l+1}}, o_1, o_2)$ then for the walk to be valid $e_{i_l}.o_2 = e_{i_{l+1}}.o_1$. If $v_j = v_i$ and $e_{i_1}.o_1 = e_{i_{m+1}}.o_2$ then the walk is called *cyclic*. A walk on the bi-directed graph is referred to as a *bi-directed walk*. We define an orientation function $\mathcal{O} : V^2 \rightarrow \{\triangleright, \triangleleft\}^2$ which gives the orientation of the bi-directed edge between a pair of vertices – if one exists . For instance if $(v_i, v_j, \triangleleft, \triangleright)$ is a bi-directed edge between v_i and v_j then $\mathcal{O}(v_i, v_j) = \triangleleft-\triangleright$. An edge which is adjacent on a vertex with an orientation $\triangleright$ ($\triangleleft$) is called an *incoming* (*outgoing*) edge. The incoming(outgoing) degree of a vertex v is denoted by $d_{in}(v)$ ($d_{out}(v)$). A vertex v is called *balanced* iff $d_{in}(v) - d_{out}(v) = 0$. A vertex is called *imbalanced* iff $|d_{in}(v) - d_{out}(v)| > 0$. The imbalance of a vertex is called *positive* iff $d_{in}(v) - d_{out}(v) > 0$. Similarly a vertex is *negative* imbalanced iff $d_{in}(v) - d_{out}(v) < 0$. A bi-directed graph is called *connected* iff every pair of vertices have a bi-directed walk between them.

A de Bruijn graph $D^k(s)$ of the order k on a given string s is defined as follows. The vertex set V of $D^k(s)$ is defined as the $k-$spectrum of s (i.e., $V = \mathbb{S}(s, k)$). We use the notation $\mathrm{suf}(v_i, l)(\mathrm{pre}(v_i, l))$ to denote the suffix(prefix) of length l in string v_i. The symbol . denotes concatenation between two strings. Finally the set of directed edges E of $D^k(s)$ is defined as follows $E = \{(v_i, v_j)|\operatorname{suf}(v_i, k-1) = \operatorname{pre}(v_j, k-1) \wedge v_i[1].\operatorname{suf}(v_i, k-1).v_j[k] \in \mathbb{S}(s, k+1)\}$. We can further generalize the definition of a de Bruijn graph $B^k(S)$ on a set $S = \{s_1, s_2 \ldots s_n\}$ of strings, $V = \cup_{i=1}^n \mathbb{S}(s_i, k)$ and $E = \{(v_i, v_j)|\operatorname{suf}(v_i, k-1) = \operatorname{pre}(v_j, k-1) \wedge \exists l : v_i[1].\operatorname{suf}(v_i, k-1).v_j[k] \in \mathbb{S}(s_l, k+1)\}$.

To model the double strandedness of the DNA molecules we should also consider the reverse compliments ($\bar{S} = \{\bar{s}_1, \bar{s}_2 \ldots \bar{s}_n\}$) while we build the de Bruijn graph. To address this a bi-directed de Bruijn graph $BD^k(S \cup \bar{S})$ has been suggested in [6]. The set of vertices V of $BD^k(S \cup \bar{S})$ consists of all the possible $k-$molecules from Σ^k. For every $k+1-$mer $z \in S \cup \bar{S}$, if x, y are the two $k-$mer's of z then an edge is introduced between the $k-$molecules (v_i, v_j) corresponding to x and y. The orientations of the arrow heads on the edges is chosen as follows. If both x, y are the positive strands in v_i, v_j an edge $(v_i, v_j, \triangleright, \triangleright)$ is introduced. If x is a positive strand in v_i and y is a negative strand in v_j an edge $(v_i, v_j, \triangleright, \triangleleft)$ is introduced. Finally if x is a negative strand in v_i and y is a positive strand in v_j an edge $(v_i, v_j, \triangleleft, \triangleright)$ is introduced.

3   Problem Definitions

A *Chinese Postman walk* in a bi-directed graph is a bi-directed walk which visits every edge at least once. A *cyclic Chinese Postman walk* of minimum cost on a weighted bi-directed graph is denoted as CPW. The problem of finding a CPW is referred to as CPP. The problem of finding a CPW on an un-weighted bi-directed de Bruijn graph (of order k) constructed from a set of reads is called the *Shortest Double stranded DNA* string (SDDNA) problem. In this paper we give algorithms for the cyclic CPP and SDDNA problems.

4   Single Source Shortest Path Algorithm on a Bi-directed de Bruijn Graph

We first present an algorithm for the single source shortest path problem on a bi-directed de Bruijn graph. The bi-directed de Bruijn graph in the context of sequence assembly has non-negative weights on the edges. This makes it possible to extend the classic Dijkstra's single source shortest path algorithm to these graphs. In our algorithm we attach two labels for each vertex in the bi-directed graph.

Given a source vertex s, the algorithm initializes all the labels similar to Dijkstra's algorithm. In each stage of the algorithm a label with the smallest cost is picked and some of labels corresponding to adjacent nodes are updated. The only major difference between Dijkstra's algorithm and our algorithm is the way we update the labels. Dijkstra's algorithm updates all the labels/nodes which are adjacent to the smallest label/node currently picked. However our algorithm updates only those labels/nodes which are consistent with the bi-directed walk property. We now give details of our algorithm and prove its correctness.

Let $G = (V, E)$ be the bi-directed graph of interest. Also let s be the source and t be the destination. We are interested in finding a *shortest bi-directed* walk from s to t. We introduce two labels $dist^+[u]$, $dist^-[u]$ for every vertex $u \in V$. The algorithm first initializes labels corresponding to the source (i.e. $dist^+[s]$ and $dist^-[s]$) to zero. Along with these labels of all the nodes adjacent to s are also initialized with the corresponding edge weight. The orientation of the edge determines the label we use for initialization. For instance, if (s, v) is a bi-directed edge with $\triangleright\!-\!\triangleleft$ as the orientation, the label $dist^-[v]$ is initialized with $w_{s,v}$ and $dist^+[v]$ is left uninitialized. In contrast, if the orientation of the edge is $\triangleright\!-\!\triangleright$ then $dist^+[v]$ is initialized to $w_{s,v}$ and $dist^-[v]$ is left uninitialized. All the uninitialized labels contains ∞ by default.

In each iteration of the algorithm a label with the minimum cost is picked. Since we have two types of labels, the minimum label can come from either $dist^+$ or $dist^-$. In the first case let u^+ be the node corresponding to the minimum label during the iteration. This intuitively means that we have a path from s to u^+ and the orientation of the edge adjacent to u^+ in this path is either $\triangleleft\!-\!\triangleright$ or $\triangleright\!-\!\triangleright$. We are going to prove this fact later in the correctness. On the other hand if u^+ is different from the destination t, then u^+ may possibly appear as an internal node in the shortest bi-directed walk between s and t. In this case the path through u^+ should satisfy *bi-directed walk constraint*. Thus we should explore only those node(s) adjacent to u^+ with an edge(s) orientated as $\triangleright\!-\!\triangleleft$ or $\triangleright\!-\!\triangleright$. The orientation of the edge determines the type of the label we need to update – similar to the label initialization. For instance let (u^+, v) be an edge adjacent on u^+ with an orientation of $\triangleright\!-\!\triangleleft$. In this case we should use label $dist^-[v]$ to make an update. Similarly if the orientation of the same edge is $\triangleright\!-\!\triangleright$ then $dist^+[v]$ is used in the update process. Consistent with the classical terminology of the Dijkstra's algorithm, we refer to the minimum cost label picked in each iteration as the *permanent label*. For instance if a label $dist^-[v]$ is picked to be the minimum label in an iteration then we call $dist[v]$ as the *permanent label* of node v. Now to prove the correctness of the algorithm. It is sufficient to show that the cost on the permanent label of a node in each iteration is the weight of the shortest bi-directed path from s to that node.

Theorem 1. *The permanent label of a node $u \in V$ in each iteration of **Algorithm 1** is the weight of the shortest bi-directed path from s to u.*

Proof. We prove the statement by induction on the number (n) of iterations in Algorithm 1. We now prove the base case when $n = 1$. Since we have initialized $dist^+[s] = dist^-[s] = 0$ and the values of the remaining both initialized and uninitialized nodes are > 0; the first iteration picks s and zero is trivially the cost of shortest bi-directed path form s to s.

Assume that the statement is true for $n = 1 \ldots k$. As per the induction hypothesis the permanent labels $dist[s], dist[v_{i_2}] \ldots dist[v_{i_k}]$ correspond to the costs of the shortest bi-directed paths between s and $s, v_{i_2} \ldots v_{i_k}$.

Now let $dist'[v_{i_{k+1}}] < dist[v_{i_{k+1}}]$ be the cost of the shortest bi-directed walk from s to $v_{i_{k+1}}$. Also let $s, v_{j_2} \ldots v_{j_k}, v_{i_{k+1}}$ be the path corresponding to the cost $dist'[v_{i_{k+1}}]$. Note that v_{j_k} cannot be one of the nodes with a permanent label. If

---

Algorithm 1. Algorithm to find the shortest bi-directed path from s to t

---

INPUT   : Bi-directed graph $G = (V, E)$ and two vertices $s, t \in V$
OUTPUT: Cost of the shortest bi-directed path between s and t

1
2 $dist^+[s] = dist^-[s] = 0$
3 $dist^+[v] = dist^-[v] = \infty \ \forall v \in V \wedge v \neq s$
4
5 **while** $dist^+ \neq \phi$ *or* $dist^- \neq \phi$ **do**
6       $u^+ = \min_u\{dist^+\}$
7       $u^- = \min_u\{dist^-\}$
8
9   **if** $u^+ = t$ *or* $u^- = t$ **then**
10 | **return** $\min\{dist^+[u^+], dist^-[u^-]\}$
11
12
13   **if** $dist^+[u^+] < dist^-[u^-]$ **then**
14        $U^+ = \{v|(u^+, v) \in E \wedge (\mathcal{O})(u^+, v) = \vartriangleleft\text{-}\vartriangleleft)\}$
15        $U^- = \{v|(u^+, v) \in E \wedge (\mathcal{O})(u^+, v) = \vartriangleleft\text{-}\vartriangleright)\}$
16        $dist[u^+] = dist^+[u^+]$
17        $dist^+ = dist^+ - \{u^+\}$
18   **else**
19        $U^+ = \{v|(u^-, v) \in E \wedge (\mathcal{O})(u^-, v) = \vartriangleright\text{-}\vartriangleleft)\}$
20        $U^- = \{v|(u^-, v) \in E \wedge (\mathcal{O})(u^+, v) = \vartriangleright\text{-}\vartriangleright)\}$
21        $dist[u^-] = dist^-[u^-]$
22        $dist^- = dist^- - \{u^-\}$
23
24
25   **foreach** $u \in dist^+$ **do**
26 | $dist^+[u] = \min\{dist^+[u], dist^+[u^+] + w[u^+, u]\}$
27   **foreach** $u \in dist^-$ **do**
28 | $dist^-[u] = \min\{dist^-[u], dist^-[u^-] + w[u^-, u]\}$
29
30
31 return ∞

---

not, we would have $dist'[v_{i_{k+1}}] = dist[v_{i_{k+1}}]$ (because we should have updated v_{k+1} when the v_{j_k} was given a permanent label) which is a contradiction. Now let $dist'[v_{j_k}]$ be the cost of the shortest path from s to v_{j_k}. Clearly, $dist'[v_{j_k}] < dist[v_{i_{k+1}}]$ and this means that none of the nodes $v_{j_2}, v_{j_3} \ldots v_{j_k}$ haa a permanent label. Since in the iteration $n = 1$ the algorithm updated the labels adjacent to all the nodes this means that either $dist^+[v_{j_2}]$ or $dist^-[v_{j_2}]$ should have a cost $0 < w_{s,j_2}$ and $dist'[v_{i_{k+1}}] \geq w_{s,j_2}$. In each iteration from $n = 1, \ldots, (k + 1)$ we picked the globally minimum label $dist[v_{i_{k+1}}] < w_{s,j_2} \leq dist'[v_{i_{k+1}}]$ which is a contradiction. $\square$

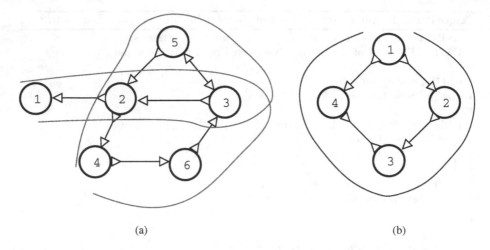

(a)   (b)

Fig. 1. (a) node 4 contains two bi-directed walks from node 1, the green colored path is the shortest.**(b)** the walk starting from node 1 and ending at node 1 is a Chinese walk but not a cyclic Chinese walk.

We now give a simple example to illustrate the algorithm. Consider the bi-directed graph in Figure 1(a), with a unit weight on every edge. Let $s = 1$ and $t = 4$ for instance. From Figure 1(a) we see two bi-directed walks – *red, green*. The *green* path is the shortest path of length 4 units.

5   Terminal Oriented Shortest Bi-directed Walks

In the previous section we have seen how to find a shortest bi-directed walk between two nodes in a given bi-directed graph. We now define a *terminal oriented bi-directed walk* as follows. Let $w(v_i, v_j) = v_i, e_{i_1}, v_{i_1}, e_{i_2}, v_{i_2} \ldots v_{i_m}, e_{i_{m+1}}, v_j$ be any bi-directed walk between two nodes v_i and v_j in a bi-directed graph. Then this bi-directed walk $w(v_i, v_j)$ is called *terminal oriented bi-directed walk* iff $e_{i_1}.o_1 = \triangleright$ and $e_{i_{m+1}}.o_2 = \triangleright$. For example in Figure 1(a) there are two bi-directed walks between nodes 4 and 1 – marked with green and red. However only the green bi-directed walk is terminally oriented. A terminal oriented bi-directed walk w is called the *shortest terminal oriented bi-directed walk* iff there is no other terminal oriented bi-directed walk shorter than w.

5.1   An Algorithm for Finding a Terminal Oriented Shortest Bi-directed Walk

It is easy to modify Algorithm 1 to find a terminal oriented shortest path between s and t. We only have to modify the initialization step and the step which checks if the target node has been reached. During the initialization at line 2 of Algorithm 1 we make $dist^+[s] = 0$ and $dist^-[s] = \infty$. This avoids the exploration of bi-directed walks which does not start with $\triangleright$. In line 9, we stop our exploration only if $u^+ = t$. These changes ensure that the bi-directed walk at s starts with $\triangleright$ and ends with $\triangleright$ at t.

6 A Sufficient Condition for an Eulerian Tour on a Bi-directed Graph

The following Lemma 1 [6] is a sufficient condition for a cyclic Eulerian tour in a bi-directed graph. A bi-directed graph which has a cyclic Eulerian tour is called an Eulerian bi-directed graph.

Lemma 1. *A connected bi-directed graph is Eulerian if and only if every vertex is balanced.*

Note that if a bi-directed graph is Eulerian then a cyclic CP walk is the same as a cyclic Eulerian walk. We emphasize the cyclic adjective for the following reason. Figure 1(b) has a CP walk starting and ending at vertex 1. However the CP walk is not cyclic because the walk starts with $\triangleright$ and ends with $\triangleleft$. The bi-directed graph in Figure 1(b) is not balanced. If the bi-directed graph is not Eulerian, the key strategy to find a cyclic CP walk is to make it Eulerian by introducing multi-edges into the original graph. The hope is that introducing multi-edges would make the bi-directed graph balanced. Thus a cyclic Eulerian walk on a balanced multi-edge bi-directed graph would give a cyclic CP walk on the original graph. Since we are interested in finding a shortest cyclic CP walk, we would like to minimize the number of multi-edges we introduce in the original graph.

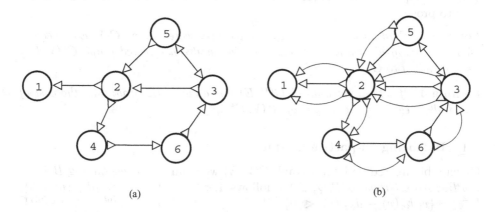

(a)   (b)

Fig. 2. (a) a simple bi-directed graph, (b) a multi-bi-directed graph. Notice that orientations of the multi-edges is the same as the orientation of the original edge.

7 A Deterministic Algorithm to Find a Cyclic CP Walk on a Bi-directed Graph

We now describe our deterministic algorithm to find a cyclic CP walk on a weighted bi-directed graph. First we define a *multi-bi-directed graph* as a bi-directed graph in which an edge between two nodes is overlaid at least once, without changing its orientation. Figure 2(a) shows a bi-directed graph; Figure 2(b) shows a valid multi-bi-directed graph. Notice that while overlaying the edge we did not change its orientation. Since the orientation of the multi-edges is the same as the original edges, any bi-directed walk involving multi-edges is consistent with the bi-directed walk in the original graph. Another important property of the multi-bi-directed graphs is their ability to make the nodes balanced. Notice that the vertex 3 in the original bi-directed graph is positively imbalanced – $d_{in}(v_3) = 2, d_{out}(v_3) = 1$. However in the multi-bi-directed graph in Figure 2(b) we are able to balance vertex 3 by introducing some multi-edges into the original graph. Given a bi-directed graph $G = (V, E)$, let $G^m = (V, E^m)$ be some multi-bi-directed graph corresponding to G. The following Lemma 2 gives a characterization for G to have a cyclic CP walk.

Lemma 2. *A non Eulerian bi-directed graph $G = (V, E)$ has a cyclic Chinese Postman walk $\iff \exists$ a corresponding multi-bi-directed graph $G^m = (V, E^m)$ which is Eulerian.*

Given a multi-bi-directed graph $G^m(V, E^m)$ corresponding to some bi-directed graph $G = (V, E)$, we define the *multi-bi-directed graph weight* as $W(G^m) = \sum_{e \in E^m} c(e)$, where $c : e \in E \to \mathbb{R}^+$ is a cost function on the bi-directed graph $G(V, E)$. We denote $G^*(V, E^*)$ as the minimum weight Eulerian multi-bi-directed graph corresponding to $G(V, E)$ if at all one exists. The following Lemmas are easy to prove.

Lemma 3. *Finding a cyclic CP walk on a bi-directed graph $G(V, E)$ is equivalent to finding a minimum weight Eulerian multi-bi-directed graph $G^*(V, E^*)$ corresponding to G.*

Lemma 4. *If a bi-directed-graph $G(V, E)$ has a cyclic CP walk then the cost of that walk is equal to the weight of $G^*(V, E^*)$.*

7.1 Balancing Bi-partite Graph

Given a bi-directed de Bruijn graph $G(V, E)$ we define a corresponding *Balancing Bi-partite Graph*, $B(P, Q, E^b)$ as follows. Let $V^+ = \{v | \, d_{in}(v) - d_{out}(v) > 0\}$, $V^- = \{v | \, d_{in}(v) - d_{out}(v) < 0\}$. $P = \cup_{p \in V^+} \{p^{(1)}, p^{(2)} \dots p^{(|d_{in}(p) - d_{out}(p)|)}\}$, $Q = \cup_{q \in V^-} \{q^{(1)}, q^{(2)} \dots q^{(|d_{in}(q) - d_{out}(q)|)}\}$. We now introduce an edge between $p^{(i)} \in P$ and $q^{(j)} \in Q$ iff $p, q \in V$ are connected by a *terminal oriented bi-directed walk* from p to q. Let $dist^t(p, q)$ be the weight of this walk. Then $E^b = \{(p^{(i)}, q^{(j)}) | \, dist^t(p, q) \neq \infty \land p, q \in V\}$. The weight of the edge $(p^{(i)}, q^{(j)}) \in E^b$ is the weight of terminal oriented bi-directed walk $dist^t(p, q)$.

Lemma 5. *A non Eulerain bi-directed graph $G(V, E)$ has a cyclic CP walk $\Longleftrightarrow$ the balancing bi-partite graph $B(P, Q, E^b)$ has a perfect match.*

7.2   Constructing a Family of Eulerian Multi-bi-directed Graphs

We now give a construction for generating Eulerian multi-bi-directed graphs corresponding to a given non Eulerian bi-directed graph which has a cyclic CP walk. We call this a *Balancing Match Family* denoted by $\mathcal{F}$. Lemma 5 can be used to generate $\mathcal{F}$. Assume that $G(V, E)$ is a non Eulerian bi-directed graph that has a cyclic CP walk. The following construction generates a family of Eulerian multi-bi-directed graphs corresponding to $G(V, E)$.

- STEP-1: Create a balancing bi-partite graph $B(P, Q, E^m)$ corresponding to $G(V, E)$ by choosing some terminal oriented bi-directed walk between $p^{(i)} \in P$ and $q^{(j)} \in Q$.
- STEP-2: Find a perfect match M_b in $B(P, Q, E^m)$. For each edge in M_b overlay the corresponding terminal oriented bi-directed walk on $G(V, E)$. This generates a Eulerian multi-bi-directed graph $G^m(V, E^m)$.

The following Lemma 6 is easy to see.

Lemma 6. *If $G(V, E)$ is a non Eulerian bi-directed graph that has a cyclic CP walk, then every corresponding Eulerian multi-bi-directed graph $G^m(V, E^m)$ belongs to the family $\mathcal{F}$.*

The following Lemma gives an expression for the weight of any $G^m(V, E^m) \in \mathcal{F}$.

Lemma 7. *Let $G(V, E, c)$ be a non Eulerian weighted bi-directed graph which has a cyclic CP walk $c : E \to \mathbb{R}^+$. Let $G^m(V, E^m, c) \in \mathcal{F}$ be some Eulerian multi-bi-directed graph. Then, $\mathcal{W}(G^m) = \sum_{e \in E} c(e) + \sum_{(p^{(i)}, q^{(j)}) \in M_b} dist^t(p, q)$, where M_b is a perfect match in $B(P, Q, E^b)$.*

7.3   An Algorithm for Finding an Optimal Cyclic CP Walk

We now put together all the results in the preceding sub-section(s) to give an algorithm to find $G^*(V, E^*)$. The algorithm is summarized in the following steps.

- STEP-1: We first identify positive and negative imbalanced nodes in G. Let $V^+ = \{v | d_{in}(v) - d_{out}(v) > 0\}$, $V^- = \{v | d_{in}(v) - d_{out}(v) < 0\}$
- STEP-2: Find the cost of a *terminal oriented shortest bi-directed walk* between every pair $(v, u) \in V^+ \times V^-$. Let this cost be denoted as $dist^t(v, u)$.
- STEP-3: Create a *balancing bi-partite graph* $B(P, Q, E^b)$ as follows. Let $P = \cup_{v \in V^+} \{v^{(1)}, v^{(2)}, \ldots, v^{(|d_{in}(v) - d_{out}(v)|)}\}$, $Q = \cup_{u \in V^-} \{u^{(1)}, u^{(2)}, \ldots, u^{(|d_{in}(u) - d_{out}(u)|)}\}$, $E = \{(v^{(i)}, u^{(j)}) | v^{(i)} \in P \wedge u^{(j)} \in Q\}$. The cost of an edge $c(v^{(i)}, u^{(j)}) = dist^t(v, u)$.

- STEP-4: Find a minimum cost perfect match in B. Let this match be M_b. If B does not have a perfect match then G does not have a cyclic CP walk.
- STEP-5: For each edge $(v^{(i)}, u^{(j)}) \in M_b$, overlay the terminal oriented shortest bi-directed walk between v and u in the $G(V, E)$. After overlaying all the terminal oriented bi-directed walks from M_b on to $G(V, E)$ we obtain $G^*(V, E^*)$. We will prove that it is optimal in Theorem 2.

Theorem 2. *If $G(V, E)$ is a bi-directed graph that has a cyclic CP walk, then the cost of this cyclic CP walk is equal to $\mathcal{W}(G^*) = \sum_{e \in E} c(e) + \sum_{(v^{(i)}, u^{(j)}) \in M_b} dist^t(v, u)$. Here M_b is the min-cost perfect match in the balancing bi-partite graph B.*

7.4   Runtime Analysis of the Algorithm to Find a Cyclic CP Walk

Let $p = \max\{|V^+|, |V^-|\}$ and $d_{max} = \max_{v \in V}\{|d_{in}(v) - d_{out}(v)|\}$. STEP-2 of the algorithm runs in $\Theta(p(|V| + |E|) \log(|V|))$ time to compute $dist^t(v, u)$. In STEP-3 $|P| \leq d_{max}p$, $|Q| \leq d_{max}p$. For STEP-4 Hungarian method can be applied to solve the weighted matching problem in $\Theta((d_{max}p)^3)$ time. So the total runtime of this deterministic algorithm is $\Theta(p(|V| + |E|) \log(|V|) + (d_{max}p)^3)$. As mentioned before if p is much smaller than $|V|$ this algorithm performs better than the bi-directed flow algorithm.

7.5   Runtime Analysis of the Algorithm to Find SDDNA

Since SDDNA runs on a bi-directed de Bruijn graph which is un-weighted, STEP-2 of the algorithm runs in $\Theta(p(|V| + |E|))$ time – because we don't need to use a Heap, we just do a BFS on the bi-directed graph. The rest of the analysis for the runtime remains the same and the total run time of the algorithm is $\Theta(p(|V| + |E|) + (d_{max}p)^3)$.

8   Dealing with Practical Bi-directed de Bruijn Graphs with no Cyclic CP Walks

As we have mentioned earlier most of the bi-directed de Bruijn graphs constructed from the reads do not satisfy the sufficient condition for cyclic CP walks. In such cases our algorithm can still be used, by modifying it to find a *maximum* match in the balancing bi-partite graph rather than perfect match. We can introduce a hypothetical node h and connect all the un-matched nodes in the balancing bi-partite graph to h with appropriate bi-directed edges and thus make all the original nodes balanced. We can now find a cyclic CP walk in this hypothetical graph. Every sub-walk in the cyclic CP walk that starts from h and ends at h can be reported as a *contig*. Thus our algorithm is capable of handling cases when the bi-directed graph cannot have a cyclic CP walk.

9   Experimental Results

As we have mentioned in the previous sections the asymptotic complexity of our algorithm depends on p – the maximum of positively and negatively imbalanced nodes. In the case of de Bruijn graphs $d_{max} \leq |\Sigma|$, where $|\Sigma|$ is the size the alphabet from which the strings are drawn. In our case this is exactly four. So we can safely ignore d_{max} in the case of de Bruijn graphs and just concentrate on p. In the rest of the discussion we would like to refer to p as the number of imbalanced nodes. It is clear that p is a random variable with support in $[0, |V|]$. So we would like to estimate the expected number of imbalanced nodes in a graph with $|V|$ bi-directed edges. We estimated the mean of the random variable $\frac{p}{|V|}$ from several samples of bi-directed de Bruijn graphs constructed from reads from a plant genome. A simple t–test is applied to to estimate the 95% confidence interval of $\frac{p}{|V|}$. See Table 1 for the details of the samples used. Notice that as we increase the size of k (de Bruijn graph order) from 21 to 25, the number of imbalanced nodes in columns corresponding to $|V^+|$ and $|V^-|$ reduces. This is because increasing k reduces the number of edges which may reduce the number of imbalanced nodes. On the other hand for a fixed value of k the number of imbalanced nodes increases consistently with the nodes. However the rate of growth is very slow compared to the rate of growth of the number of nodes. Finally we use this evidence to hypothesize that the number of imbalanced nodes in practical bi-directed graphs is only between 0.087% to 0.133% of the number of nodes in the graph, with a probability of 95%.

Table 1. The value of p on short read data from a plant genome sequencing data from CSHL

READS	k	NODES	P-IMBAL	N-IMBAL	BAL-BI-GRAPH			
			$|V^+|$	$|V^-|$	$|P|$	$|Q|$	p	$\frac{p \times 100}{|V|}$
102400	21	1588569	1157	1133	1186	1173	1186	0.075
153600	21	2353171	2240	2141	2298	2211	2298	0.098
204800	21	3097592	3509	3492	3601	3590	3601	0.116
256000	21	3825101	4953	5004	5074	5131	5131	0.134
307200	21	4538734	6719	6748	6878	6912	6912	0.152
358400	21	5235821	8586	8603	8789	8802	8802	0.168
409600	21	5917489	10665	10693	10914	10934	10934	0.185
102400	25	1202962	569	521	588	540	588	0.049
153600	25	1788533	1104	1026	1139	1062	1139	0.064
204800	25	2362981	1744	1708	1788	1759	1788	0.076
256000	25	2927656	2521	2523	2579	2592	2592	0.089
307200	25	3484849	3370	3414	3451	3517	3517	0.101
358400	25	4032490	4333	4369	4441	4485	4485	0.111
409600	25	4571554	5390	5467	5518	5613	5613	0.123

$\left[\bar{x} - z_{\frac{\alpha}{2}} \frac{S}{\sqrt{n}} , \bar{x} + z_{-\frac{\alpha}{2}} \frac{S}{\sqrt{n}} \right]$: 95% C.I for average $\frac{p \times 100}{|V|}$ is $[0.0872\%, 0.1330\%]$

9.1 Implementation and Data

An implementation of the algorithms discussed is available at `http://trinity.engr.uconn.edu/~vamsik/fast_cpp.tgz`.

10 Conclusion and Further Research

In this paper we have given an algorithm for cyclic Chinese Postman walk on a bi-directed de Bruijn graph. Our algorithm is based on identifying shortest bi-directed walks and weighted matching. This algorithm performs asymptotically better than the bi-directed flow algorithm when the number of imbalanced nodes are much smaller than the nodes in the bi-directed graph. On the other hand this algorithm can also handle the instances of bi-directed graphs which does not have a cyclic CP walk and provide a minimal set of walks, cyclic walks which cover every edge in the bi-directed graph at least once.

There are several research directions which can be pursued. Firstly, we need to address how the addition of paired reads may impose new constraints on the cyclic CPP walk. Secondly, while Eulerization of the bi-directed graph we have chosen the shortest path bi-directed path, however this may not correspond to the repeating region in the genome. Other strategy to make the graph Eulerian is to choose the path with maximum read multiplicity. This on other hand may increase the length of the Chinese walk, can we simultaneously optimize these two objectives ?.

Acknowledgements. This work has been supported in part by the following grants: NSF 0326155, NSF 0829916 and NIH 1R01GM079689-01A1.

References

1. Lander, E.S., Linton, L.M., Birren, B., Nusbaum, C., Zody, M.C.e.a.: Initial sequencing and analysis of the human genome. Nature 409, 860–921 (2001)
2. Craig Venter, J., Adams, M.D., Myers, E.W., Li, P.W., Mural, R.J.e.: The sequence of the human genome. Science 291, 1304–1351 (2001)
3. Zerbino, D.R., Birney, E.: Velvet: Algorithms for de novo short read assembly using de bruijn graphs. Genome research 18, 821–829 (2008)
4. Pevzner, P.A., Tang, H., Waterman, M.S.: An eulerian path approach to dna fragment assembly. Proceedings of the National Academy of Sciences of the United States of America 98, 9748–9753 (2001)
5. Myers, E.W.: The fragment assembly string graph. Bioinformatics 21, ii79–ii85 (2005)
6. Medvedev, P., Georgiou, K., Myers, G., Brudno, M.: Computability of models for sequence assembly. In: Giancarlo, R., Hannenhalli, S. (eds.) WABI 2007. LNCS (LNBI), vol. 4645, pp. 289–301. Springer, Heidelberg (2007)

On the Hardness and Inapproximability of Optimization Problems on Power Law Graphs

Yilin Shen, Dung T. Nguyen, and My T. Thai

Department of Computer Information Science and Engineering
University of Florida, Gainesville, FL, 32611
{yshen,dtnguyen,mythai}@cise.ufl.edu

Abstract. The discovery of power law distribution in degree sequence (i.e. the number of vertices with degree i is proportional to $i^{-\beta}$ for some constant β) of many large-scale real networks creates a belief that it may be easier to solve many optimization problems in such networks. Our works focus on the hardness and inapproximability of optimization problems on power law graphs (PLG). In this paper, we show that the MINIMUM DOMINATING SET, MINIMUM VERTEX COVER and MAXIMUM INDEPENDENT SET are still APX-hard on power law graphs. We further show the inapproximability factors of these optimization problems and a more general problem (ρ-MINIMUM DOMINATING SET), which proved that a belief of $(1 + o(1))$-approximation algorithm for these problems on power law graphs is not always true. In order to show the above theoretical results, we propose a general cycle-based embedding technique to embed any d-bounded graphs into a power law graph. In addition, we present a brief description of the relationship between the exponential factor β and constant greedy approximation algorithms.

Keywords: Theory, Complexity, Inapproximability, Power Law Graphs.

1 Introduction

In real life, the remarkable discovery shows that many large-scale networks follow a power law distribution in their degree sequences, ranging from biological networks, the Internet, the WWW to social networks [19] [20]. That is, the number of vertices with degree i is proportional to $i^{-\beta}$ for some constant β in these graphs, which is called power law graphs (PLG). The observations show that the exponential factor β ranges between 1 and 4 for most real-world networks [8]. Intuitively, the following theoretical question is raised: What are the differences in terms of complexity and inapproxamability of several optimization problems between on general graphs and on PLG?

Many experimental results on random power law graphs give us a belief that the problems might be much easier to solve on PLG. Eubank *et al.* [12] claimed that a simple greedy algorithm leads to a $1 + o(1)$ approximation ratio on MINIMUM DOMINATING SET (MDS) problem (without any formal proof) although MDS has been proved NP-hard to be approximated within $(1 - \epsilon) \log n$ unless

W. Wu and O. Daescu (Eds.): COCOA 2010, Part I, LNCS 6508, pp. 197–211, 2010.

NP=ZPP. The approximating result on MINIMUM VERTEX COVER (MVC) was also much better than the 1.366-inapproximability on general graphs [10]. In [22], Gopal claimed that there exists a polynomial time algorithm that guarantees a $1 + o(1)$ approximation of the MVC problem with probability at least $1 - o(1)$. However, there is no such formal proof for this claim either. Furthermore, several papers also have some theoretical guarantees for some problems on PLG. Gkantsidis et al. [14] proved the flow through each link is at most $O(n \log^2 n)$ on power law random graphs (PLRG) where the routing of $O(d_u d_v)$ units of flow between each pair of vertices u and v with degrees d_u and d_v. In [14], the authors take advantage of the property of power law distribution by using the structural random model [1],[2] and show the theoretical upper bound with high probability $1 - o(1)$ and the corresponding experimental results. Likewise, Janson et al. [16] gave an algorithm that approximated MAXIMUM CLIQUE within $1 - o(1)$ on PLG with high probability on the random poisson model $G(n, \alpha)$ (i.e. the number of vertices with degree at least i decreases roughly as n^{-i}). Although these results were based on experiments and random models, it raises an interest in investigating hardness and inapproximability of classical optimization problems on PLG.

Recently, Ferrante et al. [13] had an initial attempt to show that MVC, MDS and MAXIMUM INDEPENDENT SET (MIS) ($\beta > 0$), MAXIMUM CLIQUE (CLIQUE) and MINIMUM GRAPH COLORING (COLORING) ($\beta > 1$) still remain NP-hard on PLG. Unfortunately, there is a minor error in the proof of their Lemma 5 which makes the proof of NP-hardness of MIS, MVC, MDS with $\beta < 1$ no longer hold. Indeed, it is not trivial to fix that error and thus we present in Appendix A another way to show the NP-hardness of these problems when $\beta < 1$.

Our Contributions: In this paper, we show the APX-hardness and the inapproximability of MIS, MDS, and MVC according to a general Cycle-Based Embedding Technique which embeds any d-bounded graph into a power law graph with the exponential factor β. The inapproximability results of the above problems on PLG are shown in Table 1 with some constant c_1, c_2 and c_3. Then, the further inapproximability results on CLIQUE and COLORING are shown by taking advantage of the reduction in [13]. We also analyze the relationship between β and constant greedy approximation algorithms for MIS and MDS.

In addition, recent studies on social networks have led to a new problem of spreading the influence through a social network [18] [17] by initially influencing a minimum small number of people. By formulating this problem as ρ-Minimum Dominating Set (ρ-MDS), we show that ρ-MDS is Unique Game-hard to be approximated within $2 - (2 + o_d(1)) \log \log d / \log d$ factor on d-bounded graphs and further leading to the following inapproximability result on PLG (shown in Table 1).

Organization: In Section 2, we introduce some problem definitions, the model of PLG, and corresponding concepts. In Section 3, the general embedding technique are introduced by which we can use to show the hardness and inapproximability of MIS, MDS, MVC in Section 4 and Section 5 respectively. In addition, the

Table 1. Inapproximability Results on Power Law Graph with Exponential Factor β

Problem	Inapproximability Factor	Condition
MDS	$1 + 2\left(\log c_3 - O(\log \log c_3) - 1\right)/((c_3 + 1)\zeta(\beta))$	$NP \not\subseteq DTIME\left(n^{O(\log \log n)}\right)$
MIS	$1 - 2\left(c_1 - O\left(\log^2 c_1\right)\right)/(c_1(c_1 + 1)\zeta(\beta))$	Unique Game Conjecture
MVC	$1 + 2\left(1 - (2 + o_{c_2}(1))\frac{\log \log c_2}{\log c_2}\right)/((c_2 + 1)\zeta(\beta))$	Unique Game Conjecture
ρ-MDS	$1 + \left(1 - (2 + o_{c_2}(1))\frac{\log \log c_2}{\log c_2}\right)/((c_2 + 1)\zeta(\beta))$	Unique Game Conjecture
CLIQUE	$O\left(n^{1/(\beta+1)-\epsilon}\right)$	$NP \neq ZPP$
COLORING	$O\left(n^{1/(\beta+1)-\epsilon}\right)$	$NP \neq ZPP$

inapproximability result of CLIQUE and COLORING are also shown in Section 5. In Section 6, we analyze the relationship between β and constant approximation algorithms, which further proves that the integral gap is typically small for optimization problems on PLG than that on general bounded graphs. We fix the *NP*-hardness proof for $\beta < 1$ presented in [13] in Appendix A.

2 Preliminaries

This section provides several parts. First, we recall the definition of the new optimization problem ρ-Minimum Dominating Set. Next, the power law model and some corresponding concepts are proposed. Finally, we introduce some special graphs which will be used in the analysis throughout the paper.

2.1 Problem Definitions

The ρ-Minimum Dominating Set is defined as general version of MDS problem. In the context of influence spreading, the ρ-MDS problem says that given a graph modeling a social network, where each vertex v has a fix threshold $\rho|N(v)|$ such that the vertex v will adopt a new product if $\rho|N(v)|$ of its neighbors adopt it. Thus our goal is to find a small set DS of vertices such that targeting the product to DS would lead to adoption of the product by a large number of vertices in the graph in t propagations. To be simplified, we define ρ-MDS problem in the case that $t = 1$.

Definition 1 (ρ-Minimum Dominating Set). *Given an undirected graph* $G = (V, E)$, *find a subset* $DS \subseteq V$ *with the minimum size such that for each vertex* $v_i \in V \setminus DS$, $|DS \cap N(v_i)| \geq \rho|N(v_i)|$, *where* $0 < \rho \leq 1/2$.

2.2 Power Law Model and Concepts

A great number of models [5] [6] [1] [2] [21] on power law graphs are emerging in the past recent years. In this paper, we do the analysis based on the general (α, β) model, that is, the graphs only constrained with the distribution on the number of vertices with different degrees.

Definition 2 ((α, β) Power Law Graph Model). *A graph $G_{(\alpha,\beta)} = (V, E)$ is called a (α, β) power law graph where multi-edges and self-loops are allowed if the maximum degree is $\Delta = \lfloor e^{\alpha/\beta} \rfloor$ and the number of vertices of degree i is:*

$$y_i = \begin{cases} \lfloor e^{\alpha}/i^{\beta} \rfloor, & \text{if } i > 1 \text{ or } \sum_{i=1}^{\Delta} \lfloor e^{\alpha}/i^{\beta} \rfloor \text{ is even} \\ \lfloor e^{\alpha} \rfloor + 1, & \text{otherwise} \end{cases} \tag{1}$$

Definition 3 (d-Bounded Graph). *Given a graph $G = (V, E)$, G is a d-bounded graph if the degree of any vertex is upper bounded by an integer d.*

Definition 4 (Degree Set). *Given a power law graph $G_{(\alpha,\beta)}$, let $D_i(G_{(\alpha,\beta)})$ be the set of vertices with degree i on graph $G_{(\alpha,\beta)}$.*

2.3 Special Graphs

Definition 5 (Cubic Cycle CC_n). *A cubic cycle CC_n is composed of two cycles. Each cycle has n vertices and two i^{th} vertices in each cycle are adjacent with each other. That is, Cubic Cycle CC_n has $2n$ vertices and each vertex has degree 3. An example CC_8 is shown in Figure 1.*

Then a Cubic Cycle CC_n can be extended into a d-Regular Cycle RC_n^d with the given vector d. The definition is as follows.

Definition 6 (d-Regular Cycle RC_n^d). *Give a vector $d = (d_1, \ldots, d_n)$, a d-Regular Cycle RC_n^d is composed of a two cycles. Each cycle has n vertices and two i^{th} vertices in each cycle are adjacent with each other by $d - 2$ multi-edges. That is, d-Regular Cycle RC_n^d has $2n$ vertices and the two i^{th} vertex has degree d_i. An example RC_8^d is shown in Figure 3.*

Definition 7 (d-Cycle C_n^d). *Give a vector $d = (d_1, \ldots, d_n)$, a d-Cycle C_n^d is a cycle with a even number of vertices n such that each vertex has degree d_i with $(d_i - 2)/2$ self-loops. An example C_8^d is shown in Figure 4.*

Definition 8 (κ-Branch-d-Cycle κ-BC_n^d). *Given a d-Cycle and a vector $\kappa = (\kappa_1, \ldots, \kappa_m)$, the κ-Branch-d-Cycle is composed of $|\kappa|/2$ branches appending C_n^d, where $|\kappa|$ is a even number. An example is shown in Figure 5.*

Fact 1. *κ-Branch-d-Cycle has $|\kappa|$ even number of vertices with odd degrees.*

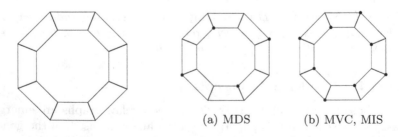

(a) MDS   (b) MVC, MIS

Fig. 1. CC_8 **Fig. 2.** Solutions on CC_8

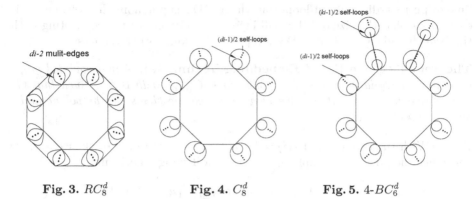

Fig. 3. RC_8^d **Fig. 4.** C_8^d **Fig. 5.** $4\text{-}BC_6^d$

3 General Cycle-Based Embedding Technique

In this section, we present *General Cycle-Based Embedding Technique* on (α, β) power law graph model with $\beta > 1$. The idea on *Cycle-Based Embedding Technique* is to embed an arbitrary d-bounded graph into PLG with $\beta > 1$ with a d_1-Regular Cycle, a κ-Branch-d_2-Cycle and a number of cliques K_2, where d_1, d_2 and κ are defined by α and β. Since the classical problems can be polynomially solved in both d-Regular Cycles and κ-Branch-d-Cycle according to Corollary 1 and Lemma 2, Cycle-Based Embedding Technique helps to prove the complexity of such problem on PLG according to the complexity result of the same problem on bounded graphs.

Lemma 1. *MDS, MVC and MIS is polynomially solvable on Cubic Cycle.*

Proof. Here we just prove MDS problem is polynomially solvable on Cubic Cycle. The algorithm is simple. First we arbitrarily select a vertex, then select the vertex on the other cycle in two hops. The algorithm will terminate until all vertices are dominated. Now we will show that this gives the optimal solution. Let's take CC_8 as an example. As shown in Fig. 2(a), the size of MDS is 4. Notice that each node can dominate exact 3 vertices, that is, 4 vertices can dominate exactly 12 vertices. However, in CC_8, there are altogether 16 vertices, which have to be dominated by at least 4 vertices apart from the vertices in MDS. That is, the algorithm returns an optimal solution. Moreover, MVC and MIS can be proved similarly as shown in Fig. 2(b).

Corollary 1. *MDS, MVC and MIS is polynomially solvable on d-Regular Cycle and d-Cycle.*

Lemma 2. *MDS, MVC and MIS is polynomially solvable on κ-Branch-d-Cycle.*

Proof. Let us take the MDS as an example. First we select the vertices connecting both the branches and the cycle. Then by removing the branches, we will have a line graph regardless of self-loops, on which MDS is polynomially solvable. It is easy to see that the size of MDS will increase if any one vertex connecting both the branch and the cycle in MDS is replaced by some other vertices.

Theorem 1 (Cycle-Based Embedding Technique). *Any d-bounded graph G_d can be embedded into a power law graph $G_{(\alpha,\beta)}$ with $\beta > 1$ such that G_d is a maximal component and the above classical problems can be polynomially solvable on $G_{(\alpha,\beta)} \setminus G_d$.*

Proof. With the given β and $\tau(i) = \lfloor e^\alpha/i^\beta \rfloor - n_i$ where $n_i = 0$ when $i > d$, we construct the power law graph $G_{(\alpha,\beta)}$ as the following algorithm:

1. Choose a number α such that $e^\alpha = \max_{1 \leq i \leq d}\{n_i \cdot i^\beta\}$ and $e^{\alpha/\beta} \geq d$;
2. For the vertices with degree 1, add $\lfloor \tau(1)/2 \rfloor$ number of cliques K_2;
3. For $\tau(2)$ vertices with degree 2, add a cycle with the size $\tau(2)$;
4. For all vertices with degree larger than 2 and smaller than $\lfloor e^{\alpha/\beta} \rfloor$, construct a d_1-Regular Cycle where d_1 is a vector composed of $2\lfloor \tau(i)/2 \rfloor$ number of i elements for all i satisfying $\tau(i) > 0$;
5. For all leftover isolated vertices L such that $\tau(i) - 2\lfloor \tau(i)/2 \rfloor = 1$, construct a d_2^1-Branch-d_2^2-Cycle, where d_2^1 is a vector composed of the vertices in L with odd degrees and d_2^2 is a vector composed of the vertices in L with even degrees.

The last step holds since the number of vertices with odd degrees has to be even. Therefore, $e^\alpha = \max_{1 \leq i \leq d}\{n_i \cdot i^\beta\} \leq n$, that is, the number of vertices in graph $G_{(\alpha,\beta)}$ $N = \zeta(\beta)n = \Theta(n)$ meaning that N/n is a constant. According to Corollary 1 and Lemma 2, since $G_{(\alpha,\beta)} \setminus G_d$ is composed of a d_1-Regular Cycle and a k-Branch-d_2-Cycle, it can be polynomially solvable.

4   Hardness of Optimization Problems on PLG

In this section, we prove that MIS, MDS, MVC are *APX*-hard on PLG.

Theorem 2 (Alimonti *et al.* [3]). *MDS is APX-hard on cubic graphs.*

Theorem 3. *MDS is APX-hard on PLG.*

Proof. According to Theorem 1, we use the *Cycle-Based Embedding Technique* to show $\mathcal{L}$-reduction from MDS on d-bounded graph G_d to MDS on power law graph $G_{(\alpha,\beta)}$. Let ϕ and φ be a feasible solution on G_d and $G_{(\alpha,\beta)}$ respectively.

We first consider MDS on different graphs. Notice that MDS on a K_2 is 1, $n/4$ on a d-Regular Cycle according to Lemma 1 and $n/3$ on a cycle. Therefore, for a solution ϕ on G_d, we have a solution φ on $G_{(\alpha,\beta)}$ is $\varphi = \phi + n_1/2 + n_2/3 + n_3/4$, where n_1, n_2 and n_3 corresponds to $\tau(1)$, $\tau(2)$ and all leftover vertices in Theorem 1. Correspondingly, we have $OPT(\varphi) = OPT(\phi) + n_1/2 + n_2/3 + n_3/4$.

On one hand, for a d-bounded graph with vertices n, the optimal MDS is lower bounded by $n/(d+1)$. Thus, we know

$$OPT(\varphi) = OPT(\phi) + n_1/2 + n_2/3 + n_3/4$$
$$\leq OPT(\phi) + (N - n)/2 \leq OPT(\phi) + (\zeta(\beta) - 1)n/2$$
$$\leq OPT(\phi) + (\zeta(\beta) - 1)(d+1)OPT(\phi)/2 = [1 + (\zeta(\beta) - 1)(d+1)/2] OPT(\phi)$$

where N is the number of vertices in $G_{(\alpha,\beta)}$.

On the other hand, with $|OPT(\phi) - \phi| = |OPT(\varphi) - \varphi|$, we proved the $\mathcal{L}$-reduction with $c_1 = 1 + (\zeta(\beta) - 1)(d+1)/2$ and $c_2 = 1$.

Theorem 4. *MVC is* APX-*hard on PLG.*

Proof. In this proof, we construct as *Cycle-Based Embedding Technique*, according to Theorem 1, to show $\mathcal{L}$-reduction from MVC on d-bounded graph G_d to MVC on power law graph $G_{(\alpha,\beta)}$. Let ϕ be a feasible solution on G_d and φ be a feasible solution on $G_{(\alpha,\beta)}$.

However, MVC on K_2, cycle, d-Regular Cycle and κ-Branch-d-Cycle is $n/2$. Therefore, for a solution ϕ on G_d, we have a solution φ on $G_{(\alpha,\beta)}$ is $\varphi = \phi + (N - n)/2$. Correspondingly, we have $OPT(\varphi) = OPT(\phi) + (N - n)/2$.

On one hand, for a d-bounded graph with vertices n, the optimal MVC is lower bounded by $n/(d+1)$. Therefore, similarly as the proof in Theorem 3,

$$OPT(\varphi) \leq [1 + (\zeta(\beta) - 1)(d+1)/2] OPT(\phi)$$

On the other hand, with $|OPT(\phi) - \phi| = |OPT(\varphi) - \varphi|$, we proved the $\mathcal{L}$-reduction with $c_1 = 1 + (\zeta(\beta) - 1)(d+1)/2$ and $c_2 = 1$.

Corollary 2. *MIS is* APX-*hard on PLG.*

5   Inapproximability of Optimization Problems on PLG

5.1   MDS, MIS, MVC

Theorem 5 (P. Austrin *et al.* [4]). *For every sufficiently large integer d, MIS on a graph d-bounded G is UG-hard to approximate within a factor $O\left(d/\log^2 d\right)$.*

Theorem 6 (P. Austrin *et al.* [4]). *For every sufficiently large integer d, MVC on a graph d-bounded G is UG-hard to approximate within a factor $2 - (2 + o_d(1))\log\log d/\log d$.*

Theorem 7 (M. Chlebík *et al.* [9]). *For every sufficiently large integer d, there is no $(\log d - O(\log\log d))$-approximation for MDS on d-bounded graphs unless $NP \subseteq DTIME\left(n^{O(\log\log n)}\right)$.*

Theorem 8. *MIS is* UG-*hard to approximate to within a factor* $1 - \frac{2\left(c_1 - O\left(\log^2 c_1\right)\right)}{c_1(c_1+1)\zeta(\beta)}$ *on PLG.*

Proof. In this proof, we construct the power law graph based on *Cycle-Based Embedding Technique* in Theorem 1 and show the Gap-Preserving from MIS on d-bounded graph G_d to MIS on power law graph $G_{(\alpha,\beta)}$. Let ϕ be a feasible solution on G_d and φ be a feasible solution on $G_{(\alpha,\beta)}$. We show *Completeness* and *Soundness* with $m' = m + (N-n)/2$.

- If $OPT(\phi) = m \Rightarrow OPT(\varphi) = m'$
  Let $OPT(\phi) = m$ be the MIS on graph G_d, we have $OPT(\varphi)$ which is composed of several parts: (1) $OPT(\phi) = m$; (2) MIS on clique K_2, cycle and d-Regular Cycle are all exactly half number of all vertices. Therefore, MIS on $G_{(\alpha,\beta)} \setminus G_d$ is $(N-n)/2$, where N and n are respectively the number of vertices on $G_{(\alpha,\beta)}$ and G_d. We have $OPT(\varphi) = OPT(\phi) + (N-n)/2$. That is, $OPT(\varphi) = m'$ where $m' = m + (N-n)/2$.

- If $OPT(\phi) < O\left(\log^2 d/d\right) m \Rightarrow OPT(\varphi) < \left(1 - \frac{2\left(c_1 - O\left(\log^2 c_1\right)\right)}{c_1(c_1+1)\zeta(\beta)}\right) m'$

$$OPT(\varphi) = OPT(\phi) + \frac{N-n}{2} < O\left(\frac{\log^2 d}{d}\right) m + \frac{N-n}{2}$$

$$= \left(1 - \frac{\left(1 - O\left(\frac{\log^2 d}{d}\right)\right) m}{m + \frac{N-n}{2}}\right) m' < \left(1 - \frac{\left(1 - O\left(\frac{\log^2 d}{d}\right)\right)}{\frac{N}{2m}}\right) m'$$

$$< \left(1 - \frac{1 - O\left(\frac{\log^2 d}{d}\right)}{\frac{(d+1)N}{2n}}\right) m' < \left(1 - \frac{2n\left(1 - O\left(\frac{\log^2 d}{d}\right)\right)}{N(d+1)}\right) m'$$

$$= \left(1 - \frac{2n\left(1 - O\left(\frac{\log^2 d}{d}\right)\right)}{\zeta(\beta)(d+1)n}\right) m' \leq \left(1 - \frac{2\left(c_1 - O\left(\log^2 c_1\right)\right)}{c_1(c_1+1)\zeta(\beta)}\right) m'$$

where c_1 is the minimum integer d satisfying Theorem 5.

Equation (1) holds since $1 \leq OPT(\phi) < O\left(\frac{\log_2 d}{d}\right) m$. Since G_d is a d-bounded graph, $m \geq n/(d+1)$. The last step holds since it is easy to see that function $f(x) = (x - O\left(\log^2 x\right))/(x(x+1))$ is monotonously decreasing when $f(x) > 0$ for any $x > 0$.

Theorem 9. *MVC is UG-hard to be approximated within* $1 + \frac{2\left(1-(2+o_{c_2}(1))\frac{\log\log c_2}{\log c_2}\right)}{(c_2+1)\zeta(\beta)}$ *on PLG.*

Proof. The proof is similar to the inapproximability of MIS. We only show the *Soundness* here.

$$OPT(\varphi) = OPT(\phi) + \frac{N-n}{2} > \left(1 + \frac{1 - (2 + o_d(1))\frac{\log\log d}{\log d}}{1 + \frac{N-n}{2m}}\right) m'$$

$$> \left(1 + \frac{2n\left(1 - (2 + o_d(1))\frac{\log\log d}{\log d}\right)}{(d+1)\zeta(\beta)n}\right) m' > \left(1 + \frac{2\left(1 - (2 + o_{c_2}(1))\frac{\log\log c_2}{\log c_2}\right)}{(c_2+1)\zeta(\beta)}\right) m'$$

where c_2 is the minimum integer d satisfying Theorem 6 and $m' = (N - n)/2$. The inequality holds since function $f(x) = (1-(2+o_x(1))\log\log x/\log x)/(x+1)$ is monotonously decreasing when $f(x) > 0$ for all x.

Theorem 10. *There is no* $1 + \frac{2(\log c_3 - O(\log\log c_3) - 1)}{(c_3+1)\zeta(\beta)}$ *-approximation for Minimum Dominating Set on PLG unless* $NP \subseteq DTIME\left(n^{O(\log\log n)}\right)$.

Proof. In this proof, we construct the power law graph based on *Cycle-Based Embedding Technique* in Theorem 1 and show the Gap-Preserving from MDS on d-bounded graph G_d to MDS on power law graph $G_{(\alpha,\beta)}$. Let ϕ and φ be feasible solutions on G_d and $G_{(\alpha,\beta)}$. We show *Completeness* and *Soundness*.

- If $OPT(\phi) = m \Rightarrow OPT(\varphi) = m'$
  Let $OPT(\phi) = m$ be the MDS on graph G_d, we have $OPT(\varphi)$ which is composed of several parts: (1) $OPT(\phi) = m$; (2) MDS on a K_2 is 1, $n/4$ on a d-Regular Cycle according to Lemma 1 and $n/3$ on a cycle. That is, $OPT(\varphi) = m'$ where $m' = m + n_1/2 + n_2/3 + n_3/4$, where n_1, n_2 and n_3 corresponds to $\tau(1)$, $\tau(2)$ and all leftover vertices in Theorem 1.
- If $OPT(\phi) > (\log d - O(\log\log d)) m \Rightarrow OPT(\varphi) > \left(1 + \frac{2(\log c_3 - O(\log\log c_3) - 1)}{(c_3+1)\zeta(\beta)}\right) m'$

$$OPT(\varphi) = OPT(\phi) + n_1/2 + n_2/3 + n_3/4$$
$$> \left(1 + \frac{((\log d - O(\log\log d)) - 1)}{1 + (N - n)/(2m)}\right) m' > \left(1 + \frac{2(\log c_3 - O(\log\log c_3) - 1)}{(c_3 + 1)\zeta(\beta)}\right) m'$$

where $c_3 = \max\{\gamma_1, \gamma_2\}$, where γ_1 is the minimum integer d satisfying Theorem 7 and γ_2 satisfying $\frac{df(x)}{dx} = 0$ with function $f(x) = (\log x - O(\log\log x) - 1)/(x + 1)$. Why we choose such c_3 is that γ_2 is the maxima of $f(x)$.

5.2 ρ-Dominating Set Problem

Theorem 11. *ρ-PDS is UG-hard to be approximated into* $2 - (2 + o_d(1))\frac{\log\log d}{\log d}$ *on d-bounded graphs.*

Proof. In this proof, we show the Gap-Preserving from MVC on (d/ρ)-bounded graph $G = (V, E)$ to ρ-PDS on d-bounded graph $G' = (V', E')$. w.l.o.g., we assume that d and d/ρ are integers. We construct a graph $G' = (V', E')$ by adding new vertices and edges to G. For each edge $(u, v) \in E$, create k new vertices $uv_1, \ldots, uv_k$ where $1 \leq k \leq \lfloor 1/\rho \rfloor$ and $2k$ new edges (uv_i, u) and (uv_i, u) for all $i \in [1, k]$ as shown in Fig. 6. Clearly, $G' = (V', E')$ is a d-bounded graph.

Let ϕ and φ be solutions to MVC on G and G' respectively. We claim that $OPT(\phi) = OPT(\varphi)$.

On one hand, if $\{v_1, v_2, \ldots, v_j\} \in V$ is minimum vertex cover on G. Then $\{v_1, v_2, \ldots, v_j\}$ is a ρ-PDS on G' because every old vertex in V has ρ of all neighbors in MVC and every new vertex in $V' \setminus V$ has at least one of two neighbors in MVC. Thus $OPT(\phi) \geq OPT(\varphi)$. One the other hand, we can prove that $OPT(\varphi)$ does not contain new vertices, that is, $V' \setminus V$. Consider a

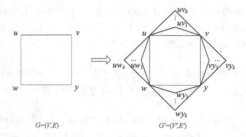

$G=(V,E)$ $G'=(V',E')$

Fig. 6. Reduction from MVC to ρ-MDS

vertex $u \in V$, if $u \in OPT(\varphi)$, the new vertices uv_i for all $v \in N(u)$ and all $i \in [1,k]$ are not needed to be selected. If $u \notin OPT(\varphi)$, it has to be dominated by rho proportion of its all neighbors. That is, for each edge (u,v) incident to u, either v or all uv_i has to be selected since every uv_i has to be selected or dominated. If all uv_i are selected in $OPT(\varphi)$ for some edge (u,v), v is still not dominated by enough vertices if there are some more edges incident to v and the number of vertices uv_i k is great than 1, that is, $\lfloor 1/\rho \rfloor \geq 1$. In this case, therefore, v will be selected to dominate uv. Thus, $OPT(\varphi)$ does not contain new vertices. Since the verices in V selected is a solution to ρ-MDS, that is, for each vertex u in graph G, u will be selected or at least the number of neighbors of u will be selected. Therefore, the vertices in $OPT(\varphi)$ consist a Vertex Cover in G. Thus $OPT(\phi) \leq OPT(\varphi)$. Then we present the *Completeness* and *Soundness*.

- If $OPT(\phi) = m \Rightarrow OPT(\varphi) = m$
- If $OPT(\phi) > \left(2 - (2 + o_d(1))\frac{\log\log(d/2)}{\log(d/2)}\right) m \Rightarrow OPT(\varphi) > \left(2 - (2 + o_d(1))\frac{\log\log d}{\log d}\right) m$

$$OPT(\varphi) > \left(2 - (2 + o_d(1))\frac{\log\log(d/\rho)}{\log(d/\rho)}\right) m > \left(2 - (2 + o_d(1))\frac{\log\log d}{\log d}\right) m$$

since the function $f(x) = 2 - \log x/x$ is monotonously increasing for any x.

Theorem 12. ρ-PDS is UG-hard to be approximated into $1 + \dfrac{2\left(1 - (2 + o_{c_2}(1))\frac{\log\log c_2}{\log c_2}\right)}{(c_2+1)\zeta(\beta)}$ on PLG.

Proof. In this proof, we will show the Gap-Preserving from ρ-MDS on bounded degree graph G_d to ρ-MDS on power law graph $G_{(\alpha,\beta)}$.

We use the same construction as in Theorem 8. Let ϕ be a solution on G'_d and φ be a solution on $G_{(\alpha,\beta)}$, we prove the *Completeness* and *Soundness*.

- If $OPT(\phi) = m \Rightarrow OPT(\varphi) = m'$
    Let $OPT(\phi) = m$ be the ρ-MDS on graph G_d, we have $OPT(\varphi)$ which is composed of several parts: (1) $OPT(\phi) = m$; (2) MDS on a K_2 is 1, $g(\rho)n$ on a d-Regular Cycle according to Lemma 1 and $f(\rho)n$ on a cycle, where

$$f(\rho) = \begin{cases} \frac{1}{4}, & \rho \leq \frac{1}{3} \\ \frac{1}{3}, & \frac{1}{3} < \rho \leq \frac{1}{2} \end{cases} \quad \text{and } g(\rho) = \frac{1}{3} \text{ for all } \rho \leq \frac{1}{2}.$$

Therefore, ρ-MDS on $G_{(\alpha,\beta)}$ to be m' where $m' = m + n_1/2 + f(\rho)n_2 + g(\rho)n_3$, where n_1, n_2 and n_3 corresponds to $\tau(1)$, $\tau(2)$ and all leftover vertices in Theorem 1.

- If $OPT(\phi) > \left(2 - (2 + o_d(1))\frac{\log\log d}{\log d}\right) m \Rightarrow OPT(\varphi) > \left(1 + \frac{1 - (2 + o_{c_2}(1))\frac{\log\log c_2}{\log c_2}}{(c_2 + 1)\zeta(\beta)}\right) m'$

$$OPT(\varphi) = OPT(\phi) + n_1/2 + f(\rho)n_2 + g(\rho)n_3$$
$$> \left(1 + \frac{2n\left(1 - (2 + o_d(1))\frac{\log\log d}{\log d}\right)}{(d + 1)\zeta(\beta)n}\right) m' > \left(1 + \frac{2\left(1 - (2 + o_{c_2}(1))\frac{\log\log c_2}{\log c_2}\right)}{(c_2 + 1)\zeta(\beta)}\right) m'$$

Again, c_2 is the minimum integer d satisfying Theorem 6. The inequality holds since function $f(x) = (1 - (2 + o_\tau(1))\log\log x / \log \tau) / (x + 1)$ is monotonously decreasing when $f(x) > 0$ for any x.

5.3   Maximum Clique, Minimum Coloring

Theorem 13 (Hastad [15]). *There is no $n^{1-\epsilon}$-approximation on Maximum Clique problem unless NP=ZPP.*

Lemma 3 (Ferrante et al. [13]). *Let $G = (V, E)$ be a simple graph with n vertices and $\beta \geq 1$. Let $\alpha \geq \max\{4\beta, \beta\log n + \log(n + 1)\}$. Then, $G_2 = G \setminus G_1$ is a bipartite graph.*

Lemma 4. *Given a function $f(x)$ $(x \in \mathbb{Z}, f(x) \in \mathbb{Z}^+)$ monotonously decreases, $\sum_x f(x) \leq \int_x f(x)$.*

Corollary 3. $e^\alpha \sum_{i=1}^{e^{\alpha/\beta}} \left(\frac{1}{d}\right)^\beta < (e^\alpha - e^{\alpha/\beta})/(\beta - 1)$.

Theorem 14. *Maximum Clique cannot be approximated within $O\left(n^{1/(\beta+1)-\epsilon}\right)$ on large PLG with $\beta > 1$ and $n > 54$ for any $\epsilon > 0$ unless NP=ZPP.*

Proof. In [13], the authors proved the hardness of Maximum Clique problem on power law network. Here we use the same construction. According to Lemma 3, $G_2 = G \setminus G_1$ is a bipartite graph when $\alpha \geq \max\{4\beta, \beta\log n + \log(n+1)\}$ for any $\beta \geq 1$. Let ϕ be a solution on general graph G and φ be a solution on power law graph G_2. We show the *Completeness* and *Soundness*.

- If $OPT(\phi) = m \Rightarrow OPT(\varphi) = m$
   If $OPT(\phi) \leq 2$ on graph G, we can solve Clique problem in polynomial time by iterating the edges and their end vertices one by one, where G is not a general graph in this case. w.l.o.g, assuming $OPT(\phi) > 2$, then $OPT(\varphi) = OPT(\phi) > 2$ since the maximum clique on bipartite graph is 2.
- If $OPT(\phi) \leq m/n^{1-\epsilon} \Rightarrow OPT(\varphi) < O\left(1/(N^{1/(\beta+1)-\epsilon'})\right) m$
   In this case, we consider the case that $4\beta < \beta\log n + \log(n + 1)$, that is, $n > 54$. According to Lemma 3, let $\alpha = \beta\log n + \log(n+1)$. From Corollary 3, we have

$$N = e^\alpha \sum_{d=1}^{e^{\alpha/\beta}} \left(\frac{1}{d}\right)^\beta < \frac{e^\alpha - e^{\alpha/\beta}}{\beta - 1} = \frac{n^\beta(n+1) - n(n+1)^{1/\beta}}{\beta - 1} < \frac{2n^{\beta+1} - n}{\beta - 1}$$

Therefore, $OPT(\varphi) = OPT(\phi) \leq m/n^{1-\epsilon} < O\left(m/\left(N^{1/(\beta+1)-\epsilon'}\right)\right)$.

Corollary 4. *The Minimum Coloring problem cannot be approximated within* $O\left(n^{1/(\beta+1)-\epsilon}\right)$ *on large PLG with* $\beta > 1$ *and* $n > 54$ *for any* $\epsilon > 0$ *unless* NP=ZPP.

6   Relationship between β and Approximation Hardness

As shown in previous sections, many hardness results depend on β. In this section, we analyze the hardness of some optimization problems based on the value of β by showing that trivial greedy algorithms can achieve constant guarantee factor on MIS and MDS.

Lemma 5. *When* $\beta > 2$, *the size of MDS of a power law graph is greater than* Cn *where* n *is the number of vertices,* C *is some constant depended only on* β.

Proof. Let $MDS = (v_1, v_2, \ldots, v_t)$ with degrees $d_1, d_2, \ldots, d_t$ be the MDS of power-law graph $G = (V, E)$. The total of degrees of vertices in dominating set must be at least the number of vertices outside the dominating set. Thus $\sum_{i=1}^{i=t} d_i \geq |V \backslash DS|$. With a given total degrees, a set of vertices has minimum size when it includes highest degree vertices. With $\beta > 2$ the function $\zeta(\beta-1) = \sum_{i=1}^\infty \frac{1}{i^{\beta-1}}$ is converged, there exists a constant $t_0 = t_0(\beta)$ such that

$$\sum_{i=t_0}^{\lfloor e^{\alpha/\beta} \rfloor} i \left\lfloor \frac{e^\alpha}{i^\beta} \right\rfloor \leq \sum_{i=1}^{t_0} \left\lfloor \frac{e^\alpha}{i^\beta} \right\rfloor$$

where α is any large enough constant. Thus the size of MDS is at least

$$\sum_{i=t_0}^{\lfloor e^{\alpha/\beta} \rfloor} \left\lfloor \frac{e^\alpha}{i^\beta} \right\rfloor \approx \left(\zeta(\beta) - \sum_{i=1}^{t_0} \frac{1}{i^\beta}\right) e^\alpha \approx C|V|$$

where $C = (\zeta(\beta) - \sum_{i=1}^{t_0} \frac{1}{i^\beta})/(\zeta(\beta))$.

Consider the greedy algorithm which selects vertices from the highest degree vertices to lowest one. In the worst case, it selects all vertices with degree greater than 1 and a half of vertices with degree 1 to form a dominating set. The approximation factor of this simple algorithm is a constant.

Corollary 5. *Given a power law graph with* $\beta > 2$, *the greedy algorithm that selects vertices in decreasing order of degrees provides a dominating set of size at most* $\sum_{i=2}^{\lfloor e^{\alpha/\beta} \rfloor} \lfloor e^\alpha/i^\beta \rfloor + \frac{1}{2}e^\alpha \approx (\zeta(\beta) - 1/2)e^\alpha$. *Thus the approximation ratio is* $(\zeta(\beta) - \frac{1}{2})/(\zeta(\beta) - \sum_{i=1}^{t_0} 1/i^\beta)$.

Let us consider a maximization problem MIS, we propose a greedy algorithm Power-law-Greedy-MIS as follows. Sort the vertices in non-increasing order then start checking from the lowest degree vertex, if the vertex is not adjacent to any selected vertex, it is selected. The set of selected vertices forms an independent set with the size at least a half the number of vertices with degree 1 which is $e^{\alpha}/2$. The size of MIS is at most a half of number of vertices, then we have

Lemma 6. *Power-law-Greedy-MIS has factor* $1/(2\zeta(\beta))$ *on PLG with* $\beta > 1$.

Acknowledgment

This work is partially supported by NSF Career Award ♯ 0953284 and DTRA, Young Investigator Award, Basic Research Program ♯ HDTRA1-09-1-0061.

References

1. Aiello, W., Chung, F., Lu, L.: A random graph model for massive graphs. In: STOC 2000, pp. 171–180. ACM, New York (2000)
2. Aiello, W., Chung, F., Lu, L.: A random graph model for power law graphs. Experimental Math. 10, 53–66 (2000)
3. Alimonti, P., Kann, V.: Hardness of approximating problems on cubic graphs. In: Bongiovanni, G., Bovet, D.P., Di Battista, G. (eds.) CIAC 1997. LNCS, vol. 1203, pp. 288–298. Springer, Heidelberg (1997)
4. Austrin, P., Khot, S., Safra, M.: Inapproximability of vertex cover and independent set in bounded degree graphs. In: CCC 2009, pp. 74–80 (2009)
5. Barabási, A.-L., Albert, R.: Emergence of scaling in random networks. Science 286, 509–512 (1999)
6. Bianconi, G., Barabási, A.L.: Bose-einstein condensation in complex networks (2000)
7. Bondy, J., Murty, U.: Graph theory with applications. MacMillan, London (1976)
8. Bornholdt, S., Schuster, H.G. (eds.): Handbook of Graphs and Networks: From the Genome to the Internet. John Wiley & Sons, Inc., New York (2003)
9. Chlebík, M., Chlebíková, J.: Approximation hardness of dominating set problems in bounded degree graphs. Inf. Comput. 206(11), 1264–1275 (2008)
10. Dinur, I., Safra, S.: On the hardness of approximating minimum vertex cover. Annals of Mathematics 162, 2005 (2004)
11. Erdos, P., Gallai, T.: Graphs with prescribed degrees of vertices. Mat. Lapok 11, 264–274 (1960)
12. Eubank, S., Kumar, V.S.A., Marathe, M.V., Srinivasan, A., Wang, N.: Structural and algorithmic aspects of massive social networks. In: SODA 2004, pp. 718–727. Society for Industrial and Applied Mathematics, Philadelphia (2004)
13. Ferrante, A., Pandurangan, G., Park, K.: On the hardness of optimization in power-law graphs. Theoretical Computer Science 393(1-3), 220–230 (2008)
14. Gkantsidis, C., Mihail, M., Saberi, A.: Conductance and congestion in power law graphs. SIGMETRICS Perform. Eval. Rev. 31(1), 148–159 (2003)
15. Hastad, J.: Clique is hard to approximate within $n^{1-\epsilon}$. In: FOCS 1996, Washington, DC, USA, p. 627. IEEE Computer Society, Los Alamitos (1996)
16. Janson, S., Luczak, T., Norros, I.: Large cliques in a power-law random graph (2009)

17. Kempe, D., Kleinberg, J., Tardos, É.: Influential nodes in a diffusion model for social networks. In: Caires, L., Italiano, G.F., Monteiro, L., Palamidessi, C., Yung, M. (eds.) ICALP 2005. LNCS, vol. 3580, pp. 1127–1138. Springer, Heidelberg (2005)
18. Kempe, D., Kleinberg, J., Tardos, E.: Maximizing the spread of influence through a social network. In: KDD, pp. 137–146. ACM Press, New York (2003)
19. Kleinberg, J.: The small-world phenomenon: An algorithmic perspective. In: 32nd STOC, pp. 163–170 (2000)
20. Kumar, R., Raghavan, P., Rajagopalan, S., Sivakumar, D., Tomkins, A., Upfal, E.: Stochastic models for the web graph. In: FOCS 2000, p. 57 (2000)
21. Norros, I., Reittu, H.: On a conditionally poissonian graph process. In: Advances in Applied Probability, pp. 38–59 (2006)
22. Pandurangan, G., http://www.cs.purdue.edu/homes/gopal/powerlawtalk.pdf

Appendix A : Embedding Construction with $\beta < 1$

Ferrante *et. al.* [13] proved the *NP*-hardness of MIS, MDS, and MVC where $\beta < 1$ based on Lemma 7 which is invalid. A counter-example is as follows. Let $D_1 =< 3, 2, 2, 1 >$ and $D_2 =< 7, 6, 5, 4, 3, 2, 2, 1 >$ then D_1 is eligible and $Y_1 =< 1, 2, 1 ><$ $Y_2 =< 1, 2, 1, 1, 1, 1, 1 >$ but D_2 is NOT eligible with $f_{D_2}(4) < 0$. In this part, we present an alternative lemma to prove the hardness of these problems on power-law graphs with $\beta < 1$.

Definition 9 (d-Degree Sequence). *Given a graph $G = (V, E)$, the d-degree sequence of G is a sequence $D =< d_1, d_2, \ldots, d_n >$ of vertex degrees in non-increasing order.*

Definition 10 (y-Degree Sequence). *Given a graph $G = (V, E)$, the y-degree sequence of G is a sequence $Y =< y_1, y_2, \ldots, y_m >$ where m is the maximum degree of G and $y_i = |\{u|u \in V$ and $degree(u) = i\}|$.*

Definition 11 (Eligible Sequences). *A sequence of integers $S =< s_1, \ldots, s_n >$ is eligible if $s_1 \geq s_2 \geq \ldots \geq s_n$ and, for all $k \in [n]$, $f_S(k) \geq 0$, where*

$$f_S(k) = k(k - 1) + \sum_{i=k+1}^{n} min\{k, s_i\} - \sum_{i=1}^{k} s_i$$

Lemma 7 (Invalid Lemma, [13]). *Let Y_1 and Y_2 be two y-degree sequences with m_1 and m_2 elements respectively such that (1) $Y_1(i) \leq Y_2(i)$, $\forall 1 \leq i \leq m_1$, and (2) two corresponding d-degree sequences D_1 and D_2 are contiguous. If D_1 is eligible then D_2 is eligible.*

Erdős and Gallai [11] showed that a sequence of integers to be graphic - d-degree sequence of an graph, *iff* it is eligible and the total of all elements is even. Then Havel and Hakimi [7] gave an algorithm to construct a simple graph from a degree sequence.

Lemma 8 ([7]). *A sequence of integers $D =< d_1, \ldots, d_n >$ is graphic if and only if it is non-increasing, and the sequence of values $D' =< d_2 - 1, d_3 - 1, \ldots, d_{d_1+1} - 1, d_{d_1+2}, \ldots, d_n >$ when sorted in non-increasing order is graphic.*

We now prove the following lemma, which can substitute Lemma 7 for the NP-hardness proof in [13].

Lemma 9. *Given an undirected graph $G = (V, E)$, $0 < \beta < 1$, there exists polynomial time algorithm to construct power-law graph $G' = (V', E')$ of exponential factor β such that G is a set of maximal components of G'.*

Proof. To construct G', we choose $\alpha = max\{\beta \ln(n-1) + \ln(n+2), 3\ln 2\}$ then $\lfloor e^{\alpha}/((n-1)^{\beta}) \rfloor > n+2$, i.e. if there are a least 2 vertices of G' having degree d, there are at least 2 vertices of $G' \backslash G$ having degee d. According to the definition, the total degrees of all vertices in G' and G are even. Therefore, the lemma will follow if we prove that the degree sequence D of $G' \backslash G$ is eligible.

In D, the maximum degree is $\lfloor e^{\alpha/\beta} \rfloor$. There is only one vertex of degree i if $1 \leq e^{\alpha}/i^{\beta} < 2$ and furthermore $e^{\alpha/\beta} \geq i > e^{(\alpha - \ln 2)/\beta} = (e^{\alpha}/2)^{1/\beta}$.

We check $f_D(k)$ in two cases:

1. **Case 1:** $k \leq \lfloor e^{\alpha/\beta}/2 \rfloor$

$$
\begin{aligned}
f_D(k) &= k(k-1) + \sum_{i=k+1}^{n} min\{k, d_i\} - \sum_{i=1}^{k} d_i \\
&> k(k-1) + \sum_{i=k}^{T-k} k + \sum_{i=B}^{k-1} i + \sum_{i=1}^{B-1} 2 - \sum_{i=1}^{k} (T-k+1) \\
&= k(T-k) + (k-B)(k-1+B)/2 + B(B-1) - k(2T-k+1)/2 \\
&= (B^2 - B)/2 - k
\end{aligned}
$$

where where $T = \lfloor e^{\alpha/\beta} \rfloor$ and $B = \lfloor (e^{\alpha}/2)^{1/\beta} \rfloor + 1$. Note that with $\alpha > 3\ln 2$, $\alpha/\beta > \ln 2 (2/\beta + 1)$. Hence $(\lfloor (e^{\alpha}/2)^{1/\beta} \rfloor + 1)(\lfloor (e^{\alpha}/2)^{1/\beta} \rfloor) > \lfloor e^{\alpha/\beta} \rfloor \geq 2k$, so $f_D(k) > 0$.

2. **Case 2:** $k > \lfloor e^{\alpha/\beta}/2 \rfloor$

$$
f_D(k+1) \geq f_D(k) + 2k - 2d_{k+1} \geq f_D(k) \geq \ldots \geq f_D\left(\left\lceil e^{\alpha/\beta}/2 \right\rceil\right) > 0
$$

Cyclic Vertex Connectivity of Star Graphs*

Zhihua Yu[1], Qinghai Liu[2], and Zhao Zhang[1],**

College of Mathematics and System Sciences, Xinjiang University
Urumqi, Xinjiang, 830046, People's Republic of China
hxhzz@163.com

Abstract. For a connected graph G, a vertex subset $F \subset V(G)$ is a cyclic vertex-cut of G if $G - F$ is disconnected and at least two of its components contain cycles. The cardinality of a minimum cyclic vertex-cut of G, denoted by $\kappa_c(G)$, is the cyclic vertex-connectivity of G. In this paper, we show that for any integer $n \geq 4$, the n-dimensional star graph SG_n has $\kappa_c(SG_n) = 6(n - 3)$.

Keywords: star graph; cyclic vertex-connectivity.

1 Introduction

Let $G = (V(G), E(G))$ be a simple connected graph, where $V(G)$, $E(G)$ are the vertex set and the edge set, respectively. A vertex subset $F \subseteq V(G)$ is a *cyclic vertex-cut* of G if $G - F$ has at least two connected components containing cycles. Vertices in F are called *faulty*, and vertices in $V(G) - F$ are said to be *good*. If G has a cyclic vertex-cut, then the *cyclic vertex-connectivity* of G, denoted by $\kappa_c(G)$, is the minimum cardinality over all cyclic vertex-cuts of G. When G has no cyclic vertex-cut, the definition of $\kappa_c(G)$ can be found in [15] using Betti number. The cyclic edge-connectivity $\lambda_c(G)$ can be defined similarly, changing 'vertex' to 'edge' (see for example [13,14]).

The concepts of cyclic vertex- and edge-connectivity date to Tait (1880) in attacking Four Color Conjecture [16]. Since then, they are used in many classic fields of graph theory such as integer flow conjectures [21], n-extendable graphs [9,12], etc.

In [18], the authors showed that $\lambda_c(G)$ coincides with $\lambda^2(G)$, where $\lambda^k(G)$ is a kind of conditional connectivity [7] defined as follows: for a connected graph G, an edge subset $F \subset V(G)$ is a R^k-*edge-cut* if $G - F$ is disconnected and each vertex in $V(G) - F$ has at least k good neighbors in $G - F$ (or equivalently, $\delta(G - F) \geq 2$, where δ is the minimum degree of the graph). The R^k-*edge connectivity* of G, denoted by $\lambda^k(G)$ is the cardinality of a minimum R^k-vertex-cut of G. Thus many results obtained for $\lambda^2(G)$ can be directly transformed to those of $\lambda_c(G)$, for example, results in [11,20].

* This research is supported by NSFC (10971255), the Key Project of Chinese Ministry of Education (208161), Program for New Century Excellent Talents in University, and The Project-sponsored by SRF for ROCS, SEM.
** Corresponding author.

W. Wu and O. Daescu (Eds.): COCOA 2010, Part I, LNCS 6508, pp. 212–221, 2010.
© Springer-Verlag Berlin Heidelberg 2010

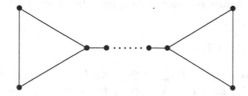

Fig. 1. $\kappa_c(G) = 1 < n - 6 = \kappa^2(G)$, where n is the number of vertices in G

However, the story is different for $\kappa_c(G)$. Changing 'edge' to 'vertex', we obtain the definition of R^k-*vertex-connectivity* $\kappa^k(G)$. Since every graph with minimum degree at least 2 has a cycle, we have $\kappa_c(G) \le \kappa^2(G)$ as long as both $\kappa_c(G)$ and $\kappa^2(G)$ exist. The following example shows that the strict inequality may hold and the gap between $\kappa_c(G)$ and $\kappa^2(G)$ can be arbitrarily large.

In this paper, we determine κ_c for star graphs. Let S_n be the *symmetric group* of order n, that is, the set of all permutations of $\{1, 2, ..., n\}$. The n-*dimensional star graph* SG_n is the graph with vertex set $V(SG_n) = S_n$, two vertices u, v are adjacent in SG_n if and only if $v = u(1i)$, for some $2 \le i \le n$. We say that the *label* on the edge uv is $(1i)$. Star graphs have been shown to have many desirable properties such as high connectivity, small diameter ect., which makes it favorable as a network topology (see for example [2,8]).

We will show in this paper that $\kappa_c(SG_n) = 6(n - 3)$ for $n \ge 4$. In [17], Wan and Zhang proved that for any integer $n \ge 4$, $\kappa^2(S_n) = 6(n - 3)$. We guess that this is not an accidental coincidence, which deserves further study.

2   Some Preliminaries

Terminologies not defined here are referred to [3].

For a graph G, a subgraph G_1 of G, and a vertex $u \in V(G)$, we use $N_{G_1}(u) = \{v \in V(G_1) \mid v$ is adjacent with u in $G\}$ to denote the *neighbor set of u in G_1*. In particular, if $G_1 = G$, then $N_G(u)$ is the neighbor set of u in G, and $d_G(u) = |N_G(u)|$ is the *degree* of vertex u in G. The *minimum degree* of G is $\delta(G) = \min\{d_G(u) \mid u \in V(G)\}$. For a vertex subset $U \subseteq V(G)$, let $N_{G_1}(U) = (\bigcup_{u \in U} N_{G_1}(u)) - U$ be the neighbor set of U in G_1. For simplicity of notation, we sometimes use a subgraph and its vertex set interchangeably, for example, $N_G(G_1)$ is used to denote $N_G(V(G_1))$ where G_1 is a subgraph of G, and $N_A(U)$ is used to denote $N_{G[A]}(U)$ where A, U are two vertex sets and $G[A]$ is the *subgraph of G induced by A*.

It is known that SG_n is $(n - 1)$-regular, bipartite, vertex transitive, and edge transitive [1]. We will also use the following result given by Cheng and Lipman.

Lemma 1 ([4]). *For $n \ge 4$, let T be a vertex subset of SG_n with $|T| \le 2n - 4$. Then one of the following occurs:*

   (i) $SG_n - T$ *is connected;*

   (ii) $SG_n - T$ *has two connected components, one of which is a singleton;*

(iii) $SG_n - T$ *has two connected components, one of which is an edge uv, furthermore, $T = N_{SG_n}(uv)$.*

As a corollary of Lemma 1, we have

Corollary 1. *For $n \geq 4$, $\kappa^1(SG_n) = 2n - 4$. Furthermore, if T is a minimum R^1-vertex-cut of SG_n, then $T = N_{SG_n}(uv)$ for some edge $uv \in E(SG_n)$.*

The *girth* of a graph G is the length of the shortest cycle in G. The following lemma characterizes the structure of shortest cycles of SG_n.

Lemma 2 ([17]). *The girth of SG_n is 6. Any 6-cycle in SG_n has the form $u_1 u_2 u_3 u_4 u_5 u_6 u_1$, where $u_2 = u_1(1i), u_3 = u_2(1j), u_4 = u_3(1i), u_5 = u_4(1j), u_6 = u_5(1i), u_1 = u_6(1j)$ for some i, j with $i \neq j$.*

Lemma 2 shows that any 6-cycle of SG_n has its edges labeled with $(1i)$ and $(1j)$ alternately for some $i, j \in \{2, ..., n\}$ and $i \neq j$. As a consequence, we see that

Corollary 2. *Any two 6-cycles of SG_n have at most one common edge.*

Proof. Suppose $C_1 = u_1 u_2 u_3 u_4 u_5 u_6 u_1$ and $C_2 = u_1 u_2 v_3 v_4 v_5 v_6 u_1$ are two 6-cycles of SG_n having a common edge $u_1 u_2$, the label on $u_1 u_2$ is $(1i)$, and the label on $u_2 u_3$ is $(1j)$ for $j \neq i$. By Lemma 2, the label on $u_2 v_3$ is $(1k)$ for some $k \neq i, j$. Then the common edges of C_1 and C_2 must have label $(1i)$. Notice that $v_3, v_6 \notin V(C_1)$ since the girth of SG_n is 6. Hence $v_3 v_4$ and $v_5 v_6$, which are the only two other edges on C_2 with label $(1i)$, do not belong to C_1. Thus $u_1 u_2$ is the only common edge of C_1 and C_2. $\square$

Let S_n^i be the subset of S_n that consists of all permutations with element i in the rightmost position, and let SG_{n-1}^i be the subgraph of SG_n induced by S_n^i. Clearly SG_{n-1}^i is isomorphic to SG_{n-1}, and thus we call it a *copy* of SG_{n-1}. It is easy to see that SG_n can be decomposed into n copies of SG_{n-1}, namely $SG_{n-1}^1, SG_{n-1}^2, ..., SG_{n-1}^n$. For any copy SG_{n-1}^i and any vertex $u \in V(SG_{n-1}^i)$, there is exactly one neighbor of u outside of SG_{n-1}^i, namely the vertex $u(1n)$. We call it the *outside neighbor* of u and use u' to denote it.

The following property was proved in Lemma 3 of [17], though we state it in a different way to suit the needs of this paper.

Lemma 3 ([17]). *For any path $P = u_0 u_1 u_2$ which is contained in some copy, the outside neighbors u_0', u_1', u_2' are in three different copies. As a consequence, for any edge $u_1 u_2$ in some copy, u_1' and u_2' are in different copies.*

The next result can also be found in [17].

Lemma 4 ([17]). *For any $i \in \{1, 2, ..., n\}$, $N_{SG_n}(SG_{n-1}^i)$ is an independent set of cardinality $(n-1)!$, and $|N_{SG_{n-1}^j}(SG_{n-1}^i)| = (n-2)!$ for any $j \neq i$.*

3   Main Result

In this section, we determine the value of $\kappa_c(SG_n)$ for $n \geq 4$.

Lemma 5. *Let C be a 6-cycle of SG_n ($n \geq 4$). Then $N_{SG_n}(C)$ is a cyclic vertex-cut of SG_n.*

Proof. Clearly, $SG_n - N_{SG_n}(C)$ is disconnected which contains cycle C as a connected component. Hence to prove the lemma, it suffices to show that the subgraph $\widetilde{G} = SG_n - N_{SG_n}(C) - C$ has a cycle. In fact, we can prove a stronger property $\delta(\widetilde{G}) \geq 2$ as follows.

Suppose $C = u_1 u_2 \dots u_6 u_1$. By Lemma 2, there exist two indices $i, j \neq n$ such that the labels on the edges of C are $(1i)$ and $(1j)$ alternately. If $\delta(\widetilde{G}) \leq 1$, then there exists a vertex $v \in V(\widetilde{G})$ which has at least $n - 2 \geq 2$ neighbors in $N_{SG_n}(C)$ (recall that SG_n is $(n-1)$-regular). Let v_1, v_2 be two distinct vertices in $N_{SG_n}(v) \cap N_{SG_n}(C)$. Suppose, without loss of generality, that v_1 is a neighbor of u_1. Since SG_n is bipartite, there is no odd cycle in SG_n. Hence v_2 can only be a neighbor of vertex u_3 or u_5, say u_3. But then $C' = v v_1 u_1 u_2 u_3 v_2 v$ is a 6-cycle of SG_n which have two common edges $u_1 u_2$, $u_2 u_3$ with the 6-cycle C, contradicting Corollary 2. Thus $\delta(\widetilde{G}) \geq 2$.

Since every graph with minimum degree at least 2 has a cycle, the lemma is proved.   □

Theorem 1. *For any integer $n \geq 4$, $\kappa_c(S_n) = 6(n-3)$.*

Proof. Let C be a 6-cycle in SG_n and $F = N_{SG_n}(C)$. Since the girth of SG_n is 6, no two vertices on C have a common neighbor in $N_{SG_n}(C)$. Thus $|F| = 6(n-3)$. By Lemma 5, F is a cyclic vertex-cut. Hence $\kappa_c(SG_n) \leq |F| \leq 6(n-3)$.

To prove the converse, let F be a minimum cyclic vertex-cut of SG_n. Suppose $|F| < 6(n-3)$, we are to derive a contradiction. For $i \in \{1, 2, \dots, n\}$, denote $F_i = F \cap SG_{n-1}^i$, and U_i the set of isolated vertices of $SG_{n-1}^i - F_i$.

Claim 1. If $|F_i| \leq 2n - 7$, then $|U_i| \leq 1$.

Otherwise, let u, v be two vertices in U_i. Since SG_{n-1}^i has girth 6, we see that u, v have at most one common neighbor. Hence $|N_{SG_{n-1}^i}(\{u, v\})| \geq 2(n-2) - 1 > 2n - 7$. Since U_i is an independent set, we see that $N_{SG_{n-1}^i}(U_i) \supseteq N_{SG_{n-1}^i}(\{u, v\})$. It follows that $|F_i| \geq |N_{SG_{n-1}^i}(U_i)| \geq |N_{SG_{n-1}^i}(\{u, v\})| > 2n - 7$, contradicting that $|F_i| \leq 2n - 7$.

Claim 2. If $|F_i| \leq 2n - 7$, then $SG_{n-1}^i - (F_i \cup U_i)$ is connected.

Suppose this is not true, then $F_i \cup U_i$ is a R^1-vertex-cut of SG_{n-1}^i. By Corollary 1, $|F_i \cup U_i| \geq \kappa^1(SG_{n-1}^i) = 2n - 6$. Combining this with $|U_i| \leq 1$ (by Claim 1) and $|F_i| \leq 2n - 7$, we see that $|U_i| = 1$ and $|F_i \cup U_i| = \kappa^1(SG_{n-1}^i)$ (thus $F_i \cup U_i$ is a minimum R^1-vertex-cut of SG_{n-1}^i). Again by Corollary 1, $F_i \cup U_i = N_{SG_{n-1}^i}(vw)$ for some edge $vw \in E(SG_{n-1}^i)$. Let u be the unique vertex in U_i. Then u is adjacent with either v or w, contradicting that u is an isolated vertex in $SG_{n-1}^i - F_i$. Thus Claim 2 is proved.

Let $I = \{i : |F_i| \geq 2n - 6\}$. Since $|F| < 6(n-3)$, we have $|I| \leq 2$.

Claim 3. Let G_1 be the subgraph of SG_n induced by $\bigcup_{i \notin I} V(SG_{n-1}^i - (F_i \cup U_i))$. Then G_1 is connected.

By Claim 2, $SG_{n-1}^i - (F_i \cup U_i)$ is connected for any $i \notin I$. Hence to prove Claim 3, it suffices to show that for any two indices $i, j \notin I$, there is a path in G_1 connecting $SG_{n-1}^i - (F_i \cup U_i)$ and $SG_{n-1}^j - (F_j \cup U_j)$. For such i, j, $|U_i|, |U_j| \leq 1$ by Claim 1.

If there is an edge between $SG_{n-1}^i - (F_i \cup U_i)$ and $SG_{n-1}^j - (F_j \cup U_j)$, then we are done. Hence we suppose that there is no edge between $SG_{n-1}^i - (F_i \cup U_i)$ and $SG_{n-1}^j - (F_j \cup U_j)$. Then

$$N_{SG_{n-1}^j}(SG_{n-1}^i - (F_i \cup U_i)) \subseteq F_j \cup U_j, \tag{1}$$

and thus $|N_{SG_{n-1}^j}(SG_{n-1}^i - (F_i \cup U_i))| \leq |F_j| + 1$. We will show that this inequality can be refined to

$$|N_{SG_{n-1}^j}(SG_{n-1}^i - (F_i \cup U_i))| \leq |F_j|. \tag{2}$$

Suppose (2) is not true, then $N_{SG_{n-1}^j}(SG_{n-1}^i - (F_i \cup U_i)) = F_j \cup U_j$ and $|U_j| = 1$. Let u be the unique vertex in U_j. Since u is an isolated vertex in $SG_{n-1}^j - F_j$, it has a neighbor v in F_j. By $N_{SG_{n-1}^j}(SG_{n-1}^i - (F_i \cup U_i)) = F_j \cup U_j$, we see that u and v have outside neighbors u' and v' in $SG_{n-1}^i - (F_i \cup U_i)$, respectively. Since the girth of SG_n is 6 and u and v are neighbors to each other, $u' \neq v'$. Since either u and v has another neighbor in SG_{n-1}^i, by the latter part of Lemma 3, we have a contradiction.

Next, we show that

$$|N_{SG_{n-1}^j}(F_i \cup U_i)| \leq |F_i|. \tag{3}$$

Since each vertex has exactly one outside neighbor, we have $|N_{SG_{n-1}^j}(F_i \cup U_i)| \leq |F_i \cup U_i| \leq |F_i| + 1$. If equality holds, then $|U_i| = 1$ and every vertex in $F_i \cup U_i$ has its outside neighbor in SG_{n-1}^j. Similar to the above, the unique vertex $u \in U_i$ has a neighbor v in F_i, and thus the outside neighbors u', v' can not be both in the same copy. This contradiction establishes inequality (3).

By Lemma 4 and inequalities (2), (3),

$$\begin{aligned}
(n-2)! &= |N_{SG_{n-1}^j}(SG_{n-1}^i)| \\
&= |N_{SG_{n-1}^j}(SG_{n-1}^i - (F_i \cup U_i))| + |N_{SG_{n-1}^j}(F_i \cup U_i)| \\
&\leq |F_i| + |F_j| \\
&\leq 2(2n-7).
\end{aligned}$$

This is impossible for $n \geq 6$. Thus $n = 4$ or 5, in which case the above inequalities become equalities, and thus

$$|F_i| = |F_j| = 2n - 7, \tag{4}$$

$$|N_{SG_{n-1}^j}(SG_{n-1}^i - (F_i \cup U_i))| = |F_j|, \tag{5}$$

$$|N_{SG_{n-1}^i}(SG_{n-1}^j - (F_j \cup U_j))| = |F_i|. \tag{6}$$

We can show that

$$N_{SG_{n-1}^j}(SG_{n-1}^i - (F_i \cup U_i)) = F_j. \tag{7}$$

Suppose this is not true, then by (1) and (5), we see that

$$N_{SG_{n-1}^j}(SG_{n-1}^i - (F_i \cup U_i)) = \{u\} \cup (F_j \setminus \{v\}), \tag{8}$$

where u is the unique vertex in U_j and v is some vertex in F_j. Since u is an isolated vertex in $SG_{n-1}^j - F_j$, it has at least two neighbors in F_j (recall that SG_{n-1} is $(n-2)$-regular and $n \geq 4$). Thus there exists a vertex $w \in F_j$ such that w is adjacent with u and $w \neq v$. By Lemma 3, the outside neighbors u' and w' can not be both in SG_{n-1}^i, contradicting (8). Thus (7) is proved.

In the case that $n = 4$ and $|I| = 2$, suppose, without loss of generality, that $I = \{1, 2\}$. By (4), $|F_3| = |F_4| = 2n - 7$. Hence $|F| \geq 2(2n-7) + 2(2n-6) = 6$, contradicting that $|F| < 6(n-3) = 6$. Hence $n = 5$, or $n = 4$ and $|I| = 1$. In these cases, there exists an index $k \notin I$ and $k \neq i, j$ (recall that $|I| \leq 2$). Since (7) says that every faulty vertex in SG_{n-1}^j has its outside neighbor in SG_{n-1}^i, we see that vertices in $N_{SG_{n-1}^j}(SG_{n-1}^k)$ are all good. Hence in the case that $n = 5$, by $|N_{SG_{n-1}^j}(SG_{n-1}^k - (F_k \cup U_k))| \geq (n-2)! - |F_k \cup U_k| \geq (n-2)! - (2n-7) - 1 = 2$ and $|U_j| \leq 1$, we see that $SG_{n-1}^k - (F_k \cup U_k)$ has a good neighbor in $SG_{n-1}^j - (F_j \cup U_j)$. In the case that $n = 4$, we must have $U_k = \emptyset$. Otherwise U_k has a unique vertex u by Claim 1. Since $N_{SG_{n-1}^k}(u) \subseteq F_k$, we have $n - 2 = |N_{SG_{n-1}^k}(u)| \leq |F_k| \leq 2n - 7$, and thus $n \geq 5$, contradicting $n = 4$. Similarly, $U_j = \emptyset$. Then by $|N_{SG_{n-1}^j}(SG_{n-1}^k - F_k)| \geq (n-2)! - |F_k| \geq (n-2)! - (2n-7) = 1$, we see that $SG_{n-1}^k - F_k$ has a good neighbor in $SG_{n-1}^j - F_j$. In any case, there is an edge between $SG_{n-1}^k - (F_k \cup U_k)$ and $SG_{n-1}^j - (F_j \cup U_j)$. Symmetrically, it can be shown that there is an edge between $SG_{n-1}^k - (F_k \cup U_k)$ and $SG_{n-1}^i - (F_i \cup U_i)$. Then $SG_{n-1}^i - (F_i \cup U_i)$ is connected to $SG_{n-1}^j - (F_j \cup U_j)$ through $SG_{n-1}^k - (F_k \cup U_k)$ (all the vertices on the path connecting $SG_{n-1}^i - (F_i \cup U_i)$ and $SG_{n-1}^j - (F_j \cup U_j)$ belong to G_1).

Claim 3 is proved.

By Claim 3, we may assume that G_1 is contained in a connected component $\widetilde{C}$ of $SG_n - F$. If $I = \emptyset$, then $V(SG_n - F - \widetilde{C}) \subseteq \bigcup_{i=1}^n U_i$. For each vertex $u \in \bigcup_{i=1}^n U_i$, if its outside neighbor is good, then $d_{SG_n - F}(u) = 1$, otherwise $d_{SG_n - F}(u) = 0$. It follows that $\delta(SG_n - F - \widetilde{C}) \leq 1$ and thus $SG_n - F - \widetilde{C}$ has no cycle, contradicting that F is a cyclic vertex-cut. Hence $1 \leq |I| \leq 2$. Let G_2 be the subgraph of SG_n induced by $\bigcup_{i \in I}(SG_{n-1}^i - F_i)$.

Claim 4. Let C be a connected component of G_2 which contains at least one cycle. Then there is an edge between C and G_1.

Suppose this is not true, then $N_{SG_n}(C) \subseteq F \cup U$, where $U = \bigcup_{j \notin I} U_j$. It follows that for any index $j \notin I$, $N_{SG_{n-1}^j}(C) \subseteq F_j \cup U_j$, and for any index $i \in I$, $N_{SG_{n-1}^i}(C) \subseteq F_i$. As a consequence, using Claim 1,

$$|N_{SG_{n-1}^j}(C)| \leq |F_j| + |U_j| \leq |F_j| + 1 \text{ for } j \notin I, \text{ and} \tag{9}$$

$$|N_{SG_{n-1}^i}(C)| \leq |F_i| \text{ for } i \in I. \tag{10}$$

We can further refine (9) to

$$|N_{SG_{n-1}^j}(C)| \leq |F_j| \text{ for } j \notin I. \tag{11}$$

Suppose (11) is not true, then $N_{SG_{n-1}^j}(C) = F_j \cup U_j$ and $|U_j| = 1$. Let u be the unique vertex in U_j, and v, w be two neighbors of u in F_j. By Lemma 3, the outside neighbors u', v', w' should be in three different copies. But this is impossible since C has non-empty intersection with at most two copies, namely the copies corresponding to I. Thus (11) is proved.

Combining inequalities (10) and (11), we have

$$|N_{SG_n}(C)| \leq |F|. \tag{12}$$

In the following, we count $|N_{SG_n}(C)|$ and derive contradictions to (12).

Case 1. $|I| = 1$.

Suppose, without loss of generality, that $I = \{1\}$. In this case,

$$C \text{ is contained in } SG_{n-1}^1 - (F_1 \cup U_1). \tag{13}$$

Let D be a shortest cycle in C, and $u_1, ..., u_6$ be six sequential vertices on D. Since the girth of SG_n is 6 and there is no odd cycle in SG_n, we see that if $u_1 u_6$ is an edge, then no vertices of $\{u_1, ..., u_6\}$ can have a common neighbor outside of D; if $u_1 u_6$ is not an edge, then the only pairs of vertices of $\{u_1, ..., u_6\}$ that may have a common neighbor outside of D are $\{u_1, u_5\}$ and $\{u_2, u_6\}$. Furthermore, we see from Corollary 2 that if u_1, u_5 have a common neighbor outside of D, then u_2, u_6 cannot have common neighbor outside of D, and vice versa. Denote $Y = N_{SG_{n-1}^1}(D)$. By the above analysis, we see that $|Y| \geq 6(n-4) = 6n - 24$ if $u_1 u_6$ is an edge, and $|Y| \geq 4(n-4) + 2(n-3) - 1 = 6n - 23 > 6n - 24$ otherwise.

Let $Y' = Y \cap V(C)$ and $Y'' = Y \setminus Y'$. Clearly, each vertex $y \in Y''$ is in $N_{SG_n}(C)$. For each vertex $y \in Y'$, since its outside neighbor $y' \notin V(C)$ (by (13)), we have $y' \in N_{SG_n}(C)$. Since the outside neighbors of vertices in a same copy are all different, we have

$$|N_{SG_n}(C)| \geq |Y''| + |\{y' \mid y \in Y'\}| + |\{u_1', ..., u_6'\}| = |Y| + 6 \geq 6n - 18 > |F|,$$

contradicting (12).

Case 2. $|I| = 2$.

Suppose, without loss of generality, that $I = \{1, 2\}$. Then C is contained in $(SG_{n-1}^1 - F_1) \cup (SG_{n-1}^2 - F_2)$. If C is completely contained in SG_{n-1}^1 or SG_{n-1}^2, then a contradiction can be obtained as in Case 1. Thus we assume $V(C) \cap V(SG_{n-1}^i) \neq \emptyset$ for $i = 1, 2$. In this case, there exists an edge of C

between SG_{n-1}^1 and SG_{n-1}^2. Let uv be such an edge, and let P be a path of C on 6 vertices which passes through uv. Denote $X_1 = V(P) \cap V(SG_{n-1}^1)$ and $X_2 = V(P) \cap V(SG_{n-1}^2)$. Then $|X_1|, |X_2| \geq 1$, and thus $|X_1|, |X_2| \leq 5$ by $|X_1| + |X_2| = 6$.

For $i = 1, 2$, let $Y_i = N_{SG_{n-1}^i}(X_i)$. Since SG_{n-1}^i has girth 6, we have

$$|Y_1| + |Y_2| = \begin{cases} 6n - 21 & \text{if one of } X_1 \text{ and } X_2 \text{ is a path on five vertices} \\ & \text{the ends of which have a common neighbor,} \\ 6n - 20 & \text{otherwise.} \end{cases} \quad (14)$$

For $i = 1, 2$, let $n_i = N_{SG_{n-1}^{3-i}}(X_i \cup Y_i) \cap V(C)$. We claim that

$$1 \leq n_i \leq \begin{cases} 1, & \text{for } |X_i| = 1, \\ |X_i| - 1, & \text{for } 2 \leq |X_i| \leq 5. \end{cases} \quad (15)$$

The left hand side $n_i \geq 1$ is obvious because of the edge uv. For the case that $|X_1| = 5$, assume X_1 induces a path $u_1 u_2 u_3 u_4 u_5$ in SG_{n-1}^1, where $u = u_5$. Then u_5 has its outside neighbor v in SG_{n-1}^2. By Lemma 3, vertices in $N_{SG_{n-1}^1}(\{u_5, u_4\})$ do not have their outside neighbors in SG_{n-1}^2; at most one vertex in $N_{SG_{n-1}^1}(\{u_3\})$ has its outside neighbor in SG_{n-1}^2; for $i = 1, 2$, at most one vertex in $N_{SG_{n-1}^1}(\{u_i\}) \cup \{u_i\}$ has its outside neighbor in SG_{n-1}^2. Thus $n_i \leq 4 = |X_i| - 1$. The other cases can be proved similarly.

Suppose $|Y_1| + |Y_2| - n_1 - n_2 \geq 6n - 24$. For $i = 1, 2$, denote $Y_i' = Y_i \cap V(C)$ and $Y_i'' = Y_i \setminus Y_i'$. Then $Y_i'' \subseteq N_{SG_n}(C)$, and each vertex y in $Y_1' \cup X_1$ (resp. $Y_2' \cup X_2$) whose outside neighbor y' is not in SG_{n-1}^2 (resp. SG_{n-1}^1) has $y' \in N_{SG_n}(C)$. Hence

$$|N_{SG_n}(C)| \geq |Y_1''| + |\{y' \mid y \in Y_1' \cup X_1\}| - n_1 + |Y_2''| + |\{y' \mid y \in Y_2' \cup X_2\}| - n_2$$
$$= |Y_1| + |Y_2| - n_1 - n_2 + 6 \geq 6n - 18 > |F|$$

(observe that since the outside neighbors counted in the above inequality are not in $V(SG_{n-1}^1 \cup SG_{n-1}^2)$, no two of them can coincide), which contradicts (12).

Next we consider the case that

$$\text{for any edge } uv \text{ between } SG_{n-1}^1 \text{ and } SG_{n-1}^2 \text{ and any path } P \quad (16)$$
$$\text{taken as above, } |Y_1| + |Y_2| - n_1 - n_2 \leq 6n - 25.$$

Combining this with (14), we have

$$n_1 + n_2 \geq \begin{cases} 5, & \text{if } |Y_1| + |Y_2| = 6n - 20, \\ 4, & \text{if } |Y_1| + |Y_2| = 6n - 21. \end{cases} \quad (17)$$

If both $|X_1| \geq 2$ and $|X_2| \geq 2$, then by (15) and the fact $|X_1| + |X_2| = 6$, we have $n_1 + n_2 \leq 4$. Then by (17), we see that $n_1 + n_2 = 4$ and $|Y_1| + |Y_2| = 6n - 21$. But by (14), one of $|X_1|$ and $|X_2|$ must be 1, a contradiction. Hence suppose

$$\text{for any edge } uv \text{ between } SG_{n-1}^1 \text{ and } SG_{n-1}^2 \text{ and any path } P \quad (18)$$
$$\text{taken as above, } |X_1| = 5, |X_2| = 1, \text{ or vice versa.}$$

Suppose $P = u_1u_2u_3u_4u_5u_6$ is such a path, where u_6 is the only vertex in X_2. Then $n_2 = 1$ and $n_1 \geq 3$ by (17). By the deduction in proving (15), we see that besides u_6, the only outside neighbors which can contribute to n_1 are in $N_{SG^1_{n-1}}(u_3)$, $N_{SG^1_{n-1}}(\{u_2\}) \cup \{u_2\}$, $N_{SG^1_{n-1}}(\{u_1\}) \cup \{u_1\}$, and at most one from each of the three sets. If there is a vertex $u_7 \in N_{SG^1_{n-1}}(u_3)$ whose outside neighbor $u'_7 \in SG^2_{n-1} \cap V(C)$, then $u'_7u_7u_3u_4u_5u_6$ is a path contradiction (18). If $u'_2 \in SG^2_{n-1} \cap V(C)$, then $u'_2u_2u_3u_4u_5u_6$ is a path contradiction (18). Hence in order that $n_1 \geq 3$, there must be a vertex $u_7 \in N_{SG^1_{n-1}}(u_2)$ such that $u'_7 \in SG^2_{n-1} \cap V(C)$. By Lemma 3, $u'_1 \notin SG^2_{n-1}$. Hence in order that $n_1 \geq 3$, there must be a vertex $u_8 \in N_{SG^1_{n-1}}(u_1)$ such that $u'_8 \in SG^2_{n-1} \cap V(C)$. But then $u'_8u_8u_1u_2u_7u'_7$ is a path contradiction (18).

Claim 4 is proved.

As a consequence of Claim 4, every connected component of G_2 which contains a cycle is in $\widetilde{C}$. Thus $SG_n - F - \widetilde{C}$ consists of some vertices in U and some acyclic connected components of G_2. Since every vertex in U has degree at most 1 in $SG_n - F$, we see that $SG_n - F - \widetilde{C}$ does not contain cycle, contradicting that F is a cyclic vertex-cut. The theorem is proved.   □

4   Conclusion and Future Work

In this paper, we determined the cyclic vertex-connectivity κ_c of the n-dimensional star graph SG_n. Generally, κ_c is different from κ^2. For SG_n, these two parameters coincide. Is there something deeper under the coincidence? This is the focus of our future research.

References

1. Akers, S.B., Harel, D., Krishnamurthy, B.: The star graph: an attractive alternative to the n-cube. In: Proc. Int. Conf. Parallel Processing, pp. 393–400 (1987)
2. Akers, S.B., Krishnamurthy, B.: A group-theoretic model for symmetric interconnection networks. IEEE Transactions on Computers 38, 555–566 (1989)
3. Bondy, J.A., Murty, U.S.R.: Graph theory with application. Macmillan, London (1976)
4. Cheng, E., Lipman, M.J.: Increasing the connectivity of the star graphs. Networks 40, 165–169 (2002)
5. Day, K., Tripathi, A.: A comparative study of topological properties of hypercubes and star graphs. IEEE Trans. Comp. 5, 31–38 (1994)
6. Esfahanian, A.H.: Generalized measures of fault tolerance with application to n-cube networks. IEEE Trans. Comp. 38, 1586–1591 (1989)
7. Harary, F.: Conditional connectivity. Networks 13, 347–357 (1983)
8. Heydemann, M.C., Ducourthial, B.: Cayley graphs and interconnection networks. In: Hahn, G., Sabidussi, G. (eds.) Graph Symmetry, Montreal, PQ. NATO Advanced Science Institutes Series C, Mathematica and Physical Sciences, vol. 497, pp. 167–224. Kluwer Academic Publishers, Dordrecht (1996)
9. Holton, D.A., Lou, D., Plummer, M.D.: On the 2-extendability of plannar graphs. Discrete Math. 96, 81–99 (1991)

10. Hu, S.C., Yang, C.B.: Fault tolerance on star graphs. In: Proceedings of the First Aizu International Symposium on Parallel Algorithms/Architecture Synthesis, pp. 176–182 (1995)
11. Latifi, S., Hegde, M., Pour, M.N.: Conditional connectivity measures for large multiprocessor systems. IEEE Trans. Comp. 43, 218–222 (1994)
12. Lou, D., Holton, D.A.: Lower bound of cyclic edge connectivity for n-extendability of regular graphs. Discrete Math. 112, 139–150 (1993)
13. Nedela, R., Skoviera, M.: Atoms of cyclic connectivity in cubic graphs. Math. Slovaca 45, 481–499 (1995)
14. Plummer, M.D.: On the cyclic connectivity of planar graphs. Lecture Notes in Mathematics, vol. 303, pp. 235–242 (1972)
15. Robertson, N.: Minimal cyclic-4-connected graphs. Trans. Amer. Math. Soc. 284, 665–684 (1984)
16. Tait, P.G.: Remarks on the colouring of maps. Proc. Roy. Soc., Edinburgh 10, 501–503 (1880)
17. Wan, M., Zhang, Z.: A kind of conditional vertex connectivity of star graphs. Appl. Math. Letters 22, 264–267 (2009)
18. Wang, B., Zhang, Z.: On cyclic edge-connectivity of transitive graphs. Discrete Math. 309, 4555–4563 (2009)
19. Watkins, M.E.: Connectivity of transitive graphs. J. Combin. Theory 8, 23–29 (1970)
20. Xu, J.M., Liu, Q.: 2-restricted edge connectivity of vertex-transitive graphs. Australasian Journal of Combinatorics 30, 41–49 (2004)
21. Zhang, C.Q.: Integer flows and cycle covers of graphs. Marcel Dekker Inc., New York (1997)

The Number of Shortest Paths
in the (n, k)-Star Graphs

Eddie Cheng[1], Ke Qiu[2], and Zhi Zhang Shen[3]

[1] Dept. of Mathematics and Statistics
Oakland University
Rochester, MI 48309, U.S.A.
[2] Department of Computer Science
Brock University
St. Catharines, Ontario, L2S 3A1 Canada
[3] Dept. of Computer Science and Technology
Plymouth State University
Plymouth, NH 03264, U.S.A.

Abstract. We enumerate all of the shortest paths between any vertex v and the identity vertex in an (n, k)-star graph by enumerating the minimum factorizations of v in terms of the transpositions corresponding to edges in that graph. This result generalizes a previous one for the star graph, and can be applied to obtain the number of the shortest paths between a pair of vertices in some of the other similar structures. It also implies an algorithm to enumerate all such paths.

Keywords: parallel computing, networks and graphs, minimum factorization, combinatorics, shortest paths, algorithms.

1 Introduction

Given a graph G, a well-known problem is to find out *the number of the shortest paths*, not necessarily disjoint, between a pair of vertices in G. A solution to this counting problem can serve as an important topological property for an interconnection network in terms of strong connectivity, effective fault-tolerance, lower communication cost and desired routing flexibility [15]. In fact, a parameter related to the number of shortest paths, referred to as *semigirth*, has been defined, and made use of, in [6,9,13,18].

This number of shortest paths between two vertices has been calculated for several vertex symmetric structures. For example, it is easy to see that there are $H(u \oplus v)!$ different shortest paths from u to v, two vertices in a hypercube graph, where $H(\cdot)$ is the Hamming weight function, and '$\oplus$' the exclusive-or operation. A closed-form formula for this quantity has also been obtained for the hexagonal networks [7], and for the star graphs [11], respectively.

Since a star graph [1] is a special case of the (n, k)-star graph [3], it is natural to consider whether there is a closed-form expression for the corresponding problem in the (n, k)-star graph, as well. Such a result is first reported in [2], obtained via

W. Wu and O. Daescu (Eds.): COCOA 2010, Part I, LNCS 6508, pp. 222–236, 2010.
© Springer-Verlag Berlin Heidelberg 2010

a many-to-one mapping from shortest paths in an (n,k)-star graph to a suitable star graph by making use of a combinatorial result as established in [11].

In this paper, we further investigate this shortest path problem as related to the (n,k)-star graphs by generalizing the argument given in [11]. An (n,k)-star graph, denoted as $S_{n,k}$ in this paper, is defined on the relationship between k-permutations. It turns out that, each and every shortest path between a vertex, v, and $\mathcal{E}_k\,(=12\cdots k)$, the *identity vertex* in such a graph can be characterized as a unique minimum factorization of (the cyclic form of) v in terms of (n,k)-*star transpositions*, corresponding to edges in such a graph. Hence, the number of the shortest paths between v and $\mathcal{E}_k$ in $S_{n,k}$ equals the number of minimum factorizations of v in terms of (n,k)-star transpositions.

The general problem of counting the minimum factorizations of a permutation, especially for the *full cycle*, i.e., $(1,2,\ldots,n)$, with *prescribed cycle types*, has been studied for quite some time [8,10]. In particular, the minimum factorization counting problem has been studied in the past for various allowed transpositions, starting with Denes' classic work [5], with no restriction being placed on the type of transpositions. Later on, Stanley studied the same problem, where the allowed transpositions form a Coxeter group [17], i.e., those in the form of $\{(i,i+1)|i\in[1,n)\}$, corresponding to edges in a bubblesort graph [12]; and Irving *et al*, while further extending a combinatorial result achieved by Pak [14], solved this problem for the *star transpositions* [11], namely, those in the form of $\{(1,i)|i\in[2,n]\}$, corresponding to edges in a star graph. To continue with this line of work, we count, in this paper, the number of minimum factorizations of a vertex in the (n,k)-star graph in terms of the (n,k)-star transpositions, which, when $k=n-1$, become star transpositions, by generalizing a bijection between the star transpositions and a collection of bi-colored trees as proposed in [11] to that between the (n,k)-star transpositions and a collection of bi-colored trees. Therefore, our results generalize those from [11]. The technique used can be applied to derive such quantity in other similar structures. In addition, our discussion implies an algorithm that enumerates all such minimum factorizations, and consequently, all the shortest paths between an arbitrary vertex and the identity.

The rest of this paper proceeds as follows: In the next section, we characterize the cycle structures of a vertex in an (n,k)-star graph, and discuss, in Section 3, the structure of minimum factorizations of a vertex in $S_{n,k}$. We then derive a general formula for the number of the shortest paths between any vertex and the identity vertex in an (n,k)-star graph in Section 4. We finally conclude this paper in Section 5.

2 The (n,k)-Star Graph and the Cycle Structures of Its Vertices

Let $\langle n\rangle$ stand for $\{1,2,\ldots,n\}$, in an (n,k)-*star graph*, $S_{n,k}(V,E)$, $k\in[1,n)$, $S_{n,k}$ for short, V is the collection of k-permutations on $\langle n\rangle$; and, for any $u,v\in V$, (u,v) is an edge in $S_{n,k}$ iff v can be obtained from $u=u_1u_2\cdots u_k$ by either 1) applying

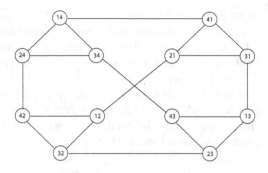

Fig. 1. A $(4, 2)$-star $S_{4,2}$

a transposition $(1, i)$ to u, $i \in [2, k]$; or 2) for some $x \in \langle n \rangle - \{u_i | i \in [1, k]\}$, replacing u_1 with x in u. Figure 1 shows a $(4, 2)$-star.

An n-star or S_n [1], on the other hand, is defined over Γ_n, the set of all permutations on $<n>$, where (u, v) is an edge iff v can be obtained from u by applying a star transposition $(1, i)$, $2 \leq i \leq n$. It is easy to see that $S_{n,n-1}$ is isomorphic to the n-star.

Let $v = v_1 v_2 \cdots v_k$ be a vertex of $S_{n,k}$. For all $i \in [1, k]$, we refer to v_i as *a symbol occurring in* v, and i its *position*. Following the terms adopted in [3], we refer to v_i as an *internal symbol*, if $v_i \in [1, k]$; and as an *external symbol* if $v_i \in (k, n]$. Similarly, we refer to a position i as an *internal*, or *external*, position, respectively, depending on whether $i \in [1, k]$.

We now associate each k-permutation v with a permutation v' on $\{1, 2, \ldots, n\}$ called the *extended permutation of* v. Suppose v is $v_1 v_2 \cdots v_k$. We define its extended permutation $v'_1 v'_2 \ldots v'_n$ of v as follows: $v'_i = v_i$ for $i = 1, 2, \ldots, k$; for $j \in \{k + 1, k + 2, \ldots, n\}$, we consider two cases. Note that the k-permutation $v = v_1 v_2 \cdots v_k$ is a function from $\{1, 2, \ldots, k\}$ to $\{1, 2, \ldots, n\}$ with $v(l) = a_l$ for $l \in \{1, 2, \ldots, k\}$. For $j \in \{k + 1, k + 2, \ldots, n\}$, if $j \notin \{v_1, v_2, \ldots, v_k\}$, that is, j is not in the image of v, define $v'_j = j$. If $v_i = j$ for some i, then the definition is more involved. Let $r = j$. Repeat $r := v^{-1}(r)$ until r is not in the image of v. Let $v'_j = r$.

For example, for $v = 2968134 \in S_{9,7}$, we extend it to a 9-permutation as follows: besides the first 7 symbols as shown in v, for the external position 8, as 8 occurs in position 4, which occurs in position 7, but 7 does not occur in v, we set $v'_8 = 7$. Similarly, we set $v'_9 = 5$ to get $v' = 296813475$.

From now on, when talking about any node $v \in S_{n,k}$, we refer to its extended permutation.

This distinction of internal and external symbols naturally leads to a distinction between *internal cycles* and *external cycles* that might occur in the cyclic representation of a vertex in an (n, k)-star graph. The definition of an *internal cycle* for $v \in S_{n,k}$ is given as usual:

$$C = (c_1, \cdots, c_l),$$

such that 1) for all $j \in [1, l]$, $c_j \in [1, k]$; 2) for $j \in [1, l-1]$, the position of c_{j+1} in v is c_j; and 3) the position of c_1 in v is c_l.

On the other hand, let e_m be an external symbol occurring in an internal position of $v \in S_{n,k}$, $m \geq 1$, E_{e_m}, the external cycle associated with e_m in v, is defined as follows:

$$E_{e_m} = (e_m; e_0, \cdots, e_{m-1}),$$

where 1) for all $j \in [0, m-1]$, $e_j \in [1, k]$; 2) for $j \in [1, m]$, the position of e_j in v is e_{j-1}; 3) the position of e_0 in v is e_m, an external position.

We note, in the above definition, although e_0 is located in an external position, thus not part of a vertex in $S_{n,k}$, it is now included in the associated external cycle.

The following result [16, Proposition 2.1] is parallel to the general one for the symmetric group:

Proposition 1. *Every vertex $v \in S_{n,k}, v \neq \mathcal{E}_k$, can be factorized into the following product of disjoint cycles, each containing at least two symbols:*

$$v = E_{e_{m_1}} \cdots E_{e_{m_p}} C_1 \cdots C_r, p + r > 0,$$

where $E_{e_{m_i}}, i \in [1, p]$, are the external cycles and $C_j, j \in [1, r]$, the internal ones. This factorization is unique, except for the order in which the cycles are written.

For example, for $v = 2968134 \in S_{9,7}$, since its extended permutation is 296813475, it is immediate that its equivalent cyclic form is $(9; 5, 1, 2)(8; 7, 4)(3, 6)$, where the first two cycles are external and the last one is internal.

We can thus identify any vertex v of a given (n, k)-star graph with its unique cyclic factorization, which we will refer to as its *cycle structure*, denoted as $\mathcal{C}(v)$, in the rest of this paper. We will also use $b(v)$ to refer to the total number of symbols in $\mathcal{C}(v)$, $b_I(v)$ $(b_E(v))$ the total number of symbols in the internal (external) cycles of $\mathcal{C}(v)$, $g(v)$ the number of all the cycles in $\mathcal{C}(v)$, and $g_I(v)$ $(g_E(v))$ the number of internal (external) cycles in $\mathcal{C}(v)$. We will drop the parameter v in these expressions when the context is clear.

We note that, when factorizing a cycle of $\mathcal{C}(v), v \in S_{n,k}$, in terms of the (n, k)-star transpositions, corresponding to the edges in $S_{n,k}$, any such a transposition has to be in the form of $(1, j), j \in [2, n]$, by the structural definition of the (n, k)-star graph. Since such a transposition is in the same format as that for the star transposition as defined for the vertex decomposition for the star graph [11], we call this collection of transpositions allowed for the (n, k)-star graph, referred to earlier as the (n, k)-*star transpositions*, the *extended star transpositions* in this paper.

Because of this central role played by the position 1, we refer to any cycle that contains position 1 a *primary cycle*, and *normal* otherwise. A primary cycle could be either internal or external.

In general, a cycle, $\sigma = (c_1, c_2, \ldots, c_l), l \geq 2$, can be factorized into a product of $l - 1$ transpositions: $(c_1, c_2) \cdots (c_1, c_l)$, *when applied from left to right. Such a factorization puts every symbol occurring in σ to its correct position in $\mathcal{E} = 12 \cdots n$, the identify permutation in Γ_n. The following result [5, Lemma 1] shows that this decomposition is a shortest one of this nature.

Lemma 1. *A cycle of degree $k, k \geq 2$, can not be represented by a product of less than $k - 1$ transpositions.*

When a cycle is primary, without loss of generality, $c_1 = 1$, any transposition in the above shortest product is in the form of $(1, j), j \in [2, n]$, thus giving us the desired minimum transposition sequence of $\mathcal{C}(v)$. Otherwise, it is easy to see that the product $(1, c_1)(1, c_2) \cdots (1, c_l)(1, c_1)$ is a minimum factorization of $\sigma = (c_1, c_2, \ldots, c_l), l \geq 2, c_i \neq 1$, with its length being $l + 1$.

From now on, by a *minimal factorization* of $\mathcal{C}(v), v \in S_{n,k}$, denoted as $f(v)$, we mean a *minimum factorization of $\mathcal{C}(v)$ in terms of extended star transpositions*, $(1, j), j \in [2, n]$. Let $f(v) = f_1 \cdots f_m$ be such a minimum factorization, we refer to each $f_i, i \in [1, m]$, a *factor* of $f(v)$, and, if f_i is associated with a cycle σ_j, we also say f_i *meets* σ_j, following terminology adopted in [11].

Clearly, the length of a minimum factorization of $v \in S_{n,k}$ in terms of the extended star transpositions equals the distance between v and $\mathcal{E}_k$ in $S_{n,k}$, since such a minimum factorization immediately leads to a path between v and $\mathcal{E}_k$ in $S_{n,k}$. On the other hand, if there were a shorter path from v to $\mathcal{E}_k$, we would have an even shorter factorization. This latter distance between a vertex v and $\mathcal{E}_k$ in $S_{n,k}$ is a known result [3,16], given as follows:

Theorem 1. *The distance between $\mathcal{E}_k$ and v in $S_{n,k}$ can be expressed as follows:*

1. If v does not contain any external cycle, then

$$d_{S_{n,k}}(\mathcal{E}_k, v) = \begin{cases} b_I + g_I, & \text{if none of the internal cycles is primary,} \\ b_I + g_I - 2. & \text{otherwise.} \end{cases}$$

2. Otherwise,

$$d_{S_{n,k}}(\mathcal{E}_k, v) = \begin{cases} b + g_I + 1 & \text{if none of the cycles is primary,} \\ b + g_I - 1, & \text{otherwise.} \end{cases}$$

We use $\mathcal{F}(v)$ to denote the collection of all the minimum factorizations of $\mathcal{C}(v), v \in S_{n,k}$. In the rest of this paper, we will calculate the number of such minimum factorizations of a vertex $v \in S_{n,k}$, namely, $|\mathcal{F}(v)|$.

3   The Structural Characterization of a Minimum Factorization

We first characterize the structure for those extended star transpositions associated with the same internal cycle.

Lemma 2. *Let σ be an internal cycle in $\mathcal{C}(v), v \in S_{n,k}$, $f(\sigma)$ be a minimum factorization of σ. Then*

1. if σ is primary, i.e., $\sigma = (1, b_2, \ldots, b_l)$, then for all $j \in [2, l]$, $(1, b_j)$ occurs exactly once in $f(\sigma)$, with the order of the transpositions in $f(\sigma)$ being $(1, b_2)$, $\cdots$, $(1, b_l)$, when applied from left to right;

2. *otherwise, $\sigma = (a_1, a_2, \ldots, a_l)$, such that for no $j \in [1, l], a_j = 1$, then for some $j \in [1, l]$, $(1, a_j)$ occurs exactly twice in $f(\sigma)$ and for all other $i \neq j$, $(1, a_j)$ occurs exactly once in $f(\sigma)$, with the order of the transpositions, for this choice of j, being $(1, a_j), (1, a_{j+1}), \cdots, (1, a_{j-l}), (1, a_j)$, again when applied from left to right.*

We omit the proof of the above result as it is the same as that of a result for the star graph [11, Lemma 3], since no external symbols occur in internal cycles.

We now discuss the role played by all the external cycles as contained in $\mathcal{C}(v), v \in S_{n,k}$, in the minimum factorization of v. Let $\sigma_1^E, \sigma_2^E, \ldots, \sigma_{g_E}^E$ be the external cycles, such that $\sigma_i^E = (e_{m_i}^i; e_0^i, \ldots, e_{m_i-1}^i)$.

We can certainly factorize such external cycles in the way as we did with the internal cycles, but, the length of such a resulted factorization will be

$$\begin{cases} b + g, & \text{if none of the cycles is primary;} \\ b + g - 2, & \text{otherwise.} \end{cases}$$

When compared with the distance result as contained in Theorem 1, it is easy to see that such a factorization won't be minimum when $\mathcal{C}(v)$ contains more than one external cycles. On the other hand, the above factorization will place all the symbols, including those external symbols that occur in internal positions, of a vertex v to their correct positions in $\mathcal{E}_n$. For external symbols, this is certainly not necessary since the correct positions for those external symbols are external, thus invisible.

The above consideration leads to a minimum routing algorithm as suggested in [3,16], for vertices containing (an) external cycle(s), which only places all the internal symbols to their correct internal positions in $\mathcal{E}_k$, while placing external symbols in some external positions, but not necessary the correct, external, ones.

We first discuss the case when one of the external cycles is primary. Without loss of generality, let σ_1^E be primary, i.e., for some $j \in [1, m-1]$, $\sigma_1^E = (1, e_{j+1}^1, \ldots, e_{m-1}^1, e_m^1; e_0^1, \ldots, e_{j-1}^1)$, then a factorization corresponding to these external cycles could be the following:

$$(1, e_{j+1}^1)(1, e_{j+2}^1) \cdots (1, e_{m_1-1}^1)(1, e_{m_2}^2)(1, e_0^2) \cdots (1, e_{m_2-1}^2)$$

$$\cdots$$

$$(1, e_{m_{g_E}}^{g_E})(1, e_0^{g_E}) \cdots (1, e_{m_{g_E}-1}^{g_E})(1, e_{m_1}^1)(1, e_0^1) \cdots (1, e_{j-1}^1),$$

which is effectively a factorization of the following *type-1 aggregated external cycle:*

$$(1, e_{j+1}^1, \ldots, e_{m_1-1}^1)\sigma_2^E \cdots \sigma_{g_E}^E (e_{m_1}^1; e_0^1, \ldots, e_{j-1}^1),$$

where $\sigma_2^E \cdots \sigma_{g_E}^E$ stands for the concatenation of the symbols in $\sigma_i^E, i \in [2, g_E]$.

As an example, for $v = 2968134 \in S_{9,7}$, i.e., $\mathcal{C}(v) = (9; 5, 1, 2)(8; 7, 4)(3, 6)$, a factorization associated with the two external cycles is the following:

$$(1, 2)(1, 8)(1, 7)(1, 4)(1, 9)(1, 5).$$

Indeed, when applying the above product to v,

$$296813\underline{475} \overset{(1,2)}{\to} 926813\underline{475} \overset{(1,8)}{\to} 726813\underline{495} \overset{(1,7)}{\to} 426813\underline{795} \overset{(1,4)}{\to} 826413\underline{795}$$
$$\overset{(1,9)}{\to} 526413\underline{798} \overset{(1,5)}{\to} 126453\underline{798},$$

where the last two underlined digits represent the two invisible digits in $S_{9,7}$.

We caution that the relative order of the positions in each external cycle is important, thus should remain the same. In fact, if we rearrange the positions in an external cycle σ_i^E as $(e_j^i, e_{j+1}^i, \ldots, e_{j-1}^i)$ such that $j \neq m_i, i \in [2, g_E]$, the external symbol e_j^i, located in position e_{j-1}^i, will not be placed in the correct position e_j^i, but the very first position as contained in σ_{i+1}^E; or the external position $e_{m_1}^1$ in case $i = g_E$.

As shown in [16, Corollary 3.1], as a result of applying the above factorization, all the internal symbols, e_j^i, $j \in [0, m_i] - \{m_{i-1}\}, i \in [1, g_E]$, will be moved to its correct position while the external symbols, $e_{m_i}^i$, located in position $e_{m_i-1}^i$, $i \in [1, g_E]$, will be moved to $e_{m_{i+1}}^{i+1}$, i.e., cyclically shifted to the right.

Since the length of this factorization for the external cycles is $b_E - 1$, the length of a resulting factorization for v is $b + g_I - 1$, thus minimum by Theorem 1.

We note that any permutation of $\sigma_2^E, \cdots, \sigma_{g_E}^E$, when the order of the symbols in each cycle stays the same, will also lead to a minimum factorization of $\mathcal{C}(v)$, which again will place all the internal symbols in their respective correct positions, while placing each external symbol in some external position in the extension of v.

When none of the external cycles in $\mathcal{C}(v)$ is primary, we can follow a similar argument to obtain the following minimum factorization for $\mathcal{C}(v)$:

$$(1, e_{m_1}^1)(1, e_0^1) \cdots (1, e_{m_1-1}^1)(1, e_{m_2}^2)(1, e_0^2) \cdots (1, e_{m_2-1}^2)$$
$$\cdots (1, e_{m_{g_E}}^{g_E})(1, e_0^{g_E}) \cdots (1, e_{m_{g_E}-1}^{g_E})(1, e_{m_1}^1),$$

which is effectively a factorization of the following *type-2 aggregated external cycle:*

$$(e_{m_1}^1; e_0^1, \ldots, e_{m_1-1}^1)\sigma_2^E \cdots \sigma_{g_E}^E,$$

where the order of the symbols in each cycle has to stay the same, since, for this case, the relative order of the positions in the above type-2 aggregated external cycle is also important.

We summarize the above discussion with the following result.

Lemma 3. *Let $\sigma^E = \{\sigma_1^E, \sigma_2^E, \ldots, \sigma_{g_E}^E\}$ be the external cycles in $\mathcal{C}(\pi) \in S_{n,k}$, $g_E \geq 1$, then*

1. *if, without loss of generality, σ_1^E is primary, then $f(\sigma_{E,1})$, a minimum factorization of the external cycles when σ_1^E is primary, is the same as that for the following* type-1 aggregated external cycle:

$$\sigma^{E,1} = (1, e_{j+1}^1, \ldots, e_{m_1-1}^1)\sigma_2^E \cdots \sigma_{g_E}^E (e_{m_1}^1; e_0^1, \ldots, e_{j-1}^1).$$

where $\sigma_2^E \cdots \sigma_{g_E}^E$ stands for the concatenation of all the elements in those cycles subject to permutation of the cycles, while the order of the positions in each $\sigma_i^E, i \in [2, g_E]$ stays the same.

2. *Otherwise, $f(\sigma_{E,2})$, a minimum factorization of the external cycles, when none of them is primary, is the same as that for the following type-2 aggregated external cycle:*

$$\sigma^{E,2} = (e_{m_1}^1; e_0^1, \ldots, e_{m_1-1}^1)\sigma_2^E \cdots \sigma_{g_E}^E,$$

where $\sigma_2^E \cdots \sigma_{g_E}^E$ stands for the concatenation of all the elements in those cycles subject to permutation of the cycles, while the order of the positions in each $\sigma_i^E, i \in [2, g_E]$ stays the same.

3. *In both cases, there are $(g_E - 1)!$ ways to construct an aggregated external cycle, which contains exactly b_E symbols.*

We now look at the relationship between transpositions associated with distinct cycles. Assume σ_1 is a primary cycle, with its associated factorization being $(1, c_2) \cdots (1, c_{l_1})$, and σ_2 is a normal cycle with factorization being $(1, a_1)(1, a_2) \cdots (1, a_{l_2})(1, a_1)$.

If some transposition(s) associated with σ_1 are embedded in those for σ_2 in a $f(v)$, i.e., for some $j \in [2, l], 1 \le i_0 \le j_0 \le l_1$, and transposition products $f_1(v)$ and $f_2(v)$,

$$f(v) = f_1(v)(1, a_1)(1, a_2) \cdots (1, a_j)(1, c_{i_0}) \cdots (1, c_{j_0})(1, a_{j+1}) \cdots (1, a_{l_2})(1, a_1)f_2(v),$$

then, the symbol a_{j+1}, located in position a_j, after being placed in position 1 by $(1, a_j)$, instead of being placed in its correct position a_{j+1}, will be placed in position c_{i_0}, and will get stuck there, since σ_1 is primary thus the transposition $(1, c_{i_0})$ occurs only once in $f(v)$. Thus, transpositions corresponding to a primary cycle cannot be embedded within those corresponding to a normal cycle.

On the other hand, assume $c_i, c_{i+1} \in \sigma_2$, and $a_{j-1}, a_j \in \sigma_3$, both normal, and

$$f(v) = f_1(v)(1, a_{j-1})(1, c_i)(1, a_j)(1, c_{i+1})f_2(v).$$

After the transposition $(1, a_{j-1})$ places the symbol a_j to position 1, $(1, c_i)$ will place c_{i+1} to position 1 and a_j to position c_i. then $(1, a_j)$ is to place c_{i+1} in position a_j, and the symbol a_{j+1} to position 1. Finally, the transposition $(1, c_{i+1})$ is to place the symbol a_{j+1} to position c_{i+1}. Since neither position c_i nor a_j is the position 1, it takes at least two more occurrences of $(1, a_j)$ in $f_2(v)$ to restore the symbol a_j to its correct position a_j in e, which requires the transposition $(1, a_j)$ appear at least three times in $f(v)$. But, by Lemma 2, $(1, a_j)$ occurs at most twice.

In other words, if transpositions of a normal cycle are embedded by transpositions of another cycle, either normal or primary, all such transpositions in the former normal cycles have to be completely embedded by transpositions of the latter cycle.

This proves the following result about the relationship of those transpositions associated with different cycles, which is similar to that for the star graph [11, Lemma 3].

Lemma 4. *Let σ_1 and σ_2 be two different cycles in $\mathcal{C}(v)$, $v \in S_{n,k}$, and $f(v)$ a minimum factorization of v. Suppose that for $a < c < b$, the factors at positions a and b are associated with σ_1, and that at position c is associated with σ_2. Then*

- *σ_2 is not primary, and*
- *all the transpositions associated with σ_2 are embedded between factors at position a and b in $f(v)$.*

It turns out that the combination of Lemma 2, Lemma 3 and Lemma 4 is also sufficient for a collection of extended star transpositions to be a minimum factorization of a vertex in $S_{n,k}$. Due to the space limit, we have to omit the sufficiency proof.

We note that Irving *et al* showed in [11] that the collection of minimum transitive factorizations of a vertex in a star graph are characterized by Lemmas 2 and 4. A factorization, f, of a vertex in S_n is *transitive* if the orbit of some element of $\langle n \rangle$, when acted on by the group, as generated by the factors in f, equals $\langle n \rangle$. Thus, e.g., given $v = 13245 = (2,3) \in S_5$, the factorization $(1,2)(1,3)$ of v, although minimum, is not transitive, since the orbits of 1, 2 and 3 all equal $\{1,2,3\}$, orbit of 4 equals $\{4\}$, and that of 5 equals $\{5\}$. On the other hand, $(1,2)(1,3)(1,4)(1,4)(1,5)(1,5)$ is a minimum transitive factorization of v, since, with the corresponding group, the orbit of 1 equals $\{1,2,3,4,5\}$.

It is pointed out in [11] that the length of a minimum transitive factorization of a vertex v in S_n, containing m cycles, is $n + m - 2$, regardless whether the symbol 1 constitutes a fixed point.

4   The Number of Minimum Factorizations of a Vertex in (n, k)-Star

Irving *et al*, in [11], enumerated the number of the minimum transitive factorizations of a vertex in a star graph, S_n, in terms of star transpositions, by setting up a bijection between such factorizations and a collection of certain bi-colored trees, via an intermediate structures of words.

More specifically, given a vertex $v \in S_n$, such that $\mathcal{C}(v) = \sigma_1 \sigma_2 \cdots \sigma_m, |\sigma_i| = l_i$, the resulting bi-colored tree, $T(v)$, has the following properties: 1) The root of $T(v)$ is white; 2) the non-root white vertices are labeled $2, \ldots, m$; 3) the white vertex labeled i has $l_i - 1$ black children, and; 4) all black children are leaves.

Let $\mathcal{T}(v)$ be the collection of all such bi-colored trees, Irving *et al* derived the following fundamental result in [11]: The total number of such bi-colored trees is the following:

$$|\mathcal{T}(v)| = \frac{(n + m - 2)!}{n!} l_1,$$

where the factor l_1 counts the arrangements of the $l_1 - 1$ black children for a given structure of the factors corresponding to the $m - 1$ sub-trees.

As a result, Irving *et al* gave the number of the minimum transitive factorizations of a vertex $v \in S_n$ as follows [11]:

$$|\mathcal{F}_1(v)| = |\mathcal{T}(v)| \times l_2 \times \cdots \times l_m = \frac{(n+m-2)!}{n!} l_1 \times \cdots \times l_m.$$

When a cycle structure $\mathcal{C}(v)$ contains no fixed points other than 1, a minimum transitive factorization is the same as a minimum factorization in terms of the extended star transpositions. This is not the case when $\mathcal{C}(v)$ does contain such fixed points. For example, for $u = 21, \mathcal{C}(u) = (1, 2)$, thus the minimum factorization of $\mathcal{C}(u)$ is simply $(1, 2)$, giving a shortest path from u to $\mathcal{E}_2$ in $S_{4,2}$, as shown in Figure 1. On the other hand, since u contains both 3 and 4 as fixed points, one of the minimum transitive factorizations of u is $(1, 3)(1, 4)(1, 4)(1, 3)(1, 2)$, which leads to a path $21 - 31 - 41 - 31 - 21 - 12$ from 21 to 12 in Figure 1 of length 5. In fact, since $u = 21\underline{34}$ contains four symbols and three cycles, the length of all the minimum transitive factorizations of u is 5.

In general, for each f, a fixed point greater than 1, the transitivity requirement adds a pair of $(1, f)$ into a minimum transitive factorization which plays no role in routing at all, thus a minimum transitive factorization of a vertex in $S_{n,k}$ does not always lead to a shortest path. Since what we want to count is the number of the shortest paths, we must remove this transitivity requirement from our consideration. On the other hand, it turns out that Irving *et al*'s process that bijectively maps a minimum transitive factorization of a vertex in S_n to a bi-colored tree is also a bijection from the collection of the minimum factorizations of a vertex v in $S_{n,k}$, when we aggregate all the external cycles in $\mathcal{C}(v)$ into one aggregated external cycle within Γ_n, to the collection of the bi-colored trees where every white vertex, except the root, contains at least one black child, when we ignore all the fixed points greater than 1 in v.

For $v = 2968134 \in S_{9,7}$, we convert $\mathcal{C}(v) = (9; 5, 1, 2)(8; 7, 4)(3, 6)$ into the following structure while combining the two external cycles, one of them primary, into a type-1 aggregation external cycle $\sigma^{E,1}$:

$$\mathcal{C}'(v) = (1, 2, 8, 7, 4, 9, 5)(3, 6) = \sigma^{E,1}\sigma_1^I = \sigma_1\sigma_2.$$

By Lemmas 2, 3, and 4, one of the minimum factorizations of $\mathcal{C}'(v)$ is the following:

$$f(v) = (1, 2)(1, 8)(1, 3)(1, 6)(1, 3)(1, 7)(1, 4)(1, 9)(1, 5).$$

Then, we have

$$\Phi(f(v)) = w(v); 3 = (1, 1, 2, 2, 2, 1, 1, 1, 1); 3.$$

The label 3 indicates that, in this particular case, we decompose σ_1^I, i.e., σ_2, into $(1, 3)(1, 6)(1, 3)$. Thus, $(1, 3)$ is the leftmost factor of $f(v)$ that meets σ_1^I. By Lemma 3, another way to decompose σ_2 is certainly $(1, 6)(1, 3)(1, 6)$.

We notice that in the above example, the sub-tree rooted at 2 can be placed to the left of the first black child, or to the right of the last one, or in between any adjacent black children. There are seven such arrangements, exactly that of l_1. If we also consider the two choices of decomposing $(3, 6)$, we would have fourteen different such bi-colored trees, corresponding to fourteen minimum factorizations of v.

In general, the aforementioned Φ is a bijection from $\mathcal{F}(v)$, the collection of minimum factorizations as specified by Lemmas 2, 3 and 4, to a collection of words, $\mathcal{W}(v)$, and a list of factors with which the factorization meets their respective, normal, cycles.

Furthermore, $w(v) = (1, 1, 2, 2, 2, 2, 1, 1, 1, 1)$ is uniquely mapped to the bi-colored tree as shown in Figure 2.

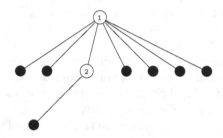

Fig. 2. An example of a bi-colored tree construction

In general, this procedure provides a bijection between the collection of such words, $\mathcal{W}(v)$, and a collection of bi-colored trees, $\mathcal{T}(v)$ as characterized earlier. Therefore, the problem of enumerating the number of minimum factorizations of a vertex $v \in S_{n,k}$ in terms of the extended star transpositions also reduces to the bi-colored tree enumeration problem, which Irving *et al* solved in [11], where each white child contains zero or more black children.

We are now ready to derive a closed-form formula for the number of the shortest paths between $v \in S_{n,k}$ and the identity vertex in $S_{n,k}$. We carry out calculations in different cases as given in Theorem 1.

– If $\mathcal{C}(v)$ contains no external cycle, all the symbols occurring in v are taken from $[1, k]$, and the only kind of transpositions we need to factorize v are those in the form of $(1, j), j \in [2, k]$, i.e., those for the star graph S_k. Hence, the number of minimum factorizations of v in $S_{n,k}$ in this case is simply that of v in S_k.

Let v contain m cycles, trivial or not, and p fixed points in $[2, k.]$ By a known result [11,2].

$$|\mathcal{F}(v)| = \frac{((k-p)+(m-p)-2)!}{(k-p)!} l_1 \times \cdots \times l_{g(v)}. \qquad (1)$$

We note that, in this case, all the external symbols are fixed points in v, as a vertex in $S_{n,k}$.

We now express Eq. 1 in our terminology.

- If the symbol 1 is also a fixed point, then 1, as a symbol that occurs in a trivial cycle, is not counted in $b(v)$, similarly, the trivial cycle (1) is not counted in $g(v)$. Hence, we have $b(v) = k - p - 1$, and $g(v) = m - p - 1$.
- Otherwise, we have $b(v) = k - p$ and $g(v) = m - p$.

Thus, when $\mathcal{C}(v)$ does not contain external cycles,

$$|\mathcal{F}(v)| = \begin{cases} \frac{(b(v)+g(v))!}{(b(v)+1)!} l_1 \times \cdots \times l_{g(v)}, & \text{if } v_1 = 1; \\ \frac{(b(v)+g(v)-2)!}{b(v)!} l_1 \times \cdots \times l_{g(v)}, & \text{otherwise.} \end{cases} \tag{2}$$

– Otherwise, let $\mathcal{C}(v) = \sigma_1^I \cdots \sigma_{g_I}^I \sigma_1^E \cdots \sigma_{g_E}^E$. We first convert $\mathcal{C}(v)$ to $\mathcal{C}'(v) = \sigma_1^I \cdots \sigma_{g_I}^I \sigma^E$, where σ^E is an aggregated external cycle.

- if $\mathcal{C}'(v)$ contains a primary cycle, i.e., $v_1 \neq 1$, without loss of generality, we assume $\mathcal{C}'(v) = \sigma_1 \sigma_2 \cdots \sigma_{g_I+1}$, where σ_1 is primary, either an internal cycle or an aggregated external cycle. We now follow Irving et al's construction to obtain $T(v)$, where there are $g_I + 1$ white vertices, $b_I + b_E = b$ vertices in total, thus the total number of such trees, for a given σ^E is

$$|T(v)| = \frac{((b + (g_I + 1)) - 2)!}{b!} l_1 = \frac{(b + g_I - 1)!}{b!} l_1.$$

By Lemma 4, there are $(g_E - 1)!$ ways to generate an aggregated external cycle, and an aggregated external cycle contains exactly b_E symbols. Moreover, by Lemmas 2, the order of the factors associated with a normal cycle is circularly equivalent.

  * If $\sigma_1 = l_1^I$ is an internal cycle, then

$$|\mathcal{F}(v)| = \frac{(b + g_I - 1)!}{b!} \times l_1^I \times \cdots \times l_{g_I}^I \times b_E \times (g_E - 1)!. \tag{3}$$

  * Otherwise, σ_E is a primary cycle, i.e., a type-1 aggregated external cycle, thus,

$$|\mathcal{F}(v)| = \frac{(b + g_I - 1)!}{b!} \times b_E \times l_1^I \times \cdots \times l_{g_I}^I \times (g_E - 1)!. \tag{4}$$

- Otherwise, if $\mathcal{C}'(v)$ does not contain a primary cycle, $v_1 = 1$, the resulting $T(v)$ contains an extra white vertex, i.e., the root vertex labeled 1. Thus, $T(v)$ contains $g_I + 2$ white vertices, a total of $b_I + b_E + 1 = b + 1$ vertices. Hence, the total number of such trees, for a given type-2 aggregated external cycle, is the following:

$$|T(v)| = \frac{((b + 1) + (g_I + 2) - 2)!}{(b + 1)!} l_1^I = \frac{(b + g_I + 1)!}{(b + 1)!},$$

as $l_1^I = 1$.

Again, by Lemma 4, there are $(g_E - 1)!$ ways to generate a type-2 aggregated external cycle, containing b_E symbols; and by Lemmas 2 and 3, the order of the factors associated with normal internal cycles are circularly equivalent, we have the following result:

$$|\mathcal{F}(v)| = \frac{(b + g_I + 1)!}{(b + 1)!} l_1^I \times \cdots \times l_{g_I}^I \times b_E \times (g_E - 1)!. \tag{5}$$

We note that, in this case, when $v_1 = 1$, $|\sigma_1^I| = l_1^I = 1$. But, we still keep this l_1^I term to achieve a uniformity.

To reiterate, By Eqs. 2, 3, 4 and 5, we have obtained the following main result of this paper:

Theorem 2. *Let* $v \in S_{n,k}$ *such that* $\mathcal{C}(v) = \sigma_1^I \cdots \sigma_{g_I}^I \sigma_1^E \cdots \sigma_{g_E}^E, g_E \geq 0$, *and* $\mathcal{F}(v)$ *be the collection of the minimum factorizations of* v. *Then*

1. *If* v *does not contain any external cycle, then*

$$|\mathcal{F}(v)| = \begin{cases} \frac{(b(v)+g(v))!}{(b(v)+1)!} l_1 \times \cdots \times l_{g(v)}, & \text{if } v_1 = 1; \\ \frac{(b(v)+g(v)-2)!}{b(v)!} l_1 \times \cdots \times l_{g(v)}, & \text{otherwise.} \end{cases}$$

2. *Otherwise,*

$$|\mathcal{F}(v)| = \begin{cases} \frac{(b(v)+g_I(v)+1)!}{(b(v)+1)!} l_1^I \times \cdots \times l_{g_I(v)}^I \times b_E \times (g_E(v) - 1)!, & \text{if } v_1 = 1; \\ \frac{(b(v)+g_I(v)-1)!}{b(v)!} l_1^I \times \cdots \times l_{g_I(v)}^I \times b_E \times (g_E(v) - 1)!, & \text{otherwise.} \end{cases}$$

Corollary 1. *Let* $v \in S_{n,k}, v \neq \mathcal{E}_k$, *and let* $\mathcal{P}(v)$ *be the collection of all the shortest paths between* v *and* e_k, *then* $|\mathcal{P}(v)| = |\mathcal{F}(v)|$, *where* $\mathcal{F}(v)$ *is given in Theorem 2.*

The result as shown in the above corollary agrees with the one as obtained in [2], which, as mentioned in the introductory section, was obtained via the structural relationship between the star graphs and the (n, k)-star graphs. Moreover, since (n, k)-star graph is vertex symmetric, Corollary 1 applies to any two vertices in $S_{n,k}$.

We end our discussion with an example. Again, consider $v = 2968134 \in S_{9,7}$, $\mathcal{C}(v) = (9; 5, 1, 2) \ (8; 7, 4)(3, 6)$, we have that

$$\mathcal{C}'(v) = (1, 2, 8, 7, 4, 9, 5)(3, 6) = \sigma^E \sigma_1^I.$$

Since $v_1 \neq 1$, $g_I = 1, l_1^I = 2, b = 9$, and $b_E = 7$, by Corollary 1, the number of the shortest paths from v to e_7, the identity vertex in $S_{9,7}$, is

$$|\mathcal{F}(v)| = \frac{(9 + 1 - 1)!}{9!} \times 2 \times 7 = 14,$$

which agrees with our earlier combinatorial analysis.

5 Concluding Remarks

In this paper, we further extended earlier works done on the number of minimum factorizations of vertices in terms of star transpositions to the number of such minimum factorizations in terms of the general (n, k)-star transpositions, and derived the number of the shortest paths between any two vertices in the (n, k)-star graph.

We believe the approach that we reported in this paper, continuing the line of enumerating minimum factorizations of a permutation in terms of a certain class of allowed transpositions, suggests a general scheme to calculate the number of shortest paths between vertices in a structure defined on permutations as these two quantities are equal to each other. The results as related to the (n, k)-star graph that we reported in this work provides a demonstrating example in this regard. As another example, it turns out that the collection of minimum transpositions of a permutation in terms of the *arrangement transpositions,* corresponding to the edges in the arrangement graphs [4], can be bijectively mapped to a collection of the ordered forests of bi-colored trees. We are currently investigating this arrangement transposition case, and will report our results in a separate paper.

References

1. Akers, S.B., Krishnamurthy, B.: A Group Theoretic Model for Symmetric Interconnection Networks. IEEE Trans. on Computers 38(4), 555–566 (1989)
2. Cheng, E., Grossman, J., Lipták, L., Qiu, K., Shen, Z.: Distance Formula and Shortest Paths for the (n, k)-Star Graphs. Information Sciences 180, 1671–1680 (2010)
3. Chiang, W., Chen, R.: The (n, k)-Star graph: A Generalized Star Graph. Information Processing Letters 56, 259–264 (1995)
4. Day, K., Tripathi, A.: Arrangement Graphs: A Class of Generalized Star Graphs. Information Processing Letter 42, 235–241 (1992)
5. Denés, J.: The Representation of Permutation as the Product of a Minimal Number of Transpositions and its Connection with the Theory of Graphs. Magyar Tudományos Akadémia. Matematikai Kutatóintézet 4, 63–71 (1959)
6. Fàbrega, J., Fiol, M.A.: Maximally Connected Digraphs. J. Graph Theory 13(6), 657–668 (1989)
7. García, F., Solano, J., Stojmenović, I., Stojmenović, M.: Higher Dimensional Hexagonal Networks. J. Parallel and Distributed Computing 63, 1164–1172 (2003)
8. Goupil, A., Schaeffer, G.: Factoring N-Cycles and Counting Maps of Given Genus. Europ. J. Combinatorics 19, 819–834 (1998)
9. Gross, J., Yellen, J. (eds.): Handbook of Graph Theory. CRC Press, Boca Raton (2003)
10. Irving, J.: On the Number of Factorizations of a Full Cycle. J. Combinatorial Theory, Series A 113, 1549–1554 (2006)
11. Irving, J., Rattan, A.: Factorizations of Permutations into Star Transpositions. Discrete Mathematics 309(6), 1435–1442 (2009)
12. Latifi, S., Srimani, P.: Transposition Networks as a Class of Fault-Tolerant Robust Networks. IEEE Trans. on Computers 45(2), 230–238 (1996)

13. Marcote, X., Balbuena, C., Pelayo, I.: Diameter, Short Paths and Superconnectivity in Diagraphs. Discrete Mathematics 288, 113–123 (2004)
14. Pak, I.: Reduced Decompositions of Permutations in Terms of Star Transpositions, Generalized Catalan Numbers and k-ary Trees. Discrete Mathematics 204, 329–335 (1999)
15. Schwiebert, L.: There is no Optimal Routing Policy for the Torus. Information Processing Letters 83, 331–336 (2002)
16. Shen, Z., Qiu, K., Cheng, E.: On the Surface Area of the (n, k)-Star Graph. Theoretical Computer Science 410, 5481–5490 (2009)
17. Stanley, R.P.: On the Number of Reduced Decompositions of Elements of Coxeter Group. Europ. J. Combinatorics 5, 359–372 (1984)
18. Tang, J., Balbuena, C., Lin, Y., Miller, M.: An Open Problem: Superconnectivity of Regular Digraphs with Respect to Semigirth and Diameter. In: Program of International Workshop on Optimal Network Topologies (IWONT 2007), Pilsen-Černice, Czech Republic (September 2007), http://iti.zcu.cz/iwont2007/iwont2007.pdf

Complexity of Determining the Most Vital Elements for the 1-median and 1-center Location Problems

Cristina Bazgan, Sonia Toubaline, and Daniel Vanderpooten

Université Paris-Dauphine, LAMSADE,
Place du Maréchal de Lattre de Tassigny, 75775 Paris Cedex 16, France
{bazgan,toubaline,vdp}@lamsade.dauphine.fr

Abstract. We consider the k most vital edges (nodes) and min edge (node) blocker versions of the 1-median and 1-center location problems. Given a weighted connected graph with distances on edges and weights on nodes, the k most vital edges (nodes) 1-median (respectively 1-center) problem consists of finding a subset of k edges (nodes) whose removal from the graph leads to an optimal solution for the 1-median (respectively 1-center) problem with the largest total weighted distance (respectively maximum weighted distance). The complementary problem, min edge (node) blocker 1-median (respectively 1-center), consists of removing a subset of edges (nodes) of minimum cardinality such that an optimal solution for the 1-median (respectively 1-center) problem has a total weighted distance (respectively a maximum weighted distance) at least as large as a specified threshold. We show that k most vital edges 1-median and k most vital edges 1-center are NP-hard to approximate within a factor $\frac{7}{5} - \epsilon$ and $\frac{4}{3} - \epsilon$ respectively, for any $\epsilon > 0$, while k most vital nodes 1-median and k most vital nodes 1-center are NP-hard to approximate within a factor $\frac{3}{2} - \epsilon$, for any $\epsilon > 0$. We also show that the complementary versions of these four problems are NP-hard to approximate within a factor 1.36.

Keywords: most vital edges and nodes, 1-median, 1-center, complexity, approximation.

1 Introduction

For problems of security or reliability, it is important to assess the ability of a system to resist to a destruction or a failure of a number of its entities. This amounts to identifying critical entities which can be determined with respect to a measure of performance or a cost associated to the system. In this paper we focus on simple location problems. Consider for instance the following problem. We aim at locating one hospital or one supermarket in order to serve n areas. Each area is characterized by a population which represents a potential demand. The areas are connected by roads with a given distance. The objective for locating this hospital or supermarket is not the same. Indeed, for the hospital,

W. Wu and O. Daescu (Eds.): COCOA 2010, Part I, LNCS 6508, pp. 237–251, 2010.
© Springer-Verlag Berlin Heidelberg 2010

we aim at finding the location that minimizes the maximum distance weighted by population from the hospital to all areas while for the supermarket we aim at finding the location that minimizes the total weighted distance from the supermarket to all areas. However, there may occur incidents such as works on road or floods that make some roads inaccessible. In this case several problems may arise. We can aim at detecting the critical roads whose failure causes the largest increase in the weighted distance. Alternatively, wa can aim at determining the maximum number of damaged roads which still ensures a certain quality of service level. Modeling the considered network by a weighted connected graph with distances on edges and weights on nodes, where roads are edges and areas are nodes, these problems consist either of finding among all subset of edges or nodes, a subset whose removal from the graph generates the largest increase in the total or maximum weighted distance or of determining a subset of edges or nodes of minimal cardinality such that, when we remove this subset from the graph, the total or maximum weighted distance is at least as large as a specified threshold. In the literature these problems are referred respectively to as the k *most vital edges/nodes* and the *min edge/node blocker* problems.

The k most vital edges/nodes and min edge/node blocker versions have been studied for several problems, including shortest path, minimum spanning tree, maximum flow, maximum matching and independent set. The k most vital edges problem with respect to shortest path was proved NP-hard [2]. Later, k most vital edges/nodes shortest path (and min edge/node blocker shortest path, respectively) were proved not 2-approximable (not 1.36-approximable, respectively) if $P \neq NP$ [8]. For minimum spanning tree, k most vital edges is NP-hard [6] and $O(\log k)$-approximable [6]. In [11] it is proved that k most vital edges maximum flow is NP-hard. For maximum matching, min edge blocker is NP-hard even for bipartite graphs [12], but polynomial for grids and trees [10]. In [3], the k most vital nodes and min node blocker versions with respect to independent set for bipartite graphs remain polynomial on the unweighted graphs and become NP-hard for weighted graphs. For bounded treewidth graphs and cographs these versions remain polynomial [3]. Concerning the approximation on bipartite weighted graphs, k most vital nodes with respect to independent set has no ptas [3].

In this paper the k most vital edges (nodes) and min edge (node) blocker versions for the 1-median and 1-center problems are studied.

After introducing some preliminaries in Section 2, we prove in Section 3 that k MOST VITAL EDGES (NODES) 1-MEDIAN (1-CENTER) and MIN EDGE (NODE) BLOCKER 1-MEDIAN (1-CENTER) are not constant approximable for some constants, unless $P=NP$. Final remarks are provided in Section 4.

2   Basic Concepts and Definitions

Consider $G = (V, E)$ a connected weighted graph with $|V| = n$ and $|E| = m$. Let $d_{v_i v_j}$ be the distance between v_i and v_j for $(v_i, v_j) \in E$ and w_{v_i} be the weight associated to node v_i for $i = 1, \ldots, n$ (w_{v_i} represents the demand occurring at node v_i). Denote by $d(v_i, v_j)$ the minimum distance between two nodes v_i and

v_j of G. The 1-median (respectively 1-center) problem consists of locating the median (respectively the center) of a graph G, that is the node v which minimizes the total weighted distance (respectively the maximum weighted distance) to all nodes of the graph given by $\sum\limits_{v_i \in V} w_{v_i} \, d(v, v_i)$ (respectively $\max\limits_{v_i \in V} w_{v_i} \, d(v, v_i)$).

Denote by $G - R$ the graph obtained from G by removing the subset R of edges or nodes.

We consider in this paper the k most vital edges (nodes) and min edge (node) blocker versions of the 1-median and 1-center problems. These problems are defined as follows:

k Most Vital Edges 1-median (1-center)
Input: A connected graph $G = (V, E)$ weighted by two functions $d : E \to N$ and $w : V \to N$ and a positive integer k.
Output: A subset $S^* \subseteq E$, with $|S^*| = k$, whose removal generates an optimal solution for the 1-median (1-center) problem in the graph $G - S^*$ of maximal value.

k Most Vital Nodes 1-median (1-center)
Input: A connected graph $G = (V, E)$ weighted by two functions $d : E \to N$ and $w : V \to N$ and a positive integer k.
Output: A subset $N^* \subseteq V$, with $|N^*| = k$, whose removal generates an optimal solution for the 1-median (1-center) problem in the graph $G - N^*$ of maximal value.

Min Edge blocker 1-median (1-center)
Input: A connected graph $G = (V, E)$ weighted by two functions $d : E \to N$ and $w : V \to N$ and a positive integer U.
Output: An *edge blocker* $S^* \subseteq E$ of minimal cardinality where an edge blocker is a subset of edges such that the value of an optimal solution for the 1-median (1-center) problem in the graph $G - S^*$ is greater than or equal to U.

Min Node blocker 1-median (1-center)
Input: A connected graph $G = (V, E)$ weighted by two functions $d : E \to N$ and $w : V \to N$ and a positive integer U.
Output: A *node blocker* $N^* \subseteq V$ of minimal cardinality where a node blocker is a subset of nodes such that the value of an optimal solution for the 1-median (1-center) problem in the graph $G - N^*$ is greater than or equal to U.

Given an NPO optimization problem and an instance I of this problem, we use $|I|$ to denote the size of I, $opt(I)$ to denote the optimum value of I, and $val(I, S)$ to denote the value of a feasible solution S of instance I. The *performance ratio* of S (or *approximation factor*) is $r(I, S) = \max\left\{ \frac{val(I,S)}{opt(I)}, \frac{opt(I)}{val(I,S)} \right\}$. The *error* of S, $\varepsilon(I, S)$, is defined by $\varepsilon(I, S) = r(I, S) - 1$.

For a function f, an algorithm is an $f(n)$-*approximation*, if for every instance I of the problem, it returns a solution S such that $r(I, S) \leq f(|I|)$.

The notion of a *gap*-reduction was introduced in [1] by Arora and Lund. A minimization problem Π is called *gap-reducible* to a maximization problem Π' with parameters (c, ρ) and (c', ρ'), if there exists a polynomial time computable function f such that f maps an instance I of Π to an instance I' of Π', while satisfying the following properties.

- If $opt(I) \leq c$ then $opt(I') \geq c'$
- If $opt(I) > c\rho$ then $opt(I') < \frac{c'}{\rho'}$

Parameters c and ρ are function of $|I|$ and parameters c' and ρ' are function of $|I'|$. Also, $\rho, \rho' \geq 1$.

The interest of a *gap*-reduction is that if Π is not approximable within a factor ρ then Π' is not approximable within a factor ρ'.

The notion of an *E*-reduction (*error-preserving* reduction) was introduced in [9] by Khanna et al. A problem Π is called *E-reducible* to a problem Π', if there exist polynomial time computable functions f, g and a constant β such that

- f maps an instance I of Π to an instance I' of Π' such that $opt(I)$ and $opt(I')$ are related by a polynomial factor, i.e. there exists a polynomial p such that $opt(I') \leq p(|I|)opt(I)$,
- g maps any solution S' of I' to one solution S of I such that $\varepsilon(I, S) \leq \beta\varepsilon(I', S')$.

An important property of an *E*-reduction is that it can be applied uniformly to all levels of approximability; that is, if Π is *E*-reducible to Π' and Π' belongs to $\mathcal{C}$ then Π belongs to $\mathcal{C}$ as well, where $\mathcal{C}$ is a class of optimization problems with any kind of approximation guarantee (see also [9]).

3 *NP*-Hardness of Approximation

We first prove that k MOST VITAL EDGES (NODES) 1-MEDIAN and k MOST VITAL EDGES (NODES) 1-CENTER are not constant approximable for some constants, unless $P=NP$. For this, we construct, in theorems 1 and 2, *gap*-reductions from MIN VERTEX COVER restricted to tripartite graphs. This problem is shown *NP*-hard in [7] where Garey *et al.* prove that it is *NP*-hard to find a minimum vertex cover in graphs of maximum degree 3, considering also that these graphs, with the exception of the clique K_4, are 3-colorable [4].

Theorem 1. k MOST VITAL EDGES 1-MEDIAN *and* k MOST VITAL EDGES 1-CENTER *are NP-hard to approximate within a factor* $\frac{7}{5} - \epsilon$ *and* $\frac{4}{3} - \epsilon$ *respectively, for any* $\epsilon > 0$.

Proof. We first consider k MOST VITAL EDGES 1-MEDIAN.

Let I be an instance of MIN VERTEX COVER formed by a graph $G = (V, E)$ with a tripartition $V = V_1 \cup V_2 \cup V_3$ and $|V| = n$. We construct an instance I' of k MOST VITAL EDGES 1-MEDIAN consisting of a graph $G' = (V', E')$ with $k < n$ as follows (see Figure 1). We associate for each node $v_\ell^i \in V_i$, two nodes

$v_{\ell,1}^i$ and $v_{\ell,2}^i$ in V' and connect them in E', for $i = 1, 2, 3$ and $\ell = 1, \ldots, |V_i|$. We add for each edge $(v_\ell^i, v_r^j) \in E$, with $i < j$, the edge $(v_{\ell,2}^i, v_{r,1}^j)$ to E'. We also add four nodes x_1, x_2, x_2', x_3 connected by the path $(x_1, x_2'), (x_2', x_2), (x_2, x_3)$. We connect x_1 to $v_{\ell,1}^1$ for $\ell = 1, \ldots, |V_1|$, x_2' to $v_{\ell,1}^2$ and x_2 to $v_{\ell,2}^2$ for $\ell = 1, \ldots, |V_2|$ and x_3 to $v_{\ell,2}^3$ for $\ell = 1, \ldots, |V_3|$. We assign a distance 1 to edges (x_1, x_2'), $(x_1, v_{\ell,1}^1)$, $(x_2', v_{j,1}^2)$, $(x_2, v_{j,2}^2)$ and (x_2', x_3) for $\ell = 1, \ldots, |V_1|$ and $j = 1, \ldots, |V_2|$, a distance 2 for the edge (x_2', x_2) and a distance 0 for all the other edges in E'. We set $w_{x_1} = 8$, $w_{x_2} = w_{x_3} = 1$ and assign a weight 0 to all other nodes in V'. We replace all edges of E', except the edges $(v_{\ell,1}^i, v_{\ell,2}^i)$, for $i = 1, 2, 3$ and $\ell = 1, \ldots, |V_i|$, by the gadget given in Figure 2. For each edge to be replaced, one chooses indifferently the vertex playing the role of i in Figure 2, except for all edges incident to x_1 for which we take x_1 as i. We show in the following that:

1. $opt(I) \le k \Rightarrow opt(I') \ge 7$
2. $opt(I) > k \Rightarrow opt(I') \le 5$

which proves that k MOST VITAL EDGES 1-MEDIAN is NP-hard to approximate within a factor $\frac{7}{5} - \epsilon$, for any $\epsilon > 0$.

First observe that there exists at least one optimal solution of k MOST VITAL EDGES 1-MEDIAN containing only edges among the edges $(v_{\ell,1}^i, v_{\ell,2}^i)$, for $i = 1, 2, 3$ and $\ell = 1, \ldots, |V_i|$. Indeed, if a solution contains edges from a gadget

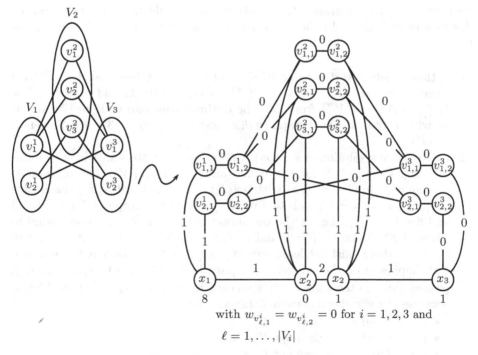

with $w_{v_{\ell,1}^i} = w_{v_{\ell,2}^i} = 0$ for $i = 1, 2, 3$ and
$\ell = 1, \ldots, |V_i|$

Fig. 1. Construction of G' from G

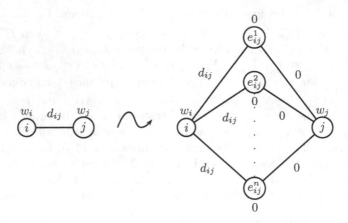

Fig. 2. The replacement gadget of an edge $e = (i, j) \in E'$

corresponding to an initial edge (i, j), it must contain at least n edges from this gadget in order to have a chance to increase the solution value by suppressing communication between i and j. Therefore, since $k < n$, it is at least as good to select k edges among those which do not belong to the gadgets.

Observe also that G' is designed so as to ensure that x_1 will always be the optimal 1-median node. Indeed, since the weight of vertex x_1 is 8 and all edges incident to x_1 have distance 1, any other node would have a total weighted distance of at least 8. In the following, x_1 has always a total distance of at most 7.

1. If there exists a vertex cover $V' \subseteq V$ of cardinality less than k in G then consider any set of vertices $V'' \supset V'$ of cardinality k, and remove $S'' = \{(v_{\ell,1}^i, v_{\ell,2}^i) : v_\ell^i \in V''\}$ from G'. The optimal 1-median node in $G' - S''$ is x_1 with a total weighted distance $d(x_1, x_2) + d(x_1, x_3) = 3 + 4 = 7$. Hence, $opt(I') \geq 7$.

2. Let S^* be any solution of k MOST VITAL EDGES 1-MEDIAN which contains only edges $(v_{\ell,1}^i, v_{\ell,2}^i)$, for $i = 1, 2, 3$ and $\ell = 1, \ldots, |V_i|$. The optimal 1-median node in $G' - S^*$ is x_1 with $opt(I') = d(x_1, x_2) + d(x_1, x_3)$. Each edge $(v_{\ell,1}^i, v_{\ell,2}^i)$ of S^* corresponds to a node $v_\ell^i \in V_i$ in the graph G, for $i = 1, 2, 3$ and $\ell = 1, \ldots, |V_i|$. Let N^* be the subset of nodes in G that correspond to edges of S^*. Since $|N^*| = k$ and $opt(I) > k$, N^* is not a vertex cover in G. Thus, there exists at least one edge $(v_\ell^i, v_r^j) \in E$ which is not covered. This implies in G' the existence of a path from x_i (or x_i') to x_j, with $i < j$, passing through the gadget corresponding to the edge $(v_{\ell,2}^i, v_{r,1}^j)$, enabling a decrease of some shortest path distances. Hence,
   - if $i = 1$ and $j = 2$ then $opt(I') \leq 6$
   - if $i = 1$ and $j = 3$ then $opt(I') \leq 3$
   - if $i = 2$ and $j = 3$ then $opt(I') \leq 6$

   Therefore, $opt(I') \leq 6$.

We consider now k MOST VITAL EDGES 1-CENTER. We use the same construction as above. We show that:

1. $opt(I) \leq k \Rightarrow opt(I') \geq 4$
2. $opt(I) > k \Rightarrow opt(I') \leq 3$

which proves that k MOST VITAL EDGES 1-CENTER is NP-hard to approximate within a factor $\frac{4}{3} - \epsilon$, for any $\epsilon > 0$.

Similarly as above, there exists at least one optimal solution of k MOST VITAL EDGES 1-CENTER containing only edges among the edges $(v_{\ell,1}^i, v_{\ell,2}^i)$, for $i = 1, 2, 3$ and $\ell = 1, \ldots, |V_i|$. Moreover, as before, x_1 will always be the optimal 1-center node.

1. If there exists a vertex cover $V' \subseteq V$ of cardinality less than k in G then consider any set of vertices $V'' \supset V'$ of cardinality k, and remove $S'' = \{(v_{\ell,1}^i, v_{\ell,2}^i) : v_\ell^i \in V''\}$ from G'. The optimal 1-center node in $G' - S''$ is x_1 with a maximum weighted distance $\max\{d(x_1, x_2), d(x_1, x_3)\} = 4$. Hence, $opt(I') \geq 4$.
2. Let S^* be any solution of k MOST VITAL EDGES 1-CENTER which contains only edges $(v_{\ell,1}^i, v_{\ell,2}^i)$, for $i = 1, 2, 3$ and $\ell = 1, \ldots, |V_i|$. The optimal 1-center node in $G' - S^*$ is x_1 with $opt(I') = \max\{d(x_1, x_2), d(x_1, x_3)\}$. Each edge $(v_{\ell,1}^i, v_{\ell,2}^i)$ of S^* corresponds to a node $v_\ell^i \in V_i$ in the graph G, for $i = 1, 2, 3$ and $\ell = 1, \ldots, |V_i|$. Let N^* be the subset of nodes of G corresponding to edges in S^*. Since $|N^*| = k$ and $opt(I) > k$, N^* is not a vertex cover in G. Thus, there exists at least one edge $(v_\ell^i, v_r^j) \in E$ which is not covered. This implies in G' the existence of a path from x_i (or x_i') to x_j, with $i < j$, passing through the gadget corresponding to the edge $(v_{\ell,2}^i, v_{r,1}^j)$. Hence,
   - if $i = 1$ and $j = 2$ then $opt(I') \leq 3$
   - if $i = 1$ and $j = 3$ then $opt(I') \leq 3$
   - if $i = 2$ and $j = 3$ then $opt(I') \leq 3$

   Therefore, $opt(I') \leq 3$. $\qquad\square$

Theorem 2. k MOST VITAL NODES 1-MEDIAN and k MOST VITAL NODES 1-CENTER are NP-hard to approximate within a factor $\frac{3}{2} - \epsilon$, for any $\epsilon > 0$.

Proof. We first consider k MOST VITAL NODES 1-MEDIAN.

Let I be an instance of MIN VERTEX COVER formed by a graph $G = (V, E)$ with a tripartition $V = V_1 \cup V_2 \cup V_3$ and $|V| = n$. We construct an instance I' of k MOST VITAL NODES 1-MEDIAN consisting of a graph $G' = (V', E')$ with $k < n$ as follows (see Figure 3). G' is a copy of G to which we add complete graphs K_n^i with n nodes $x_i^1, \ldots, x_i^n$ for $i = 1, 2, 3$. We connect each node $v_\ell^i \in V_i$ with each node x_i^r, for $i = 1, 2, 3$, $\ell = 1, \ldots, |V_i|$ and $r = 1, \ldots, n$. We connect also each node x_i^r to each node x_{i+1}^r for $i = 1, 2$ and $r = 1, \ldots, n$. We assign a distance 2 to edges (x_i^r, x_{i+1}^r) for $i = 1, 2$ and $r = 1, \ldots, n$, a distance 1 to edges (x_1^r, v_ℓ^1) for $\ell = 1, \ldots, |V_1|$ and $r = 1, \ldots, n$ and a distance 0 to all other edges in E'. We set $w_{x_1^r} = 7$ and $w_{x_2^r} = w_{x_3^r} = 1$ for $r = 1, \ldots, n$, and $w_{v_\ell^i} = 0$ for $i = 1, 2, 3$, $\ell = 1, \ldots, |V_i|$. We show in the following that:

1. $opt(I) \leq k \Rightarrow opt(I') \geq 6n$
2. $opt(I) > k \Rightarrow opt(I') \leq 4n$

which proves that k MOST VITAL NODES 1-MEDIAN is NP-hard to approximate within a factor $\frac{3}{2} - \epsilon$, for any $\epsilon > 0$.

First observe that there exists at least one optimal solution of k MOST VITAL NODES 1-MEDIAN containing only nodes of V. Indeed, if a solution contains nodes from K_n^i for some i, it must contain all nodes of K_n^i in order to have a chance to increase the solution value by disconnecting these nodes from the graph. Therefore, since $k < n$, it is at least as good to select k nodes in V only.

Observe also that G' is designed so as to ensure that any node x_1^r for $r = 1, \ldots, n$ will always be an optimal 1-median node. Indeed, since the weight of a vertex x_1^r is 7 and all edges incident to x_1^r, except the edges (x_1^r, x_1^j) for $j = 1, \ldots, n$ and $j \neq r$ have distance at least 1, any other node would have a total weighted distance of at least 7, while any node x_1^r has always a total weighted distance of at most 6. We consider arbitrarily in the following that x_1^1 is the selected optimal 1-median node.

1. If there exists a vertex cover $V' \subseteq V$ of cardinality less than k in G then consider any set of vertices $V'' \supset V'$ of cardinality k, and remove V'' from G'. Taking x_1^1 as the optimal 1-median node in $G' - V''$, we get a total weighted distance $\sum_{j=1}^{n}(d(x_1^1, x_2^j) + d(x_1^1, x_3^j)) = \sum_{j=1}^{n}(2 + 4) = 6n$. Hence, $opt(I') \geq 6n$.

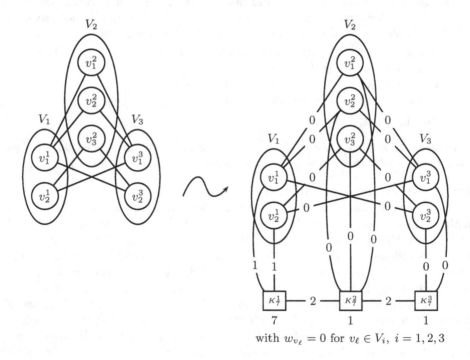

with $w_{v_\ell} = 0$ for $v_\ell \in V_i$, $i = 1, 2, 3$

Fig. 3. Construction of G' from G

2. Let $N^* \subseteq V$ be any solution of k MOST VITAL NODES 1-MEDIAN which contains only nodes of V. Taking x_1^1 as the optimal 1-median node in $G' - N^*$, we get $opt(I') = \sum_{\ell=1}^{n}(d(x_1^1, x_2^\ell) + d(x_1^1, x_3^\ell))$. Since $|N^*| = k$ and $opt(I) > k$, N^* is not a vertex cover in G. Thus, there exists at least one edge $(v_i, v_j) \in E$ which is not covered. This implies in G' the existence of a path from each x_i^r to each x_j^r for $r = 1, \ldots, n$, passing through the edge (v_i, v_j). Hence,
- if $i = 1$ and $j = 2$ then $opt(I') \leq \sum_{\ell=1}^{n}(1 + 3) = 4n$
- if $i = 1$ and $j = 3$ then $opt(I') \leq \sum_{\ell=1}^{n}(2 + 1) = 3n$
- if $i = 2$ and $j = 3$ then $opt(I') \leq \sum_{\ell=1}^{n}(2 + 2) = 4n$

Consequently, $opt(I') \leq 4n$.

We consider now k MOST VITAL NODES 1-CENTER. We use the same construction as above, but we modify the distance associated to the edges (x_2^r, x_3^r) for $r = 1, \ldots, n$ for which we assign a distance 1. We show that:

1. $opt(I) \leq k \Rightarrow opt(I') \geq 3$
2. $opt(I) > k \Rightarrow opt(I') \leq 2$

which proves that k MOST VITAL NODES 1-CENTER is NP-hard to approximate within a factor $\frac{4}{3} - \epsilon$, for any $\epsilon > 0$.

As previously, we can show that only the nodes of V can be removed. We observe as above that any node x_1^r for $r = 1, \ldots, n$ will always be an optimal 1-center node. We consider arbitrarily in the following that x_1^1 is the selected optimal 1-center node.

1. If there exists a vertex cover $V' \subseteq V$ of cardinality less than k in G then consider any set of vertices $V'' \supset V'$ of cardinality k, and remove V'' from G'. Taking x_1^1 as the optimal 1-center node in $G' - V''$, we get a maximum weighted distance $\max \{ \max_{j=1,\ldots,n} d(x_1^1, x_2^j), \max_{j=1,\ldots,n} d(x_1^1, x_3^j) \} = 3$. Hence, $opt(I') \geq 3$.
2. Let $N^* \subseteq V$ be any solution of k MOST VITAL NODES 1-CENTER which contains only nodes of V. Taking x_1^1 as the optimal 1-center node in $G' - N^*$, we get $opt(I') = \max \{ \max_{\ell=1,\ldots,n} d(x_1^1, x_2^\ell), \max_{\ell=1,\ldots,n} d(x_1^1, x_3^\ell) \}$. Since $|N^*| = k$ and $opt(I) > k$, N^* is not a vertex cover in G. Thus, there exists at least one edge $(v_i, v_j) \in E$ which is not covered. This implies in G' the existence of a path from each x_i^r to each x_j^r for $r = 1, \ldots, n$, passing through the edge (v_i, v_j). Hence,
   - if $i = 1$ and $j = 2$ then $opt(I') = \max\{d(x_1^1, x_2^1), d(x_1^1, x_3^1)\} \leq 2$
   - if $i = 1$ and $j = 3$ then $opt(I') = d(x_1^1, x_2^1) \leq 2$
   - if $i = 2$ and $j = 3$ then $opt(I') = \max\{d(x_1^1, x_2^1), d(x_1^1, x_3^1)\} \leq 2$.

   Therefore, $opt(I') \leq 2$.

$\square$

We prove now that the four problems MIN EDGE (NODE) BLOCKER 1-MEDIAN and MIN EDGE (NODE) BLOCKER 1-CENTER are not 1.36 approximable, unless $P=NP$. These results, stated in theorems 3 and 4, are obtained by constructing E-reductions from MIN VERTEX COVER shown NP-hard to approximate within a factor 1.36 [5].

Theorem 3. MIN EDGE BLOCKER 1-MEDIAN *and* MIN EDGE BLOCKER 1-CENTER *are* NP-*hard to approximate within a factor 1.36.*

Proof. We first consider MIN EDGE BLOCKER 1-MEDIAN.

Let I be an instance of MIN VERTEX COVER consisting of a graph $G = (V, E)$ with $V = \{v_1, \ldots, v_n\}$. We construct an instance I' of MIN EDGE BLOCKER 1-MEDIAN formed by a graph $G' = (V', E')$ and a positive integer U as follows (see Figure 4). We associate for each node $v_i \in V$ two nodes v_i and v_i' in V' and connect them in E' for $i = 1, \ldots, n$. We add for each edge $(v_i, v_j) \in E$, with $i < j$, an edge (v_i', v_j) to E'. We also add $2n$ nodes $x_1, x_1', x_2, x_2', \ldots, x_n, x_n'$ connected by the path $(x_1, x_1'), (x_1', x_2), (x_2, x_2'), (x_2', x_3), \ldots, (x_{n-1}', x_n), (x_n, x_n')$. Finally, we connect x_i to v_i and x_i' to v_i' for $i = 1, \ldots, n$. We assign the following distances to the edges of E': $d_{v_i v_i'} = 0$, $d_{x_i v_i} = d_{x_i' v_i'} = 1$ and $d_{x_i x_i'} = 2$ for $i = 1, \ldots, n$, $d_{x_i' x_{i+1}} = 0$ for $i = 1, \ldots, n-1$ and $d_{v_i' v_j} = 2(j-i) - 1$ for $(v_i, v_j) \in E$ and $i < j$. We set $w_{x_1} = 2n^2 + 1$, $w_{x_i} = 1$ for $i = 2, \ldots, n$, $w_{x_i'} = 1$ and $w_{v_i} = w_{v_i'} = 0$ for $i = 1, \ldots, n$ and we consider that $U = 2n^2$. We replace each edge of E', except the edges (v_i, v_i') for $i = 1, \ldots, n$, by the gadget given in Figure 2 where each edge is replaced by $n+1$ instead of n disjoint paths of length 2 (for edges (x_1, v_1) and (x_1, x_1'), x_1 plays the role of i in Figure 2).

Observe that G' is designed so as to ensure that x_1 will always be the optimal 1-median node. Indeed, since the weight of vertex x_1 is $2n^2 + 1$ and all edges incident to x_1 have distance at least 1, any other node would have a total weighted distance of at least $2n^2 + 1$. In the following, x_1 has always a total distance of at most $2n^2$.

We prove first that $opt(I') \leq opt(I)$. Let $V^* \subseteq V$ be a minimum vertex cover of G. Let us consider $S^* = \{(v_i, v_i') : v_i \in V^*\}$. By removing the edges

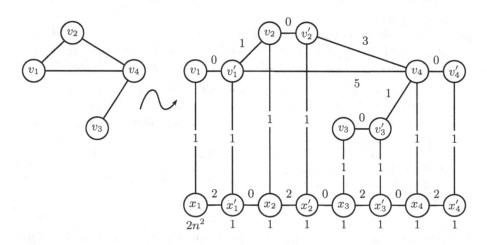

with $w_{v_\ell} = w_{v_\ell'} = 0$ for $\ell = 1, \ldots, 4$

Fig. 4. Construction of G' from G with $n = 4$ nodes

in S^* from G', the optimal 1-median node is x_1 with a total weighted distance $\sum_{i=1}^n w_{x_i} d(x_1, x_i) + \sum_{i=1}^n w_{x_i'} d(x_1, x_i') = 2(\sum_{i=1}^{n-1} i + \sum_{i=1}^n i) = 2n^2 = U$. Hence, $opt(I') \leq |S^*| = opt(I)$.

When we remove all edges (v_i, v_i'), for $i = 1, \ldots, n$ from G', the optimal 1-median node in the resulting graph is x_1 with value U. Hence, $opt(I') \leq n$. Let $S \subseteq E'$ be an edge blocker for G'. If S contains an edge (i, e_{ij}^ℓ) or (e_{ij}^ℓ, j) from a gadget corresponding to an initial edge (i, j), it must contain at least $n + 1$ edges from this gadget in order to suppress the communication between i and j, otherwise the value of an optimal solution for the 1-median problem in $G' - S$ is the same as in $G' - (S \backslash \{(i, e_{ij}^\ell)\})$ or $G' - (S \backslash \{(e_{ij}^\ell, j)\})$. Therefore, since $opt(I') \leq n$, we can consider in the following that S contains only edges among the edges (v_i, v_i'), $i \in \{1, \ldots, n\}$.

Let us consider $N = \{v_i : (v_i, v_i') \in S\}$ where S is an edge blocker. We prove, by contradiction, that N is a vertex cover in G. Suppose that there exists an edge $(v_i, v_j) \in E$ such that $v_i \notin N$, $v_j \notin N$ and $i < j$. We show in the following that by removing S from G', the value of an optimal solution for the 1-median problem in the remaining graph is strictly less than $2n^2$. Indeed, x_1 is the optimal 1-median node in $G' - S$. Let $D(x_1)$ be the total weighted distance associated to x_1 in $G' - S$. We have $D(x_1) = \sum_{\ell=1}^n d(x_1, x_\ell') + \sum_{\ell=1}^n d(x_1, x_\ell) = \sum_{\ell=1}^{j-1} d(x_1, x_\ell') + d(x_1, x_j') + \sum_{\ell=j+1}^n d(x_1, x_\ell') + \sum_{\ell=1}^n d(x_1, x_\ell)$. Then, $D(x_1) \leq 2 \sum_{\ell=1}^{j-1} \ell + d(x_1, x_j') + 2 \sum_{\ell=j+1}^n \ell + 2 \sum_{\ell=1}^{n-1} \ell = 2 \sum_{\ell=1}^n \ell - 2j + d(x_1, x_j') + 2 \sum_{\ell=1}^{n-1} \ell = 2n^2 - 2j + d(x_1, x_j')$. The edge (v_i, v_j) being not covered, this implies the existence of a path from x_1 to x_j' using a subpath from x_1 to x_i and joining x_i to x_j' by a subpath passing through the gadget associated to the edge (v_i, v_j). We have $d(x_1, x_j') \leq 2(i-1) + 1 + 2(j-i) - 1 + 1 = 2j - 1$. Thus, we have $D(x_1) \leq 2n^2 - 1 < 2n^2$, contradicting the assumption that S is an edge blocker. Therefore, N is a vertex cover in G such that $val(I, N) = val(I', S)$. Consequently, $\varepsilon(I, N) = \frac{val(I,N)}{opt(I)} - 1 \leq \frac{val(I',S)}{opt(I')} - 1 = \varepsilon(I', S)$, which achieves the proof.

We consider now MIN EDGE BLOCKER 1-CENTER.

We use the same construction as above with $U = 2n$. As above, G' is designed so as to ensure that x_1 will always be the optimal 1-center node.

We show first that $opt(I') \leq opt(I)$. Let $V^* \subseteq V$ be a minimum vertex cover in G. Let us consider $S^* = \{(v_i, v_i') : v_i \in V^*\}$. By removing the edges of S^* from the graph G', the optimal 1-center node is x_1 with a maximum weighted distance $d(x_1, x_n') = 2n = U$. Hence, $opt(I') \leq |S^*| = opt(I)$.

Let $S \subseteq E'$ be an edge blocker. We can assume, similarly to the 1-median problem, that S contains only edges among the edges (v_i, v_i'), $i \in 1, \ldots, n$. Let us consider $N = \{v_i : (v_i, v_i') \in S\}$. In the following, we show by contradiction that N is a vertex cover in G. Suppose that there exists an edge $(v_i, v_j) \in E$ such that $v_i \notin N$, $v_j \notin N$ and $i < j$. Then x_1 is the optimal 1-center node in $G' - S$ with a maximum weighted distance $D_{max}(x_1) = d(x_1, x_n)$. The edge (v_i, v_j) being not covered, this implies the existence of a path from x_1 to x_j' using a subpath from x_1 to x_i and joining x_i to x_j' by a subpath passing through the gadget associated to the edge (v_i, v_j). Then $D_{max}(x_1) \le 2(i-1)+1+2(j-i) - 1+1+2(n-j) = 2n-1 < 2n$, contradicting the assumption that S is an edge blocker. Therefore N is a vertex cover in G such that $val(I, N) = val(I', S)$. Consequently, $\varepsilon(I, N) = \frac{val(I,N)}{opt(I)} - 1 \le \frac{val(I',S)}{opt(I')} - 1 = \varepsilon(I', S)$, which achieves the proof.   □

Theorem 4. MIN NODE BLOCKER 1-MEDIAN *and* MIN NODE BLOCKER 1-CENTER *are* NP-*hard to approximate within a factor* 1.36.

Proof. We consider first MIN NODE BLOCKER 1-MEDIAN.

Let I be an instance of MIN VERTEX COVER consisting of a graph $G = (V, E)$ with $V = \{v_1, \ldots, v_n\}$. We construct an instance I' of MIN NODE BLOCKER 1-MEDIAN formed by a graph $G' = (V', E')$ and a positive integer U as follows (see Figure 5). G' is a copy of G to which we add one node x_1 and complete graphs K_{n+1}^i with $n + 1$ nodes $x_i^1, \ldots, x_i^{n+1}$ for $i = 2, \ldots, n$. We connect x_1 to v_1 and x_2^r for $r = 1, \ldots, n + 1$, and each node x_i^r to v_i for $i = 2, \ldots, n$ and $r = 1, \ldots, n + 1$. We also connect each node x_i^r to each node x_{i+1}^r for $r = 1, \ldots, n + 1$ and $i = 2, \ldots, n - 1$. We assign a distance 1 to the edge (x_1, v_1), a distance 2 to the edges (x_1, x_2^r) and (x_i^r, x_{i+1}^r) for $i = 2, \ldots, n - 1$ and $r = 1, \ldots, n + 1$, and a distance 0 to all other edges in E'. Let us set $w_{x_1} = n^3$, $w_{x_i^r} = 1$ for $i = 2, \ldots, n$ and $r = 1, \ldots, n + 1$ and $w_{v_i} = 0$ for $i = 1, \ldots, n$. Finally, we set $U = n(n^2 - 1)$.

Observe that G' is designed so as to ensure that x_1 will always be the optimal 1-median node. Indeed, since the weight of vertex x_1 is n^3 and all edges incident to x_1 have distance at least 1, any other node would have a total weighted distance of at least n^3. In the following, x_1 has always a total distance of at most $n(n^2 - 1)$.

We show first that $opt(I') \le opt(I)$. Let $V^* \subseteq V$ be a minimum vertex cover in G. By removing V^* from G', the optimal 1-median node is x_1 with a total

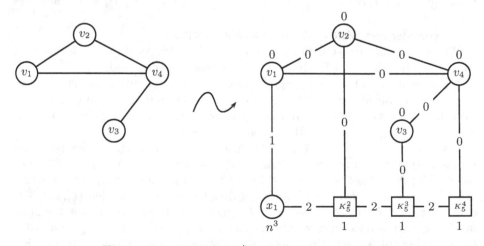

Fig. 5. Construction of G' from G with $n = 4$ nodes

weighted distance $\sum_{\ell=2}^{n} \sum_{r=1}^{n+1} d(x_1, x_\ell^r) = 2(n+1) \sum_{i=1}^{n-1} i = n(n-1)(n+1) = U$. Hence, $opt(I') \leq |V^*| = opt(I)$.

Let $N \subseteq V'$ be a node blocker. According to the construction of G', in order to obtain an optimal solution for the 1-median problem in the graph $G' - N$ of a value at least U, N must be included in V. We show, by contradiction, that N is a vertex cover in G. Suppose that there exists an edge $(v_i, v_j) \in E$ such that $v_i \notin N$, $v_j \notin N$ and $i < j$. The optimal 1-median node in $G' - N$ is x_1 with value strictly less than $n(n-1)(n+1)$. Indeed, let $D(x_1)$ be the total weighted distance associated to x_1 in $G' - N$. Hence, $D(x_1) = \sum_{\ell=2}^{n} \sum_{r=1}^{n+1} d(x_1, x_\ell^r) = \sum_{\ell=2}^{j-1} \sum_{r=1}^{n+1} d(x_1, x_\ell^r) + \sum_{r=1}^{n+1} d(x_1, x_j^r) + \sum_{r=1}^{n+1} \sum_{\ell=j+1}^{n} d(x_1, x_\ell^r))$. We distinguish two cases:

- If $v_i = v_1$ then $d(x_1, x_j^r) = d_{x_1 v_1} + d_{v_1 v_j} + d_{v_j x_j^r} = 1$ for $r = 1, \ldots, n+1$. Hence, we obtain $D(x_1) \leq 2(n+1) \sum_{\ell=1}^{j-2} \ell + (n+1) + 2(n+1) \sum_{\ell=j}^{n-1} \ell < 2(n+1) \sum_{\ell=1}^{j-2} \ell + 2(j-1)(n+1) + 2(n+1) \sum_{\ell=j}^{n-1} \ell = n(n-1)(n+1)$, contradiction.
- If $v_i \neq v_1$ then $d(x_1, x_j^r) = d(x_1, x_i^1) + d_{x_i^1 v_i} + d_{v_i v_j} + d_{v_j x_j^r} = d(x_1, x_i^1)$ for $r = 1, \ldots, n+1$. Hence, we obtain $D(x_1) \leq 2(n+1) \sum_{\ell=1}^{j-2} \ell + 2(i-1)(n+1) + 2(n+1) \sum_{\ell=j}^{n-1} \ell < 2(n+1) \sum_{\ell=1}^{j-2} \ell + 2(j-1)(n+1) + 2(n+1) \sum_{\ell=j}^{n-1} \ell = n(n-1)(n+1)$, contradiction.

Therefore N is a vertex cover in G such that $val(I, N) = val(I', N)$. Consequently, $\varepsilon(I, N) = \frac{val(I,N)}{opt(I)} - 1 \leq \frac{val(I',N)}{opt(I')} - 1 = \varepsilon(I', N)$, which achieves the proof.

We consider now MIN NODE BLOCKER 1-CENTER.

We use the same construction as above with $U = 2(n-1)$. Here again, we observe that G' is designed so as to ensure that x_1 will always be the optimal 1-center node.

We show first that $opt(I') \leq opt(I)$. Let $V^* \subseteq V$ be a minimum vertex cover in G. By deleting the nodes of V^* from G', the optimal 1-center node in the remaining graph is x_1 with a maximum weighted distance $d(x_1, x_n^r) = 2(n-1) = U$ for any $r = 1, \ldots, n+1$. Hence, $opt(I') \leq |V^*| = opt(I)$.

When we remove all nodes v_i, $i = 1, \ldots, n$ from G', the optimal 1-center node in the resulting graph is x_1 with value U. Hence, $opt(I') \leq n$. Let $N \subseteq V'$ be a node blocker. According to the construction of G', in order to obtain an optimal 1-center node in $G' - N$ of value at least U, N cannot contain x_1. If N contains nodes x_i^ℓ for a given i and ℓ, then N must contains all the $n+1$ nodes x_i^r for $r = 1, \ldots, n+1$, otherwise the value of an optimal solution for the 1-center problem in $G' - N$ is the same as in $G' - (N \setminus \{x_i^\ell\})$. Therefore, since $opt(I') \leq n$, we can consider in the following that N is included in V. In the following, we prove by contradiction that N forms a vertex cover in G. Suppose that there exists an edge $(v_i, v_j) \in E$ such that $v_i \notin N$, $v_j \notin N$ and $i < j$. By removing N from G', the optimal 1-center node is x_1 with a maximum weighted distance $D_{max}(x_1) = d(x_1, x_n^r)$ for any $r = 1, \ldots, n$. We distinguish two cases:

- if $v_i = v_1$ then $D_{max}(x_1) = d_{x_1 v_1} + d_{v_1 v_j} + d_{v_j x_j^1} + d(x_j^1, x_n^r) \le 1 + 0 + 0 + 2(n - j) \le 1 + 2n - 4 < 2(n - 1)$, contradiction.
- if $v_i \ne v_1$ then $D_{max}(x_1) \le d(x_1, x_i^1) + d_{x_i^1 v_i} + d_{v_i v_j} + d_{v_j x_j^1} + d(x_j^1, x_n^r) \le 2(i - 1) + 0 + 0 + 0 + 2(n - j) = 2(n - 1) - 2(j - i) < 2(n - 1)$, contradiction.

Therefore N is a vertex cover in G such that $val(I, N) = val(I', N)$. Consequently, $\varepsilon(I, N) = \frac{val(I,N)}{opt(I)} - 1 \le \frac{val(I',N)}{opt(I')} - 1 = \varepsilon(I', N)$, which achieves the proof. $\qquad\square$

4   Conclusion

We established in this paper negative results concerning the approximation of k most vital edges (nodes) and min edge (node) blocker versions of the 1-median and 1-center location problems. An interesting open question would be to establish positive results concerning the approximability of these problems. Another interesting perspective is to find efficient exact algorithms to solve them.

References

1. Arora, S., Lund, C.: Hardness of approximations. In: Approximation Algorithms for NP-hard Problems, pp. 399–446. PWS Publishing Company (1996)
2. Bar-Noy, A., Khuller, S., Schieber, B.: The complexity of finding most vital arcs and nodes. Technical Report CS-TR-3539, Department of Computer Science, University of Maryland (1995)
3. Bazgan, C., Toubaline, S., Tuza, Z.: Complexity of most vital nodes for independent set in graphs related to tree structures. In: Proceedings of the 21st International Workshop on Combinatorial Algorithms (IWOCA 2010). Springer, Heidelberg (2010) (to appear in LNCS)
4. Brooks, R.L.: On colouring the nodes of a network. Mathematical Proceeing of the Cambridge Philosophical Society 37(2), 194–197 (1941)
5. Dinur, I., Safra, S.: On the hardness of approximating minimum vertex cover. Annals of Mathematics 162(1), 439–485 (2005)
6. Frederickson, G.N., Solis-Oba, R.: Increasing the weight of minimum spanning trees. In: Proceedings of the 7th Annual ACM-SIAM Symposium on Discrete Algorithms (SODA 1996), pp. 539–546 (1996)
7. Garey, M.R., Johnson, D.S., Stockmeyer, L.: Some simplified NP-complete graph problems. Theoretical Computer Science 1(3), 237–267 (1976)
8. Khachiyan, L., Boros, E., Borys, K., Elbassioni, K., Gurvich, V., Rudolf, G., Zhao, J.: On short paths interdiction problems: total and node-wise limited interdiction. Theory of Computing Systems 43(2), 204–233 (2008)
9. Khanna, S., Motwani, R., Sudan, M., Vazirani, U.: On syntactic versus computational views of approximability. In: Proceedings of the 35th Annual IEEE Annual Symposium on Foundations of Computer Science (FOCS 1994), pp. 819–830 (1994); Also published in SIAM Journal on Computing 28(1), 164-191 (1999)

10. Ries, B., Bentz, C., Picouleau, C., de Werra, D., Costa, M., Zenklusen, R.: Blockers and transversals in some subclasses of bipartite graphs: When caterpillars are dancing on a grid. Discrete Mathematics 310(1), 132–146 (2010)
11. Wood, R.K.: Deterministic network interdiction. Mathematical and Computer Modeling 17(2), 1–18 (1993)
12. Zenklusen, R., Ries, B., Picouleau, C., de Werra, D., Costa, M., Bentz, C.: Blockers and transversals. Discrete Mathematics 309(13), 4306–4314 (2009)

PTAS for Minimum Connected Dominating Set with Routing Cost Constraint in Wireless Sensor Networks

Hongwei Du[1], Qiang Ye[1], Jioafei Zhong[2], Yuexuan Wang[3],
Wonjun Lee[4], and Haesun Park[5]

[1] Dept. of Computer Science and Information Science,
University of Prince Edward Island, Canada
hongwei.ddu@gmail.com
[2] Department of Computer Science, University of Texas at Dallas,
Richardson, TX 75080, USA
{weiliwu,fayzhong}@utdallas.edu
[3] Institute for Theoretical Computer Science, Tsinghua University,
Beijing, 100084, P.R. China
wangyuexuan@tsinghua.edu.cn
[4] Dept. of Computer Science and Engineering, Korea University,
Seoul, Republic of Korea
wlee@korea.ac.kr
[5] School of Computational Science and Engineering,
George Institute of Technology, USA
hpark@cc.gatech.edu

Abstract. To reduce routing cost and to improve road load balance, we study a problem of minimizing size of connected dominating set D under constraint that for any two nodes u and v, the routing cost through D is within a factor of α from the minimum, the cost of the shortest path between u and v. We show that for $\alpha \geq 5$, this problem in unit disk graphs has a polynomial-time approximation scheme, that is, for any $\varepsilon > 0$, there is a polynomial-time $(1 + \varepsilon)$-approximation.

1 Introduction

Given a graph $G = (V, E)$, a node subset $D \subseteq V$ is called a *dominating set* if every node not in D has a neighbor in D. A dominating set is said to be a *connected dominating set* (CDS) if it induces a connected subgraph.

Due to applications in wireless networks, the MCDS problem, i.e., computing the minimum connected dominating set (MCDS) for a given graph, has been studied extensively since 1998 [15,1,10,11,12,14,2,3].

Guha and Khuller [7] showed that the MCDS in general graph has no polynomial time approximation with performance ratio $\rho \ln \delta$ for $0 < \rho < 1$ unless $NP \subseteq DTIME(n^{O(\log \log n)})$ where δ is the maximum node degree of input graph. They also gave a polynomial-tme $(3 + \ln \delta)$-approximation. Ruan *et al.* [9] and Du *et al.* [6] made improvements.

W. Wu and O. Daescu (Eds.): COCOA 2010, Part I, LNCS 6508, pp. 252–259, 2010.
© Springer-Verlag Berlin Heidelberg 2010

Recently, motivated from reducing routing cost [4] and from improving road load balancing [13], the following problem has been proposed:

> MOC-CDS: Given a connected graph $G = (V, E)$, compute a connected dominating set D with minimum cardinality under condition that for every two nodes $u, v \in V$, there exists a shortest path between u and v such that all intermediate nodes belong to D.

Ding *et al.* [4] showed that MOC-CDS has no polynomial time approximation with performance ratio $\rho \ln \delta$ for $0 < \rho < 1$ unless $NP \subseteq DTIME(n^{O(\log \log n)})$ where δ is the maximum node degree of input graph G. They also gave a polynomial time distributed approximation algorithm with performance ratio $H\left(\frac{\delta(\delta-1)}{2}\right)$ where H is the hamonic function, i.e., $H(k) = \sum_{i=1}^{k} \frac{1}{i}$.

However, some example shows that the solution of MOC-CDS may be much bigger than the solution of MCDS. Thus, to reach the minimum routing cost, the size of CDS may be increased too much. Motivated from this situation, Du *et al.* [5] proposed the following problem for any constant $\alpha \geq 1$.

> αMOC-CDS: Given a graph, compute the minimum CDS D such that for any two nodes u and v, $m_D(u, v) \leq \alpha \cdot m(u, v)$ where $m_D(u, v)$ is the number of intermediate nodes on a shortest path connecting u and v through D and $m(u, v) = m_G(u, v)$.

1MOC-CDS is exactly MOC-CDS. For $\alpha > 1$, the constraint on routing cost is relaxed and hence the CDS size becomes smaller. Du *et al.* [5] showed that for any $\alpha \geq 1$, αMOC-CDS in general graphs is APX-hard and hence has no PTAS unless $NP = P$. Liu *et al.* [8] showed that αMOC-CDS in unit disk graphs is NP-hard for $\alpha \geq 4$.

In this paper, we show that for $\alpha \geq 5$, αMOC-CDS in unit disk graphs has PTAS, that is, for any $\varepsilon > 0$. αMOC-CDS has polynomial-time $(1 + \varepsilon)$-approximation.

2   Main Result

First, let us quote a lemma in [5], simplifying the routing cost constraint. For convenience of the reader, we also include their proof here.

Lemma 1. *Let G be a connected graph and D a dominating set D of G. Then, for any two nodes u and v,*

$$m_D(u, v) \leq \alpha m(u, v),$$

if and only if for any two nodes u and v with $m(u, v) = 1$,

$$m_D(u, v) \leq \alpha. \tag{1}$$

Proof. It is trivial to show the "only if" part. Next, we show the "if" part. Consider two nodes u and v. If $m(u, v) = 0$, it is clear that $m_D(u, v) = 0 = \alpha m(u, v)$. Next, assume $m(u, v) \geq 1$. Consider a shortest path $(u, w_1, ..., w_k, v)$ where $k = m(u, v) \geq 1$. Let us assume k is even. For odd k, the proof is similar.

Note that $m(u, w_2) = m(w_2, w_4) = \cdots = m(w_{k-2}, w_k) = 1$. By (1), there exist paths $(u, s_{1,1}, s_{1,2}, ..., s_{1,h_1}, w_2)$, $(w_2, s_{3,1}, s_{3,2}, ..., s_{3,h_3}, w_4)$, ..., $(w_{k-2}, s_{k-1,1}, s_{k-1,2}, ..., s_{k-1,h_{k-1}}, w_k)$ such that $1 \leq h_i \leq \alpha$ for all $i = 1, 3, ..., k-1$ and $s_{i,j} \in D$ for all $i = 1, 3, ..., k-1$ and $j = 1, 2, ..., h_i$. Now, note that $m(s_{1,h_1}, s_{3,1}) = \cdots = m(s_{k-1,h_{k-1}}, v) = 1$. By (1), there exist paths $(s_{1,h_1}, s_{2,1}, s_{2,2}, ..., s_{2,h_2}, s_{3,1})$, ..., $(s_{k-1,h_{k-1}}, s_{k,1}, s_{k,2}, ..., s_{k,h_k}, v)$ such that $1 \leq h_i \leq \alpha$ for $i = 2, 4, ..., k$ and $s_{i,j} \in D$ for $i = 2, 4, ..., k$ and $j = 1, 2, ..., h_i$. Therefore, there is a path $(u, s_{1,1}, ..., s_{1,h_1}, s_{2,1}, ..., s_{k,h_k}, v)$ with $h_1 + h_2 + \cdots + h_k$ $(\leq \alpha k)$ intermediate nodes all in D. Thus, $m_D(u, v) \leq \alpha \cdot m(u, v)$.   □

Next, the following lemma indicates that a dominating set satisfying condition (1) must be feasible for αMOC-CDS.

Lemma 2. *In a connected graph, a dominating set D satisfying condition (1) must be a connected dominating set.*

Proof. If $|D| = 1$, then the subgraph induced by D consists of only single node, which is clearly a connected subgraph.

Next, assume $|D| \geq 2$. For any two nodes $u, v \in D$, by Lemma 1, there is a path between u and v with all intermediate nodes in D. Therefore, D induces a connected subgraph.   □

Now, we start to construct a PTAS for unit disk graphs.

First, we put input unit disk graph $G = (V, E)$ in the interior of the square $[0, q] \times [0, q]$. Then construct a grid $P(0)$ as shown in Fig. 1. $P(0)$ divides the sequare $[0, pa] \times [0, pa]$ into p^2 cells where $a = 2(\alpha + 2)k$ for a positive integer k and $p = 1 + \lceil q/a \rceil$. Each cell e is a $a \times a$ square, including its left boundary and its lower boundary, so that all cells are disjoint and their union covers the interior of the square $[0, q] \times [0, q]$.

For each cell e, construct a $(a+4) \times (a+4)$ square and a $(a+2\alpha+4) \times (a+2\alpha+4)$ square with the same center as that of e (Fig. 2). The closed area bounded by the first square is called the *central area* of cell e, denoted by e^c. The area between the second square and cell e, including the boundary of e and excluding the boundary of the second square, is called the *boundary area* of cell e, denoted by e^b. The union of the boundary area and the central area is the open area bounded by the second square, denoted by e^{cb}.

Now, for each cell e, we study the following problem.

LOCAL(e): Find the minimum subset D of nodes in $V \cap e^{cb}$ such that (a) D dominates all nodes in $V \cap e^c$, and (b) for any two nodes $u, v \in V \cap e^c$ with $m(u, v) = 1$ and $\{u, v\} \cap e \neq \emptyset$, $m_D(u, v) \leq \alpha$.

Lemma 3. *Suppose $\alpha \geq 5$ and $|V \cap e^{cb}| = n_e$. Then the minimum solution of LOCAL(e) problem can be computed in time $n_e^{O(a^4)}$.*

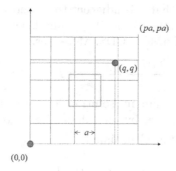

Fig. 1. Grid $P(0)$

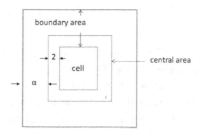

Fig. 2. Central Area e^c and Boundary Area e^b

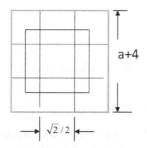

Fig. 3. Decompsition of Central Area e^c

Proof. Cut e^c into $\lceil(a+4)\sqrt{2}\rceil^2$ small squares with edge length at most $\sqrt{2}/2$ (Fig. 3). Then for each (closed) small square s, if $V \cap s \neq \emptyset$, then choose one which would dominates all nodes in $V \cap s$. Those nodes form a set D dominating $V \cap e^c$ and $|D| \leq \lceil(a+4)\sqrt{2}\rceil^2$.

For any two nodes $u, v \in D$ with $m(u,v) \leq 3$, connect them with a shortest path between u and v. Namely, let $M(u,v)$ denote the set of all intermediate nodes on a shortest path between u and v. Define

$$C = D \cup \left(\cup_{u,v\in D:m(u,v)\leq 3}M(u,v)\right).$$

We show that C is a feasible solution of LOCAL(e) problem. For any two nodes $u, v \in V \cap e^c$ with $m(u,v) = 1$ and $\{u,v\} \neq \emptyset$, since D dominates $V \cap e^c$, there are $u', v' \in D$ such that u is adjacent to u' and v is adjacent to v'. Thus, $m(u', v') \leq 3$. This implies that $M(u, v) \subseteq C$ and hence $m_C(u, v) \leq 5$. Therefore, C is a feasible solution of αMOC-CDS. Moreover,

$$|C| \leq |D| + 3 \cdot \frac{|D|(|D| - 1)}{2} \leq 1.5|D|^2 \leq 1.5 \cdot \lceil (a + 4)\sqrt{2} \rceil^4.$$

This means that the minimum solution of LOCAL(e) has size at most $1.5 \cdot \lceil (a + 4)\sqrt{2} \rceil^4$. Therefore, by an exhausting search, we can compute the minimum solution of LOCAL(e) in time $n_e^{O(a^4)}$. $\qquad \square$

Let D_e denote the minimum solution for the LOCAL(e) problem. Define $D(0) = \cup_{e \in P(0)} D_e$ where $e \in P(0)$ means that e is over all cells in partition $P(0)$.

Lemma 4. $D(0)$ *is a feasible solution of* α*MOC-CDS and* $D(0)$ *can be computed in time* $n^{O(a^4)}$ *where* $n = |V|$.

Proof. Since every node in V belongs to some e^c, $D(0)$ is a dominating set. Moreover, for every two nodes $u, v \in V$ with $m(u, v) = 1$, we have $u \in e$ for some cell e, which implies that $u, v \in e^c$. Hence, $m_{D_e}(u, v) \leq \alpha$. If follows that $m_{D(0)}(u, v) \leq \alpha$. By Lemma 2, $D(0)$ is feasible for αMOC-CDS.

Note that each node may appear in e^{cb} for at most four cells e. Therefore, by Lemma 3, $D(0)$ can be computed in time

$$\sum_{e \in P(0)} n_e^{O(a^4)} \leq (4n)^{O(a^4)} = n^{O(a^4)}$$

where $n = |V|$. $\qquad \square$

To estimate $|D(0)|$, we consider a minimum solution D^* of αMOC-CDS. Let $P(0)^b = \cup_{e \in P(0)} e^b$.

Lemma 5. $|D(0)| \leq |D^*| + 4|D^* \cap P(0)^b|$.

Proof. We claim that $D^* \cap e^{cb}$ is feasible for LOCAL(e) problem. In fact, it is clear that $D^* \cap e^{cb}$ dominates $V \cap e^c$. For any two nodes $u, v \in e^c$ with $m(u, v) = 1$ and $\{u, v\} \cap e \neq \emptyset$, the path between u and v with at most α intermediate nodes must lie inside of e^{cb} and $m_{D^*}(u, v) \leq \alpha$ implies $m_{D^* \cap e^{cb}}(u, v) \leq \alpha$. This completes the proof of our claim.

Our claim implies that $|D_e| \leq |D^* \cap e^{cb}|$. Thus

$$|D(0)| \leq \sum_{e \in P(0)} |D_e|$$

$$\leq \sum_{e \in P(0)} |D^* \cap e^{cb}|$$

$$\leq \sum_{e \in P(0)} |D^* \cap e| + \sum_{e \in P(0)} |D^* \cap e^b|$$

$$\leq |D^*| + 4|D^* \cap P(0)^b|. \qquad \square$$

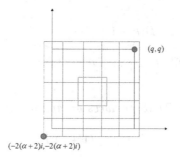

Fig. 4. Grid $P(i)$

Now, we shift partition $P(0)$ to $P(i)$ as shown in Fig. 4 such that the left and lower corner of the grid is moved to point $(-2(\alpha + 2)i, -2(\alpha + 2)i)$. For each $P(i)$, we can compute a feasible solution $D(i)$ in the same way as $D(0)$ for $P(0)$. Then we have

(a) $D(i)$ is a feasible solution of αMOC-CDS.
(b) $D(i)$ can be computed in time $n^{O(a^4)}$.
(c) $|D(i)| \leq |D^*| + 4|D^* \cap P(i)^b|$.

In addition, we have

Lemma 6. $|D(0) + |D(1)| + \cdots + |D(k-1)| \leq (k+8)|D^*|$.

Proof. Note that $P(i)^b$ consists of a group of horizontal strips and a group of vetical strips (Fig. 5). All horizontal strips in $P(0)^b \cup P(1)^b \cup \cdots \cup P(k-1)^b$ are disjoint and all vertical strips in $P(0)^b \cup P(1)^b \cup \cdots \cup P(k-1)^b$ are also disjoint. Therefore,

$$\sum_{i=0}^{k-1} |D^* \cap P(i)^b| \leq 2|D^*|.$$

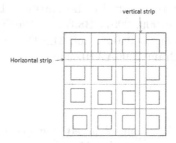

Fig. 5. Horizontal and Vertical Strips

Hence,

$$\sum_{i=0}^{k-1} |D(i)| \leq (k+8)|D^*|. \qquad \square$$

Set $k = \lceil 1/(8\varepsilon) \rceil$ and run the following algorithm.

Algorithm PTAS

Compute $D(0), D(1), ..., D(k-1)$;

Choose i^*, $0 \leq i^* \leq k-1$ such that

$|D(i^*)| = \min(|D(0)|, |D(1)|, ..., |D(k-1)|)$;

Output $D(i^*)$.

Theorem 1. *Algorithm PTAS produces an approximation solution for αMOC-CDS with size*

$$|D(i^*)| \leq (1+\varepsilon)|D^*|$$

and runs in time $n^{O(1/\varepsilon^4)}$.

Proof. It follows from Lemmas 4 and 6.   □

3   Conclusion

We showed that for $\alpha \geq 5$, αMOC-CDS has PTAS and leave the problem open for $1 \leq \alpha < 5$. Actually, how to connect a dominating set into a feasible solution for αMOC-CDS is the main difficulty for $1 \leq \alpha < 5$. So far, no good method has been found without increasing too much number of nodes.

Acknowledgement

This research was jointly supported in part by MEST, Korea under WCU (R33-2008-000-10044-0), by the KOSEF grant funded by the Korea government (MEST) (No. R01- 2007-000-11203-0), by KRF Grant funded by (KRF-2008- 314-D00354), and by MKE, Korea under ITRC IITA-2009- (C1090-0902-0046) and IITA-2009-(C1090-0902-0007). This research was also supported in part by National Science Foundation of USA under grants CNS0831579 and CCF0728851 and supported in part by the National Basic Research Program of China Grant 2007CB807900, 2007CB807901, the National Natural Science Foundation of China Grant 60604033, and the Hi-Tech Research & Development Program of China Grant 2006AA10Z216.

References

1. Bharghavan, V., Das, B.: Routing in ad hoc networks using minimum connected dominating sets. In: International Conference on Communication, Montreal, Canada (June 1997)
2. Cardei, M., Cheng, M.X., Cheng, X., Du, D.-Z.: Connected domination in ad hoc wireless networks. In: Proc. the Sixth International Conference on Computer Science and Informatics (CS&I 2002) (2002)
3. Cheng, X., Huang, X., Li, D., Wu, W., Du, D.-Z.: A polynomial-time approximation scheme for minimum connected dominating set in ad hoc wireless networks. Networks 42, 202–208 (2003)

4. Ding, L., Gao, X., Wu, W., Lee, W., Zhu, X., Du, D.-Z.: Distributed Construction of Connected Dominating Sets with Minimum Routing Cost in Wireless Network. To appear in the 30th International Conference on Distributed Computing Systems, ICDCS 2010 (2010)
5. Du, H., Ding, L., Wu, W., Willson, J., Lee, W., Du, D.-Z.: Approximation for minimum vitual backbone with routing cost constraint in wireless networks (manuscript)
6. Du, D.-Z., Graham, R.L., Pardalos, P.M., Wan, P.-J., Wu, W., Zhao, W.: Analysis of Greedy Approximations with Nonsubmodular Potential Functions. In: Proceedings of the 19th Annual ACM-SIAM Symposium on Dicrete Algorithms (SODA), San Francisco, USA, January 20-22, pp. 167–175 (2008)
7. Guha, S., Khuller, S.: Approximation algorithms for connected dominating sets. Algorithmica 20, 374–387 (1998)
8. Liu, Q., Zhang, Z., Willson, J., Ding, L., Wu, W., Lee, W., Du, D.-Z.: Approximation for minimum connected dominating set with routing cost constraint in unit disk graphs (manuscript)
9. Ruan, L., Du, H., Jia, X., Wu, W., Li, Y., Ko, K.-I.: A greedy approximation for minimum connected dominating set. Theoretical Computer Science 329, 325–330 (2004)
10. Salhieh, A., Weinmann, J., Kochha, M., Schwiebert, L.: Power Efficient topologies for wireless sensor networks. In: ICPP 2001, pp.156–163 (2001)
11. Sivakumar, R., Das, B., Bharghavan, V.: An improved spine-based infrastructure for routing in ad hoc networks. In: IEEE Symposium on Computer and Communications, Athens, Greece (June 1998)
12. Stojmenovic, I., Seddigh, M., Zunic, J.: Dominating sets and neighbor elimination based broadcasting algorithms in wireless networks. In: Proc. IEEE Hawaii Int. Conf. on System Sciences (January 2001)
13. Willson, J., Gao, X., Qu, Z., Zhu, Y., Li, Y., Wu, W.: Efficient Distributed Algorithms for Topology Control Problem with Shortest Path Constraints (submitted)
14. Wan, P., Alzoubi, K.M., Frieder, O.: Distributed construction of connected dominating set in wireless ad hoc networks. In: Proc. 3rd ACM Int. Workshop on Discrete Algorithms and Methods for Mobile Computing and Communications, pp. 7–14 (1999)
15. Wu, J., Li, H.: On Calculating Connected Dominating Set for Efficient Routing in Ad Hoc Wireless Networks. In: Proceedings of the 3rd ACM International Workshop on Discrete Algorithms and Methods for Mobile Computing and Communications, pp. 7–14 (1999)

A Primal-Dual Approximation Algorithm for the Asymmetric Prize-Collecting TSP

Viet Hung Nguyen

LIP6, Université Pierre et Marie Curie Paris 6, 4 place Jussieu, Paris, France

Abstract. We present a primal-dual $\lceil \log(n) \rceil$-approximation algorithm for the version of the asymmetric prize collecting traveling salesman problem, where the objective is to find a directed tour that visits a subset of vertices such that the length of the tour plus the sum of penalties associated with vertices not in the tour is as small as possible. The previous work on the problem [9] is based on the Held-Karp relaxation and heuristic methods such as the Frieze et al.'s heuristic [6] or the recent Asadpour et al.'s heuristic for the ATSP [2]. Depending on which of the two heuristics is used, it gives respectively $1 + \lceil \log(n) \rceil$ and $3 + 8\frac{\log(n)}{\log(\log(n))}$ as an approximation ratio. Our approximation ratio $\lceil \log(n) \rceil$ outperforms the first in theory and the second in practice. Moreover, unlike the method in [9], our algorithm is combinatorial.

1 Introduction

Let $G = (V, A)$ be a complete directed graph with the vertex set $V = \{1, 2, \ldots, n\}$ and the arc set A. We associate with each arc $e = (i, j)$ a cost c_e and with each vertex $i \in V$ a nonnegative penalty π_i. The arc costs are assumed to satisfy the triangle inequality, that is, $c_{(i,j)} \leq c_{(i,k)} + c_{(k,j)}$ for all $i, j, k \in V$. In this paper, we consider a simplified version of the *Asymmetric Prize Collecting Traveling Salesman Problem* (APCTSP), namely, to find a tour that visits a subset of the vertices such that the length of the tour plus the sum of penalties of all vertices not in the tour is as small as possible. Note that in the general version of APCTSP, introduced by Balas [4], the arc costs are not assumed to satisfy the triangle inequality. Furthermore, in [4] associated with each vertex there is a certain reward or prize, and in the optimization problem one must choose a subset of vertices to be visited so that the total reward is at least a given a parameter W_0.

The *Symmetric Prize Collecting Traveling Salesman Problem* (SPCTSP) which is the symmetric version of APCTSP (i.e. when $c_{(i,j)} = c_{(j,i)}$ for all $i, j \in V$) has been studied intensively since the work of Balas [4], especially on the design of approximation algorithms. The first constant approximation algorithm was given by Bienstock et al. [5] achieves a ratio $\frac{5}{2}$. This algorithm is based on the solution of a linear programming problem. The second approximation algorithm, developped by Goemans and Willamson [8], is purely combinatorial. They presented a general approximation technique for constrained forest problems, that can be extended to SPCTSP with 2 as approximation ratio. Recently, Archer et al. [1] and Goemans [7] respectively improve the primal-dual algorithm of Goemans and Williamson to 1.990283 and 1.91456.

For APCTSP, though exact algorithms was developed in [3], there was no work on approximation algorithm until the recent Nguyen et al.'s work [9]. The latter is based

W. Wu and O. Daescu (Eds.): COCOA 2010, Part I, LNCS 6508, pp. 260–269, 2010.

on the Held-Karp relaxation and heuristic methods such as the Frieze et al.'s heuristic [6] or the recent Asadpour et al.'s heuristic for the ATSP [2]. Depending on which of the two heuristics is used, it gives respectively $1 + \lceil \log(n) \rceil$ and $3 + 8\frac{\log(n)}{\log(\log(n))}$ as an approximation ratio. In this paper, we present a primal-dual $\lceil \log(n) \rceil$-approximation algorithm for APCTSP. This ratio obviously improves $1 + \lceil \log(n) \rceil$ in theory. It also improves the second ratio in practice since $3 + 8\frac{\log(n)}{\log(\log(n))}$ is asymptotically better than $\lceil \log(n) \rceil$ but for realistic values of n, $3 + 8\frac{\log(n)}{\log(\log(n))}$ is at least nearly $\frac{3}{2}$ times the value of $\lceil \log(n) \rceil$ (for example when $n = 10^{20}$, $3 + 8\frac{\log(n)}{\log(\log(n))} \approx 90$ and $\lceil \log(n) \rceil = 67$). Moreover, unlike the method in [9], our algorithm is combinatorial.

The paper is organized as follows. In Section 2, we give an integer formulation for the problem. Section 3 describes a primal-dual algorithm based on the linear relaxation of the formulation in Section 2. In Section 4, we prove that the algorithm outputs a solution for APCTSP which is at most $\lceil \log(n) \rceil$ times the optimal solution.

2 Integer Formulation

Let $G = (V, A)$ be directed graph with $|V| = n$ and $|A| = m$. Each arc $a \in A$ is associated to a cost c_a. Each vertex $v \in V$ is associated to a penalty π_v. The arc cost c is assumed to satisfy the triangle inequality. Our aim is to find a tour T which minimizes $\sum_{a \in T} c_a + \sum_{v \notin T} \pi_v$. We consider the following integer formulation inspired from the undirected version in [5] for APCTSP$_j$, the subproblem of APCTSP when we impose a specific vertex j to be in T. For every $i \in V$, for every arc $a \in A$, let

$$y_i = \begin{cases} 1 \text{ if } i \in T \\ 0 \text{ otherwise} \end{cases} \text{ and } x_a = \begin{cases} 1 \text{ if } a \in T \\ 0 \text{ otherwise} \end{cases}$$

For every subset S, let $\delta^+(S)$ be the set of arcs with tail in S and head in $V \setminus S$ and $\delta^-(S)$ be the set of arcs with head in S and tail in $V \setminus S$. Then APCTSP$_j$ can be formulated as follows.

$$\min Z_j = \sum_{e \in A} c_e x_e + \sum_{i \in V} \pi_i (1 - y_i)$$

$$\text{subject to } x(\delta^+(i)) = x(\delta^-(i)) = y_i \; \forall i \in V, \tag{1}$$

$$x(\delta^+(S)) \geq y_i \; \forall S \subset V \setminus \{j\}, \; |S| \geq 2 \text{ and } \forall i \in S \tag{2}$$

$$x(\delta^-(S)) \geq y_i \; \forall S \subset V \setminus \{j\}, \; |S| \geq 2 \text{ and } \forall i \in S \tag{3}$$

$$y_j = 1, \tag{4}$$

$$0 \leq x_e \leq 1 \text{ and integer}, \text{ and } 0 \leq y_i \leq 1 \text{ and integer}.$$

The constraints (1) ensure that T is Eulerian and the constraints (2) are the subtour elimination constraints. Note that one of the two families of constraints (2) and (3) is unnecessary if the Eulerian constraints (1) are respected since we have always $x(\delta^+(S)) = x(\delta^-(S))$ for all $S \subset V \setminus \{j\}$. But we include both of the constraints (2) and (3) in the formulation since we will work on a relaxation called (R) where the Eulerian constraints (1) will be ignored. Precisely, the constraints (1) and (4) are relaxed to

$$x(\delta^+(i)) \geq y_i \text{ for all } i \in V \setminus \{j\},$$
$$x(\delta^-(i)) \geq y_i \text{ for all } i \in V \setminus \{j\},$$
$$x(\delta^+(j)) \geq 1,$$
$$x(\delta^-(j)) \geq 1.$$

We can regroup the first two constraints with constraints (2) and (3) by allowing $|S| = 1$ in these constraints. Let $C = \sum_{i \in V \setminus \{j\}} \pi_i$ which is a constant, we can then write down the relaxation (R) as follows:

$$(R) \ \min Z_j = \sum_{e \in A} c_e x_e - \sum_{i \in V \setminus \{j\}} \pi_i y_i + C$$

$$\text{subject to } x(\delta^+(j)) \geq 1, \tag{5}$$
$$x(\delta^-(j)) \geq 1, \tag{6}$$
$$x(\delta^+(S)) \geq y_i, \ \emptyset \neq S \subset V \setminus \{j\} \text{ and } \forall i \in S, \tag{7}$$
$$x(\delta^-(S)) \geq y_i, \ \emptyset \neq S \subset V \setminus \{j\} \text{ and } \forall i \in S, \tag{8}$$
$$y_i \leq 1, \ i \in V \setminus \{j\} \tag{9}$$
$$y_i \geq 0, \ i \in V \setminus \{j\}$$
$$x_e \geq 0, \ e \in A$$

Let us introduce the dual variable(s):

- z_j^+ associated to the constraint (5),
- z_j^- associated to the constraint (6),
- $z_{S,i}^+$ associated to the constraints (7),
- $z_{S,i}^-$ associated to the constraints (8),
- and finally, z_i associated to the constraints (9).

then the dual program (D) of (R) can be written as follows:

$$(D) \ \max C + z_j^+ + z_j^- - \sum_{i \in V \setminus \{j\}} z_i$$

$$\text{subject to } \sum_{S \subset V \setminus \{j\} \text{ s.t. } i \in S} (z_{S,i}^- + z_{S,i}^+) + z_i \geq \pi_i \ \forall i \in V \setminus \{j\} \tag{10}$$

$$\sum_{S \subset V \setminus \{j\} \text{ s.t. } e \in \delta^-(S)} \sum_{i \in S} z_{S,i}^- + \sum_{S \subset V \setminus \{j\} \text{ s.t. } e \in \delta^+(S)} \sum_{i \in S} z_{S,i}^+ \leq c_e \ \forall e \in A \tag{11}$$

$$z_{S,i}^+, z_{S,i}^- \geq 0 \ \forall S \subset V \setminus \{j\} \text{ and } \forall i \in S$$
$$z_i \geq 0 \ \forall i \in V \setminus \{j\}$$
$$z_j^+, z_j^- \geq 0$$

In the sequel, we will try to find an approximation algorithm for APCTSP$_j$. An approximation algorithm for APCTSP of the same ratio can be simply deduced from approximating APCTSP$_j$ for each $j \in V$.

3   Primal-Dual Approximation Algorithm

3.1   General Idea of the Algorithm

We will present a primal-dual algorithm that has at most $\lceil \log_2(n) \rceil$ iterations of dual augmentation. The algorithm starts with a feasible dual solution of (D):

- $z_{S,i}^+ = z_{S,i}^- = 0$ for all $i \in V \setminus \{j\}$ and for all $S \subset V \setminus \{j\}$ such that $i \in S$,
- $z_j^+ = z_j^- = 0$.
- $z_i = \pi_i$.

and applies at most $\lceil \log_2(n) \rceil$ dual augmentations. In the algorithm, we maintain an arc subset denoted by T which contains the arcs that will constitute our solution for APCTSP$_j$ at the end of the algorithm. At initialization $T = \emptyset$, and at each iteration, based on the current dual feasible solution of (D), we add a set of vertex disjoint simple cycles to T. Hence, T is always a collection of strongly connected Eulerian components. At each iteration of the algorithm, we consider the graph $\bar{G}$ obtained from G by shrinking the vertex subsets of G corresponding to strongly connected Eulerian components in T. We define the cost for the arcs in $\bar{G}$ with respect to the current reduced cost. From $\bar{G}$, we build a bipartite graph B and transform the dual augmenting problem to a minimum cost assignment problem, called (A), in B with respect to the current reduced cost. We consider the classical linear programming formulation for (A) and its dual. In particular, we make the correspondence from each dual variable of (A) to some dual variable of (D). We solve (A) by any known primal-dual algorithm for the minimum cost assignement problem and assign the value of the dual optimal solution of (A) to the corresponding dual variable of (D). Note that each dual variable of (D) will be augmented at most once and after that its value will not be changed until the end of the algorithm. After each iteration of the algorithm, we add to T a set of arcs and may eliminate definitely some vertices from the solution tour T. Note that T can contain a multiplicity of an arc and if a vertex i is eliminated from T, then all the vertices that belong to the same connected component in the subgraph induced by T, will be also eliminated. Thus there will be no arc of T such that one end-vertex is in T and the other has been eliminated from T. The algorithm stops when T becomes connected. As at each iteration, for each strongly connected Eulerian component H of T, either H is merged with some other strongly connected Eulerian component of T, or the vertices in H are eliminated from T, at most $\lceil \log_2(n) \rceil$ iterations was performed. We prove that the cost of the arcs added to T in each iteration is at most C^* (C^* is the value of an optimal solution of APCTSP$_j$). In particular, for the last iteration the cost of the arcs added to T plus the total penalty associated to the vertices eliminated from T (from the first iteration to the end) is at most C^*. As T is Eulerian, T is a solution of APCTSP$_j$ and the cost of T is at most $\lceil \log_2(n) \rceil C^*$.

3.2   Computed Items and Their Meaning

Let us define the main items that will be computed in the algorithm:

- T is the set of solution arcs.
- $V_{\bar{T}} \subset V$ denotes the set of the vertices which will not be included in T.

- $V_T = V \setminus V_{\bar{T}}$ which is the set of the vertices belonging to T.
- Let G_T be the subgraph of G with vertex set V_T and arc set T.
- p denotes the counter of iterations.
- In the algorithm, we always set

$$z_i = \pi_i - \sum_{S \subset V \setminus \{j\} \text{ s.t. } i \in S} (z_{S,i}^- + z_{S,i}^+)$$

for all $i \in V \setminus \{j\}$.

3.3   Initialization

The algorithm starts with a feasible dual solution:

- $z_{S,i}^+ = z_{S,i}^- = 0$ for all $i \in V \setminus \{j\}$ and for all $S \subset V \setminus \{j\}$ such that $i \in S$,
- $z_j^+ = z_j^- = 0$ and $z_i = \pi_i - \sum_{S \subset V \setminus \{j\} \text{ s.t. } i \in S} (z_{S,i}^- + z_{S,i}^+) = \pi_i$ for all $i \in V \setminus \{j\}$.

Let us set the reduced cost $\bar{c} = c$. Set $T = \emptyset$ and $V_{\bar{T}} = \emptyset$. Set $V_T = V$. Set $p = 1$.

3.4   The pth iteration

The tranformation to an assignment problem. We build a graph $\bar{G}^p = (\bar{V}^p, \bar{A}^p)$ where the vertices in $\bar{V}^p$ correspond to the strongly connected components in G_T. Hence a vertex in $\bar{G}^p$ is either a vertex in G or a pseudo-vertex corresponding to a vertex subset $S \subseteq V_T$. Note that the subsets associated to the pseudo-vertices in $\bar{G}^p$ are pairwise disjoint. The arcs in $\bar{A}^p$ are formed as follows:

- for any two vertices i and k which are in both G and $\bar{G}^p$, if there is an arc $(i, k) \in A$ then there is also an arc $(i, k) \in \bar{A}^p$. The cost of (i, k) in $\bar{G}^p$ is the reduced cost $\bar{c}_{(i,k)}$.
- for a vertex i in both G and $\bar{G}^p$ and a pseudo-vertex s in $\bar{G}^p$ corresponding to a subset $S \subseteq V_T$, there is an arc $(s, i) \in \bar{A}^p$ if there exists at least one arc $(k, i) \in A$ with $k \in S$. Let the cost of (s, i), $\bar{c}_{(s,i)} = \min\{\bar{c}_{(k,i)} \mid k \in S \text{ and } (k, i) \in A\}$. Any arc $(k, i) \in A$ with $k \in S$ of reduced cost equal to $\bar{c}_{(s,i)}$ is called *representative arc* of (s, i).
- for two pseudo-vertices s and t corresponding to respectively the disjoint subsets S and T, there is an arc $(s, t) \in \bar{A}^p$ if there exists at least one arc $(k, i) \in A$ with $k \in S$ and $i \in T$. Let the cost of (s, t), $\bar{c}_{(s,t)} = \min\{\bar{c}_{(k,i)} \mid k \in S, i \in T \text{ and } (k, i) \in A\}$. Any arc $(k, i) \in A$ with $k \in S$ and $i \in T$ of reduced cost equal to $\bar{c}_{(s,t)}$ is called *representative arc* of (s, t).

Note that at initialization $\bar{G}^p$ is just a copy of G and the reduced costs $\bar{c}$ is equal to the original cost c and z_i is the penalty π_i for all $i \in V \setminus \{j\}$.

From $\bar{G}^p$, we form a bipartite graph $B = (V_B, A_B)$ as follows:

- For the vertex in $\bar{G}^p$ containing j, we create two vertices j^+ (positive vertex) and j^- (negative vertex) in B and an arc (j^+, j^-) of cost 0.

- For each vertex $i \in \bar{V} \setminus \{j\}$, we create four vertices i^+, M_i^+ (positive vertices) and i^-, M_i^- (negative vertices) in B. We create three arcs (M_i^+, i^-), (i^+, M_i^-) and (M_i^+, M_i^-). The cost $\bar{c}_{(M_i^+, M_i^-)}$ of (M_i^+, M_i^-) is assigned to 0. If i is a pseudo-vertex corresponding to a subset $S \subseteq V \setminus \{j\}$, then the costs $\bar{c}_{(M_i^+, i^-)}$ and $\bar{c}_{(i^+, M_i^-)}$ are assigned to the same value being equal to $z_i/2$ (note that we could also set $z_{S_i,k}^+ = \lambda z_i$ and $z_{S_i,k}^- = (1 - \lambda)z_i$ for any $0 < \lambda < 1$).
- For any two distinct vertices $i, k \in \bar{V}$, if $\bar{G}^p$ contains an arc (i, k) with cost $\bar{c}_{(i,k)}$, we create an arc (i^+, k^-) in B with the same cost. Similarly, if $\bar{G}^p$ contains an arc (k, i) with cost $\bar{c}_{(k,i)}$, we create an arc (k^+, i^-) in B with the same cost.

Remark 1. B is a bipartite graph with the positive vertices in a side and the negative vertices in the other.

Let B^+ and B^- be respectively the set of positive and negative vertices in B.

Remark 2. The number of positive vertices is equal to the number of the negative vertices, i.e. $|B^+| = |B^-|$.

We consider the assignment problem in B which aims at finding the minimum cost assignment of the positive vertices to the negative ones. The classical linear programming formulation for the assignment problem is

$$(A) \min \sum_{e \in A_B} \bar{c}_e \chi_e$$

$$\text{subject to}$$

$$\sum_{e \in \delta^+(i)} \chi_e = 1 \text{ for all } i \in B^+ \tag{12}$$

$$\sum_{e \in \delta^-(i)} \chi_e = 1 \text{ for all } i \in B^- \tag{13}$$

$$\chi_e \geq 0 \text{ for all } e \in A_B \tag{14}$$

and it's dual is

$$(DA) \min \sum_{i \in B^+} u_i + \sum_{k \in B^-} v_k$$

$$\text{subject to}$$

$$u_i + v_k \leq \bar{c}_{(i,k)} \text{ for all } (i, k) \in A_B \tag{15}$$

Applying any primal-dual algorithm for this assignment problem, we obtain a primal optimal solution χ^* and its corresponding dual optimal solution u^* and v^*.

Updating the dual feasible solution of (D) and the sets T, $V_{\bar{T}}$. In the solution χ^*, for each pair of vertices M_i^+ and M_i^- with $i \in \bar{V}^p \setminus \{j\}$, there are two possible cases:

1. M_i^+ **is assigned to** M_i^-. Then i^+ and i^- are respectively assigned to the vertices which are not M-vertices, i.e. the strongly connected component containing i will

be merged with another strongly connected component in G_T. Let S_i be the subset represented by i ($S_i = \{i\}$ when i is not a pseudo-vertex). Suppose that i^+ is assigned to some vertex h^- and i^- is assigned to some vertex k^+, we choose any representative arc $e_i^+ = (i', h')$ of (i, h) and any representative arc $e_i^- = (k'', i'')$ of (k, i). Set $z_{S_i,i'}^+ = u_i^*$ and $z_{S_i,i''}^- = v_i^*$. Set $T = T \cup \{e_i^+, e_i^-\}$. If i is pseudo-vertex and $i' \neq i''$, let $P_{i'',i'}$ be the path from i'' to i' in T, let us set $T = T \cup P_{i'',i'}$.

2. M_i^- **is assigned to** i^+ **and** M_i^+ **is assigned to** i^-. In this case, i will be eliminated from T, i.e. i will not be visited by T. If i is not a pseudo-vertex, set $V_{\bar{T}} = V_{\bar{T}} \cup \{i\}$. Otherwise, i.e. i is a pseudo-vertex in $\bar{G}^p$ representing a vertex subset S_i, all the vertices in S_i are eliminated from T, i.e. we set $V_{\bar{T}} = V_{\bar{T}} \cup S_i$. For all $k \in S_i$, let us set $z_{S_i,k}^+ = z_k/2$ and $z_{S_i,k}^- = z_k/2$.

For the vertex j, there are also two following possible cases:

1. j^+ **is assigned to** j^-. The dual variables u_j^* and v_j^* are necessary equal to 0, there is no updating operation.
2. j^+ **is assigned to some vertex** h^- **and** j^- **is assigned to some vertex** k^+. Let S_j be the subset represented by j, let $\bar{S}_j = V \setminus S_j$. Let us choose $e_j^+ = (j', h')$ any representative arc of (j, h) and $e_j^- = (k'', j'')$ any representative arc of (k, j). Set $T = T \cup \{e_j^+, e_j^-\}$. We set $z_{\bar{S}_j,h'}^- = u_j^*$ and $z_{\bar{S}_j,k''}^+ = v_j^*$. If $j'' \neq j'$, let $P_{j'',j'}$ be the path from j'' to j' in T, let us set $T = T \cup P_{j'',j'}$.

For all $i \in V \setminus \{j\}$, we update z_i by the formula:

$$z_i = \pi_i - \sum_{S \subset V \setminus \{j\} \text{ s.t. } i \in S} (z_{S,i}^- + z_{S,i}^+).$$

If T is connected, output T, z and STOP. Otherwise, set $p = p + 1$ and reiterate.

3.5   Analysis of the Algorithm

Lemma 1. *There are at most* $\lceil \log_2(n) \rceil$ *iterations.*

Proof. At p^{th} iteration, for any vertex i in $\bar{V}^p$,

- either i belongs to a (non trivial) strongly connected Eulerian component which will shrink into a pseudo-vertex in the next iteration,
- or i will be added to $V_{\bar{T}}$.

and the vertices in $\bar{V}^{p+1}$ are the pseudo vertices corresponding to the (non trivial) strongly connected components in $\bar{G}_p$. Hence, $|\bar{V}^{p+1}| \leq \frac{1}{2}|\bar{V}^p|$.

Proposition 1. *For an arc* (i, k) *in* T, *we have*

$$\sum_{S \subset V \setminus \{j\} \text{ s.t. } i \in S \text{ and } k \notin S} z_{S,i}^+ + \sum_{S \subset V \setminus \{j\} \text{ s.t. } k \in S \text{ and } i \notin S} z_{S,k}^- = c_{(i,k)}$$

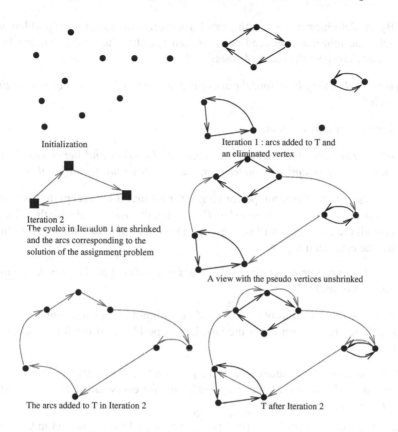

Fig. 1. An example of how arcs are added to T

Proof. Let consider the moment when (i, k) was added for the first time to T in the algorithm, let S_i (respectively S_k) be the vertex subset of the strongly connected Eulerian component containing i (respectively k). At that moment the value of the dual variables $z_{S_i,i}^+$ and $z_{S_k,k}^-$ was updated such that

$$\sum_{S \subset V \setminus \{j\} \text{ s.t. } i \in S \text{ and } k \notin S} z_{S,i}^+ + \sum_{S \subset V \setminus \{j\} \text{ s.t. } k \in S \text{ and } i \notin S} z_{S,k}^- = c_{(i,k)} \qquad (16)$$

After that, since S_i and S_k are going to be merged and so i and k are going to be in the same strongly connected Eulerian component, no dual variable $z_{S,i}^+$ such that $i \in S$ and $k \notin S$ or $z_{S,k}^-$ such that $k \in S$ and $i \notin S$ will be updated. Hence, (16) remains true until the end of the algorithm.

Proposition 2. *If a vertex i is eliminated from T then $z_i = 0$.*

Proof. Directly by construction.

Proposition 3. *At the end of each iteration, the dual variables are always feasible.*

Proof. By the definition of costs in the transformation to the assignment problem which always take the minimum reduced cost, we can see that dual augmentation in each iteration always respect the reduced costs.

Proposition 4. *In every iteration, the set of arcs added to T is a collection of disjoint simple cycles.*

Proof. Directly by construction.

Corollary 1. *The algorithm outputs a T which is Eulerian and hence can be transformed (under the assumption of metric costs) to a tour without additional cost.*

For the analysis of the algorithm performance, for all iterations except the last, we will estimate only the cost of the arcs added to T. The penalty associated with the eliminated vertices (in all the iterations) will be regrouped with the the the arcs added to T in the last iteration to be estimated together.

Lemma 2. *For every iteration except the last, the total cost of the arcs added to T at this iteration is at most C^*.*

Proof. Let z be the dual feasible solution of (D) output by the algorithm. Let us consider the pth iteration which is not the last. Let us build a dual feasible solution zp as follows.

Case 1. For each arc (i, k) added to T at the pth iteration, for every subset $S \subset V \setminus \{j\}$ such that $i \in S$ and $k \notin S$, set $zp_{S,i}^+ = z_{S,i}^+$ and for every subset $S' \subset V \setminus \{j\}$ such that $k \in S'$ and $i \notin S'$, set $zp_{S',k}^- = z_{S',k}^-$.

Case 2. All the other variables of type $zp_{S,i}^+$ or $zp_{S,i}^-$, which are not set in Case 1., are set to 0.

Case 3. Set $zp_i = \pi_i - \sum_{S \subset V \setminus \{j\} \text{ s.t. } i \in S} (zp_{S,i}^- + zp_{S,i}^+)$ for all $i \in V \setminus \{j\}$.

We can see easily that the vector zp with the components built as above is feasible for (D) and by Proposition 1 the cost of zp is equal to the cost of the arc add to T in the pth iteration. This cost is obviously at most C^*.

Lemma 3. *For the last iteration, the cost of the arcs added to T at this iteration plus the penalties associated to the vertices eliminated from T (from the first iteration) is at most C^*.*

Proof. The proof is quite similar to the one for Lemma 2. Given z, let us build a dual feasible solution zl as follows.

Case 1. For each arc (i, k) added to T at the last iteration, for every subset $S \subset V \setminus \{j\}$ such that $i \in S$ and $k \notin S$, set $zl_{S,i}^+ = z_{S,i}^+$ and for every subset $S' \subset V \setminus \{j\}$ such that $k \in S'$ and $i \notin S'$, set $zl_{S',k}^- = z_{S',k}^-$.

Case 2. For each vertex i not in T, for every subset $S \subset V \setminus \{j\}$ such that $i \in S$, set $zl_{S,i}^+ = z_{S,i}^+$ and $zl_{S,i}^- = z_{S,i}^-$.

Case 3. All the other variables of type $zl_{S,i}^+$ or $zl_{S,i}^-$, which are not set in Case 1. and Case 2., are set to 0.

Case 4. Set $zl_i = \pi_i - \sum_{S \subset V \setminus \{j\}} \text{ s.t. } _{i \in S}(zl^-_{S,i} + zl^+_{S,i})$ for all $i \in V \setminus \{j\}$.

Obviously, the vector zl with the components built as above is feasible for (D) and by Propositions 1 and 2 the cost of zl is equal to the cost of the arc added to T in the pth iteration plus the penalties associated to the vertices not in T. This cost is obviously at most C^*.

Theorem 1. *The cost of T plus the penalties associated to the vertices eliminated from T is at most $\lceil \log(n) \rceil C^*$.*

Proof. The theorem is a direct consequence of Lemmas 2 and 3.

4   Conclusions

Since the asymmetric TSP is a special case of the APCTSP, the algorithm described in this paper can be viewed as a primal-dual $\lceil \log(n) \rceil$-approximation algorithm for the asymmetric TSP. It is the first combinatorial algorithm for APCTSP achieving the same approximation ratio as the best combinatorial approximation algorithm for the asymmetric TSP by Frieze et al. [6]. This ratio improves the approximation ratio $1 + \lceil \log(n) \rceil$ given in [9] and is (for reasonable values of n) substantially smaller than the ratio $3 + 8 \frac{\log(n)}{\log(\log(n))}$ which can be obtained if in the algorithm in [9], we use the recent algorithm for asymmetric TSP by Asadpour et al. [2]. In our opinion, it is interesting for further works to derive a combinatorial algorithm for both the asymmetric TSP and the APCTSP achieving an approximation ratio which is asymptotically better than $\lceil \log(n) \rceil$.

References

1. Archer, A., Bateni, M., Hajiaghayi, M., Karloff, H.: Improved Approximation Algorithms for Prize-Collecting Steiner Tree and TSP. In: Proceedings of the 50th Annual Symposium on Foundations of Computer Science (2009)
2. Asadpour, A., Goemans, M.X., Madry, A., Oveis Gharan, S., Saberi, A.: An O(log n/log log n)-approximation algorithm for the asymmetric traveling salesman problem. In: 21st ACM-SIAM Symposium on Discrete Algorithms (2010)
3. Dell'Amico, M., Maffioli, F., Väbrand, P.: On Prize-collecting Tours and the Asymmetric Travelling Salesman Problem. Int. Trans. Opl Res. 2, 297–308 (1995)
4. Balas, E.: The prize collecting traveling salesman problem. Networks 19, 621–636 (1989)
5. Bienstock, D., Goemans, M.X., Simchi-Levi, D., Williamson, D.P.: A note on the prize collecting traveling salesman problem. Math. Prog. 59, 413–420 (1993)
6. Frieze, A.M., Galbiati, G., Maffioli, F.: On the worst case performance of some algorithms for the asymmetric traveling salesman problem. Networks 12, 23–39 (1982)
7. Goemans, M.X.: Combining approximation algorithms for the prize-collecting TSP, arXiv:0910.0553v1 (2009)
8. Goemans, M.X., Williamson: A general approximation technique for constrained forest problems. SIAM Journal on Computing 24, 296–317 (1995)
9. Nguyen, V.H., Nguyen, T.T.: Approximating the assymetric profitable tour, Proceedings of International Symposium on Combinatorial Optimization. In: Electronic Notes in Discrete Mathematics, vol. 36C, pp. 907–914 (2010)

Computing Toolpaths for 5-Axis NC Machines

Danny Z. Chen[1,*] and Ewa Misiołek[2]

[1] Department of Computer Science and Engineering, University of Notre Dame,
Notre Dame, IN 46556, USA
dchen@cse.nd.edu
[2] Mathematics Department, Saint Mary's College, Notre Dame, IN 46556, USA
misiolek@saintmarys.edu

Abstract. We present several algorithms for computing a *feasible tool-path* with desired features for sculpting a given surface using a 5-axis numerically controlled (NC) machine in computer-aided manufacturing. A toolpath specifies the orientation of a cutting tool at each point of a path taken by the tool. Previous algorithms are all heuristics with no quality guarantee of solutions and with no analysis of the running time. We present optimal quality solutions and provide time analysis for our algorithms. We model the problems using a directed, layered graph G such that a feasible toolpath corresponds to a certain path in G, and give efficient methods for solving several path problems in such graphs.

1 Introduction

In this paper, we study the *feasible toolpath* problem in computer-aided man-ufacturing. Given a surface $\mathcal{F}$ and an already specified sculpting path $\mathcal{C}$, we develop new methods for computing feasible toolpaths with desired features for manufacturing $\mathcal{F}$ using a 5-axis numerically controlled (NC) machine. A *feasible toolpath* is a sequence of tool orientations (angles) at the points on the path $\mathcal{C}$ that allow the sculpting of $\mathcal{F}$ without collision with the surface or with the ma-chine and obey a given limit on the angular changes in the orientations between consecutive points on $\mathcal{C}$ (called the *angular change constraint*).

Depending on the number of degrees of freedom, NC machines are classified as 3-, 4-, or 5-axis. For decades, before the introduction of 5-axis machines, research was focused on programming 3- and 4-axis machines. Unfortunately, the algo-rithms developed for those (older) machines, and adopted for 5-axis machines, produce toolpaths that may include rapid changes in the tool's orientations between consecutive path points that are impossible to achieve in practice by 5-axis machines. Thus, computing *feasible* toolpaths for 5-axis machines has be-come an important problem in NC machine programming. For the accuracy of the sculpting process (when using a flat-end tool), it is often desirable that the tool accesses $\mathcal{F}$ at a *best possible* angle [4,17,21], i.e., the tool's orientation vector at a point p is as close as possible to the vector normal to $\mathcal{F}$ at p. The greater

---

* The research of this author was supported in part by the National Science Foundation under Grants CCF-0515203 and CCF-0916606.

W. Wu and O. Daescu (Eds.): COCOA 2010, Part I, LNCS 6508, pp. 270–284, 2010.

the angle between these two vectors, the greater the height of the unwanted material left on the surface after sculpting. To minimize the unwanted material, this angle should be minimized at each path point. A feasible toolpath satisfying this additional requirement is called a *maximum accuracy feasible toolpath*.

The problem of finding feasible toolpaths for 5-axis NC machines has been studied by many researchers [2,8,9,16,17,18,20,21]. A common approach is based on "toolpath smoothing" techniques [2,8,9,16,17,18] that aim to reduce the angular changes of an already found toolpath. Ho *et al.* [16] used a quaternion interpolation algorithm. Jun *et al.* [17] utilized a configuration space that includes tool orientation angles. Morishige *et al.* [18] applied a configuration space to find two (forward and backward) toolpaths and pick the one with smaller angular changes. A main drawback of the smoothing techniques is that even though changes in tool orientations are made less abrupt, the resulting toolpaths may still be infeasible — these techniques do not guarantee that the angular changes respect the machine limits. Another method, by Wang and Tang [20,21], uses visibility maps to find toolpaths that obey the angular change constraint [20], and uses iso-conic partitioning to find maximum accuracy feasible toolpaths [21].

To facilitate computerized treatment of surfaces, a discretized approximate representation of the surfaces, called *free-form surfaces*, is commonly used in practice (see Section 2 for more details). A *configuration space* is utilized to capture the possible tool positions that allow the tool to sculpt the surface without collisions. However, due to the complex shapes of the tools, machines, and surfaces, analytically computing toolpaths is nearly impossible. In most computations, discrete configuration spaces, or called *discrete visibility maps*, are instead used [2,17,20,21]. Such discrete representations also help to speed up the computation and offer flexibility in adjusting the granularity of the discrete division of the configuration space domain (e.g., into certain grid structures).

Inspired by Wang and Tang's studies [20,21], we also utilize discrete visibility maps. But, our approaches for computing feasible toolpaths are quite different: We use geometric and graph models and techniques. While the methods in [20,21] are heuristics with no quality guarantee of solutions and no analysis of the running time, we ensure optimal quality solutions and analyze the time complexity for all our algorithms. Our main results are as follows.

- An $O(kn)$ time algorithm for finding a feasible toolpath, where n is the number of points on $\mathcal{C}$ and k is the visibility map size for each point $c_i \in \mathcal{C}$.
- An $O(kn)$ time algorithm for finding a maximum accuracy toolpath, if one such path exists; otherwise, an $O(ln^2 + kn^2)$ time algorithm for finding the minimum number of such paths whose union covers $\mathcal{C}$, where $1 < l \leq n$ is the minimum number of paths needed.

We use a graph model to capture the problem constraints and the optimization criterion. Our graph $G = (V, E)$ is layered, directed, and acyclic. The size of the graph depends on the number of contact points, n, on $\mathcal{C}$ and the size, k, of the *visibility map* (i.e., the discrete grid of possible tool orientations) for each point on $\mathcal{C}$: $|V| = O(nk)$ and $|E| = O(nk^2)$. The feasible toolpaths correspond to certain (shortest) paths between the first and last layers of the graph. To solve the toolpath problem variants, we present several efficient path algorithms on such layered graphs. In most cases, we significantly reduce the time bounds for computing various paths over the standard path approach. In fact, we use the graph only as a conceptual model and do not explicitly build it. Instead, we exploit the geometric structures of the graph and utilize a number of interesting geometric and graph techniques for much faster solutions. Our methods could find applications to other problems with similar geometric or graph structures.

2   Preliminaries

A *free-form surface* $\mathcal{F}$ is a surface that describes the shape of a manufactured product using a discrete representation [22]. In this representation, a surface $\mathcal{F} = (X, N)$ in $\mathbf{R}^3$ is defined by a set of points, $X = \{x_1, \ldots, x_f\}$, on $\mathcal{F}$ and the set of unit vectors, $N = \{n_1, \ldots, n_f\}$, normal to $\mathcal{F}$ at the points of X. We assume that the tool's path is already given as a piecewise-linear *contact curve* $\mathcal{C} = (c_1, \ldots, c_n)$, where $c_i \in X$, $1 \le i \le n$.

A convenient representation of $\mathcal{F}$ for modeling NC machining problems is a *spherical representation* [22]. A spherical representation $\mathcal{F}_S = (X_S, N_S)$ of $\mathcal{F} = (X, N)$, also called the *Gaussian image* of $\mathcal{F}$, is a set of points, $X_S = \{p_1, p_2, \ldots\}$, on a unit sphere S_0 centered at the origin in $\mathbf{R}^3$ that are the heads of the unit vectors, $N_S = \{p_1, p_2, \ldots\}$, starting at the origin and corresponding to the vectors in N (i.e., the direction of a vector $n_i \in N$ is the same as that of a vector $p_j \in N_S$). For example, the spherical representation of a sphere is the entire unit sphere S_0, the spherical representation of a cylinder is a great circle of S_0, and the spherical representation of a cube is a set of six points on S_0 representing the directions of the cube's six faces. The advantage of using a spherical representation of $\mathcal{F}$ is that the set of directions from each of which a point x on $\mathcal{F}$ can be accessed by a tool can also be represented by a set of points on S_0. Such a set of points (or directions) is called a *visibility map* [22] of the point x. (For more details of the spherical representations and visibility maps, see [7,13,20].) We assume that, together with $\mathcal{C}$, the set $\mathcal{A} = \{\mathcal{A}(c_1), \ldots, \mathcal{A}(c_n)\}$ is already given, where $\mathcal{A}(c_i) \subset S_0$ ($\mathcal{A}(c_i) \ne \emptyset$) is the visibility map of $c_i \in \mathcal{C}$.

Since the distance of any point $v \in \mathcal{A}(c_i)$ from the origin is 1, using the spherical coordinates, v can be uniquely represented by a pair of angles (α, β), with $\alpha \in [0, \pi]$ and $\beta \in [0, 2\pi]$. Let θ be the machine-specific precision limit on the difference in the tool's orientations between any two consecutive points on $\mathcal{C}$. Then, the feasible toolpath problem is formally defined as follows.

Feasible Toolpath (FT) Problem. Given a free-form surface $F = (X, N)$, *a contact curve* $\mathcal{C} = (c_1, \ldots, c_n)$ *with* $c_i \in X$, *a set* $\mathcal{A}$ *of nonempty visibility maps, and a machine-specific real value* θ, *determine a feasible toolpath* $\mathcal{T} = (v_1^*, \ldots, v_n^*)$, *where* v_i^* *is a tool orientation at* $c_i \in \mathcal{C}$, *for* $i = 1, \ldots, n$, *such that*

1. $\mathcal{T}$ *satisfies the* accessibility constraint, *i.e.,* $v_i^* \in \mathcal{A}(c_i)$ *for* $i = 1, \ldots, n$, *and*
2. $\mathcal{T}$ *satisfies the* angular change constraint, *i.e., if* $v_{i-1}^* = (\alpha_{i-1}^*, \beta_{i-1}^*)$ *and* $v_i^* = (\alpha_i^*, \beta_i^*)$ *are the feasible sculpting directions for* c_{i-1} *and* c_i, *then* $|\alpha_i^* - \alpha_{i-1}^*| \le \theta$ *and* $|\beta_i^* - \beta_{i-1}^*| \le \theta$, *for* $i = 2, \ldots, n$.

If a single feasible toolpath does not exist, then determine the minimum number of feasible toolpaths whose union covers C.

In the *maximum accuracy* (MaxA) toolpath problem, we seek a feasible toolpath that additionally maximizes the accuracy of the sculpting by minimizing the sum of the differences between the tool's orientation and the vector normal to the surface at each contact point of C. Let $M(v, w)$ denote the angular difference between the vectors v and w. Such angular differences can be measured using, for example, the *maximum* metric, the L_1 metric, or the absolute value of the angle between the two vectors. The selection of a metric may depend on the technical details and requirements of a specific NC machine. With this additional requirement, the MaxA toolpath problem is defined as follows.

Maximum Accuracy (MaxA) Toolpath Problem. *Let $M(v_i, p_i)$ be the angular difference between the tool's orientation v_i at c_i and the unit vector p_i normal to $\mathcal{F}$ at c_i. Compute a feasible toolpath $\mathcal{T} = (v_1^*, \dots, v_n^*)$ under the maximum accuracy criterion:*

$$(v_1^*, \dots, v_n^*) = \underset{(v_1, \dots, v_n) \in \mathcal{A}(c_1) \times \cdots \times \mathcal{A}(c_n)}{\arg \min} \left\{ \sum_{i=1}^{n} M(v_i, p_i) \right\}. \tag{1}$$

If a single optimal toolpath does not exist, then determine the minimum number of optimal toolpaths whose union covers C.

To solve the FT and MaxA problems, we use a geometric structure underlying a graph model that captures the accessibility and angular change constraints. Since a point $v \in S_0$ is uniquely represented by a pair of angles (α, β), with $\alpha \in [0, \pi]$ and $\beta \in [0, 2\pi]$, a visibility map $\mathcal{A}(c_i)$ lies on a $[0, \pi] \times [0, 2\pi]$ rectangle in the (α, β)-plane. We represent a visibility map as a discrete set of points; hence the angle ranges $[0, \pi]$ and $[0, 2\pi]$ are divided into m and $2m$ angle directions, respectively (the granularity, i.e., value of m, may depend on the representation of $\mathcal{F}$ and the physical precision of the machine). Hence, geometrically, each $\mathcal{A}(c_i)$ lies on an $m \times 2m$ rectangular grid R_i consisting of $2m^2$ cells r_i^{kl}, where $r_i^{kl} \in R_i$ corresponds to a pair of angles (α_i^k, β_i^l) and represents either an "accessible" direction of the tool at c_i ($v_i^{kl} \in \mathcal{A}(c_i)$), or is "forbidden" due to possible collisions. Let A_i be the set of cells of R_i representing the accessible directions of $\mathcal{A}(c_i)$ (see Fig. 1(a)). The sequence of accessibility rectangles, $R = \{R_1, \dots, R_n\}$, for the contact points on C forms a "stack" of rectangles (or layers), as in Fig. 1(b). Thus, the accessibility constraint for feasible toolpaths is specified by $A = \{A_1, \dots, A_n\}$. The angular change constraint is reflected in the following definition.

Definition 1. *For any $r = (\alpha_r, \beta_r) \in R_{j-1}$, the square of successors of r, $S(r)$, is the set of all cells on R_j corresponding to the directions for sculpting $\mathcal{F}$ at c_j that can be selected without violating the angular change constraint if the orientation of the tool at c_{j-1} corresponds to r. That is,*

$$S(r) = \{q = (\alpha_q, \beta_q) : q \in R_j, \ |\alpha_q - \alpha_r| \le \theta, \ and \ |\beta_q - \beta_r| \le \theta\}.$$

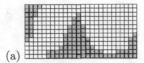 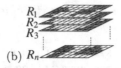 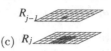

Fig. 1. (a) The accessibility rectangle R_i with accessible (shaded) and forbidden (clear) orientations of the visibility map $\mathcal{A}(c_i)$. (b) A "stack" of the R_i's. (c) The square of all possible angle directions on R_j for a cell on R_{j-1}.

The set $S(r)$ forms a square of cells on R_j (e.g., see Fig. 1(c)).

Our graph model capturing the constraints is quite intuitive. The stack of the rectangles (layers), $R = \{R_1, \ldots, R_n\}$, forms a 3-D grid for embedding a graph $G = (V, E)$. The set $A = \{A_1, \ldots, A_n\}$ defines the vertices in V (each cell $r \in A_j$ corresponds to a vertex v_r in V; each A_j defines a *layer* V_j of vertices), and the relation S defines the directed edges in E (($v_r, v_q) \in E$ if and only if $v_r \in V_{j-1}$, $v_q \in V_j$, and $q \in S(r)$). Hence, G is a layered directed acyclic (3-D grid) graph, with $|V| = O(nm^2)$ vertices and, in the worst case, $|E| = O(nm^4)$ edges, where n is the number of points on $\mathcal{C}$ and $|A_i| = O(m^2)$ for all $i = 1, \ldots, n$.

To extend the graph model to the MaxA problem, we assign weights to the vertices of G so that $w(v_j) = M(\boldsymbol{v}_j, \boldsymbol{n}_j)$ is the angular difference between the tool orientation $\boldsymbol{v}_j$ at c_j and the vector $\boldsymbol{n}_j$ normal to $\mathcal{F}$ at c_j.

Since a feasible toolpath is a sequence of feasible orientations, one for each c_i, satisfying the angular change constraint, it corresponds to a path in (the unweighted) G from a vertex of V_1 to a vertex of V_n. A feasible toolpath that additionally satisfies the maximum accuracy requirement corresponds to a *shortest* path in the respective *weighted* graph. Even though very intuitive, this graph model does not lend itself to the most efficient solutions for the problems. Instead, we exploit the geometric structures of the graph and use geometric methods.

3   The Feasible Toolpath (FT) Problem

A feasible toolpath $\mathcal{T}$ corresponds to a path from layer V_1 to layer V_n in the unweighted graph $G = (V, E)$ as defined in Section 2. If we add to G a *source vertex* s and directed unweighted edges from s to all vertices on V_1, then by using a standard method to find a path from s to any vertex on V_n, we would obtain $\mathcal{T}$ in $O(|V| + |E|) = O(nm^4)$ time. We significantly reduce this time bound by using an efficient discrete sweeping method that exploits the geometric structures of G. Our algorithm takes $O(mn + K + L)$ time, where K and L are some structural parameters of the problem and are $O(nm^2)$ in the worst case. Even in the worst case when $K, L = O(nm^2)$, our algorithm takes $O(nm^2) = O(|V|)$ time.

3.1   The Feasible Toolpath Algorithm (FTA)

For now, we assume that a single feasible toolpath $\mathcal{T}$ exists. If this is not the case, then we show in Section 3.3 how to find the minimum number of feasible toolpaths whose union covers $\mathcal{C}$.

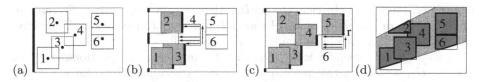

Fig. 2. (a) The six points in U_{j-1} are labeled based on the lexicographical order; the thick vertical segment is the initial (rightmost) boundary curve. (b) The first three squares have been visited and the boundary curve (thick vertical segments) updated accordingly. The arrows show the depths of the examined rows of $S(4)$. (c) Square $S(6)$ is being visited; only some of its rows are examined. (d) The diagonal band is A_j; the shaded polygonal areas within the union of squares inside A_j form U_j.

Our idea for finding $\mathcal{T}$ is similar to that of growing a "breadth-first search" tree in G rooted at the source vertex s: For each layer R_j, we determine the set of cells in R_j reachable from s, called the *feasible set* of R_j and denoted by U_j. Clearly, $U_1 = A_1$. $\mathcal{T}$ exists if $U_n \neq \emptyset$ for the last layer R_n.

By Definition 1, for any cell $r \in R_{j-1}$, we define the square of successors (cells), $S(r) \subseteq R_j$, of r. Here we extend this concept to any subset W of R_{j-1}. For a subset $W \subseteq R_{j-1}$, let $S(W) = \bigcup_{r \in W} S(r) \subseteq R_j$. For $r \in R_{j-1}$ and $W \subseteq R_{j-1}$, let $E(r) = S(r) \cap A_j$ be the *permissible image* of r, and $E(W) = S(W) \cap A_j$ be the *permissible image* of W.

The cells in $E(r)$ and $E(W)$ satisfy both the accessibility and angular change constraints as the tool goes from c_{j-1} to c_j. An easy induction on j shows that

$$U_j = \begin{cases} A_1 & \text{for } j = 1, \\ E(U_{j-1}) & \text{for } j = 2, \ldots, n. \end{cases} \tag{2}$$

Using (2), we successively compute each of $U_1, U_2, \ldots, U_n$. If $U_n \neq \emptyset$, then $\mathcal{T} = (r_1^*, \ldots, r_n^*)$ can be obtained easily by standard path reporting techniques.

Thus, the key task for our **feasible toolpath algorithm (FTA)** is: Given U_{j-1}, compute U_j efficiently, for each $j = 2, \ldots, n$. This task is performed by the **feasible set procedure (F-SET)** in the next section.

3.2 The Feasible Set Procedure (F-SET)

Given U_{j-1}, the procedure F-SET computes the feasible set U_j. By (2), $U_j = E(U_{j-1}) = A_j \cap S(U_{j-1}) = A_j \cap \left(\bigcup_{r \in U_{j-1}} S(r) \right)$. Thus, U_j is the intersection of the union of squares for all cells in U_{j-1} and the set A_j (e.g., see Fig. 2(d)).

Determining U_j is related to Klee's measure problem in 2-D [1,3,19], which computes the area of the union of a set of axis-aligned rectangles on the plane. Bentley solved this problem in optimal $O(N \log N)$ time for N rectangles [3], based on a plane sweeping approach using a segment tree. There are some differences between Klee's measure problem in 2-D and our problem. A main difference is that instead of computing the area of the union, we need to find the union itself. Furthermore, the cells in the union $S(U_{j-1})$ that are not in A_j must be excluded from $S(U_{j-1})$ to yield the set U_j. Also, our problem is on a discrete domain (a 2-D grid), and our rectangles are all squares of the same size. Therefore, we apply a different plane sweeping method than [3].

A goal of our sweeping is to construct U_j in a time proportional to $|U_j|$. To find U_j, we add to the union, one by one, each square $S(r)$, for $r \in U_{j-1}$, examine (sweep) the cells of $S(r)$, and add to U_j only those cells that belong to A_j. To achieve efficiency, during the sweep, we must avoid repeated visits to the cells that have already been visited. This is crucial since the overlap among the squares in the union may be significant. We avoid repeated examination of cells by using a careful sweeping order and the fact that all squares $S(r)$ have the same size. Our sweeping order includes both the order in which we visit the squares and the order in which we visit the cells in each square. We also keep track of the rightmost boundary, called the *boundary curve*, of the current union.

We first sort the cells of U_{j-1} in the lexicographical order (*left-to-right*, and then *top-to-bottom*) based on their coordinates: $r_1 = (\alpha_1, \beta_1) < r_2 = (\alpha_2, \beta_2)$ iff $\alpha_1 < \alpha_2$ or ($\alpha_1 = \alpha_2$ and $\beta_1 < \beta_2$). This takes $O(m + |U_{j-1}|)$ time using bucket-sort [10]. In Fig. 2(a), the six points of U_{j-1} are numbered in this order. Let $p_1, p_2, \ldots, p_{k_{j-1}}$ be the list of cells of U_{j-1} in this sorted order. In this order, we successively add the squares $S(p_1), S(p_2), \ldots, S(p_{k_{j-1}})$ to the union $S(U_{j-1})$. When a square $S(p_i)$ is being added, we visit its cells row by row, starting from the *bottom row* and up. Each row is visited *from right to left*. If a visited cell $q_{i_h} \in S(p_i)$ is in A_j, it is added to U_j (which is stored in an array $U_j[1 \ldots k_j]$). To avoid repeated examination of the cells, we use a *boundary curve array*, $B_a[1 \ldots m]$, which maintains the rightmost *boundary curve* of the current union in R_j; B_a holds the (α, β) positions (row and column) of the boundary curve such that $B_a[\alpha] = \beta$. When a new square $S(p)$ is added to the union and its cells are examined, the array B_a is updated by modifying the values for the rows containing "new cells" from $S(p)$: If the rightmost cell of $S(p)$ in row α_0 lies in column β_0 and is added to the union, then $B_a[\alpha_0] = \beta_0$. Due to our order of adding squares, when a new square is added, no cell to the left of the boundary curve needs to be examined. Initially, the boundary array holds a value "-1" for each row, indicating that no cell of R_j is swept yet. In Fig. 2(a), the thick vertical segment marks the initial boundary curve; in Fig. 2(b)-(c), the boundary curve is updated after adding $S(1)$, $S(2)$, $S(3)$, $S(4)$, and $S(5)$.

When a square $S(p)$ is added, it is either disjoint from the current union U_j' or overlaps with U_j'. If it is disjoint, then all its cells are examined. Otherwise, we use the boundary array to limit the sweeping to only the cells in $S(p) - U_j'$. We stop examining the cells of a row whenever we encounter a cell c that lies on the boundary curve (see Fig. 2(b)). Furthermore, if the cell c on the boundary curve is the *first* (rightmost) cell of a row of $S(p)$, we stop examining $S(p)$ altogether (see Fig. 2(c)).

The F-SET procedure also produces a *parent set* P_j, such that $par(q) \in P_j \subseteq U_{j-1}$ is a *parent cell* of a cell $q \in U_j$. The parent cells are used to report $\mathcal{T}$ (by a standard path reporting technique). For example, a parent for a cell $q \in U_j$ can be the center cell $p_i \in U_{j-1}$ of any square $S(p_i)$ containing q.

3.3   Computing the Minimum Number of Feasible Toolpaths

If a single toolpath $\mathcal{T}$ covering $\mathcal{C}$ does not exist, i.e., $U_j = \emptyset$ for some $j \leq n$, then we need to find the minimum number of feasible toolpaths, $\mathcal{T}_1, \ldots, \mathcal{T}_l$, whose union $\mathcal{T} = \mathcal{T}_1 \cup \cdots \cup \mathcal{T}_l$ covers $\mathcal{C}$. We first introduce the following definition.

Definition 2. *Let $I = (i_0 = 1, i_1, i_2, \ldots, i_{l-1}, i_l = n)$ be a sequence of indices such that $i_j \in \{1, \ldots, n\}$ and $i_{j-1} < i_j$ for all $j = 1, \ldots, l$. For any $i \leq j$, let $U_{i,j}$ be the permissible image of R_j by starting the F-SET computation at R_i (for the subsequence of contact points $\mathcal{C}_{i,j} = (c_i, \ldots, c_j) \subseteq \mathcal{C}$). Then I is called a path sequence if $U_{1,i_1}, U_{i_1+1,i_2}, \ldots, U_{i_{h-1}+1,i_h}, \ldots, U_{i_{l-1}+1,n}$ are all non-empty.*

If $U_{i_{h-1}+1,i_h} \neq \emptyset$, then by using the feasible toolpath algorithm, we can find a feasible toolpath $\mathcal{T}_h = \mathcal{T}_{i_{h-1}+1,i_h}$ for the subsequence $\mathcal{C}_{i_{h-1}+1,i_h} = (c_{i_{h-1}+1}, \ldots, c_{i_h})$ of $\mathcal{C}$. Thus, to obtain the minimum number of feasible toolpaths whose union covers $\mathcal{C}$, it suffices to find a path sequence of the minimum size.

To produce a minimum size path sequence, we apply the greedy method. We run the feasible toolpath algorithm starting at R_1 and until the first $U_j = \emptyset$ is met, for some $1 < j \leq n$, and construct the first feasible toolpath $\mathcal{T}_1 = \mathcal{T}_{i_0,i_1}$ (with $i_1 = j - 1$). We then repeat this process, starting with R_j as the first layer and until either an empty $U_{j'}$ is encountered or a non-empty U_n is reached. The union of feasible toolpaths thus computed, $\mathcal{T} = \mathcal{T}_1, \ldots, \mathcal{T}_l$, covers $\mathcal{C}$ and utilizes the minimum number of paths. If the size of the resulting path sequence is l, then the sculpting of $\mathcal{F}$ along $\mathcal{C}$ will require at least $l - 1$ repositionings of the tool (e.g., for the different segments of the toolpath $\mathcal{T}$).

3.4   Time Complexity of the Feasible Toolpath Algorithm

Let $|U_{j-1}| = k_{j-1}$ and $|S(U_{j-1})| = l_j$. Using bucket-sort [10], the cells of U_{j-1} are sorted in $O(m + k_{j-1})$ time. Since each cell of $S(U_{j-1})$ is visited only once and the operations on each visited cell can be performed in $O(1)$ time, it takes $O(l_j)$ time to examine all cells in $S(U_{j-1})$. Thus, given U_{j-1}, the F-SET procedure computes U_j in $O(m + k_{j-1} + l_j)$ time. From U_1, the feasible toolpath algorithm applies the F-SET procedure $n - 1$ times to compute $U_2, U_3, \ldots, U_n$, in $O(m + k_1 + l_2) + O(m + k_2 + l_3) + \cdots + O(m + k_{n-1} + l_n) = O(mn + K + L)$ time, where $K = \sum_{i=1}^{n-1} k_i$ and $L = \sum_{i=2}^{n} l_i$. Finally, by using the parent cells, a feasible toolpath $\mathcal{T} = (r_1^*, r_2^*, \ldots, r_n^*)$ is reported in $O(n)$ time. Hence, computing a single feasible toolpath (if one exists) takes $O(mn + K + L))$ time. If a single feasible toolpath does not exist, then the greedy algorithm produces l path segments for $\mathcal{T}$ with the smallest possible value l, yet its total running time is the same as for finding one single feasible toolpath. That is, it also takes $O(mn + K + L)$ time in this case. It should be noted that the total input size of the discrete visibility maps, $\sum_{i=1}^{n} |A_i| = \sum_{i=1}^{n} |\mathcal{A}(c_i)|$, is $O(nm^2)$ in the worst case.

We summarize these results in the next theorem.

Theorem 1. *Given a contact curve $\mathcal{C} = (c_1, \ldots, c_n)$ and a sequence $A = (A_1, \ldots, A_n)$, where $A_i \neq \emptyset$ corresponds to the discrete visibility map for the point $c_i \in \mathcal{C}$ and $|A_i| = O(m^2)$, the feasible toolpath problem can be solved in $O(mn + K + L)$ time, where $K, L = O(nm^2)$ in the worst case.*

Even in the worst case with $K, L = O(nm^2)$ for which our feasible toolpath algorithm takes $O(nm^2)$ time, it is still a significant improvement over the $O(nm^4)$ time solution applying standard path finding techniques to the graph G.

4   The Maximum Accuracy (MaxA) Toolpath Problem

A MaxA solution is a feasible toolpath $\mathcal{T}$ such that the sum of the differences between the tool's orientation at each contact point $c_i \in \mathcal{C}$ and the vector normal to $\mathcal{F}$ at c_i is minimized. It is equivalent to a minimum total weight path from layer V_1 to layer V_n in the directed graph $G = (V, E)$ with non-negative vertex weights defined in Section 2. Using standard shortest path techniques, MaxA can be solved in $O(|E| + |V|) = O(nm^4)$ time. To obtain a faster solution, we extend the feasible toolpath algorithm in Section 3 and give an $O(mn + K + L)$ time algorithm for the single toolpath case, where $K, L = O(nm^2)$ in the worst case. Again, we show first how to solve the problem when a single feasible toolpath exists (Sections 4.1 and 4.2), and then how to find the minimum number of toolpaths if a single toolpath does not exist (Section 4.3).

4.1   The Maximum Accuracy Feasible Toolpath Algorithm (MaxA-FTA)

A feasible MaxA toolpath $\check{\mathcal{T}}$ optimizes the maximum accuracy criterion (1). To capture this criterion in computing $\check{\mathcal{T}}$, to each cell $r \in A_i$, we assign a weight $w(r)$, where $w(r) = M(\boldsymbol{v_r}, \boldsymbol{p_i})$ is the angular difference between the tool's orientation $\boldsymbol{v_r}$ at c_i corresponding to the cell r and the unit vector $\boldsymbol{p_i}$ normal to $\mathcal{F}$ at c_i. Then, a feasible toolpath $\check{\mathcal{T}} = (r_1^*, \ldots, r_n^*)$ is a solution to MaxA if $w(\check{\mathcal{T}}) = \sum_{i=1}^n w(r_i^*)$, the total weight of $\check{\mathcal{T}}$, is minimized over all feasible toolpaths. We call such an optimal toolpath $\check{\mathcal{T}}$ a *minimum weight feasible toolpath* for $\mathcal{C}$. We also let $\check{\mathcal{T}}_j = (r_1^*, \ldots, r_j^*)$ denote a minimum weight feasible toolpath for $\mathcal{C}_j = (c_1, \ldots, c_j) \subseteq \mathcal{C}$ and $\check{\mathcal{T}}_j(r) = (r_1^*, \ldots, r_{j-1}^*, r)$ denote a minimum weight feasible toolpath for $\mathcal{C}_j$ that ends at a cell $r \in U_j$.

To solve MaxA, we extend the FTA algorithm in Section 3. A main extension is on how to choose a "best" parent cell for each cell in U_j. A "best" parent cell for a cell $q \in U_j$, $p = par(q)$, must be such that $w(\check{\mathcal{T}}_{j-1}(p))$ is the smallest among all possible parents of q in U_{j-1} (in a graph context, such a parent cell corresponds to q's predecessor vertex on a shortest path from the source vertex).

A straightforward way to compute $par(q)$ for each cell $q \in U_j$ is to examine *all* cells r in U_{j-1} such that $q \in S(r)$ and then let $par(q) = p$ such that $w(\check{\mathcal{T}}_{j-1}(p))$ is the smallest among all such cells in U_{j-1}. Since in the worst case, q may have $O(k_{j-1}) = O(|U_{j-1}|)$ possible parents in U_{j-1}, computing this may take $O(k_{j-1})$ time for each cell $q \in U_j$, thus significantly increasing the running time of the MaxA algorithm. Our approach for computing $par(q)$ is to model this task as a special case of the 2-D range minimum queries (RMQ) and find $par(q)$ in $O(1)$ time for each $q \in U_j$. This leads to a MaxA-FTA algorithm (for the vertex-weighted graph case) with a running time that matches with that of the FTA algorithm (for the unweighted graph case). The extended version of the F-SET procedure, called the MaxA-F-SET procedure, is given in the next section.

4.2 The MaxA Feasible Set Procedure (MaxA-F-SET)

The MaxA-F-SET procedure aims to compute the feasible set $U_j = \{q_1, \ldots, q_{k_j}\}$, the parent set $P_j = \{par(q_1), \ldots, par(q_{k_j})\}$, and the set of weights of the minimum weight feasible toolpaths ending at the cells of U_j, $W_j = \{w(\check{T}_j(q_1)), \ldots, w(\check{T}_j(q_{k_j}))\}$. We assume that U_{j-1}, P_{j-1}, and W_{j-1} are already available. U_j can be computed from U_{j-1} in exactly the same way as in F-SET. Computing W_j is easy once U_j and P_j are already computed and W_{j-1} is given: $w(\check{T}_j(q_i)) = w(q_i) + w(\check{T}_{j-1}(par(q_i)))$. Hence, the main task is to develop an efficient method for computing P_j.

The MaxA-F-SET procedure computes each $par(q_j)$ in $O(1)$ time by modeling it as a special case of the 2-D range minimum queries (RMQ) [12,23]. A well-known (1-D) RMQ algorithm by Gabow, Bentley, and Tarjan [12] preprocesses an array of values in linear time and space, such that a query on finding the minimum value in any contiguous subarray can be answered in $O(1)$ time. Very recently, Yuan and Atallah [23] presented an algorithm with linear preprocessing time and space and $O(1)$ time queries (on any cubic shaped subarrays) for the RMQ problem on a d-D array for any fixed integer $d > 1$. Actually, Yuan and Atallah's algorithm [23] could be used in our solution. However, their algorithm is quite involved and relies on some specialized data structures. We give a much simpler solution that could be easily implemented for manufacturing applications. Our method is made simple by the fact that, unlike in the general RMQ, our query range is always of the same shape — it is the shape of the square of predecessors determined by the angular change limit value θ. We call this special 2-D case of RMQ the *fixed rectangle queries* (FRQ). We preprocess the array U_{j-1} in $O(|U_{j-1}|)$ time so that any $par(\cdot)$ query can be answered in $O(1)$ time. Below we first give a general description of processing the fixed rectangle queries and then show how to apply it to our problem.

The fixed rectangle query (FRQ)

We begin with the 1-D fixed range query problem (1-D FRQ). Let $B[1 \ldots m]$ be an array of m real numbers, and $M \geq 1$ be a given integer (the fixed range size of queries). We would like to construct an array $B_{min}[1 \ldots m]$ such that $B_{min}[i]$ holds the smallest element in the M-element subarray of B beginning at $B[i]$ (special cases around the boundaries of B can be easily handled). To compute $B_{min}[1 \ldots m]$, we utilize a simple version of the 1-D FRQ algorithm by Chen, Wang, and Wu [5,6]. Their method first partitions B into $N = \lceil \frac{m}{M} \rceil$ M-element subarrays and then computes the prefix minima and suffix minima in each such subarray, in altogether $O(m)$ time. Since any fixed range query (of size M) on B spans either one or two of the subarrays in the partition, the answer to the query can be found easily in $O(1)$ time from a prefix minimum and/or a suffix minimum. Using this 1-D FRQ algorithm [5,6], we can compute $B_{min}[i]$ for all $i = 1, \ldots, m$ in $O(m)$ time (by performing m 1-D FRQ queries on B).

To handle the 2-D fixed rectangle queries (2-D FRQ), we make use of the above 1-D FRQ solution repeatedly. Let $C[1 \ldots m, 1 \ldots m]$ be a 2-D array of m^2 real numbers, and $M \geq 1$ be a given integer. We would like to build a 2-D array $C_{min}[1 \ldots m, 1 \ldots m]$ such that $C_{min}[i,j]$ holds the smallest element in the $M \times M$ subarray of C with $C[i,j]$ at its upper-left corner. The construction of $C_{min}[1 \ldots m, 1 \ldots m]$ consists of two stages. In the first stage, we apply the 1-D FRQ algorithm to each row of C and construct an intermediate array $C_{min}^r[1 \ldots m, 1 \ldots m]$. A row $C_{min}^r[i, \cdot]$ of C_{min}^r is the 1-D array of minima (as defined in the above paragraph) for the row $C[i, \cdot]$ of C, for each $i = 1, \ldots, m$. In the second stage, we apply the 1-D FRQ algorithm to each column of C_{min}^r to construct the desired array C_{min}.

Since each 1-D FRQ process takes $O(m)$ time on each row or column of the arrays involved, computing C_{min}^r from C takes $O(m^2)$ time and computing C_{min} from C_{min}^r also takes $O(m^2)$ time. Thus, the overall time for computing C_{min} from C is $O(m^2)$, i.e., C_{min} is constructed in linear time with respect to $|C|$. Using C_{min}, each 2-D FRQ query (on any $M \times M$ subarray of C) takes $O(1)$ time to answer (by simply referring to a corresponding entry in C_{min}).

The fixed rectangle queries (FRQ) applied to MaxA-F-SET

Assume that W_{j-1} is stored in a 2-D array $W_{j-1}[1 \ldots m, 1 \ldots 2m]$ such that:

$$W_{j-1}[i_1, i_2] = \begin{cases} w(\breve{T}_{j-1}(p)) & \text{if } p = (\alpha_{i_1}, \beta_{i_2}) \in U_{j-1}, \\ \infty & \text{otherwise.} \end{cases}$$

Let $S^{-1}(q) \subseteq R_{j-1}$ be the *square of predecessors* of $q \in R_j$ defined as: If $q = (\alpha_q, \beta_q) \in R_j$, then

$$S^{-1}(q) = \{r = (\alpha_r, \beta_r) \in R_{j-1} : |\alpha_q - \alpha_r| \leq \theta \text{ and } |\beta_q - \beta_r| \leq \theta\}.$$

Note that, similar to the square of successors that forms a square of cells on R_j, the square of predecessors forms a square of cells on R_{j-1}. Also, note that any possible parent of a cell $q \in U_j$ must belong to $S^{-1}(q) \cap U_{j-1}$. Hence, for any cell $q = (\alpha_{j_1}, \beta_{j_2}) \in U_j$, finding $par(q)$ corresponds to a fixed rectangle query on the $M \times M$ subarray of W_{j-1} centered at $(\alpha_{j_1}, \beta_{j_2})$, where M is a constant that depends on the angular change constraint. When the 2-D FRQ algorithm is applied to W_{j-1} to produce W_{min}, the set of finite entries in W_{min} for W_{j-1} at the positions corresponding to the cells in U_j form the needed parent set P_j, i.e., if $q = (\alpha_{i_1}, \beta_{i_2}) \in U_j$, then $par(q) = W_{min}[i_1, i_2]$.

The time for computing W_{min} from W_{j-1} using the 2-D FRQ algorithm is linear in terms of the size of W_{j-1} (i.e., $|W_{j-1}| = 2m^2$). We can lower this computation time to $O(|U_{j-1}|)$ by considering only the finite entries of W_{j-1}. This can be attained by "compressing" the finite entries of each row (or column) of W_{j-1} into a consecutive sequence. The "compression" can be easily done together for all rows (or columns) of W_{j-1} by bucket-sort. Each of the two stages of the 2-D FRQ algorithm can be modified to work on the compressed rows (or columns) without much difficulty. Note that an $O(|U_{j-1}|)$ time for computing

W_{min} from W_{j-1} (and eventually P_j) is meaningful when $k_{j-1} = o(m^2)$. Hence, it takes $O(|U_{j-1}| + |U_j|)$ time to generate $par(q)$ for all $q \in U_j$ (i.e., P_j).

4.3 Computing the Minimum Number of Maximum Accuracy Feasible Toolpaths

When a single feasible toolpath does not exist for C, we must find an optimal solution with the minimum number of MaxA toolpaths to cover C. Our greedy method for finding the minimum number of feasible toolpaths covering C for the FT problem in Section 3.3 may not yield a set of MaxA toolpaths whose total sum of weights is minimized (as required by MaxA). For example, if one greedily finds two feasible MaxA toolpaths $\check{T}_{1,j-1}$ for $C_{1,j-1}$ and $\check{T}_{j,n}$ for $C_{j,n}$, the sum of their weights, $w(\check{T}_{1,j-1}) + w(\check{T}_{j,n})$, may still be larger than the sum $w(\check{T}_{1,i-1}) + w(\check{T}_{i,n})$ for two different feasible MaxA toolpaths $\check{T}_{1,i-1}$ and $\check{T}_{i,n}$, with $i \neq j$. That is, greedily making each MaxA toolpath go as long as possible may not lead to the smallest total weight for the union of the MaxA toolpaths. We still use the greedy method to determine the minimum number l of toolpaths required to cover C. But, computing an actual set of l toolpaths for an optimal MaxA solution must be carried out differently. There are $O(n^2)$ possible toolpaths between any two layers R_i and R_j, $1 \leq i \leq j \leq n$; one may choose the needed toolpaths for an optimal MaxA solution from the $O(n^2)$ MaxA toolpaths.

Let $\check{T}_{i,j} = (r_i^*, \ldots, r_j^*)$ be a minimum weight feasible toolpath for the subsequence $C_{i,j} = (c_i, \ldots, c_j) \subseteq C$ and $w(\check{T}_{i,j}) = \sum_{k=i}^{j} w(r_k^*)$ be its total weight. To produce an optimal MaxA solution for covering C, we need to solve the following *minimum-link minimum-weight toolpath (MLMWT)* problem.

Minimum-Link Minimum-Weight Toolpath (MLMWT) Problem.
Given a set of visibility maps $A = \{A_1, \ldots, A_n\}$ for the set of accessibility rectangles $R = \{R_1, \ldots, R_n\}$, find a path sequence $I = (i_0 = 1, i_1, \ldots, i_{l-1}, i_l = n)$ and the corresponding sequence of minimum weight feasible toolpaths $\check{T}_1 = \check{T}_{1,i_1}, \check{T}_2 = \check{T}_{i_1+1,i_2}, \ldots, \check{T}_l = \check{T}_{i_{l-1}+1,n}$ for covering C, such that:

1. *The cardinality l of I is minimized, and*
2. *$\check{T} = \check{T}_1 \cup \check{T}_2 \cup \cdots \cup \check{T}_l$ has the minimum total weight $w(\check{T}) = \sum_{k=1}^{l} w(\check{T}_k)$.*

We call $\check{T}$ an l-link minimum-weight sequence for R. Finding l is easy: We just apply the greedy method for the FT problem in Section 3.3. But, $\check{T}_1, \check{T}_2, \ldots, \check{T}_l$ need to be chosen from the possible MaxA toolpaths $\check{T}_{i,j}$ for all $1 \leq i \leq j \leq n$.

To find each $\check{T}_{i,j}$ and its weight $w(\check{T}_{i,j})$, consider every pair of rectangles R_i and R_j (i.e., for a subsequence $C_{i,j} \subseteq C$), $1 \leq i \leq j \leq n$. If for some pair (i, j), $i < j$, a single feasible toolpath $\check{T}_{i,j}$ does not exist (in particular, we assume that $\check{T}_{1,n}$ does not exist), then we set $\check{T}_{i,j} = \emptyset$ and $w(\check{T}_{i,j}) = \infty$. Since $|C| = n$, there are up to $O(n^2)$ feasible toolpaths. Let $S = \{\check{T}_{i,j} \neq \emptyset : 1 \leq i \leq j \leq n\}$ be the set of all feasible MaxA toolpaths. To compute the set S, we apply the MaxA-FTA algorithm to $n - 1$ problem instances, each starting at one of $R_1, R_2, \ldots, R_{n-1}$ and ending at R_n. Thus, computing S takes $O(mn^2 + n(K + L))$ time. Now given $\mathcal{S}$, we want to find a subset $\mathcal{S}_l = \{\check{T}_{1,i_1}, \check{T}_{i_1+1,i_2}, \ldots, \check{T}_{i_{l-1}+1,n}\} \subseteq \mathcal{S}$ of cardinality l that satisfies condition 2 of the MLMWT problem.

A straightforward way to compute $\mathcal{S}_l$ is to treat the MaxA toolpaths in $\mathcal{S}$ as weighted "segments" (with rectangles R_i and R_j as the "endpoints" of each $\check{T}_{i,j}$), and construct a vertex-weighted interval graph $G_S = (V_S, E_S)$ [11,14,15], such that for each MaxA toolpath $\check{T}_{i,j} \neq \emptyset$, there is a vertex $\check{v}_{i,j} \in V_S$ with a weight $w(\check{T}_{i,j})$ and two vertices $\check{v}_{i_1,j_1}$ and $\check{v}_{i_2,j_2}$ in G_S are connected by a directed edge $(\check{v}_{i_1,j_1}, \check{v}_{i_2,j_2})$ if $i_2 = j_1 + 1$. Then $\mathcal{S}_l$ corresponds to a certain l-link shortest path in G_S. Since G_S has $O(n^2)$ vertices and $O(n^3)$ edges (a vertex $\check{v}_{i,j}$ can have up to $O(n)$ outgoing edges connecting it to the $n - j$ vertices with a starting index of $j + 1$), one could find $\mathcal{S}_l$ by applying a general l-link shortest path algorithm [10] to G_S, in $O(ln^3)$ time. Together with the construction of $\mathcal{S}$ and G_S, this approach would yield an $O(ln^3 + mn^2 + n(K + L))$ time solution. Below we give a different, faster approach for computing $\mathcal{S}_l$ from $\mathcal{S}$ without building G_S, which is based on dynamic programming.

Let $\check{T}_j^g$ denote a g-link minimum-weight toolpath for $C_{1,j}$, i.e., $\check{T}_j^g$ is the union of g feasible MaxA toolpaths covering $C_{1,j}$. Thus, $\check{T} = \check{T}_n^l$. We construct a dynamic programming table H such that $H[h, j]$ holds the weight of an h-link minimum-weight toolpath for $C_{1,j}$, $w(\check{T}_j^h)$, if $\check{T}_j^h$ exists, or ∞ if $\check{T}_j^h$ does not exist. H is an $l \times n$ array since $h = 1, \ldots, l$ and $j = 1, \ldots, n$. The entries in the first row of H are the weights of 1-link minimum-weight toolpaths for $C_{1,1}, \ldots, C_{1,n}$, and the entry $H[l, n]$ is the weight of the desired l-link minimum-weight toolpath $\check{T} = \check{T}_n^l$, which is an optimal solution to the MLMWT problem (and thus the MaxA problem). An entry $H[g, j]$ for $g > 1$ can be computed quite efficiently from the entries $H[g - 1, i]$ for $1 \leq i < j$ based on the following observation.

Lemma 1. If $\check{T}_j^g = \check{T}^* \cup \check{T}_{i,j}$, $1 < i \leq j \leq n$, is a g-link minimum-weight toolpath from R_1 to R_j, then $\check{T}^*$ is a $(g-1)$-link minimum-weight toolpath from R_1 to R_{i-1} $(\check{T}^* = \check{T}_{i-1}^{g-1})$.

Proof. Suppose $\check{T}^* \neq \check{T}_{i-1}^{g-1}$, i.e., $w(\check{T}^*) > w(\check{T}_{i-1}^{g-1})$. Then, by letting $\check{P}_j^g = \check{T}_{i-1}^{g-1} \cup \check{T}_{i,j}$, we have $w(\check{P}_j^g) < w(\check{T}_j^g)$, a contradiction to the optimality of $\check{T}_j^g$. □

Based on Lemma 1, for $j = 1, \ldots, n$ and $1 \leq g \leq l$, we have:

$$H[g, j] = \begin{cases} w(\check{T}_{1,j}) & \text{for } g = 1, j = 1, \ldots, n \\ \min_{1 < i \leq j}\{H[g - 1, i - 1] + w(\check{T}_{i,j})\} & \text{for } 1 < g \leq l, 1 < j \leq n \end{cases} \quad (3)$$

To obtain $H[l, n]$, we use the dependency relation (3) to fill in the entries of the array H. To make sure that before computing $H[g, j]$, all entries in row $g - 1$ that are in the columns preceding j are available, we fill in the entries of H by starting at the first row (and going left-to-right) and then to the next row.

4.4 Time Complexity of the Maximum Accuracy Feasible Toolpath Algorithm

The time complexity of our algorithm for solving the maximum accuracy feasible toolpath problem depends on whether a single MaxA toolpath for $\mathcal{C}$ exists.

First assume that a single maximum accuracy feasible toolpath covering $\mathcal{C}$ exists. Let $|U_j| = k_j$ and $|S(U_{j-1})| = l_j$ for $1 \leq j \leq n$. The computation of U_j, given U_{j-1}, is the same as in the F-SET procedure, hence taking $O(m+k_{j-1}+l_j)$ time. Given U_j and P_j, computing W_j takes $O(1)$ time per element of U_j, i.e., we need $O(k_j)$ time to find W_j. As discussed earlier, computing P_j takes $O(k_{j-1}+k_j)$ time. Thus, the MaxA-F-SET procedure takes $O(m+k_{j-1}+k_j+l_j)$ time for each R_j. Since the MaxA-FTA algorithm utilizes the Max-F-SET procedure at most $n-1$ times, it follows that for the case when a single feasible toolpath exists for $\mathcal{C}$, the MaxA problem is solvable in $O(mn + K + L)$ time, where $K = \sum_{i=1}^{n-1} k_i$ and $L = \sum_{i=2}^{n} l_i$.

If a single feasible toolpath for $\mathcal{C}$ does not exist, then the set $\mathcal{S}$ of $O(n^2)$ MaxA toolpaths $\check{T}_{i,j}$ can be computed by solving $n-1$ instances of the MaxA-FTA problem, each with one of $R_1, R_2, \ldots, R_{n-1}$ as the first layer. Thus $\mathcal{S}$ can be computed in $O(mn^2 + n(K + L))$ time. Computing each entry of the array $H[1 \ldots l, 1 \ldots n]$ can take up to $O(n)$ time, since it involves calculating up to $O(n)$ values. Hence, we need $O(ln^2)$ time to build H. Therefore, it takes overall $O(ln^2 + mn^2 + n(K+L))$ time to find the minimum number of MaxA toolpaths.

Our results are summarized in the following theorem.

Theorem 2. *If a single feasible toolpath for $\mathcal{C}$ exists, then the maximum accuracy toolpath problem can be solved in $O(mn + K + L)$ time. If a single feasible toolpath does not exist, then the maximum accuracy toolpath problem can be solved in $O(ln^2 + mn^2 + n(K + L))$ time. In the worst case, $K, L = O(nm^2)$.*

References

1. Agarwal, P.K., Kaplan, H., Sharir, M.: Computing the volume of the union of cubes. In: SCG 2007: Proceedings of the 23rd Annual Symposium on Computational Geometry, pp. 294–301. ACM, New York (2007)
2. Balasubramaniama, M., Sarma, S.E., Marciniak, K.: Collision-free finishing toolpaths from visibility data. Computer-Aided Design 35(4), 359–374 (2003)
3. Bentley, J.: Algorithms for Klee's rectangle problems (1977) (unpublished notes)
4. Chen, D.Z., Misiołek, E.: Free-form surface partition in 3-D. In: Hong, S.-H., Nagamochi, H., Fukunaga, T. (eds.) ISAAC 2008. LNCS, vol. 5369, pp. 520–531. Springer, Heidelberg (2008)
5. Chen, D.Z., Wang, J., Wu, X.: Image segmentation with monotonicity and smoothness constraints. In: Eades, P., Takaoka, T. (eds.) ISAAC 2001. LNCS, vol. 2223, pp. 467–479. Springer, Heidelberg (2001)
6. Chen, D.Z., Wang, J., Wu, X.: Image Segmentation with asteroidality/tubularity and smoothness constraints. International Journal of Computational Geometry and Applications 12(5), 413–428 (2002)
7. Chen, L.-L., Woo, T.C.: Computational geometry on the sphere with application to automated machining. Journal of Mechanical Design 114(2), 288–295 (1992)
8. Chiou, C.-J., Lee, Y.S.: A machining potential field approach to tool path generation for multi-axis sculptured surface machining. Computer-Aided Design 34(5), 357–371 (2002)
9. Chiou, J.C.J., Lee, Y.S.: Optimal tool orientation for five-axis tool-end machining by swept envelope approach. Journal of Manufacturing Science and Engineering 127(4), 810–818 (2005)

10. Cormen, T.H., Stein, C., Rivest, R.L., Leiserson, C.E.: Introduction to Algorithms, 2nd edn. MIT Press, Cambridge (2001)
11. Fulkerson, D., Gross, O.: Incidence matrices and interval graphs. Pacific Journal of Mathematics 15(3), 835–855 (1965)
12. Gabow, H.N., Bentley, J.L., Tarjan, R.E.: Scaling and related techniques for geometry problems. In: STOC 1984: Proceedings of the 16th Annual ACM Symposium on Theory of Computing, pp. 135–143. ACM, New York (1984)
13. Gan, J.G., Woo, T.C., Tang, K.: Spherical maps: Their construction, properties and approximation. ASME Journal of Mechanical Design 116(2), 357–363 (1994)
14. Gilmore, P.C., Hoffman, A.J.: Characterization of comparability graphs and of interval graphs. Canadian Journal of Mathematics 16(3), 539–548 (1964)
15. Golumbic, M.C.: Algorithmic Graph Theory and Perfect Graphs. Annuals of Discrete Mathematics, vol. 57. North-Holland Publishing, Co., Amsterdam (1980)
16. Ho, M.-C., Hwang, Y.-R., Hu, C.-H.: Five-axis tool orientation smoothing using quaternion interpolation algorithm. International Journal of Machine Tools and Manufacture 43(12), 1259–1267 (2003)
17. Jun, C.-S., Cha, K., Lee, Y.-S.: Optimizing tool orientations for 5-axis machining by configuration-space search method. Computer-Aided Design 35(6), 549–566 (2003)
18. Morishige, K., Takeuchi, Y., Kase, K.: Tool path generation using C-Space for 5-axis control machining. Journal of Manufacturing Science and Engineering 121(1), 144–149 (1999)
19. Overmars, M.H., Yap, C.-K.: New upper bounds in Klee's measure problem. SIAM Journal on Computing 20(6), 1034–1045 (1991)
20. Wang, N., Tang, K.: Automatic generation of gouge-free and angular-velocity-compliant five-axis toolpath. Computer-Aided Design 39(10), 841–852 (2007)
21. Wang, N., Tang, K.: Five-axis tool path generation for a flat-end tool based on iso-conic partitioning. Computer-Aided Design 40(12), 1067–1079 (2008)
22. Woo, T.C.: Visibility maps and spherical algorithms. Computer-Aided Design 26(1), 6–16 (1994)
23. Yuan, H., Atallah, M.J.: Data structures for range minimum queries in multidimensional arrays. In: SODA 2010, Proceedings of the 21st Annual ACM-SIAM Symposium on Discrete Algorithms, pp. 150–160. SIAM, Philadelphia (2010)

A Trichotomy Theorem for the Approximate Counting of Complex-Weighted Bounded-Degree Boolean CSPs

Tomoyuki Yamakami

Department of Information Science, University of Fukui
3-9-1 Bunkyo, Fukui 910-8507, Japan

Abstract. We determine the complexity of approximate counting of the total weight of assignments for complex-weighted Boolean constraint satisfaction problems (or CSPs), particularly, when degrees of instances are bounded from above by a given constant, provided that all arity-1 (or unary) constraints are freely available. All degree-1 counting CSPs are solvable in polynomial time. When the degree is more than 2, we present a trichotomy theorem that classifies all bounded-degree counting CSPs into only three categories. This classification extends to complex-weighted problems an earlier result on the complexity of the approximate counting of bounded-degree unweighted Boolean CSPs. The framework of the proof of our trichotomy theorem is based on Cai's theory of signatures used for holographic algorithms. For the degree-2 problems, we show that they are as hard to approximate as complex Holant problems.

1 Bounded-Degree Boolean #CSPs

Our overall objective is to determine the approximation complexity of *Boolean constraint satisfaction problems* (or CSPs) whose instances consist of Boolean variables and their constraints, which describe "relationships" among the variables. Boolean CSPs have found numerous applications in graph theory, database theory, and artificial intelligence as well as statistical physics. A CSP asks, for a given set of Boolean variables and a set of constraints, whether all the constraints are satisfied by certain Boolean assignments to the variables. The *satisfiability problem* (SAT) is a typical example of Boolean CSPs. Since constraints used for typical CSPs are limited to certain fixed types of allowable ones (a set of these constraints is known as *constraint language*), it seems natural to parameterize CSPs in terms of a set $\mathcal{F}$ of allowable constraints and express them as CSP($\mathcal{F}$)'s. Schaefer's [13] dichotomy theorem classifies all such CSP($\mathcal{F}$)'s into two categories: polynomial-time solvable problems (i.e., in P) and NP-complete problems.

Of all types of CSPs, there has been a great interest in a particular type, in which each individual variable appears at most d times in all given constraints. The maximal number of such d on every instance is called the *degree* of the instance. This degree has played a key role in a discussion of the complexity

W. Wu and O. Daescu (Eds.): COCOA 2010, Part I, LNCS 6508, pp. 285–299, 2010.

of CSPs; for instance, the planar read-trice satisfiability problem, which is comprised of logical formulas of degree at most 3, is known to be NP-complete, while the planar read-twice satisfiability problem, whose degree is 2, falls into P. Those CSPs whose instances have degree bounded are referred to as *bounded-degree* Boolean CSPs. Dalmau and Ford [7], for instance, showed that, for certain cases of $\mathcal{F}$, the complexity of solving $\mathrm{CSP}(\mathcal{F})$ does not change even if all instances are restricted to at most degree 3.

Apart from those CSPs, a counting CSP (or #CSP, in short) asks how many Boolean assignments satisfy all given constraints. Creignou and Herman [6] gave a classification theorem for the counting complexity of #CSPs. This result was eventually extended by Cai, Lu, and Xia [5] to complex-weighted Boolean #CSPs. They also studied the exact complexity of complex-weighted Boolean #CSPs whose maximal degree does not exceed 3. From a viewpoint of approximation complexity, Dyer, Goldberg, and Jerrum [12] showed a trichotomy theorem on the approximate counting of the number of assignments for unweighted Boolean CSPs, depending on the choice of $\mathcal{F}$. This theorem is quite different from a dichotomy theorem for the exact-counting complexity of #CSPs.

A degree bound of instances to #CSPs is also crucial in a discussion on the approximation complexity of the #CSPs. We then ask, for given a degree bound d, what set $\mathcal{F}$ of constraints make #CSPs difficult to compute. Based on the aforementioned trichotomy theorem, Dyer, Goldberg, Jalsenius, and Richerby [10] recognized four categories of unweighted Boolean #CSPs whose degrees are further bounded.

Similar to the unweighted case of Dyer et al. [10], we intend to allow any complex-weighted unary constraint for free. Notice that the free use of unweighted unary constraints were frequently made (e.g., [7]). Moreover, in a more general setting of Holant problems [1,4], complex-weighted unary constraints were also given for free.

Notationally, we use the notation $\#\mathrm{CSP}_d^*(\mathcal{F})$ to denote a problem of computing the total weight of constraints for all Boolean assignments for which (i) any complex-weighted unary Boolean constraint can be used for free, (ii) each variable appears at most d times among all given constraints, including free unary constraints, and (iii) all constraints (except for free constraints) should be taken from $\mathcal{F}$.

The main purpose of this paper is to prove the following trichotomy theorem that classifies all $\#\mathrm{CSP}_d^*(\mathcal{F})$'s into only three categories.

Theorem 1. *Let $d \geq 3$ be any degree bound. If either $\mathcal{F} \subseteq \mathcal{AF}$ or $\mathcal{F} \subseteq \mathcal{ED}$, then $\#\mathrm{CSP}_d^*(\mathcal{F})$ is in $\mathrm{FP}_\mathbb{C}$. Otherwise, if $\mathcal{F} \subseteq \mathcal{IM}$, then $\#\mathrm{DOWNSET}_\mathbb{C}^* \leq_{\mathrm{AP}}$ $\#\mathrm{CSP}_d^*(\mathcal{F}) \leq_{\mathrm{AP}} \#\mathrm{DOWNSET}_\mathbb{C}$ under approximation-preserving reducibility (or AP-reducibility). Otherwise, $\#\mathrm{SAT}_\mathbb{C}^* \leq_{\mathrm{AP}} \#\mathrm{CSP}_d^*(\mathcal{F})$.*

Here, $\#\mathrm{DOWNSET}_\mathbb{C}^*$ and $\#\mathrm{DOWNSET}_\mathbb{C}$ are two complex-weighted versions of the *counting downset problem* and $\#\mathrm{SAT}_\mathbb{C}^*$ is also a similar variant of the *counting satisfiability problem*. See [16] for their precise definitions.

Theorem 1 highlights a clear difference between unweighted Boolean constraints and complex-weighted Boolean constraints, partly because of the strong expressiveness of complex-weighted unary constraints.

Our proof of Theorem 1 is based on the trichotomy theorem of Yamakami [16], who proved the theorem using a theory of signatures (see, e.g., [2,3]) used to analyze Valiant's Holographic algorithms [14,15]. In particular, our key claim, which directly yields Theorem 1, states that, for any degree bound $d \geq 3$ and for any set $\mathcal{F}$ of complex-weighted constraints, $\#\mathrm{CSP}_d^*(\mathcal{F})$ is AP-interreducible to $\#\mathrm{CSP}^*(\mathcal{F})$; in other words, $\#\mathrm{CSP}_d^*(\mathcal{F})$ is "equivalent" to $\#\mathrm{CSP}^*(\mathcal{F})$ in approximation complexity. The most part of this paper is devoted to proving this key claim. When the degree bound d is 2, on the contrary, we will show that $\#\mathrm{CSP}_2^*(\mathcal{F})$ is "equivalent" to Holant problems restricted to the set $\mathcal{F}$ of constraints, provided that all unary constraints are freely available. In the case of degree 1, every $\#\mathrm{CSP}_1^*(\mathcal{F})$ is solvable in polynomial time.

Our argument for complex-weighted constraints is quite different from Dyer et al.'s argument for unweighted constraints and also from Cai et al.'s argument for exact counting for complex-weighted constraints. While a key technique in [10] is 3-simulatability as well as ppp-definability, our proof argument exploits a notion of *limited T-constructability*—a restricted version of T-constructability developed in [16]. With its extensive use, the proof we will present in the rest of this paper becomes clean, elementary, and intuitive.

2   Preliminaries

Let $\mathbb{N}$ denotes the set of all natural numbers (i.e., non-negative integers) and $\mathbb{N}^+$ denotes $\mathbb{N} - \{0\}$. Similarly, $\mathbb{C}$ denotes the set of all complex numbers. For succinctness, the notation $[n]$ for a number $n \in \mathbb{N}$ expresses the integer set $\{1, 2, \ldots, n\}$. To improve readability, we sometimes identify the "name" of a node in a given undirected graph with the "label" of the same node.

2.1   Signatures, #CSP, and Holant Problems

For any undirected graph $G = (V, E)$ (where V is a node set and E is an edge set) and a node $v \in V$, an *incident set* $E(v)$ of v is the set of all edges *incident* to v, and $deg(v)$ is the *degree* of v. A bipartite graph is a tuple $(V_1|V_2, E)$, where V_1 and V_2 are respectively sets of nodes on the left-hand side and right-hand side of the graph and E is a set of edges such that $V_1 \times V_2$.

Each function f from $\{0,1\}^k$ to $\mathbb{C}$ is called a *k-ary Boolean signature* or simply a *k-ary signature*. This k is called the *arity* of f. We express f as a sequence of its values (assuming a standard order of all binary strings of length k). For instance, when $k = 2$, f can be expressed as $(f(00), f(01), f(10), f(11))$. A signature f is *symmetric* if f's values depend only on the Hamming weight of inputs. When f is a symmetric function of arity k, a succinct notation $f = [f_0, f_1, \ldots, f_k]$, where each f_i is the value of f on inputs of Hamming weight i, is often used. For example, $\Delta_0 = [1, 0]$ and $\Delta_1 = [0, 1]$, and $EQ_k = [1, 0, \ldots, 0, 1]$ ($k - 1$ zeros). For convenience, we write $\mathcal{U}$ for the set of all unary signatures.

A *complex-weighted Boolean #CSP* restricted to a set $\mathcal{F}$ of signatures, simply denoted #CSP($\mathcal{F}$), takes a finite set G of signatures (which are called *constraints* in Section 1) of the form $h(x_{i_1}, x_{i_2}, \ldots, x_{i_k})$ on Boolean variables $x_1, x_2, \ldots, x_n$, where $i_1, \ldots, i_k \in [k]$ and $h \in \mathcal{F}$, and it outputs the value $\sum_{x_1, x_2, \ldots, x_n \in \{0,1\}} \prod_{h \in G} h(x_{i_1}, x_{i_2}, \ldots, x_{i_k})$. To improve readability, we often omit the set notation and write, e.g., #CSP($f, g, \mathcal{F}, \mathcal{G}$) to mean #CSP($\{f, g\} \cup \mathcal{F} \cup \mathcal{G}$).

Here, we need to address a technical issue concerning complex-valued functions. Recall that each instance to a #CSP involves a finite set of signatures. How can we compute those signatures? How can we receive them as a part of input instance for the first place? It is convenient to treat such a k-ary signature f as a "black box," which answers the complex value $f(x)$ instantly whenever one makes a query $x \in \{0, 1\}^k$. This black-box convention helps us eliminate the entire description of f (e.g., bit sequences) from the instance to a #CSP.

Given two sets $\mathcal{F}_1, \mathcal{F}_2$ of signatures, a *bipartite Holant problem* Holant($\mathcal{F}_1 | \mathcal{F}_2$) (on a Boolean domain) is defined as follows. An instance is a signature grid $\Omega = (G, \mathcal{F}_1' | \mathcal{F}_2', \pi)$ composed of a finite undirected bipartite graph $G = (V_1 | V_2, E)$, two finite subsets $\mathcal{F}_1' \subseteq \mathcal{F}_1$ and $\mathcal{F}_2' \subseteq \mathcal{F}_2$, and a labeling function $\pi : V_1 \cup V_2 \to \mathcal{F}_1' \cup \mathcal{F}_2'$ such that $\pi(V_1) \subseteq \mathcal{F}_1'$, $\pi(V_2) \subseteq \mathcal{F}_2'$, and each node $v \in V_1 \cup V_2$ is labeled by the function $\pi(v) : \{0,1\}^{deg(v)} \to \mathbb{C}$. For brevity, we sometimes write f_v for $\pi(v)$. Let $Asn(E)$ be the set of all *edge assignments* $\sigma : E \to \{0,1\}$. The bipartite Holant problem is to compute the value Holant$_\Omega$ defined as Holant$_\Omega = \sum_{\sigma \in Asn(E)} \prod_{v \in V} f_v(\sigma|E(v))$, where $\sigma|E(v)$ denotes the binary string $(\sigma(w_1), \sigma(w_2), \ldots, \sigma(w_k))$ if $E(v) = \{w_1, w_2, \ldots, w_k\}$, whose elements are sorted in a certain pre-fixed order. A general Holant problem Holant($\mathcal{F}$) uses any undirected graph G, not necessarily limited to bipartite graphs.

In fact, #CSP($\mathcal{F}$) is just another name for Holant($\{EQ_k\}_{k \geq 1} | \mathcal{F}$) by identifying variable assignments for #CSP($\mathcal{F}$) with edge assignments for Holant($\{EQ_k\}_{k \geq 1} | \mathcal{F}$). Throughout this paper, we interchangeably use these two different ways to view complex-weighted Boolean #CSP problems.

When any unary signature is allowed to use for free of charge, we conveniently write #CSP*($\mathcal{F}$) instead of #CSP($\mathcal{F}, \mathcal{U}$). For each instance to #CSP*($\mathcal{F}$), the *degree* of an instance Ω is the greatest number of times that any variable appears among all constraints; that is, the maximum degree of nodes that appear on the left-hand side of a bipartite graph in the instance Ω. For any positive integer d, we write #CSP$_d^*$($\mathcal{F}$) for the restriction of #CSP*($\mathcal{F}$) to instances of degree $\leq d$.

2.2   FP$_\mathbb{C}$ and AP-Reductions

The notation FP$_\mathbb{C}$ denotes the set of all functions, mapping binary strings to $\mathbb{C}$, which can be computed deterministically in polynomial time. Here, as we have stated before, we do not treat complex numbers as bit sequences; rather, we treat them as basic "objects" and thus we can perform "natural" operations (such as, multiplications, addition, division, etc.) on them as basic operations, each of which requires only constant time to execute. Moreover, we apply such basic operations only in a clearly monitored way so that our assumption on the constant execution time of these operations causes no harm to a later discussion on the approximate computability of #CSP($\mathcal{F}$). (See [2,3] for further justification.)

Let F be any function mapping from $\{0,1\}^*$ to $\mathbb{C}$. Intuitively, a *randomized approximation scheme* for F is a randomized algorithm (equipped with a coin-flipping mechanism) that takes a standard input $x \in \Sigma^*$ together with an error tolerance parameter $\varepsilon \in (0,1)$, and outputs values w with high probability for which both real and imaginary parts of $F(x)$ can be approximated with relative error e^ε, where e is the base of natural logarithms. See [9] for more details.

Given two functions F and G, roughly speaking, a *polynomial-time approximation-preserving reduction* (or *AP-reduction*) from F to G is a random-ized algorithm M that takes a pair $(x, \varepsilon) \in \Sigma^* \times (0,1)$ as input, uses an arbitrary randomized approximation scheme N for G as an oracle, and outputs an approx-imated value with high probability in time polynomial in $(|x|, 1/\varepsilon)$. (Since we do not need to give the details of oracle mechanism here, the interested reader should refer to [9].) In this case, we write $F \leq_{\mathrm{AP}} G$ and we also say that F is *AP-reducible* to G. If $F \leq_{\mathrm{AP}} G$ and $G \leq_{\mathrm{AP}} F$, then we write $F \equiv_{\mathrm{AP}} G$. Let $\mathcal{F}$ and $\mathcal{G}$ be any two signature sets and let $e, d \in \mathbb{N}^+$. Note that, if $\mathcal{F} \subseteq \mathcal{G}$, then #CSP($\mathcal{F}$) $\leq_{\mathrm{AP}}$ #CSP($\mathcal{G}$). Moreover, if $d \leq e$, then #CSP$_d(\mathcal{F}) \leq_{\mathrm{AP}}$ #CSP$_e(\mathcal{F})$.

2.3 Limited T-Constructability

Our starter is a set of useful notations. Let $k \in \mathbb{N}^+$, let f, f_1, f_2 be any signatures of arity k, let $i, j \in [k]$, and let $c \in \{0,1\}$. Let $x_1, \ldots, x_k$ be Boolean variables. Let $f^{x_i=c}$ denote the function g satisfying that $g(x_1, \ldots, x_{i-1}, x_{i+1}, \ldots, x_k) = f(x_1, \ldots, x_{i-1}, c, x_{i+1}, \ldots, x_k)$. If $i \neq j$, then $f^{x_j=x_i}$ denotes the function g de-fined as $g(x_1, \ldots, x_{j-1}, x_{j+1}, \ldots, x_k) = f(x_1, \ldots, x_{j-1}, x_i, x_{j+1}, \ldots, x_k)$. The no-tation $f^{x_i=*}$ expresses the function g defined as $g(x_1, \ldots, x_{i-1}, x_{i+1}, \ldots, x_k) = \sum_{x_i \in \{0,1\}} f(x_1, \ldots, x_{i-1}, x_i, x_{i+1}, \ldots, x_k)$. Moreover, the notation $f_1 \cdot f_2$ denotes the function g such that $g(x_1, \ldots, x_k) = f_1(x_1, \ldots, x_k) f_2(x_1, \ldots, x_k)$.

A technical tool used in [16] is the notion of T-constructability. Since our target is bounded-degree #CSPs, we will use its modified version—*limited T-constructability*—which plays a central role in the proof of our main theorem. Let f be any signature of arity $k \geq 1$. We say that a bipartite undirected graph $G = (V_1|V_2, E)$ (with a labeling function π) *represents* f if G consists only of k nodes labeled $x_1, \ldots, x_k$, which may have a certain number of dangling* edges, and a single node labeled f, to whom each node x_i is incident. As before, we write f_w for $\pi(w)$. We also say that a bipartite undirected graph G *realizes* f by $\mathcal{G}$ if G satisfies the following conditions: (i) $\pi(V_2) \subseteq \mathcal{G} \cup \mathcal{U}$, (ii) G contains at least k nodes labeled $x_1, \ldots, x_k$ (possibly together with nodes associated with other variables), (iii) only each node x_i may have one or more dangling edges, and (iv) $f(x_1, \ldots, x_d) = \lambda \sum_{y_1, \ldots, y_m \in \{0,1\}} \prod_{w \in V_2} f_w(z_1, \ldots, z_k)$, where $\lambda \in \mathbb{C} - \{0\}$ and $z_1, \ldots, z_k \in V_1 = \{x_1, \ldots, x_k, y_1, \ldots, y_m\}$ with distinct variables $y_1, \ldots, y_m$.

Let $d \in \mathbb{N}$. We write $f \leq_{con}^{+d} \mathcal{G}$ if the following conditions hold: there exists a finite subset $\mathcal{G}'$ of $\mathcal{G} \cup \mathcal{U}$ such that, for any number $m \geq 2$ and for any graph

---

* A *dangling* edge is obtained from an edge by deleting exactly one side of the edge.

G representing f with distinct variables $x_1, \ldots, x_k$ of degree at most m, there exists another graph G' such that (i') G' realizes f by $\mathcal{G}'$, (ii') G' has the same dangling edges as in G, (iii') the nodes labeled $x_1, \ldots, x_k$ have degree at most $m+d$, and (iv') all the other nodes on the left-hand side of G' have degree at most $\max\{3, m+d\}$. For example, if f is T-constructed from g as $f(x_2) = g^{x_1=0}(x_2)$ then $f \leq_{con}^{+0} g$, because the node x_1 does not appear in any graph representing f and this x_1 is considered as a new variable in any graph realizing f by $\{g\}$.

3   Signature Sets

We treat a *relation* of arity k as both a subset of $\{0,1\}^k$ and a signature mapping k Boolean variables to $\{0,1\}$. From this duality, we often utilize the following notation: $R(x) = 1$ ($R(x) = 0$, resp.) iff $x \in R$ ($x \notin R$, resp.), for every $x \in \{0,1\}^k$. Together with some relations defined in Section 2.1, we also use the following special relations: $XOR = [0,1,0]$, $Implies = (1,1,0,1)$, $OR_k = [0,1,\ldots,1]$ (k ones), and $NAND_k = [1,\ldots,1,0]$ (k ones), where $k \in \mathbb{N}^+$. For convenience, the notation EQ (OR and $NAND$, resp.) refers to the equality function (OR-function and $NAND$-function, resp.) of *arbitrary* arity.

The *underlying relation* of a k-ary signature f is the set $R_f = \{x \in \{0,1\}^k \mid f(x) \neq 0\}$. A relation R is said to be *affine* if it is expressed as a set of solutions to a certain system of linear equations over $GF(2)$. A relation R is in IMP (slightly different from $IM\text{-}conj$ in [10]) if it is a product of a certain positive number of relations of the form $\Delta_0(x)$, $\Delta_1(x)$, and $Implies(x,y)$. Moreover, let $DISJ$ ($NAND$, resp.) be the set of relations defined as products of a positive number of OR_k ($NAND_k$, resp.), Δ_0, and Δ_1, where $k \geq 2$ (slightly different from $OR\text{-}conj$ and $NAND\text{-}conj$ in [10]). We will use the following sets of signatures.

1. Let $\mathcal{NZ}$ denote the set of all non-zero signatures.
2. Let $\mathcal{DG}$ denote the set of all signatures f of arity k that are expressed by products of k unary functions, which are applied respectively to k variables. A signature in $\mathcal{DG}$ is called *degenerate*.
3. Let $\mathcal{ED}$ denote the set of functions expressed as products of unary signatures, the equality EQ_2, and the disequality XOR (which are possibly multiplied by constants). See [5] for its basic property.
4. Let $\mathcal{IM}$ be the set of all signatures f such that R_f is in IMP and f equals $R_f \cdot g$ for a certain signature g in $\mathcal{NZ}$.
5. Let $\mathcal{AF}$ denote the set of all signatures of the form $g(x_1, \ldots, x_k) \prod_{j:j \neq i} R_j^{(i)}(x_i, x_j)$ for a certain fixed index $i \in [k]$, where g is in $\mathcal{DG}$ and each $R_j^{(i)}$ is an affine relation.
6. Let $\mathcal{DISJ}$ ($\mathcal{NAND}$, resp.) be the set of signatures f defined as $R_f \cdot g$, where $R_f \in DISJ$ ($R_f \in NAND$, resp.) and $g \in \mathcal{NZ}$.

Note that, for any signature $f \in \mathcal{IM}$, its underlying relation R_f can be factorized into a finite number of factors as $R_f = g_1 \cdot g_2 \cdots g_m$, where each factor g_i is of the form $\Delta_0(x)$, $\Delta_1(x)$, or $Implies(x,y)$ (x and y may be the same). The list

$L = \{g_1, g_2, \ldots, g_m\}$ of such factors is said to be *imp-distinctive* if (i) no single variable appears both in Δ_c and $Implies$ in L, where $c \in \{0, 1\}$, and (ii) no variable x appears as in $Implies(x, x)$, which belongs to L. Such a distinctive list always exists for an arbitrary signature f in $\mathcal{IM}$; however, such a list may not be unique [10,16].

Similar to the notion of imp-distinctive list, we introduce another notion of "or-distinctive" list for signatures in $\mathcal{DISJ}$. Let f be any signature in $\mathcal{DISJ}$. Let L be any list of all factors, of the form $\Delta_0(x)$, $\Delta_1(x)$, and $OR_d(x_{i_1}, \ldots, x_{i_k})$, that defines R_f. This list L is called *or-distinctive* if (i) no variable appears more than once in every OR in L, (ii) no Δ_c ($c \in \{0, 1\}$) and OR in L share the same variable, (iii) no OR's variables are a subset of any other's (ignoring the variable order), and (iv) every OR has at least two variables. For the signatures in $\mathcal{NAND}$, we obtain a similar notion of *nand-distinctive list* by replacing $\mathcal{DISJ}$ with $\mathcal{NAND}$. It is important to note that, for any signature f in $\mathcal{DISJ}$, there exists a unique or-distinctive list of all factors for R_f. The same holds for nand-distinctive lists and $\mathcal{NAND}$ [10].

The *width* of a signature in $\mathcal{DISJ}$ (resp. $\mathcal{NAND}$) is the maximal arity of any factor that appears in the unique or-distinctive (resp. nand-distinctive) list of all factors for R_f. For our later use, $\mathcal{DISJ}_w$ (resp. $\mathcal{NAND}_w$) denotes the set of all signatures in $\mathcal{DISJ}$ (resp. $\mathcal{NAND}$) of width exactly w.

4   Constructing AP-Reductions to the Equality

As stated in Section 1, Dyer et al. [10] analyzed the complexity of the approximate counting of bounded-degree unweighted Boolean #CSPs and gave the first approximation classification, in which they recognized four fundamental categories of bounded-degree problems. Here, we intend to extend their classification theorem from unweighted #CSPs to complex-weighted #CSPs by employing the technique of limited T-constructability given in Section 2.3.

let us begin with a brief discussion on the polynomial-time computability of bounded-degree #CSPs. For any signature set $\mathcal{F}$, it is known from [5,16] that if $\mathcal{F} \subseteq \mathcal{AF}$ or $\mathcal{F} \subseteq \mathcal{ED}$ then #CSP*($\mathcal{F}$) belongs to FP$_\mathbb{C}$. From this computability result, since #CSP$_d^*$($\mathcal{F}$) $\leq_{\mathrm{AP}}$ #CSP*($\mathcal{F}$), the following statement is immediate.

Lemma 1. *For any signature set $\mathcal{F}$ and any index $d \geq 2$, if either $\mathcal{F} \subseteq \mathcal{AF}$ or $\mathcal{F} \subseteq \mathcal{ED}$, then #CSP$_d^*$($\mathcal{F}$) $\in$ FP$_\mathbb{C}$, and thus #CSP*($\mathcal{F}$) $\equiv_{\mathrm{AP}}$ #CSP$_d^*$($\mathcal{F}$).*

In what follows, we are mostly focused on the remaining case where $\mathcal{F} \not\subseteq \mathcal{AF}$ and $\mathcal{F} \not\subseteq \mathcal{ED}$. At this point, we are ready to describe an outline of our proof of the main theorem, Theorem 1. For convenience, $\mathcal{EQ}$ denotes the infinite set $\{EQ_k\}_{k \geq 2}$, where we do not include the equality of arity 1, because it is in $\mathcal{U}$ and is always available for free of charge. Cai et al. [5] first laid out a basic scheme of how to prove a classification theorem for complex-weighted degree-3 Boolean #CSPs. Later, Dyer et al. [10] modified this scheme to prove a classification theorem for unweighted degree-d Boolean #CSPs for any $d \geq 3$. Our proof strategy closely follows theirs.

From a technical reason, it is better for us to introduce another notation $\#\mathrm{CSP}_d^*(\mathcal{E}\mathcal{Q}\|\mathcal{F})$, which is induced from $\#\mathrm{CSP}_d^*(\mathcal{E}\mathcal{Q},\mathcal{F})$, by imposing the following extra condition: no two nodes labeled EQs (possibly having different arities) on the right-hand side of a given bipartite graph, included in each instance, are incident to the same node on the left-hand side of the graph. Similarly, we can define $\#\mathrm{CSP}_d^*(EQ_d\|\mathcal{F})$ using EQ_d instead of $\mathcal{E}\mathcal{Q}$. Our proof strategy is comprised of the following four steps.

1. Initially, we will add the equality of various arities and reduce the original $\#$CSPs to bounded-degree $\#$CSPs with the above-described condition on $\mathcal{E}\mathcal{Q}$. More precisely, we will AP-reduce $\#\mathrm{CSP}^*(\mathcal{F})$ to $\#\mathrm{CSP}_2^*(\mathcal{E}\mathcal{Q}\|\mathcal{F})$.
2. For any index $d \geq 2$ and for any signature $f \in \mathcal{F}$, we will AP-reduce $\#\mathrm{CSP}_2^*(EQ_d\|\mathcal{F})$ to $\#\mathrm{CSP}_3^*(f,\mathcal{F})$, which clearly coincides with $\#\mathrm{CSP}_3^*(\mathcal{F})$ since $f \in \mathcal{F}$. In addition, we require that this reduction should be "generic" and "efficient" so that if we can AP-reduce $\#\mathrm{CSP}_2^*(EQ_d|\mathcal{F})$ to $\#\mathrm{CSP}_3^*(\mathcal{F})$ for every index $d \geq 3$, then we obtain $\#\mathrm{CSP}_2^*(\mathcal{E}\mathcal{Q}\|\mathcal{F}) \leq_{\mathrm{AP}} \#\mathrm{CSP}_3(\mathcal{F})$.
3. By combining the above two AP-reductions, we obtain the AP-reduction: $\#\mathrm{CSP}^*(\mathcal{F}) \leq_{\mathrm{AP}} \#\mathrm{CSP}_3^*(\mathcal{F})$. Since $\#\mathrm{CSP}_3^*(\mathcal{F}) \leq_{\mathrm{AP}} \#\mathrm{CSP}_d^*(\mathcal{F}) \leq_{\mathrm{AP}} \#\mathrm{CSP}^*(\mathcal{F})$ for any index $d \geq 3$, we obtain $\#\mathrm{CSP}^*(\mathcal{F}) \equiv_{\mathrm{AP}} \#\mathrm{CSP}_d^*(\mathcal{F})$.
4. Finally, we will apply the trichotomy theorem for $\#\mathrm{CSP}^*(\mathcal{F})$ given in [16] to determine the approximation complexity of $\#\mathrm{CSP}_d^*(\mathcal{F})$ using the AP-interreduction: $\#\mathrm{CSP}^*(\mathcal{F}) \equiv_{\mathrm{AP}} \#\mathrm{CSP}_d(\mathcal{F})$.

The first step of our proof strategy is quite easy (see, e.g., [10]).

Lemma 2. *For any signature set $\mathcal{F}$, it holds that $\#\mathrm{CSP}^*(\mathcal{F}) \leq_{\mathrm{AP}} \#\mathrm{CSP}_2^*(\mathcal{E}\mathcal{Q}\|\mathcal{F})$.*

To define an AP-reduction from $\#\mathrm{CSP}_2^*(EQ_d|\mathcal{F})$ to $\#\mathrm{CSP}_3^*(\mathcal{G},\mathcal{F})$ in the second step of our strategy, it suffices to prove, as shown in the following lemma, that $EQ_d \leq_{con}^{+1} \mathcal{G}$ by a generic, efficient procedure.

Lemma 3. *Let $d, m \in \mathbb{N}$ with $d \geq 2$. If $EQ_d \leq_{con}^{+m} \mathcal{G}$, then $\#\mathrm{CSP}_2^*(EQ_d\|\mathcal{F}) \leq_{\mathrm{AP}} \#\mathrm{CSP}_{2+m}^*(\mathcal{G},\mathcal{F})$. In addition, assume that there exists a procedure of transforming any graph G' representing EQ_d into another graph G'' realizing EQ_d in time polynomial in the size of d and the size of the graph G'. Then, it holds that $\#\mathrm{CSP}_2^*(\mathcal{E}\mathcal{Q}\|\mathcal{F}) \leq_{\mathrm{AP}} \#\mathrm{CSP}_{2+m}^*(\mathcal{G},\mathcal{F})$.*

Proof. Let $\Omega = (G, \mathcal{F}_1'|\mathcal{F}_2', \pi)$ be any signature grid to $\#\mathrm{CSP}_2^*(EQ_d\|\mathcal{F})$ and let $G = (V_1|V_2, E)$ be its bipartite graph. Take each node in V_2 labeled EQ_d and take any subgraph G' of G such that G' consists only of d different nodes labeled, say, $x_{i_1}, \ldots, x_{i_d}$, which are incident to this node EQ_d. Moreover, each of these d nodes (on the left-hand side of G') should contain (at most) one dangling edge, which is originally connected to a certain other node in G. Clearly, G' represents EQ_d. Since $EQ_d \leq_{con}^{+m} \mathcal{G}$, there is another bipartite graph G'' that realizes EQ_d by $\mathcal{U} \cup \mathcal{G}$. In G, we replace G' by G''. In this replacement, for each dangling edge appearing in G', we restore its original connection to the node in G. Note that any node other than $x_{i_1}, \ldots, x_{i_d}$ are treated as new nodes and are not incident to any node outside of G''.

Let $\tilde{G}$ be the graph obtained by replacing all such subgraphs G' with their corresponding subgraphs G''. Let $\tilde{\Omega}$ be the new signature grid associated with $\tilde{G}$. The degree of each node x_{i_j} in $\tilde{G}$ is m plus the original degree in G since no two nodes labeled EQ_d in G share the same variables. It is not difficult to show that $\text{Holant}_{\tilde{\Omega}} = \text{Holant}_{\Omega}$. This implies that $\#\text{CSP}_2^*(EQ_d|\mathcal{F}) \leq_{\text{AP}} \#\text{CSP}_{2+m}^*(\mathcal{G}, \mathcal{F})$.

The second part of the lemma comes from the fact that we can construct $\tilde{\Omega}$ from Ω efficiently and robustly if there is a generic procedure that transforms G' to G'' for any degree-bound d in polynomial time. □

5 Basic AP-Reductions of Binary Signatures

Since we have shown in Section 4 that $\#\text{CSP}^*(\mathcal{F})$ can be AP-reduced to $\#\text{CSP}_2^*(\mathcal{EQ}\|\mathcal{F})$, the remaining task is to further reduce $\#\text{CSP}_2^*(\mathcal{EQ}\|\mathcal{F})$ to $\#\text{CSP}_3^*(\mathcal{F})$. For this purpose, it suffices to prove that, for any index $d \geq 2$ and for any signature $f \in \mathcal{F}$, EQ_d is limited T-constructable from f together with unary signatures with maintaining the degree-bound of 3. To be more precise, we wish to prove that there exists a finite set $\mathcal{G} \subseteq \mathcal{U}$ for which $EQ_d \leq_{\text{con}}^{+1} \mathcal{G} \cup \{f\}$. By examining the proofs of each lemma given below, we can easily notice that the obtained limited T-constructability, $EQ_d \leq_{\text{con}}^{+1} \mathcal{G} \cup \{f\}$ for any $d \geq 3$, are indeed "generic" and "efficient," as requested by Lemma 3. Therefore, we finally conclude that $\#\text{CSP}_2^*(\mathcal{EQ}\|\mathcal{F}) \leq_{\text{AP}} \#\text{CSP}_3(f, \mathcal{F})$.

This section deals only with non-degenerate signatures of arity 2, because degenerate signatures were already dealt with in Lemma 1. The first case to discuss is signatures f of the form $(0, a, b, 0)$, $(a, 0, 0, b)$, or $(a, b, 0, c)$, where $abc \neq 0$.

Lemma 4. *Let $d \geq 2$ be any index.*

1. *Let $f = (0, a, b, 0)$ with $a, b \in \mathbb{C}$. If $ab \neq 0$, then $EQ_d \leq_{\text{con}}^{+1} f$.*
2. *Let $f = (a, 0, 0, b)$ with $a, b \in \mathbb{C}$. If $ab \neq 0$, then there exists a signature $u \in \mathcal{U} \cap \mathcal{NZ}$ such that $EQ_d \leq_{\text{con}}^{+1} \{f, u\}$.*
3. *Let $f = (a, b, 0, c)$ with $a, b, c \in \mathbb{C}$. If $abc \neq 0$, then there exist two signatures $u_1, u_2 \in \mathcal{U} \cap \mathcal{NZ}$ such that $EQ_d \leq_{\text{con}}^{+1} \{f, u_1, u_2\}$. By permuting variable indices, the case $h = (a, 0, b, c)$ is similar.*

The non-degenerate non-zero signatures $f = (1, a, b, c)$ are quite special, because they appear only in the case of *complex-weighted* #CSPs. When f is a Boolean relation, by contrast, it never becomes non-degenerate.

Lemma 5. *Let $d \geq 2$ and let $f = (1, a, b, c) \notin \mathcal{DG}$. If $abc \neq 0$, then there exist two signatures $u_1, u_2 \in \mathcal{U} \cap \mathcal{NZ}$ satisfying that $EQ_d \leq_{\text{con}}^{+1} \{f, u_1, u_2\}$.*

Proof. Assume that $f = (1, a, b, c)$ with $abc \neq 0$. First, we claim that $ab \neq c$. Assuming otherwise, we have $f = (1, a, b, ab)$ and thus f is written as $f(x_1, x_2) = [1, a](x_1) \cdot [1, b](x_2)$. This means that f belongs to $\mathcal{DG}$, a clear contradiction against the premise $f \notin \mathcal{DG}$. Therefore, $ab \neq c$ follows.

Now, we set $u_1 = [1, z]$, where $z = -1/c$, and we define $g(x_1, x_2) = \sum_{y \in \{0,1\}} f(x_1, y) f(y, x_2) u_1(y)$. This gives $g = (1 + abz, a(1 + cz), b(1 + cz), ab + c^2 z)$. Since $z = -1/c$, g is of the form $(1 - ab/c, 0, 0, ab - c)$. Note that, since $ab \neq c$, the first and last entries of g are non-zero. We then apply Lemma 4(2), which requires another non-zero unary signature u_2. In short, the desired signature g' is defined as $g'(x_1, x_2) = u_2(x_1) g(x_1, x_2)$. From a graph G representing EQ_2, we construct a new graph G' by replacing EQ_2 in G with $\{f, u_1, u_2\}$. Since G' requires extra two edges, one of which is incident to y and the other is incident to x_1. Overall, the degree of each node on the left-hand side of G' increases at most 1 in comparison with the same node in G.

In a general case $d \geq 3$, for the series $x = (x_1, \ldots, x_d)$ of d variables, we define $g(x) = \sum_{y_1, \ldots, y_{d-1} \in \{0,1\}} \prod_{i=1}^{d-1} (f(x_i, y_i) u_1(y_i) f(y_i, x_{i+1}))$. Since g has the form $(a', 0, \ldots, 0, b')$, by choosing a certain signature $u' \in \mathcal{U} \cap \mathcal{NZ}$, the signature $g'(x) = u_2'(x_1) g(x)$ equals EQ_d. The degree analysis for g' is similar to the base case $d = 2$.   □

A notable case is where $f = (0, a, b, c)$ of $f = (a, b, c, 0)$ with $abc \neq 0$, which respectively extend OR_2 and $NAND_2$.

Proposition 1. *Let $f = (0, a, b, c)$ with $abc \neq 0$. There exists a signature $u \in \mathcal{U} \cap \mathcal{NZ}$ such that $EQ_d \leq_{con}^{+1} \{f, u\}$. A similar statement holds for $f = (a, b, c, 0)$ with $abc \neq 0$.*

Proposition 1 follows from two useful lemmas, Lemmas 6 and 7.

Lemma 6. *Let $d \geq 2$. Let $f_1 = (0, a, b, c)$ and $f_2 = (a', b', c', 0)$ with $a, b, c, a', b', c' \in \mathbb{C}$. If $ab \neq 0$ and $b'c' \neq 0$, then $EQ_d \leq_{con}^{+1} \{f_1, f_2\}$.*

The proof of Lemma 6 is similar to the proof of [10, Lemma 13]. The next lemma ensures that, with a help of unary signatures, we can transform a signature in $\mathcal{DISJ}$ into another in $\mathcal{NAND}$ without increasing its degree. This is a special case not seen for Boolean signatures and it clearly exemplifies a power of the *complex* unary signatures.

Lemma 7. *Let $h \in \mathcal{NZ}$ be any binary signature. There are a binary signature $h' \in \mathcal{NZ}$ and a signature $u \in \mathcal{U} \cap \mathcal{NZ}$ such that $NAND_2 \cdot h' \leq_{con} \{OR_2 \cdot h, u\}$. A similar statement holds if we exchange the roles of OR_2 and $NAND_2$.*

Proof. Let $f = OR_2 \cdot h$ for a given signature $h \in \mathcal{NZ}$ of arity 2. By normalizing, we can assume that $f = (0, a, b, 1)$ with $ab \neq 0$. Let $u = [1, z]$ and define $g(x_1, x_2) = \sum_{x_3 \in \{0,1\}} f(x_1, x_3) u(x_3) f(x_3, x_2)$, which gives $g = (abz, az, bz, ab + z)$. Here, we set $z = -ab$. This makes g equal $(-(ab)^2, -a^2 b, -ab^2, 0)$. Note that $g(x_1, x_2)$ is written as $NAND_2(x_1, x_2) h'(x_1, x_2)$ if we define $h' = (-(ab)^2, -a^2 b, -ab^2, 1)$, which is a non-zero signature. As before, from a graph G representing g, we define a new graph G' by replacing g with $\{f, u\}$. The degree of node x_3 in G' is 3 and the other variable nodes have the same degree as their original ones in G. Therefore, it follows that $g \leq_{con}^{+0} \{f, u\}$.   □

6   Signatures of Higher Arity

We have shown in Section 5 that non-degenerate binary signatures limited-T-constructs EQ of arbitrary arity. Here, we want to prove a similar result for signatures of higher arity.

Most signatures in $\mathcal{IM}$ bears a trait of *Implies*. Such a trait can be used to realize EQ_d directly. Our first goal in this section is to show the following general statement.

Proposition 2. *Let $d \geq 3$ and let f be any signature of arity ≥ 3. If $f \in \mathcal{IM}$ and $f \notin \mathcal{ED} \cup \mathcal{NZ}$, then there exists a finite subset $\mathcal{G} \subseteq \mathcal{U}$ such that $EQ_d \leq^{+1}_{con} \mathcal{G} \cup \{f\}$.*

For the proof of this proposition, we need to introduce two special notions discussed in [16]. From any given signature f, it is possible to extract from R_f all factors of the form $\Delta_0(x)$, $\Delta_1(x)$, and $EQ_d(x_1, \ldots, x_d)$; in other words, we can "factorize" R_f into them. After such an extraction, the remaining portion of R_f can be expressed by a notion of "simple form." For every k-ary signature f, let us consider its *representing Boolean matrix* M_f, whose rows are indexed by all instances $a = (a_1, a_2, \ldots, a_k)$ in R_f, columns are indexed by numbers in $[k]$, and each (a, i)-entry is the Boolean value a_i. A signature is in *simple form* if its representing Boolean matrix does not contain all-0 columns, all-1 columns, or any pair of identical columns.

To deal with EQ_d, we further define the notion of "eq-distinctiveness" for signatures whose factors are Δ_c's and EQs. A list L of factors of a relation R, each of which is of the form either EQ_d or Δ_c, is called *eq-distinctive* if (i) any EQ in L has arity at least 2, (ii) no Δ_c ($c \in \{0, 1\}$) in L shares any variable with any EQ in L, (iii) a variable set of any EQ cannot become a subset of the variable set of any other EQ in L, and (iv) no variable appears more than once in a single EQ in L.

A *greedy sweeping procedure* performs on a signature f as follows. (i) If there exists an all-0 column in M_f indexed i, then delete this column. We then have $f(x_1, \ldots, x_k) = \Delta_0(x_i)f^{x_i=0}(x_1, \ldots, x_{i-1}, x_{i+1}, \ldots, x_k)$. (ii) If there exists an all-1 column in M_f indexed i, then delete this column. We then have $f(x_1, \ldots, x_k) = \Delta_0(x_i)f^{x_i=1}(x_1, \ldots, x_{i-1}, x_{i+1}, \ldots, x_k)$. (iii) If there is a pair of identical columns. We search for all columns that are identical to each other. For simplicity, assume that $\{1, \ldots, d\}$ be a list of *all* indices whose columns are identical. We delete all such columns except for the one indexed d. In this case, we have $f(x_1, \ldots, x_k) = EQ_d(x_1, \ldots, x_d)f^{x_1=x_d, \ldots, x_{d-1}=x_d}(x_d, \ldots, x_k)$.

Lemma 8. *Let $e, d \geq 2$. Let f be any signature of arity $k \geq 3$ in $\mathcal{IM}$ but not in $\mathcal{DG} \cup \mathcal{NZ}$. Assume that a greedy sweeping procedure makes $f(x_1, \ldots, x_k) = R(x_1, \ldots, x_{m'})g(x_m, \ldots, x_k)$ for a relation R such that R consists only of Δ_c's and EQs and a signature g in simple form. If an eq-distinctive list L of all factors of R contains an EQ_e, then $EQ_d \leq^{+1}_{con} f$.*

Proof.   Assume that a greedy sweeping procedure on R_f makes $R_f(x_1, \ldots, x_k) = R(x_1, \ldots, x_{m'})g(x_m, \ldots, x_k)$ for certain indices m, m' with

$1 \leq m \leq m' \leq k$. Obviously, R is in IMP. Let L be any imp-distinctive list of all factors of R. By the premise of the lemma, L contains at least one EQ_e. Hereafter, we fix this EQ_e and, for simplicity, we assume that this EQ_e takes a variable series $(x_1, \ldots, x_e)$. By the nature of the greedy sweeping procedure, no Δ_c ($c \in \{0, 1\}$) shares a common variable with $EQ_e(x_1, \ldots, x_e)$.

Now, we assign appropriate values to all variables except for $x_1, \ldots, x_e$ as follows. If $\Delta_c(x_j)$ is in L, then we define $c_j = c$; if an $EQ(x_{i_1}, \ldots, x_{i_b})$ of arbitrary arity is in L, then we set $c_{i_j} = 0$ for all $j \in [b]$; for all the other variables x_j (except for $x_1, \ldots, x_e$), we set $c_j = 0$. With these values $\{c_{e+1}, \ldots, c_k\}$, we define $q = f^{x_{e+1}=c_{e+1}, \ldots, x_k=c_k}$. Clearly, $q(x_1, \ldots, x_e)$ equals $R^{x_{e+1}=c_{e+1}, \ldots, x_{m'}=c_{m'}}(x_1, \ldots, x_e)g^{x_{e+1}=c_{e+1}, \ldots, x_k=c_k}(x_e)$. Since g is a non-zero signature, the signature $r = g^{x_{e+1}=c_{e+1}, \ldots, x_k=c_k}$ is also non-zero. It is not difficult to show that $R_q(x_1, \ldots, x_e)$ coincides with $EQ_e(x_1, \ldots, x_e)$; more precisely, $q(x_1, \ldots, x_e) = EQ_e(x_1, \ldots, x_e)r(x_e)$. This implies that q is of the form $[a, 0, \ldots, 0, b]$ with $ab \neq 0$. We define $u = [1/a, 1/b]$ and set $s(x_1, \ldots, x_e) = u(x_1)q(x_1, \ldots, x_e)$. It follows that s equals $[1, 0, \ldots, 0, 1]$.

Next, we want to show that $EQ_d \leq_{con}^{+1} \mathcal{G} \cup \{f\}$ for a certain finite set $\mathcal{G} \subseteq \mathcal{U}$. There are two cases to consider, depending on the value d. If $e \geq d$, then let $s(x_1, \ldots, x_d) = u(x_d)q^{x_{d+1}=*, \ldots, x_e=*}(x_1, \ldots, x_d)$. This makes $s(x_1, \ldots, x_d)$ equal $EQ_d(x_1, \ldots, x_d)$. The other case $e < d$ is not difficult and is omitted here.   □

Lemma 9.   *Let $d \geq 2$. Let $f \in \mathcal{IM}$ but $f \notin \mathcal{ED} \cup \mathcal{NZ}$. If f is in simple form, then there exist two signatures $u_1, u_2 \in \mathcal{U}$ for which $EQ_d \leq_{con}^{+1} \{f, u_1, u_2\}$.*

Proof.   Since $f \in \mathcal{IM}$, f is of the form $f = R_f \cdot g$ with $g \in \mathcal{NZ}$. Since f is also in simple form, the representing Boolean matrix M_f for f has no all-0 column, no all-1 column, and no identical columns. Note that, if M_f has a factor $Implies(x_i, x_j)$ for $i \neq j$, then there is no signature of the form: $\Delta_0(z)$ and $\Delta_1(z)$ for $z \in \{x_i, x_j\}$, and $Implies(x_j, x_i)$.

Let L be any imp-distinctive list of all factors of R_f. Assume that $Implies(x_1, x_2) \in L$ but $Implies(x_j, x_1) \notin L$ for all possible indices j. Since $Implies(x_1, x_2) \in L$, the signature $h = f^{x_3=1, \ldots, x_k=1}$ has the form $h(x_1, x_2) = Implies(x_1, x_2)t(x_1, x_2)$ for a certain signature t. Since $g^{x_3=1, \ldots, x_k=1} \in \mathcal{NZ}$, t should belong to $\mathcal{NZ}$. Thus, we can assume that $h = (a, b, 0, c)$ with $a, b, c \in \mathbb{C} - \{0\}$. Finally, we apply Lemma 4(3) to obtain the desired consequence.   □

The proof of Proposition 2 follows from Lemmas 8 and 9.

Proof of Proposition 2.   Let f be any signature in $\mathcal{IM} - (\mathcal{ED} \cup \mathcal{NZ})$. By running a greedy sweeping procedure, we obtain the form $f = R \cdot g$, where $g \in \mathcal{NZ}$ and g is in simple form. Since $R \in IMP$, we consider an eq-distinctive list L of all factors of R. First, assume that R contains an EQ_e for a certain index $e \geq 2$. In this case, by Lemma 8, we obtain $EQ_d \leq_{con}^{+1} \{f, u\}$. Next, consider the case where L contains no EQ_e, where $e \geq 2$. This implies that R consists only of unary signatures, and thus it belongs to $\mathcal{DG}$. Next, we want to claim that $g \notin \mathcal{ED}$. Assume otherwise. Since $R \in \mathcal{DG}$ and $g \in \mathcal{ED}$, f belongs to $\mathcal{ED}$, a contradiction. Thus, g should not belong to $\mathcal{ED} \cup \mathcal{NZ}$. Since g is in simple form, we can apply Lemma 9 and we then obtain the lemma. □

Now, we shift our attention to signatures sitting in $\mathcal{DISJ} \cup \mathcal{NAND}$. Lemma 6 has handled binary signatures chosen from $\mathcal{DISJ} \cup \mathcal{NAND}$. We want to show that this result can be extended to higher arity.

Proposition 3. *Let $k \geq 3$ and $d \geq 2$. Let f be any k-ary signature in $\mathcal{DISJ} \cup \mathcal{NAND}$. If $f \notin \mathcal{DG}$, then there exist two signatures $u_1, u_2 \in \mathcal{U}$ such that $EQ_d \leq_{con}^{+1} \{f, u_1, u_2\}$.*

For this proposition, we first claim the following useful lemma. Similar to the proof of Lemma 8, this lemma can be obtained by assigning either 0 or 1 to all variables except for carefully-selected w variables in f.

Lemma 10. *Let $w \geq 2$. For any signature $f \in \mathcal{DISJ}_w$ ($\mathcal{NAND}_w$, resp.), then there exists a non-zero signature h of arity w such that $OR_w \cdot h \leq_{con}^{+0} f$ ($NAND_w \cdot h \leq_{con}^{+0} f$, resp.).*

With a help of this lemma, the proof of Proposition 3 is given below.

Proof of Proposition 3. Assume that $f \in \mathcal{DISJ}$ and f has arity k. In addition, we assume that f has width w for a certain number $w \geq 2$; namely, $f \in \mathcal{DISJ}_w$. Obviously, $w \leq k$. Lemma 10 ensures the existence of a signature $h \in \mathcal{NZ}$ of arity w for which $OR_w \cdot h \leq_{con}^{+1} f$.

Assume that this OR_w takes variables $x_1, \ldots, x_w$. We then choose two specific variables, x_1 and x_2, and assign 0 to all the other variables. Let f' be the signature obtained from $OR_w \cdot h$ by these pinning operations. Obviously, $f' \leq_{con}^{+0} f$. It is not difficult to show that, since $h \in \mathcal{NZ}$, $R_{f'}(x_1, x_2)$ equals $OR_2(x_1, x_2)$; in other words, f' is of the form $(0, a, b, c)$ with $abc \neq 0$.

Finally, we apply Proposition 1 and then obtain two signatures $u_1, u_2 \in \mathcal{U} \cap \mathcal{NZ}$ satisfying that $EQ_d \leq_{con}^{+1} \{f', u_1, u_2\}$. By combining this with $f' \leq_{con}^{+0} f$, we conclude that $EQ_d \leq_{con}^{+1} \{f, u_1, u_2\}$.

The case where $f \in \mathcal{NAND}$ is similarly treated. □

The remaining type of signatures to consider is ones that sit outside of $\mathcal{DISJ} \cup \mathcal{NAND} \cup \mathcal{DG}$. As a key claim for those signatures, we prove the following proposition.

Proposition 4. *Let $d \geq 2$. For any signature $f \notin \mathcal{DG}$ of arity $k \geq 2$, if $f \notin \mathcal{DISJ} \cup \mathcal{NAND}$, then there exists a finite subset $\mathcal{G} \subseteq \mathcal{U}$ satisfying that $EQ_d \leq_{con}^{+1} \mathcal{G} \cup \{f\}$.*

To prove this proposition, we note, as our starting point, a lemma on binary signatures not in $\mathcal{DISJ} \cup \mathcal{NAND} \cup \mathcal{DG}$. These signatures have already treated in Section 5 under various conditions.

Lemma 11. *Let $d \geq 2$. Let $f \notin \mathcal{DISJ} \cup \mathcal{NAND} \cup \mathcal{DG}$ be any binary signature. In this case, there exists a finite signature set $\mathcal{G} \subseteq \mathcal{U} \cap \mathcal{NZ}$ with $|\mathcal{G}| \leq 2$ such that $EQ_d \leq_{con}^{+1} \mathcal{G} \cup \{f\}$.*

The second step for the proof of Proposition 4 is made by the following lemma. For convenience, let $COMP_1(f) = \{f^{x_i=c} \mid i \in [k], c \in \{0,1\}\}$.

Lemma 12. *Let $d, k \geq 2$. For any k-ary signature $f \notin \mathcal{DISJ} \cup \mathcal{NAND} \cup \mathcal{DG}$, if $EQ_d \not\leq_{con}^{+1} \mathcal{G} \cup \{f\}$ for any finite subset $\mathcal{G} \subseteq \mathcal{U}$, then there exists another signature $g \in COMP_1$ of arity $< k$ such that $g \notin \mathcal{DISJ} \cup \mathcal{NAND} \cup \mathcal{DG}$.*

Proof. Let $f \notin \mathcal{DISJ} \cup \mathcal{NAND} \cup \mathcal{DG}$ be any k-ary signature. Assume that $EQ_d \not\leq_{con}^{+1} \mathcal{G} \cup \{f\}$ for any finite subset $\mathcal{G} \subseteq \mathcal{U} \cap \mathcal{NZ}$. By letting $g_b = f^{x_1=b}$ for each value $b \in \{0,1\}$, we obtain $f(x_1, \ldots, x_k) = \sum_{b \in \{0,1\}} \Delta_b(x_1) g_b(x_2, \ldots, x_k)$.

Now, we want to claim that there is no case where either $g_0, g_1 \in \mathcal{DISJ}$ or $g_0, g_1 \in \mathcal{NAND}$. If so, then f clearly belongs to $\mathcal{DISJ} \cup \mathcal{NAND}$, a contradiction. Moreover, if both $g_0 \in \mathcal{DISJ}$ and $g_1 \in \mathcal{NAND}$ occur, then Proposition 3 implies that $EQ_d \leq_{con}^{+1} \{g_0, g_1, u_1, u_2\}$ for certain signatures $u_1, u_2 \in \mathcal{U} \cap \mathcal{NZ}$. Since $g_0 \leq_{con}^{+0} f$ and $g_1 \leq_{con}^{+0} f$, we conclude that $EQ_d \leq_{con}^{+1} \{f, u_1, u_2\}$. This is clearly a contradiction. The same holds if we exchange the roles of g_0 and g_1. Therefore, there are only two remaining cases: (i) $g_0 \in \mathcal{DISJ} \cup \mathcal{NAND}$ and $g_1 \notin \mathcal{DISJ} \cup \mathcal{NAND}$ and (ii) $g_0 \notin \mathcal{DISJ} \cup \mathcal{NAND}$ and $g_1 \in \mathcal{DISJ} \cup \mathcal{NAND}$. In either case, we can obtain the desired conclusion of the lemma.   □

Finally, Proposition 4 can be proven by combining Lemmas 11 and 12.

7   The Classification Theorem

Throughout the previous sections, we have already established all necessary foundations for our trichotomy theorem, Theorem 1, on the approximation complexity of complex-weighted bounded-degree Boolean #CSPs. Theorem 1 is an immediate consequence of our key claim, Proposition 5, which directly bridges between unbounded-degree #CSPs and bounded-degree #CSPs.

Proposition 5. *For any index $d \geq 3$ and for any signature set $\mathcal{F}$, $\#\text{CSP}^*(\mathcal{F}) \equiv_{\text{AP}} \#\text{CSP}_d^*(\mathcal{F})$.*

The proposition comes from Lemma 1 and Propositions 2, 3, and 4. Using this proposition, the main theorem follows directly from the trichotomy theorem proven in [16].

Another immediate consequence of Proposition 5 is an equivalence between $\#\text{CSP}^*(\mathcal{F})$ and $\text{Holant}(EQ_3|\mathcal{F}, \mathcal{U})$. Notice that $\#\text{CSP}^*(\mathcal{F})$ coincides with $\text{Holant}(\{EQ_k\}_{k \geq 1}|\mathcal{F}, \mathcal{U})$ as stated in Section 2.1.

Proposition 6. *For any set $\mathcal{F}$ of signatures, it holds that $\#\text{CSP}^*(\mathcal{F}) \equiv_{\text{AP}} \text{Holant}(EQ_3|\mathcal{F}, \mathcal{U})$.*

What is the approximation complexity of $\#\text{CSP}_d^*(\mathcal{F})$ when d is less than 3? Toward the end of this section, we briefly discuss this issue. Recall that $\#\text{CSP}^*(\mathcal{F})$ is shorthand for $\#\text{CSP}(\mathcal{F}, \mathcal{U})$. Similar to $\#\text{CSP}^*(\mathcal{F})$, we write $\text{Holant}^*(\mathcal{F})$ to denote $\text{Holant}(\mathcal{F}, \mathcal{U})$.

The following proposition, which answers the above question, determines the approximation complexity of $\#\text{CSP}_d^*(\mathcal{F})$ for any $d \in \{1, 2\}$.

Proposition 7.   *1.   For any signature set $\mathcal{F}$, $\#\mathrm{CSP}_1^*(\mathcal{F})$ is in $\mathrm{FP}_{\mathbb{C}}$.*
2.   For any signature set $\mathcal{F}$, $\#\mathrm{CSP}_2^*(\mathcal{F}) \equiv_{\mathrm{AP}} \mathrm{Holant}^*(\mathcal{F})$.

Overall, we have classified all $\#\mathrm{CSP}_d^*(\mathcal{F})$'s as we have initially planned.

Acknowledgments. The author greatly appreciates Jin-Yi Cai for his kind help in promoting the author's understanding of the theory of signatures while he was visiting at the University of Wisconsin in February 2010.

References

1. Cai, J., Huang, S., Lu, P.: From Holant to #CSP and back: dichotomy for Holant^c problems. Available on line at CoRR abs/1004.0803 (2010)
2. Cai, J., Lu, P.: Holographic algorithms: from arts to science. In: Proceedings of STOC 2007, pp. 401–410 (2007)
3. Cai, J., Lu, P.: Signature theory in Holographic algorithms. In: Hong, S.-H., Nagamochi, H., Fukunaga, T. (eds.) ISAAC 2008. LNCS, vol. 5369, pp. 568–579. Springer, Heidelberg (2008)
4. Cai, J., Lu, P., Xia, M.: On Holant problems. An early version appeared in Proceedings of STOC 2009, pp.715–724 (2009)
5. Cai, J., Lu, P., Xia, M.: The complexity of complex weighted Boolean #CSP. An early version appeared in Proceedings of STOC 2009, pp.715–724 (2009)
6. Creignou, N., Hermann, M.: Complexity of generalized satisfiability counting problems. Inform. and Comput. 125, 1–12 (1996)
7. Dalmau, V., Ford, D.K.: Generalized satisfiability with limited concurrences per variable: a study through Δ-matroid parity. In: Rovan, B., Vojtáš, P. (eds.) MFCS 2003. LNCS, vol. 2747, pp. 358–367. Springer, Heidelberg (2003)
8. Dyer, M., Frieze, A., Jerrum, M.: On counting independent sets in sparse graphs. SIAM J. Comput. 31, 1527–1541 (2002)
9. Dyer, M., Goldberg, L.A., Greenhill, C., Jerrum, M.: The relative complexity of approximating counting problems. Algorithmica 38, 471–500 (2004)
10. Dyer, M., Goldberg, L.A., Jalsenius, M., Richerby, D.: The complexity of approximating bounded-degree Boolean #CSP. In: Proceedings of STACS 2010, pp. 323–334 (2010)
11. Dyer, M., Goldberg, L.A., Jerrum, M.: The complexity of weighted Boolean #CSP. SIAM J. Comput. 38, 1970–1986 (2009)
12. Dyer, M., Goldberg, L.A., Jerrum, M.: An approximation trichotomy for Boolean #CSP. J. Comput. System Sci. 76, 267–277 (2010)
13. Schaefer, T.J.: The complexity of satisfiability problems. In: Proceedings of FOCS 1978, pp. 216–226 (1978)
14. Valiant, L.G.: Quantum circuits that can be simulated classically in polynomial time. SIAM J. Comput. 31, 1229–1254 (2002)
15. Valiant, L.G.: Expressiveness of matchgates. Theor. Comput. Sci. 289, 457–471 (2002)
16. Yamakami, T.: Approximate counting for complex-weighted Boolean constraint satisfaction problems. To appear in Proceedings of WAOA 2010. LNCS. Springer, Heidelberg (2010); See also arXiv:1007.0391

A Randomized Algorithm for Weighted Approximation of Points by a Step Function

Jin-Yi Liu

School of Computer and Communication Engineering, Liaoning Shihua University,
Fushun 113001, P.R. China
j_y_liu@sina.com

Abstract. The problem considered in this paper is: given an integer $k > 0$ and a set P of n points in the plane each with a corresponding non-negative weight, find a step function f with k steps that minimize the maximum weighted vertical distance between f and all the points in P. We present a randomized algorithm to solve the problem in $O(n \log n)$ expected running time. The bound is obviously optimal for the unsorted input. The previously best known algorithm runs in $O(n \log^2 n)$ worst-case time. Another merit of the algorithm is its simplicity. The algorithm is just a randomized implementation of Frederickson and Johnson's matrix searching technique, and it only exploits a simple data structure.

Keywords: Randomized algorithm, approximation, step function, matrix searching.

1 Introduction

In this paper, we consider the problem of approximating a planar point set by a step function under the weighted measure. Precisely, we are given an integer $k > 0$ and a set P of n planar points each with a corresponding non-negative weight, our objective is to find a step function f with k steps that minimize the maximum weighted vertical distance between f and all the points in P.

This problem belongs to the large class of (min-ε) approximation problems for a point set or a polygonal curve (e.g. [2], [9]). Even if the geometric problem may have application context in geometry-related areas, it is mainly motivated in the database community for histogram construction, where the step function we seek will be used as a concise representation of a large dataset [10,11]. Here, the reason why the weighted measure has received attention is that the data may hold non-uniform significance.

If all the weights are equal to 1, the special version of the problem has been extensively researched [4,10,13,15], and finally, was solved optimally by Fournier and Vigneron [6], in $O(n \log n)$ time for the unsorted case and $O(n)$ time for the sorted case. If the weights are allowed to be any non-negative values, several upper bounds has also been achieved in recent years. Table 1 summarizes the previous results and the result of this paper. Among all the previous results, the best bound depending only on n is $O(n \log^2 n)$, which was recently obtained by

W. Wu and O. Daescu (Eds.): COCOA 2010, Part I, LNCS 6508, pp. 300–308, 2010.

Chen and Wang [2]. Just like what Fournier and Vigneron did in [6], Chen and Wang obtained their bound by generalizing Frederickson's technique for path partitioning [7], and additionally by exploiting some complicated data structures such as fractional cascading. The Frederickson's technique itself is also a complicated technique because it has to employ sub-linear feasibility test and use careful counting tricks.

Table 1. Summary of the results

References	Time bounds
Guha and Shim (2007) [10]	$O(n \log n + k^2 \log^6 n)$
Karras, Sacharidis, and Mamoulis (2007) [11]	$O(n \log L)$, where L is the largest number that the optimal value may reach
Lopez and Mayster (2008) [12]	$O(n^2)$ or $O(n \log^3 n)$ (expected)
Fournier and Vigneron (2008) [6]	$O(n \log^4 n)$
Chen and Wang (2009) [2]	$O(n \log^2 n)$
The present paper	$O(n \log n)$ (expected)

In this paper, We present a simple randomized algorithm with $O(n \log n)$ expected running time. The bound is obviously optimal if we assume the input point set is unsorted. The simplicity of the algorithm consists in the fact that it is obtained by randomizing Frederickson and Johnson's matrix searching technique [8] (which is simpler than Frederickson's technique for path partitioning [7]), and only a very simple data structure is exploited.

2   Preliminaries

In this section, we present some necessary notations and assumptions for the problem, and reformulate the problem as a min-max optimization problem to make it more intelligible. We also present some observations on the problem, which are important to our algorithms.

2.1   Notations, Assumptions, and Formulations

Let $P = \{p_1, p_2, \ldots, p_n\}$ be the input point set, with $p_i = (x_i, y_i)$, and let $w_i \geq 0$ be the weight corresponding to p_i. To simplify the exposition, we assume that no two points in P has the same x-coordinate. We also assume that the points in P are given in the order of increasing x-coordinates. The sorted input assumption does not affect the generality of our algorithm since adding a sorting process does not affect the time bound of our algorithm. With this assumption, we look at $P = \{p_1, p_2, \ldots, p_n\}$ as a sequence of points, and then, the problem of this paper is just a special version of the general MIN-MAX PARTITION problem [6,2].

MIN-MAX PARTITION: Given a sequence of n elements, a positive integer $k \leq n$, and the definition of a cost function $\theta(b,e)$ that can map any interval $[b \ldots e]$ with $1 \leq b \leq e \leq n$ to a non-negative value, compute a k-partition of interval $[1 \ldots n]$, which is defined to be a non-decreasing sequence of integers $\{i_0, i_1, \ldots, i_k\}$ with $i_0 = 0$ and $i_k = n$, such that $\max_{1 \leq j \leq k} \theta(i_{j-1} + 1, i_j)$ is minimized. Here, the elements of the input sequence may be characters, numbers, points, etc..

For the problem of this paper, the definition of $\theta(b,e)$ is

$$\theta(b,e) = \min_{y \in \mathbb{R}} \max_{b \leq i \leq e} w_i \cdot |y_i - y| \tag{1}$$

That is to say, the problem has no relation with the x-coordinates of the points in P under the assumption that P is a sorted sequence.

The output of the problem is an optimal approximating step function, which can be represented by two sequences: a k-partition $\{i_0^\star, i_1^\star, \ldots, i_k^\star\}$ that achieves optimality; and the sequence of k approximating values $\{y_1^\star, y_2^\star, \ldots, y_k^\star\}$ induced by the partition, where $y_j^\star$ is the $y^\star$ corresponding to the interval $[i_{j-1}^\star + 1, \ldots, i_j^\star]$ for $1 \leq j \leq k$. We denote by $\varepsilon^\star$ the maximum weighted vertical distance between the optimal step function and P, i.e., $\varepsilon^\star = \max_{1 \leq j \leq k} \theta(i_{j-1}^\star + 1, i_j^\star)$. Just like many other optimization problems, the finding of $\varepsilon^\star$ is the key of the problem. In the rest part of this section, we will show that, once $\varepsilon^\star$ is found, the optimal step function can be easily computed in $O(n)$ time. (See the notes following Lemma 1 and Lemma 2.) Henceforth, we only focus on how to find $\varepsilon^\star$ in the next section.

It is easy to know that the problem can be solved with dynamic programming, but it is difficult to obtain a sub-quadratic running time. Fournier and Vigneron [6] and also Chen and Wang [2] observed that the function $\theta(b,e)$ of the problem satisfies the monotonicity property, i.e., $\theta(b,e) \leq \theta(b',e')$ for any $1 \leq b' \leq b \leq e \leq e' \leq n$, so that it can be efficiently solved by some parametric search techniques. In this paper, we also abide to this paradigm.

2.2   The Decision Algorithm

To apply parametric search or its variants to solve an optimization problem, we should have a decision algorithm to conduct feasibility test on a given $\varepsilon > 0$. Fournier and Vigneron [6] and also Chen and Wang [2] exploited the generalizations of Frederickson's decision algorithm for path partitioning [7], whose efficiency depends on the efficiency of evaluating $\theta(b,e)$. In this paper, we simply use the linear-time decision algorithm by Karras et al. [11].

Lemma 1 (The decision algorithm). *Given $\varepsilon > 0$, we can decide whether $\varepsilon < \varepsilon^\star$ in $O(n)$ time.*

Given $\varepsilon > 0$, the decision algorithm of Karras et al. works as follows. We scan the point sequence P from index 1 to n only one time. During the scanning, we greedily push indices to the current interval. Suppose the current interval starts from i, and we are now considering the index j. We incrementally compute

$\cap_{i \leq \ell \leq j} \sigma_\ell$, where σ_ℓ is the vertical segment $[y_\ell - \frac{\varepsilon}{w_\ell}, y_\ell + \frac{\varepsilon}{w_\ell}]$. If the intersection does not become empty, we go to the next index; otherwise we terminate the current interval and start a new one by taking j as the first index. In the end, we return FALSE or TRUE according to whether or not the total number of partitioned intervals is larger than k.

Note 1. Another usage of the decision algorithm is that, once ε^* is found, we can obtain the optimal k-partition $\{i_0^*, i_1^*, \ldots, i_k^*\}$ in $O(n)$ time.

2.3   The Evaluation of $\theta(b, e)$

The time bounds of the algorithms in [6] and [2], as well as our randomized algorithm, all depend on how fast $\theta(b, e)$ can be evaluated. Here, we present two approaches to evaluating $\theta(b, e)$. The first approach is a simple one.

Lemma 2 (The direct evaluation method). *Without preprocessing, any $\theta(b, e)$ can be evaluated in $O(e - b) = O(n)$ time.*

Proof. Guha and Shim [11] and also Chen and Wang [2] observed that, given $1 \leq b \leq e \leq n$, the computation of $\theta(b, e)$ equals to find the lowest point of the intersection of $2(e - b + 1)$ upward halfplanes in the y-z space: $z \geq w_i(y - y_i)$ and $z \geq w_i(y_i - y)$, for $b \leq i \leq e$. This task can be accomplished in linear time since it is just the problem of simplified two-variable linear programming [14]. Once the sought point (y^*, z^*) is found, then $\theta(b, e) = z^*$. $\square$

Note 2. Once the optimal k-partion $\{i_0^*, i_1^*, \ldots, i_k^*\}$ is computed, by using the above algorithm we can obtain the optimal approximating values $\{y_1^*, y_2^*, \ldots, y_k^*\}$ induced by the partition in $O(n)$ time.

The second approach to evaluating $\theta(b, e)$ is based on preprocessing and is due to Guha and Shim [10] and Chen and Wang [2].

Lemma 3 (The evaluation method with preprocessing). *After preprocessing in $O(n \log n)$ time and $O(n \log n)$ space, any $\theta(b, e)$ can be evaluated in $O(\log^c n)$ time, where $c > 1$ is some constant.*

What the actual value of c is depends on what data structure is exploited for preprocessing. Guha and Shim [10] designed a data structure leading to $c = 4$, and Chen and Wang [2] decreased it to 3 and 2, with more complicated data structures. In our randomized algorithm, we will not apply the approach of Lemma 3 to evaluating $\theta(b, e)$. It is mentioned here only for giving an alternative deterministic algorithm.

3   The Randomized Algorithm

In this section, we present our randomized algorithm for the problem of this paper. We obtain the algorithm by randomizing Frederickson and Johnson's matrix searching technique. We first give an overview of the technique and show that the deterministic time bound $O(n \log^c n)$ in [6] and [2] can also be derived with the technique, where c is the constant in lemma 3.

3.1   The Matrix Searching Technique

In this subsection, we first recall the technique of Frederickson and Johnson [8] for (deterministically) searching in a sorted square matrix. On the aspect of solving the MIN-MAX PARTITION problem, it may be slower than Frederickson's later technique [7]. However, it is simpler and can get benefits from randomization.

Suppose there is a min-max optimization problem whose objective is to find the optimal value λ^*, the minimum among all possible λs that make $\rho(\lambda)$=TRUE, where $\rho(\lambda)$ is a monotonic function that maps a real value to TRUE or FALSE. Suppose the decision version of the optimization problem can be solved in $O(D(n))$ time, and all the possible λs can be represented by a sorted $n \times n$ matrix $M = (m_{i,j})$, where "sorted" means $m_{i,j} \leq m_{i,j+1}$ and $m_{i,j} \geq m_{i+1,j}$ for all entries of M. Then, we can find λ^* by performing efficient searching in M.

Suppose the entry of M cannot be accessed in $O(1)$ time, then Frederickson and Johnson's algorithm should work as follows. It consists of two phases: *the first phase* and *the second phase*. During the first phase, we maintain a collection $\mathcal{M}$ of submatrices of M (by recording their ranges only), and initially set $\mathcal{M} = \{M\}$. We repeat the following process until all the submatrices in $\mathcal{M}$ become singleton matrices:

1. Divide each matrix in $\mathcal{M}$ into four almost equal-size matrices, and then put the smallest (i.e. lower-leftmost) entry of each matrix into the set L, and put the largest (i.e., upper-rightmost) entry of each matrix into the set U.
2. Evaluate the entries in L and U.
3. Select the median element λ_L of L, and the median element λ_U of U.
4. Discard from $\mathcal{M}$ the matrices that cannot contain λ^*, based on the results of calling the decision procedure with λ_L and λ_U and the comparisons between an extreme entry of a matrix and λ_L or λ_U. The detailed criteria for discarding can be seen in [8] or Section 3.3 of [1].

It can be proved that, through $\lceil \lg n \rceil$ iterations, all the matrices in $\mathcal{M}$ will reach singleton, and at the end of the ith iteration, the number of matrices in $\mathcal{M}$ is $O(2^i)$. That is to say, the number of matrices in $\mathcal{M}$ can never exceed $O(n)$, and the total number of matrices being processed in the whole phase is $O(\sum_{i=0}^{\lceil \lg n \rceil} 2^i)$, which is also $O(n)$.

Then, in the second phase, we first evaluate the $O(n)$ remaining entries in the singleton matrices, and sort them, and at last conduct a binary search to find λ^* by calling the decision procedure $O(\log n)$ times.

Suppose after an $O(\pi(n))$ time preprocessing, we can evaluate any entry of M in $O(\kappa(n))$ time, and we use the worst-case linear-time algorithm for the median selections (e.g., see the textbook [3]), then the whole searching process can be accomplished in $O(\pi(n) + n \cdot \kappa(n) + D(n) \log n + n \log n)$ time, where the last item $n \log n$ corresponds to the time consumed in the sorting.

For the problem of this paper, its decision problem can be solved in $O(n)$ time by Lemma 1, and the sorted matrix M is composed of $\theta(i,j)$'s and 0's. Precisely, $m_{i,j} = \theta(i,j)$ for $1 \leq i \leq j \leq n$, and $m_{i,j} = 0$ for $1 \leq j < i \leq n$. Therefore, by using the evaluation approach of Lemma 3, we can obtain another deterministic algorithm with $O(n \log^c n)$ worst-case time. It asymptotically matches the algorithms in [6] and [2], and is simpler than them.

3.2   Solving with Randomization and Comparison

We now solve the problem in $O(n \log n)$ expected running time. We achieve this simply by accommodating both phases of the matrix searching into the paradigm of *randomized prune-and-search*. In the whole algorithm, we only employ the direct method of Lemma 2 for evaluating the entries of M.

Clearly, if we continue to rely on entry evaluation for conducting the searching, we cannot obtain a faster algorithm. Our algorithm gets benefits from the property that, the comparison between an entry of M and a real value can be efficiently computed after some "light" preprocessing.

Lemma 4 (The comparison method). *Given $\varepsilon > 0$, with $O(n)$ preprocessing time and space, we can decide whether $\theta(i, j) <, =,$ or $> \varepsilon$ in $O(\log n)$ time for any $1 \le i \le j \le n$.*

Proof. Given $\varepsilon > 0$, the data structure we construct in preprocessing is a binary interval tree, with each node corresponding to an interval $[b \dots e]$ and possessing some information related to ε. The root of the tree corresponds to the entire interval $[1 \dots n]$, and the leaf nodes correspond to the intervals of length one, i.e., $[b \dots b]$. For an inner node corresponding to the interval $[b \dots e]$ with $b < e$, its left child corresponds to the interval $[b \dots q]$ and its right child corresponds to the interval $[q + 1 \dots e]$, where $q = \lfloor (b + e)/2 \rfloor$. From the above criteria, we can easily know that there are totally $O(n)$ nodes in the tree and its height is $O(\log n)$. In addition, for each node corresponding to the interval $[b \dots e]$, we record into the node a vertical segment σ that equals to $\cap_{b \le \ell \le e} \sigma_\ell$, where σ_ℓ is the vertical segment $[y_\ell - \frac{\varepsilon}{w_\ell}, y_\ell + \frac{\varepsilon}{w_\ell}]$.

We use a recursive divide-and-conquer procedure to construct the tree. At each stage of recursion, the work of combination is to compute the intersection of the two vertical segments stored already in the children of the current node. This can be accomplished in $O(1)$ time, so that the total preprocessing time is $O(n)$.

Now, suppose we are given arbitrary i and j with $1 \le i \le j \le n$. We decide the relationship between $\theta(i, j)$ and ε by conducting searching in the interval tree. During the process of searching, we maintain a vertical segment I, and initially set $I = [-\infty, +\infty]$. At the end of the searching, we wish $I = \cap_{i \le \ell \le j} \sigma_\ell$, so that we can make decision according to whether I is empty, a point, or a segment with some length. Therefore, the goal of tree searching is to compute I by using the segments already stored in the tree, as few as possible. We can implement the searching with a recursive procedure as follows:

Procedure. SEGMENT-INT(*node*, i, j)
1 if $i = b[node]$ and $j = e[node]$
2   then $I \leftarrow I \cap \sigma[node]$, and return.

3   $q \leftarrow \lfloor (b[node] + e[node])/2 \rfloor$.
4   if $i \leq q$ and $j \geq q + 1$
5   then SEGMENT-INT($lchild[node], i, q$),
6   SEGMENT-INT($rchild[node], q + 1, j$).
7   else if $j \leq q$
8   then SEGMENT-INT($lchild[node], i, j$).
9   else $\triangleright i \geq q + 1$
10   SEGMENT-INT($rchild[node], i, j$).

The procedure is correct since we go through each possible child interval of the parent interval and each branch of the tree searching must stop at some node whose corresponding interval is a subinterval of $[i \ldots j]$. It is easy to argue that, on each level of the interval tree, there are at most four nodes being visited: the nodes whose corresponding interval includes i or j, and possibly, their sister nodes. Once the nodes in the latter type are met, the branching must stop. So, the number of visited nodes during the searching is at most four times the height of the tree, which is $O(\log n)$. Considering each visit to a node consumes $O(1)$ time, we accomplish the proof.   □

For the purpose of clear exposition, we call the preprocessing in Lemma 4 the *inner preprocessing*, and call the one in Lemma 3 the *initial preprocessing*. With the existence of Lemma 4, we can now accelerate the matrix searching by decreasing the number of evaluation calls with the paradigm of randomized prune-and-search.

Consider the first phase of the matrix searching. At each iteration, the steps are rewritten as follows: At first we randomly pick λ_L from the set L and λ_U from the set U rather than select the medians. After that, we evaluate the two entries and conduct the inner preprocessing on each of the two values. Then, the subsequent work is almost the same as the deterministic matrix searching: calling the decision procedure and performing matrix discarding. What is different is that the matrix discarding should be implemented with calling the comparison procedure.

The correctness of this phase can be easily argued by using the same loop invariant as the deterministic searching, i.e., the entry corresponding to λ^* is always included in some matrix in $\mathcal{M}$. The analysis can also follow the deterministic version: There are $\lceil \lg n \rceil$ iterations, and at the end of the ith iteration, the expected number of matrices in $\mathcal{M}$ is $O(2^i)$. Thus, in the whole phase, the expected number of calls to the evaluation procedure, the inner preprocessing procedure, and the decision procedure is all $O(\log n)$, and the expected number of calls to the comparison procedure is $O(\sum_{i=0}^{\lceil \lg n \rceil} 2^i) = O(n)$. This yields that the first phase consumes a total $O(n \log n)$ expected time.

To find λ^* among $O(n)$ remaining entries, we conduct a typical process of randomized prune-and-search in the second phase. We repeat the following steps until one entry remains: Randomly choose an entry, evaluate it and conduct the inner preprocessing on it, decide its feasibility, and then discard the impossible entries with comparisons. It is not difficult to know that this phase consumes also

$O(n \log n)$ expected time, since the expected number of comparisons performed in the whole phase is $O(n)$.

In summary, we have the following result:

Theorem 1. *By exploiting the randomized matrix searching, and relying on Lemma 1 for feasibility test, Lemma 2 for entry evaluation, and Lemma 4 for entry comparison, we can solve the problem of this paper in $O(n \log n)$ expected time and $O(n)$ space.*

By an easy reduction, Fournier and Vigneron [6] showed that the unweighted version of the problem cannot be faster than sorting under the assumption that the input point set is unsorted. Certainly, neither can the general weighted version. Hence, our bound is optimal for the unsorted case.

4 Concluding Remarks

The result of this paper is achieved by using the randomized matrix searching technique. It may be interesting to know whether there are other problems that may be solved with this framework. Especially, we care about a closely-related problem – weighted approximation of planar points with a piecewise linear function [2]. The critical point is to design a data structure allowing both efficient preprocessing and efficient comparison query as we do in Lemma 4. The author only knows reference [5] employs a similar approach.

For the problem itself, we are interested in the following two questions:

- How to implement a deterministic algorithm with $O(n \log n)$ worst-case time for an unsorted input point set?
- For a sorted input point set, can the problem be solved in (expected or worst-case) linear time?

Acknowledgments. The comments by an anonymous referee to the previous version of this paper motivated the current result.

References

1. Agarwal, P.K., Sharir, M.: Efficient Algorithms for Geometric Optimization. Computing Surveys 30, 412–458 (1998)
2. Chen, D.Z., Wang, H.: Approximating Points by a Piecewise Linear Function: I. In: Dong, Y., Du, D.-Z., Ibarra, O. (eds.) ISAAC 2009. LNCS, vol. 5878, pp. 224–233. Springer, Heidelberg (2009)
3. Cormen, T.H., Leiserson, C.E., Rivest, R.L., Stein, C.: Introduction to Algorithms. MIT Press, Cambridge (2001)
4. Díaz-Báñez, J.M., Mesa, J.A.: Fitting Rectilinear Polygonal Curves to a Set of Points in the Plane. European Journal of Operation Research 130, 214–222 (2001)
5. Eppstein, D.: Fast Construction of Planar Two-Centers. In: 8th Annual ACM-SIAM Symposium on Discrete Algorithm (SODA), pp. 131–138 (1997)

6. Fournier, H., Vigneron, A.: Fitting a Step Function to a Point Set. In: Halperin, D., Mehlhorn, K. (eds.) Esa 2008. LNCS, vol. 5193, pp. 442–453. Springer, Heidelberg (2008)
7. Frederickson, G.N.: Optimal Algorithms for Tree Partitioning. In: 2nd Annual ACM-SIAM Symposium on Discrete Algorithms (SODA), pp. 168–177 (1991)
8. Frederickson, G.N., Johnson, D.B.: Generalized Selection and Ranking: Sorted Matrices. SIAM Journal on Computing 13, 14–30 (1984)
9. Goodrich, M.: Efficient Piecewise-Linear Function Approximation Using the Uniform Metric. In: 10th Annual ACM Symposium on Computational Geometry (SOCG), pp. 322–331 (1994)
10. Guha, S., Shim, K.: A Note on Linear Time Algorithms for Maximum Error Histograms. IEEE Transactions on Knowledge and Data Engineering 19, 993–997 (2007)
11. Karras, P., Sacharidis, D., Mamoulis, N.: Exploiting Duality in Summarization with Deterministic Guarantees. In: 13th ACM SIGKDD International Conference on Knowledge Discovery and Data Mining, pp. 380–389 (2007)
12. Lopez, M.A., Mayster, Y.: Weighted Rectilinear Approximation of Points in the Plane. In: Laber, E.S., Bornstein, C., Nogueira, L.T., Faria, L. (eds.) LATIN 2008. LNCS, vol. 4957, pp. 642–657. Springer, Heidelberg (2008)
13. Mayster, Y., Lopez, M.A.: Approximating a Set of Points by a Step Function. Journal of Visual Communication and Image Representation 17, 1178–1189 (2006)
14. Preparata, F.P., Shamos, M.I.: Computational Geometry: An Introduction. Spinger, New York (1985)
15. Wang, D.P.: A New Algorithm for Fitting a Rectilinear x-Monotone Curve to a Set of Points in the plane. Pattern Recognition Letters 23, 329–334 (2002)

Approximating Multilinear Monomial Coefficients and Maximum Multilinear Monomials in Multivariate Polynomials

Zhixiang Chen and Bin Fu

Department of Computer Science
University of Texas-Pan American
Edinburg, TX 78539, USA
{chen,binfu}@cs.panam.edu

Abstract. This paper is our third step towards developing a theory of testing monomials in multivariate polynomials and concentrates on two problems: (1) How to compute the coefficients of multilinear monomials; and (2) how to find a maximum multilinear monomial when the input is a $\Pi\Sigma\Pi$ polynomial. We first prove that the first problem is #P-hard and then devise a $O^*(3^n s(n))$ upper bound for this problem for any polynomial represented by an arithmetic circuit of size $s(n)$. Later, this upper bound is improved to $O^*(2^n)$ for $\Pi\Sigma\Pi$ polynomials. We then design fully polynomial-time randomized approximation schemes for this problem for $\Pi\Sigma$ polynomials. On the negative side, we prove that, even for $\Pi\Sigma\Pi$ polynomials with terms of degree ≤ 2, the first problem cannot be approximated at all for any approximation factor ≥ 1, nor "weakly approximated" in a much relaxed setting, unless P=NP. For the second problem, we first give a polynomial time λ-approximation algorithm for $\Pi\Sigma\Pi$ polynomials with terms of degrees no more a constant $\lambda \geq 2$. On the inapproximability side, we give a $n^{(1-\epsilon)/2}$ lower bound, for any $\epsilon > 0$, on the approximation factor for $\Pi\Sigma\Pi$ polynomials. When the degrees of the terms in these polynomials are constrained as ≤ 2, we prove a 1.0476 lower bound, assuming $P \neq NP$; and a higher 1.0604 lower bound, assuming the Unique Games Conjecture.

Keywords: Multivariate polynomials; monomial testing; monomial coefficients; maximum multilinear monomials; approximation algorithms; inapproximability.

1 Introduction

1.1 Background

There is a long history in theoretical computer science with heavy involvement of studies and applications of polynomials. Most notably, low degree polynomial testing/representation and polynomial identity testing have played invaluable roles in many major breakthroughs in complexity theory. For example, low degree polynomial testing is involved in the proof of the PCP Theorem, the cornerstone

W. Wu and O. Daescu (Eds.): COCOA 2010, Part I, LNCS 6508, pp. 309–323, 2010.

of the theory of computational hardness of approximation and the culmination of a long line of research on IP and PCP (see, Arora *et al.* [3] and Feige *et al.* [12]). Polynomial identity testing has been extensively studied due to its role in various aspects of theoretical computer science (see, for example, Chen and Kao [11], Kabanets and Impagliazzo [16]) and its applications in various fundamental results such as Shamir's IP=PSPACE [23] and the AKS Primality Testing [2]. Low degree polynomial representation [20] has been sought for so as to prove important results in circuit complexity, complexity class separation and subexponential time learning of Boolean functions (see, for example, Beigel [5], Fu[13], and Klivans and Servedio [18]). These are just a few examples. A survey of the related literature is certainly beyond the scope of this paper.

The rich literature about polynomial testing, the close relation of monomial testing to key problems in complexity such as the k-path testing, satisfiability, counting and permanent computing, and many other observations have motivated us to develop a new theory of testing monomials in polynomials represented by arithmetic circuits or even simpler structures. The monomial testing problem is related to, and somehow complements with, the low degree testing and the identity testing of polynomials. We want to investigate various complexity aspects of the monomial testing problem and its variants with two major objectives. One is to understand how this problem relates to critical problems in complexity, and if so to what extent. The other is to exploit possibilities of applying algebraic properties of polynomials to the study of those critical problems.

1.2 The First Two Steps

As a first step towards testing monomials, Chen and Fu [7] have proved a series of results: The multilinear monomial testing problem for $\Pi\Sigma\Pi$ polynomials is NP-hard, even when each clause has at most three terms and each term has a degree at most 2. The testing problem for $\Pi\Sigma$ polynomials is in P, and so is the testing for two-term $\Pi\Sigma\Pi$ polynomials. However, the testing for a product of one two-term $\Pi\Sigma\Pi$ polynomial and another $\Pi\Sigma$ polynomial is NP-hard. We have also proved that testing c-monomials for two-term $\Pi\Sigma\Pi$ polynomials is NP-hard for any $c > 2$, but the same testing is in P for $\Pi\Sigma$ polynomials. Finally, two parameterized algorithms have been devised for three-term $\Pi\Sigma\Pi$ polynomials and products of two-term $\Pi\Sigma\Pi$ and $\Pi\Sigma$ polynomials. These results have laid a basis for further study about testing monomials.

In the subsequent paper, Chen *et al.* [8] present two pairs of algorithms. First, they prove that there is a randomized $O^*(p^k)$ time algorithm for testing p-monomials in an n-variate polynomial of degree k represented by an arithmetic circuit, while a deterministic $O^*(6.4^k + p^k)$ time algorithm is devised when the circuit is a formula, here p is a given prime number. Second, they present a deterministic $O^*(2^k)$ time algorithm for testing multilinear monomials in $\Pi_m\Sigma_2\Pi_t \times \Pi_k\Pi_3$ polynomials, while a randomized $O^*(1.5^k)$ algorithm is given for these polynomials. The first algorithm extends the recent work by Koutis [19] and Williams [25] on testing multilinear monomials. Group algebra is exploited in the algorithm designs, in corporation with the randomized polynomial identity testing over a finite field by Agrawal and Biswas [1], the deterministic noncommunicative polynomial identity testing by Raz and Shpilka [21] and the perfect hashing functions by Chen et al. [10]. Finally, they prove that testing some special types of multilinear monomial is W[1]-hard, giving evidence that testing for specific monomials is not fixed-parameter tractable.

1.3 Contributions

Naturally, testing for the existence of any given monomial in a polynomial can be carried out by computing the coefficient of that monomial in the sum-product expansion of the polynomial. A zero coefficient means that the monomial is not in the polynomial, while a nonzero coefficient implies that it is. Moreover, coefficients of monomials in a polynomial have their own implications and are closely related to central problems in complexity. As we shall exhibit later, the coefficients of multilinear monomials correspond to counting perfect matchings in a bipartite graph and to computing the permanent of a matrix.

Consider a $\Pi\Sigma\Pi$ polynomial F. F may not have a multilinear monomial in its sum-product expansion. However, one can always find a multilinear monomial via selecting terms from some clauses of F, unless all the terms in each clause of F are not multilinear or F is simply empty. Here, the real challenging is how to find a longest multilinear from the product of a subset of clauses in F. This problem is closely related to the maximum independent set, MAX-k-2SAT and other important optimization problems in complexity.

Because of the above characteristics of monomial coefficients, we concentrate on two problems in this paper:

1. How to compute the coefficients of multilinear monomials in the sum-product expansion of a polynomial?
2. How to find/approximate a maximum multilinear monomial when the input is a $\Pi\Sigma\Pi$ polynomial?

For the first problem, we first prove that it is #P-hard and then devise a $O^*(3^n s(n))$ time algorithm for this problem for any polynomial represented by an arithmetic circuit of size $s(n)$. Later, this $O^*(3^n s(n))$ upper bound is improved to $O^*(2^n)$ for $\Pi\Sigma\Pi$ polynomials. Two easy corollaries are derived directly from this $O^*(2^n)$ upper bound. One gives an upper bound that matches the best known $O^*(2^n)$ deterministic time upper bound, which was due to Ryser [22] in early 1963, for computing the permanent of an $n \times n$ matrix. The other gives an upper bound that matches the best known $O^*(1.415^n)$ deterministic time upper bound, which was also due to Ryser [22], for counting the number of perfect matchings in the a bipartite graph.

We then design three fully polynomial-time randomized approximation schemes. The first approximates the coefficient of any given multilinear monomial in a $\Pi\Sigma$ polynomial. The second approximates the sum of coefficients of all the multilinear monomials in a $\Pi\Sigma$ polynomial. The third finds an ϵ-approximation to the coefficient of any given multilinear monomial in a $\Pi_k\Sigma_a\Pi_t \times \Pi_m\Sigma_s$ polynomial with a being a constant ≥ 2.

On the negative side, we prove that, even for $\Pi\Sigma\Pi$ polynomials with terms of degree ≤ 2, the first problem cannot be approximated *at all* regardless of the approximation factor ≥ 1. We then consider *"weak approximation"* in a much relaxed setting, following our previous work on inapproximability about exemplar breakpoint distance and exemplar conserved interval distance of two genomes [9,6]. We prove that, assuming $P \neq NP$, the first problem cannot be approximated in polynomial time within any approximation factor $\alpha(n) \geq 1$ along with any additive adjustment $\beta(n) \geq 0$, where $\alpha(n)$ and $\beta(n)$ are polynomial time computable.

For the second problem, we first present a polynomial time λ-approximation algorithm for $\Pi\Sigma\Pi$ polynomials with terms of degrees no more a constant $\lambda \geq 2$. On the inapproximability side, we give a $n^{(1-\epsilon)/2}$ lower bound, for any $\epsilon > 0$, on the approximation factor for $\Pi\Sigma\Pi$ polynomials. When the degrees of the terms in these polynomials are constrained as ≤ 2, we prove a 1.0476 lower bound, assuming $P \neq NP$. We also prove a higher 1.0604 lower bound, assuming the Unique Games Conjecture.

2   Notations and Definitions

For variables $x_1, \ldots, x_n$, let $\mathcal{P}[x_1, \cdots, x_n]$ denote the communicative ring of all the n-variate polynomials with coefficients from a finite field $\mathcal{P}$. For $1 \leq i_1 < \cdots < i_k \leq n$, $\pi = x_{i_1}^{j_1} \cdots x_{i_k}^{j_k}$ is called a monomial. The degree of π, denoted by $\deg(\pi)$, is $\sum_{s=1}^{k} j_s$. π is multilinear, if $j_1 = \cdots = j_k = 1$, i.e., π is linear in all its variables $x_{i_1}, \ldots, x_{i_k}$. For any given integer $\tau > 1$, π is called a τ-monomial, if $1 \leq j_1, \ldots, j_k < \tau$. In the setting of the *MAX-Multilinear Problem* in Section 7, we need to consider the length of the monomial $\pi = x_{i_1}^{j_1} \cdots x_{i_k}^{j_k}$ as $|\pi| = \sum_{\ell=1}^{k} \log(1 + j_\ell)$. (Strictly speaking, $|\pi|$ should be $\sum_{\ell=1}^{k} \log(1 + j_\ell) \, \log n$. But, the common $\log n$ factor can be dropped for ease of analysis.) When π is multilinear, $|\pi| = k$, i.e., the number of variables in it.

For any polynomial $F(x_1, \ldots, x_n)$ and any monomial π, we let $c(F, \pi)$ denote the coefficient of π in the sum-product of F, or in F for short. If π is indeed in F, then $c(F, \pi) > 0$. If not, then $c(F, \pi) = 0$. We also let $S(F)$ denote the sum of the coefficients of all the multilinear monomials in F. When it is clear from the context, we use $c(\pi)$ to stand for $c(F, \pi)$.

An arithmetic circuit, or circuit for short, is a direct acyclic graph with + gates of unbounded fan-ins, $\times$ gates of two fan-ins, and all terminals corresponding to variables. The size, denoted by $s(n)$, of a circuit with n variables is the number of gates in it. A circuit is called a formula, if the fan-out of every gate is at most one, i.e., its underlying direct acyclic graph is a tree.

By definition, any polynomial $F(x_1, \ldots, x_n)$ can be expressed as the sum of a list of monomials, called the sum-product expansion. The degree of the polynomial is the largest degree of its monomials in the expansion. With this expression, it is trivial to see whether $F(x_1, \ldots, x_n)$ has a multilinear monomial (or a monomial with any given pattern) along with its coefficient. Unfortunately, this expression is essentially problematic and infeasible to realize, because a polynomial may often have exponentially many monomials in its expansion.

In general, a polynomial $F(x_1, \ldots, x_n)$ can be represented by a circuit or some even simpler structure as defined in the following. This type of representation is simple and compact and may have a substantially smaller size, say, polynomially in n, in comparison with the number of all monomials in the sum-product expansion. The challenge is how to test whether F has a multilinear monomial, or some other needed monomial, efficiently without unfolding it into its sum-product expansion? The challenge applies to finding coefficients of monomials in F.

Throughout this paper, the $O^*(\cdot)$ notation is used to suppress poly(n) factors in time complexity bounds.

Definition 1. Let $F(x_1, \ldots, x_n) \in \mathcal{P}[x_1, \ldots, x_n]$ be any given polynomial. Let $m, s, t \geq 1$ be integers.

- $F(x_1, \ldots, x_n)$ is said to be a $\Pi_m \Sigma_s \Pi_t$ polynomial, if $F(x_1, \ldots, x_n) = \prod_{i=1}^{t} F_i$, $F_i = \sum_{j=1}^{r_i} X_{ij}$ and $1 \leq r_i \leq s$, and X_{ij} is a product of variables with $\deg(X_{ij}) \leq t$. We call each F_i a clause. Note that X_{ij} is not a monomial in the sum-product expansion of $F(x_1, \ldots, x_n)$ unless $m = 1$. To differentiate this subtlety, we call X_{ij} a term.
- In particular, we say $F(x_1, \ldots, x_n) = \prod_{i=1}^{t} F_i$ is a $\Pi_m \Sigma_s$ polynomial, if it is a $\Pi_m \Sigma_s \Pi_1$ polynomial. Here, each clause F_i is a linear addition of single variables. In other word, each term in F_i has degree 1.
- $F(x_1, \ldots, x_n)$ is called a $\Pi_m \Sigma_s \Pi_t \times \Pi_k \Sigma_\ell$ polynomial, if $F(x_1, \ldots, x_n) = F_1 \cdot F_2$ such that F_1 is a $\Pi_m \Sigma_s \Pi_t$ polynomial and F_2 is a $\Pi_k \Sigma_\ell$ polynomial.

When no confusion arises from the context, we use $\Pi \Sigma \Pi$ and $\Pi \Sigma$ to stand for $\Pi_m \Sigma_s \Pi_t$ and $\Pi_m \Sigma_s$, respectively.

Throughout the rest of the paper, we will focus on nonnegative integer coefficients in polynomials.

3 Multilinear Monomial Coefficients, Perfect Matchings and Permanents

Theorem 1. Let $F(x_1, \ldots, x_n)$ be any given $\Pi_m \Sigma_s \Pi_2$ polynomial. It is #P-hard to compute the coefficient of any given multilinear monomial in the sum-product expansion of F.

Proof. We shall reduce the counting problem of a bipartite graph to the problem of computing coefficient of a multilinear monomial in a polynomial. Let $G = (V_1 \cup V_2, E)$ be any given bipartite graph. We construct a polynomial F as follows.

Assume that $V_1 = \{v_1, \cdots, v_t\}$ and $V_2 = \{u_1, \cdots, u_t\}$. Each vertex $v_i \in V_1$ is represented by a variable x_i, so is $u_i \in V_2$ by a variable y_i. For every vertex $v_i \in V_1$, let $F_i = \sum_{(v_i, u_j) \in E} x_i y_j$. Define a polynomial for the graph G as $F(G) = F_1 \cdots F_t$. Let $n = 2t$, $m = t$, and s be maximum degree of the vertices in V_1. It is easy to see that $F(G)$ is a n-variate $\Pi_m \Sigma_s \Pi_2$ polynomial.

Now, suppose that G has a perfect matching $(x_1, y_{i_1}), \ldots, (x_t, y_{i_t})$. Then, we can choose $\pi_j = x_j y_{i_j}$ from F_j, $1 \leq j \leq t$. Thus, $\pi = \pi_1 \cdot \pi_2 \cdots \pi_t =$

$x_1 x_2 \cdots x_t y_1 y_2 \cdots y_t$ is a multilinear monomial in $F(G)$. Hence, the number of perfect matchings in G is at most $c(\pi)$, i.e., the coefficient of π in $F(G)$. On the other hand, suppose that $F(G)$ has a multilinear monomial $\pi = \pi_1' \cdots \pi_t' = x_1 x_2 \cdots x_t y_1 y_2 \cdots y_t$ in its sum-product expansion with π_j' being a term from F_j, $1 \leq j \leq t$. By the definition of F_j, $\pi_j' = x_j y_{i_j}$, meaning that vertices v_j and u_{i_j} are directly connected by the edge (j, i_j). Since π' is multilinear, $y_{i_1}, \ldots, y_{i_t}$ are distinct. Hence, $(x_1, y_{i_1}), \ldots, (x_t, y_{i_t})$ constitute a perfect matching in G. Hence, the coefficient $c(\pi)$ of π in $F(G)$ is at most the number of perfect matchings in G. Putting the above analysis together, we have that G has a perfect matching iff $F(G)$ has a copy of the multilinear monomial $\pi = x_1 x_2 \cdots x_t y_1 y_2 \cdots y_t$ in its sum-product expansion. Moreover, G has $c(\pi) \geq 0$ many perfect matchings iff the multilinear monomial π has a coefficient $c(\pi)$ in the expansion. Therefore, by Valiant's #P-hardness of counting the number of perfect matchings in a bipartite graph [24], computing the coefficient of π in $F(G)$ is #P-hard.

Theorem 2. *There is a $O^*(s(n)3^n)$ time algorithm to compute the coefficients of all multilinear monomials in a polynomial $F(x_1, \ldots, x_n)$ represented by an arithmetic circuit C of size $s(n)$.*

Proof. We consider evaluating F from C via a bottom-up process. Notice that at most 2^n many multilinear monomials can be formed with n variables. For each addition gate g in C with fan-ins $f_1, \ldots, f_s$, we may assume that each f_i is a sum of multilinear terms, i.e., products of distinct variables. This assumption is valid, because we can discard all the terms in f_i that are not multilinear since we are only interested in multilinear monomials in the sum-product expansion of F. We simply add $f_1 + \cdots + f_s$ via adding the coefficients of the same terms together. Since there are at most 2^n many multilinear monomials (or terms), this takes $O(n2^n)$ times.

Now we consider a multiplication gate g' in C with fan-ins h_1 and h_2. As for the addition gates, we may assume that h_i is a sum of multilinear terms, $i = 1, 2$. For each term π with degree ℓ in h_1, we only need to multiply it with terms in h_2 whose degrees are at most $n - \ell$. If the multiplication yields a non-multilinear term then that term is discarded, because we are only interested in multilinear terms in the expansion of F. This means that a term π of degree ℓ in h_1 can be multiplied with at most $2^{n-\ell}$ possible terms in h_2. Let m_i denote the number of terms in h_1 with degree i, $1 \leq i \leq n$. Then, evaluating $h_1 \cdot h_2$ for the multiplication gate g' takes time at most

$$O(n \, (m_1 \, 2^{n-1} + m_2 \, 2^{n-2} + \cdots + m_{n-1} \, 2^1)). \tag{1}$$

Since there are at most $\binom{n}{i}$ terms with degree i with respect to n variables, expression (1) is at most

$$O(n \, [\binom{n}{1} \, 2^{n-1} + \binom{n}{2} \, 2^{n-2} + \cdots + \binom{n}{n-1} \, 2^{n-n}])$$
$$= O(n \sum_{i=1}^{n} \binom{n}{i} 2^{n-i}) = O(n \, 3^n).$$

Since C has $s(n)$ gates, the total time for the entire evaluation of F for finding all its multilinear monomials with coefficients is $O(ns(n)3^n) = O^*(s(n)3^n)$.

Theorem 3. *Let $F(x_1, \ldots, x_n)$ be any given $\Pi_m \Sigma_s \Pi_t$ polynomial. One can find coefficients of all the multilinear monomials in the sum-product expansion of F in $O^*(2^n)$ time.*

Proof. (**Idea**) Similar to the analysis for Theorem 2 with the understanding that each clause in F has $\leq s$ terms.

Corollary 1. *There is a $O^*(1.415^n)$ time algorithm to compute the exact number of perfect matchings in a bipartite graph $G = (V_1 \cup V_2, E)$ with $n = 2|V_1| = 2|V_2|$ vertices.*

Proof. (**Sketch**)Let $m = n/2$, $V_1 = \{v_1, \ldots, v_m\}$ and $V_2 = \{u_1, \ldots, u_m\}$. For each vertex $u_i \in V_2$, we define a variable x_i. For each vertex $v_i \in V_1$, construct a polynomial $H_i = x_{i_1} + x_{i_2} + \cdots + x_{i_{\ell_i}}$, where $(v_i, u_{i_j}) \in E$ for $j = 1, \cdots, \ell_i$ and v_i has exactly ℓ_i adjacent vertices in G. Define $H(G) = H_1 \cdots H_{n/2}$. Apply Theorem 3 to $H(G)$.

Corollary 2. *The permanent of any given $n \times n$ matrix is computable in time $O^*(2^n)$.*

Proof. (**Sketch**) Let $A = (a_{ij})_{n \times n}$ be an $n \times n$ matrix with nonnegative entries a_{ij}, $1 \leq i, j \leq n$. Design a variable x_i for row i and define polynomials in the following:

$$R_i = (a_{i1}x_1 + \cdots + a_{in}x_n),$$
$$P(A) = R_1 \cdots R_n.$$

Apply Theorem 3 to $P(A)$.

4   Fully Polynomial-Time Approximation Schemes for $\Pi\Sigma$ Polynomials

In this section, we show that in contrast to Theorems 2 and 3, fully polynomial-time randomized approximation schemes ("FPRAS") exist for solving the problem of finding coefficients of multilinear monomials in a $\Pi\Sigma$ polynomial and some variants of this problem as well. An FPRAS $\mathcal{A}$ is a randomized algorithm, when given any n-variate polynomial F and a monomial π together with an accuracy parameter $\epsilon \in (0, 1]$, outputs a value $\mathcal{A}(F, \pi, \epsilon)$ in time poly$(n, 1/\epsilon)$ such that with high probability

$$(1 - \epsilon)c(\pi) \leq \mathcal{A}(F, \pi, \epsilon) \leq (1 + \epsilon)c(\pi).$$

Theorem 4. *There is an FPRAS for finding the coefficient of any given multilinear monomial in a $\Pi_m \Sigma_s$ polynomial $F(x_1, \ldots, x_n)$.*

Proof. Let $F(x_1, \ldots, x_n) = \prod_{i=1}^{m} F_i$ such that $F_i = \sum_{j=1}^{s_i} x_{ij}$ with $s_i \le s$. Notice that any monomial in the sum-product expansion of F will have exactly one variable from each clause F_i. This allows us to focus on multilinear monomials with exactly m variables. Let $\pi = x_{i_1} \cdots x_{i_m}$ be such a multilinear monomial. We consider how to test whether π is in F, and if so, how to find its coefficient $c(\pi)$.

For each F_i, we eliminate all the variables that are not included in π and let F_i' be the resulting clause and $F' = F_1' \cdots F_m'$. If one clause F_i' is empty, then we know that π must not be a in the expansion of F', nor in F. Now suppose that all clauses F_i', $1 \le i \le m$, are not empty. We shall reduce F' to a bipartite graph $G = (V_1 \cup V_2, E)$ as follows. Define $V_1 = \{v_1, \ldots, v_m\}$ and $V_2 = \{u_1, \ldots, u_m\}$. Here, each vertex v_i corresponds to the clause F_i', and each vertex u_j corresponds to the variable x_j. Define an edge (v_i, u_j) in E if x_j is in F_i.

Suppose that π is a multilinear monomial in F (hence in F'). Then, each x_{i_j} in π is in a distinct clause F_{t_j}, $1 \le j \le m$. This implies that edges (v_{t_j}, u_{i_j}), $1 \le j \le m$, constitute a perfect matching in G. On the other hand, if edges (v_{t_j}, u_{i_j}), $1 \le j \le m$ form a perfect matching in G, then we have that x_{i_j} is in the clause F_{t_j}. Hence, $\pi = x_{i_1} \cdots x_{i_m}$ is a multilinear monomial in F' (hence in F). This equivalence relation further implies that the number of perfect matchings in G is the same as the coefficient of the multilinear monomial π in F. Thus, the theorem follows from any fully polynomial-time randomized approximation scheme for computing the number of perfect matchings in a bipartite graph, and such an algorithm can be found in Jerrum *et al.* [15]. $\blacksquare$

In the following we shall consider how to compute the sum $S(F)$ of the coefficients of all the multilinear monomials in a $\Pi\Sigma$ polynomial F.

Theorem 5. *There is an FPRAS, when given any n-variate $\Pi_m\Sigma_s$ polynomial $F(x_1, \ldots, x_n)$, computes $S(F)$.*

Proof. Let $F(x_1, \ldots, x_n) = \prod_{i=1}^{m} F_i$ such that $F_i = \sum_{j=1}^{s_i} x_{ij}$ with $s_i \le s$. Since every monomial in the sum-product expansion of F consists of exactly one variable from each clause F_j, if $m > n$ then F must not have any multilinear in its expansion. Thus, we may assume that $m \le n$, because otherwise F will have no multilinear monomials. Let $H = (x_1 + \cdots + x_n)$. Define

$$F'(x_1, \ldots, x_n) = F \cdot H^{n-m} = F_1 \cdots F_m \cdot H^{n-m}.$$

Then, F' is a $\Pi_n\Sigma_n$ polynomial. For any given multilinear monomial

$$\pi = x_{i_1} \cdots x_{i_m}$$

in F with x_{i_j} belonging to the clause F_j, $1 \le j \le m$, let $x_{i_{m+1}}, \ldots, x_{i_{n-m}}$ be the $n - m$ variables that are not included in π, then

$$\pi' = x_{i_1} \cdots x_{i_m} \cdot x_{i_{m+1}} \cdots x_{i_{n-m}} = x_1 x_2 \cdots x_n$$

is a multilinear monomial in F'. Because F' have n clauses with n variables, the only multilinear monomial that may be possibly contained in F' is the multilinear monomial $\pi' = x_1 x_2 \cdots x_n$. If F' indeed has the multilinear monomial π' with x_{i_j} in the clause F_j, $1 \leq j \leq m$, then $\pi = x_{i_1} \cdots x_{i_m}$ is a multilinear monomial in F. This relation between π and π' is also reflected by the relation between the coefficient $c(\pi)$ of π in the expansion of F and the efficient $c(\pi')$ of π' in the expansion of F'. Precisely, the coefficient $c(\pi)$ of π in F implies that there are $c(\pi)$ copies of $x_{i_1} \cdots x_{i_m}$ for the choices of the first m variables in π'. Each additional variable x_{i_j}, $m + 1 \leq j \leq n - m$, is selected from one copy of the clause H. Since $H = (x_1 + \cdots x_n)$, there are $(n - m)!$ ways to select these $(n - m)$ variables from $(n - m)$ copies of H in F'. Hence, π contributes a value of $c(\pi)(n - m)!$ to the coefficient of π' in F'. Adding the contributions of all the multilinear monomials in F to π' in F' together, we have that the coefficient of π in F' is $S(F) \cdot (n - m)!$. By Theorem 4, there is an FPRAS to compute the coefficient of π' in F'. Dividing the output of that algorithm by $(n - m)!$ gives the needed approximation to $S(F)$.

Theorem 6. *Let $F(x_1, \ldots, x_n)$ be $\Pi_k \Sigma_a \Pi_t \times \Pi_m \Sigma_s$ polynomial with $a \geq 2$ being a constant. There is a $O(a^k poly(n, 1/\epsilon))$ time FPRAS that finds an ϵ-approximation for the coefficient of any given multilinear monomial π in the sum-product F if π is in F, or returns "no" otherwise. Here, $0 \leq \epsilon < 1$ is any given approximation factor.*

Proof. (**Idea**) Let $F = F_1 \cdot F_2$ such that F_1 is a $\Pi_k \Sigma_a \Pi_t$ polynomial, while F_2 is a $\Pi_m \Sigma_s$ polynomial. Apply Theorem 5 to the product of each monomial in the expansion of F_1 with F_2.

5   Inapproximability

We consider a relaxed setting of approximation in comparison with the ϵ-approximation in the previous section. Given any n-variate polynomial F and a monomial π together with an approximation factor $\gamma \geq 1$, we say that an algorithm $\mathcal{A}$ approximates the coefficient $c(\pi)$ in F within an approximation factor γ, if it outputs a value $\mathcal{A}(F, \pi)$ such that

$$\frac{1}{\gamma} c(\pi) \leq \mathcal{A}(F, \pi) \leq \gamma c(\pi).$$

We may also refer $\mathcal{A}$ as a γ-approximation to $c(\pi)$.

Theorem 7. *No matter what approximation factor $\gamma \geq 1$ is used, there is no polynomial time approximation algorithm for the problem of computing the coefficient of any given multilinear monomial in the sum-product expansion of a $\Pi_m \Sigma_3 \Pi_2$ polynomial, unless P=NP.*

Proof. (**Sketch**) Let $F(x_1, \ldots, x_n) = \prod_{i=1}^{m} F_i$ be a $\Pi_m \Sigma_3 \Pi_2$ polynomial. With loss of generality, we may assume that every term T_{ij} in each clause F_i is a product of two variables. Let $H = (x_1 + x_2 + \cdots x_n)$ and define

$$F' = F_1 \cdot F_2 \cdot H^{(n-2m)} \tag{2}$$

Then, F has a multilinear monomial iff F' has the only multilinear monomial ψ with its coefficient $c(\psi) = S(F)(n - 2m)!$. Hence, the theorem follows from the NP-completeness of the multilinear monomial testing problem for F that has been proved in Chen and Fu [7].

Theorem 8. *Assuming $P \neq NP$, given any n-variate $\Pi_m \Sigma_3 \Pi_2$ polynomial F and any approximation factor $\gamma \geq 1$, there is no polynomial time approximation algorithm for computing within a factor of γ the sum $S(F)$ of the coefficients of all the multilinear monomials in the sum-product expansion of F.*

Proof. (**Idea**) Similar to the proof for Theorem 7.

6   Weak Inapproximability

In this section, we shall relax the γ-approximation further in a much weak setting. Here, we allow the computed value to be within a factor of the targeted value along with some additive adjustment. Weak approximation has been first considered in our previous work on approximating the exemplar breakpoint distance [9] and the exemplar conserved interval distance [6] between two genomes. Assuming $P \neq NP$, it has been shown that the first problem does not admit any factor approximation along with a linear additive adjustment [9], while the latter has no approximation within any factor along with a $O(n^{1.5})$ additive adjustment [6]. We shall strengthen the inapproximability results of Theorems 7 and 8 to weak inapproximability for computing the coefficient of any given multilinear monomial in a $\Pi \Sigma \Pi$ polynomials. But first let us define the weak approximation.

Definition 2. *Let Z be the set of all nonnegative integers. Given four functions $f(x), h(x), \alpha(x)$ and $\beta(x)$ from Z to Z with $\alpha(x) \geq 1$, we say that $h(x)$ is a weak $(\alpha(x), \beta(x))$-approximation to $f(x)$, if*

$$max\left\{0, \frac{f(x) - \beta(x)}{\alpha(x)}\right\} \leq h(x) \leq \alpha(x) \, f(x) + \beta(x). \tag{3}$$

Theorem 9. *Let $\alpha(x) \geq 1$ and $\beta(x)$ be any two polynomial time computable functions from Z to Z. There is no polynomial time weak $(\alpha(x), \beta(x))$-approximation algorithm for computing the coefficient of any given multilinear monomial in an n-variate $\Pi_m \Sigma_3 \Pi_2$ polynomial, unless $P=NP$.*

Proof. Let $F(x_1, \ldots, x_n) = \prod_{i=1}^{m} F_i$ be a $\Pi_m \Sigma_3 \Pi_2$ polynomial. Like in the proof of Theorem 7, we assume without loss of generality that every term in each clause F_i is a product of two variables. We further assume that $2m > n$, because otherwise there are no multilinear monomials in F.

Choose k such that $k! > 2\alpha(n+k)\beta(n+k)+\beta(n+k)$. Notice that finding such a $k \leq 2n$ is possible when n is large enough, because both α and β are polynomial time computable. Let $H = (x_1 + x_2 + \cdots x_n)$ and $G = (y_1 + y_2 + \cdots y_k)$ with y_i being new variables. Define

$$F' = F \cdot H^{n-2m} \cdot G^k = F_1 \cdots F_m \cdot H^{n-2m} \cdot H^k. \tag{4}$$

It is easy to see from the above expression (4) that F has a multilinear monomial iff F' has one. Furthermore, the only multilinear monomial that F' can possibly have is $\psi = x_1 \cdots x_n \cdot y_1 \cdots y_k$.

Now consider that F has a multilinear monomial π with its coefficient $c(\pi) > 0$. Since the degree of π is $2m$, let $x_{i_1}, \ldots, x_{i_{n-2m}}$ be the variables that are not included in π. Then, the concatenation of π with each permutation of $x_{i_1}, \ldots, x_{i_{n-2m}}$ selected from H^{n-2m} and each permutation of $y_1, \ldots, y_k$ chosen from G^k will constitute a copy of the only multilinear monomial ψ in F'. Thus, π contributes $c(\pi)(n - 2m)! \, k!$ to the coefficient $c(\psi)$ of ψ in F'. When all the possible multilinear monomials in F are considered, the coefficient of $c(\psi)$ in F' is $S(F)(n - 2m)!k!$. If F' has a multilinear monomial, i.e., the only one ψ, then F has at least one multilinear monomial. In this case, the above analysis also yields $c(\psi) = S(F)(n - 2m)!k!$ in F'.

Assume that there is a polynomial time weak (α, β)-approximation algorithm $\mathcal{A}$ to compute the coefficient of any given the multilinear monomial in a $\Pi_m \Sigma_3 \Pi_2$ polynomial. Apply $\mathcal{A}$ to F' for the multilinear monomial ψ. Let $\mathcal{A}(\psi)$ be the coefficient returned by $\mathcal{A}$ for ψ. Then, by expression (3) we have

$$\mathcal{A}(\psi) \leq \alpha(n + k) \, c(\psi) + \beta(n + k)$$
$$= \alpha(n + k) \, S(F) \, (n - 2m)! \, k! + \beta(n + k), \tag{5}$$

$$\mathcal{A}(\psi) \geq \frac{c(\psi) - \beta(n + k)}{\alpha(n + k)}$$
$$= \frac{S(F) \, (n - 2m)! \, k! - \beta(n + k)}{\alpha(n + k)}. \tag{6}$$

When F does not have any multilinear monomials, then F' does not either, implying $S(F) = 0$. In this case, by the relation (5), we have

$$\mathcal{A}(\psi) \leq \beta(n + k). \tag{7}$$

When F has multilinear monomials, then F' does as well. By the relation (6), we have

$$\mathcal{A}(\psi) \geq \frac{S(F) \, (n - 2m)! \, k! - \beta(n + k)}{\alpha(n + k)}$$
$$\geq \frac{k! - \beta(n + k)}{\alpha(n + k)} > \frac{(2\alpha(n + k)\beta(n + k) + \beta(n + k)) - \beta(n + k)}{\alpha(n + k)}$$
$$= 2\beta(n + k). \tag{8}$$

Since there is a clear gap between $(-\infty, \beta(n+k)]$ and $(2\beta(n+k), +\infty)$, inequalities (7) and (8) provide us with a sure way to test whether F has a multilinear monomial or not: If $\mathcal{A}(\psi) > 2\beta(n + k)$, then F has multilinear monomials. If $\mathcal{A}(\psi) \leq \beta(n + k)$ then F does not. Since $\mathcal{A}$ runs in polynomial time, $\beta(n + k)$ is polynomial time computable and $k \leq 2n$, this implies that one can test whether F has a multilinear monomial in polynomial time. Since it has been proved in Chen and Fu [7] that the problem of testing multilinear monomials a $\Pi_m \Sigma_3 \Pi_2$ polynomial is NP-complete, such an algorithm $\mathcal{A}$ does not exist unless P=NP.

Combining the analysis for proving Theorems 8 and 9, we have the following weak inapproximability for computing the sum of coefficients of all the multilinear monomials in a $\Pi\Sigma\Pi$ polynomial.

Theorem 10. *Let $\alpha(x) \geq 1$ and $\beta(x)$ be any two polynomial time computable functions from Z to Z. Assuming $P \neq NP$, there is no polynomial time weak $(\alpha(x), \beta(x))$-approximation algorithm for computing the sum $S(F)$ of the coefficients of all the multilinear monomials in the sum-product expansion of a $\Pi_m\Sigma_3\Pi_2$ polynomial F.*

7   The Maximum Multilinear Problem and Its Approximation

Given any $\Pi\Sigma\Pi$ polynomial $F(x_1,\ldots,x_n) = F_1\cdots F_m$, F may not have any multilinear monomial in its sum-product expansion. But even if this is the case, one can surely find a multilinear monomial by selecting terms from a proper subset of the clauses in F, unless all the terms in F are not multilinear or F is simply empty. In this section, we consider the problem of finding the largest (or longest) multilinear monomials from subsets of the clauses in F. We shall investigate the complexity of approximating this problem.

Definition 3. *Let $F(x_1,\ldots,x_n) = F_1\cdots F_m$ be a $\Pi_m\Sigma_s\Pi_t$ polynomial. Define MAX-SIZE(F) as the maximum length of multilinear monomials $\pi = \pi_{i_1}\cdots\pi_{i_k}$ with π_{i_j} in F_{i_j}, $1 \leq j \leq k$ and $1 \leq i_1 < \cdots < i_k$. Let MAX-MLM($F$) to be a multilinear monomial π such that $|\pi| =$ MAX-SIZE(F), and we call such a multilinear monomial as a MAX-multilinear monomial in F.*

The MAX-MLM problem for an n-variate $\Pi\Sigma\Pi$ polynomial F is to find MAX-MLM(F). Sometimes, we also refer the MAX-MLM problem as the problem of finding MAX-SIZE(F). We say that an algorithm $\mathcal{A}$ is an approximation scheme within a factor $\gamma \geq 1$ for the MAX-MLM problem if, when given any $\Pi\Sigma\Pi$ polynomial F, $\mathcal{A}$ outputs a multilinear monomial denoted as $\mathcal{A}(F)$ such that MAX-SIZE(F) $\leq \gamma|\mathcal{A}(F)|$.

Theorem 11. *Let $\lambda \geq 2$ be a constant integer. Let F be any given n-variate $\Pi_m\Sigma_s\Pi_\lambda$ polynomial with $s \geq 2$. There is a polynomial time approximation algorithm that approximates the MAX-MLM problem for F within a factor of λ.*

Proof. (**Sketch**) Let $F(x_1,\ldots,x_n) = F_1\cdots F_m$ such that each clause F_i has at most s terms with degrees at most λ. Let $M = M_1 \cdot M_2\cdots M_k$ be a MAX-multilinear monomial in F. Without loss of generality, assume $|M_1| \geq |M_2| \geq \cdots |M_k|$. We shall devise a simple greedy strategy to find a multilinear monomial π to approximate M.

We first find the longest term π_1 from a clause F_{i_1}. Mark the clause F_{i_1} off in F. Let $\pi = \pi_1$. From all the unmarked clauses in F, find the longest term π_2 from a clause F_{i_2} such that π_2 has no common variables in π. Mark F_{i_2} off and let $\pi = \pi_1 \cdot \pi_2$. Repeat this process until no more terms can be found. At this point, we obtain a multilinear monomial $\pi = \pi_1 \cdot \pi_2 \cdots \pi_\ell$.

Notice that each term in F has at most λ variables. Each π_i may share certain common variables with some terms in M. If this is the case, then π_i will share common variables with at most λ terms in M. This means that we can select at least $\ell \geq \lceil \frac{k}{\lambda} \rceil$ terms for π. The greedy strategy implies that

$$\text{MAX-SIZE}(F) = |M| \leq \lambda |\pi|.$$

Theorem 12. *Let $F(x_1, \ldots, x_n)$ be any given n-variate $\Pi_m \Sigma_s \Pi_t$ polynomial. Unless $P = NP$, there can be no polynomial time algorithm that approximates MAX-MLM(F) within a factor of $n^{(1-\epsilon)/2}$, for any $\epsilon > 0$.*

Proof. (**Sketch**) We shall reduce the maximum independent set problem to the MAX-MLM problem. Let $G = (V, E)$ be any given indirected graph with $V = \{v_1, \ldots, v_n\}$. For each edge $(v_i, v_j) \in E$, we design a variable x_{ij} representing this edge. For each vertex $v_i \in V$, let $d(v_i)$ denote the number of edges connecting to it and define a term $T(v_i)$ as follows:

$$T(v_i) = \begin{cases} \prod_{(v_i,v_j)\in E} x_{ij}, & \text{if } d(v_i) = n-1, \\ \left(\prod_{(v_i,v_j)\in E} x_{ij}\right) \cdot \left(\prod_{j=1}^{n-1-d(v_i)} y_{ij}\right), & \text{if } d(v_i) < n-1. \end{cases}$$

We now define a polynomial $F(G)$ for the graph G as

$$F(G) = (T(v_1) + \cdots + T(v_n))^n.$$

It follows from the above definitions that G has a maximum independent set of size $\mathcal{K}$ iff $F(G)$ has a MAX-multilinear monomial of length $\mathcal{K}(n-1)$. The lower bound given in the theorem follows from Zuckerman's inapproximability lower bound of $n^{1-\epsilon}$ [26] on the maximum independent set problem.

Håstad [14] proved that there is no polynomial time algorithm to approximate the MAX-2-SAT problem within a factor of $\frac{22}{21}$. By this result, we can derive the following inapproximability about the MAX-MLM problem for the $\Pi_m \Sigma_2 \Pi_2$. Notice that Chen and Fu proved [7] that testing multilinear monomials in a $\Pi \Sigma_2 \Pi$ polynomial can be done in quadratic time.

Theorem 13. *Unless P=NP, there is no polynomial time algorithm to approximate MAXM-MLM(F) within a factor 1.0476 for any given $\Pi_m \Sigma_2 \Pi_2$ polynomial F.*

Proof. (**Sketch**) We reduce the MAX-2-SAT problem to the MAX-MLM problem for $\Pi_m \Sigma_2 \Pi_2$ polynomials. Let $F = F_1 \wedge \cdots \wedge F_m$ be a 2SAT formula. Without loss of generality, we assume that every variable x_i in F appears at most three times, and if x_i appears three times, then x_i itself occurs twice and $\bar{x}_i$ once. (It is easy to see that a simple preprocessing procedure can transform any 2SAT formula to satisfy these properties.) The reduction is similar to, but with subtle differences from, the one that was used in [7] to reduce a 3SAT formula to a $\prod_m \sum_3 \prod_2$ polynomial.

If x_i (or $\bar{x}_i$) appears only once in F then we replace it by $y_{i1}y_{i2}$. When x_i appears twice, then we do the following: If x_i (or $\bar{x}_i$) occurs twice, then replace the first occurrence by $y_{i1}y_{i2}$ and the second by $y_{i3}y_{i4}$. If both x_i and $\bar{x}_i$ occur, then replace both occurrences by $y_{i1}y_{i2}$. When x_i occurs three times with x_i appearing twice and $\bar{x}_i$ once, then replace the first x_i by $y_{i1}y_{i2}$ and the second by $y_{i3}y_{i4}$, and replace $\bar{x}_i$ by $y_{i1}y_{i3}$.

Let $G = G_1 \cdots G_m$ be the polynomial resulted from the above replacement process. Here, G_i corresponds to F_i with boolean literals being replaced. Then, the maximum number of the clauses in F can be satisfied by any true assignment is $\mathcal{K}$ iff a MAX-multilinear monomial in G has length $2\mathcal{K}$. The lower bound in the theorem follows from Håstad's inapproximability lower bound on the MAX-2-SAT problem [14].

Khot *et al.* [17] proved that assuming the Unique Games Conjecture, there is no polynomial time algorithm to approximate the MAX-2-SAT problem within a factor of $\frac{1}{0.943}$. Notice that $\frac{1}{0.943} > 1.0604 > \frac{22}{21} > 1.0476$. This tighter lower bound and the analysis in the proof of Theorem 13 implies the following tighter lower bound on the inapproximability of the MAX-MLM problem.

Theorem 14. *Assuming the Unique Games Conjecture, there is no polynomial time algorithm to approximate MAXM-MLM(F) within a factor 1.0604 for any given $\prod_m \sum_2 \prod_2$ polynomial F.*

Acknowledgments

We thank Yang Liu and Robbie Schweller for many valuable discussions during our weekly seminar. Bin Fu's research is supported by an NSF CAREER Award, 2009 April 1 to 2014 March 31.

References

1. Agrawal, M., Biswas, S.: Primality and Identity Testing via Chinese Remaindering. Journal of the ACM 50(4), 429–443 (2003)
2. Agrawal, M., Kayal, N., Saxena, N.: PRIMES Is in P. Ann. of Math. 160(2), 781–793 (2004)
3. Arora, S., Lund, C., Motwani, R., Sudan, M., Szegedy, M.: Proof Verification and the Hardness of Approximation Problems. Journal of the ACM 45(3), 501–555 (1998)
4. Aspvall, B., Plass, M.F., Tarjan, R.E.: A Linear-Time Algorithm for Testing the Truth of Certain Quantified Boolean Formulas. Information Processing Letters 8(3), 121–123 (1979)
5. Beigel, R.: The Polynomial Method in Circuit Compplexity. In: Proceedings of the Eighth Conference on Structure in Complexity Theory, pp. 82–95 (1993)
6. Chen, Z., Fowler, R.H., Fu, B., Zhu, B.: On the Inapproximability of the Exemplar Conserved Interval Distance Problem of Genomes. J. Comb. Optim. 15(2), 201–221 (2008)

7. Chen, Z., Fu, B.: The Complexity of Testting Monomials in Multivariate Polynomials. In: Electronic Colloquium on Computational Complexity, ECCC-TR10-114 (2010)

8. Chen, Z., Fu, B., Liu Y., Schweller, R.: Algorithms for Testing Monomials in Multivariate Polynomials. In: Electronic Colloquium on Computational Complexity, ECCC-TR10-122 (2010)

9. Chen, Z., Fu, B., Zhu, B.: The Approximability of the Exemplar Breakpoint Distance Problem. In: Cheng, S.-W., Poon, C.K. (eds.) AAIM 2006. LNCS, vol. 4041, pp. 291–302. Springer, Heidelberg (2006)

10. Chen, J., Lu, S., Sze, S.H., Zhang, F.: Improved Algorithms for Path, Matching, and Packing Problems. In: Proceesings of SODA 2007, pp. 298–307 (2007)

11. Chen, Z.Z., Kao, M.Y.: Reducing Randomness via Irrational Numbers. SIAM J. Comput. 29(4), 1247–1256 (2000)

12. Feige, U., Goldwasser, S., Lovász, L., Safra, S., Szegedy, M.: Interactive Proofs and the Hardness of Approximating Cliques. Journal of the ACM 43(2), 268–292 (1996)

13. Fu, B.: Separating PH from PP by Relativization. Acta Math. Sinica 8(3), 329–336 (1992)

14. Håstad, J.: Some Optimal Inapproximability Results. Journal of the ACM 48(4), 798–859 (2001)

15. Jerrum, M., Sinclaire, A., Vigoda, E.: A Polynomial-Time Appriximation Algorithm for the Permanent of a Matrix with Nonnegative Entries. Journal of the ACM 51(4), 671–697 (2004)

16. Kabanets, V., Impagliazzo, R.: Derandomizing Polynomial Identity Tests Means Proving Circuit Lower Bounds. In: Proceedings of STOC 2003, pp. 355–364 (2003)

17. Khot, S., Kindler, G., Mossel, E., O'Donnell, R.: Optimal Inapproximability Results for MAX-CUT and Other 2-Variable CSPs? In: Proceedings of FOCS 2004, pp. 146–154 (2004)

18. Klivans, A., Servedio, R.A.: Learning DNF in Time $2^{\tilde{O}(n^{1/3})}$. In: Proceedings of STOC 2001, pp. 258–265 (2001)

19. Koutis, I.: Faster Algebraic Algorithms for Path and Packing Problems. In: Aceto, L., Damgård, I., Goldberg, L.A., Halldórsson, M.M., Ingólfsdóttir, A., Walukiewicz, I. (eds.) ICALP 2008, Part I. LNCS, vol. 5125, pp. 575–586. Springer, Heidelberg (2008)

20. Minsky, M., Papert, S.: Perceptrons (Expanded Edition 1988). MIT Press, Cambridge (1968)

21. Raz, R., Shpilka, A.: Deterministic Polynomial Identity Testing in Non-commutative Models. Computational Complexity 14(1), 1–19 (2005)

22. Ryser, H.J.: Combinatorial Mathematics. The Carus Mathematical Monographs, vol. 14. The Mathematical Association of America (1963)

23. Shamir, A.: IP = PSPACE. Journal of the ACM 39(4), 869–877 (1992)

24. Valiant, L.G.: The Complexity of Computing the Permanent. Theoretical Computer Science 8(2), 189–201 (1979)

25. Williams, R.: Finding Paths of Length k in $O^*(2^k)$ Time. Information Processing Letters 109, 315–318 (2009)

26. Zuckerman, D.: Linear Degree Extractors and the Inapproximability of Max Clique and Chromatic Number. Theory of Computing 3, 103–128 (2007)

The Union of Colorful Simplices Spanned by a Colored Point Set

André Schulz[1,*] and Csaba D. Tóth[2,**]

[1] Institut für Mathematsche Logik und Grundlagenforschung, Universität Münster, Germany
andre.schulz@uni-muenster.de
[2] Department of Mathematics and Statistics, University of Calgary, AB, Canada
cdtoth@ucalgary.ca

Abstract. A simplex spanned by a colored point set in Euclidean d-space is *colorful* if all vertices have distinct colors. The union of all full-dimensional colorful simplices spanned by a colored point set is called the *colorful union*. We show that for every $d \in \mathbb{N}$, the maximum combinatorial complexity of the colorful union of n colored points in $\mathbb{R}^d$ is between $\Omega(n^{(d-1)^2})$ and $O(n^{(d-1)^2} \log n)$. For $d = 2$, the upper bound is known to be $O(n)$, and for $d = 3$ we present an upper bound of $O(n^4 \alpha(n))$, where $\alpha(\cdot)$ is the extremely slowly growing inverse Ackermann function. We also prove several structural properties of the colorful union. In particular, we show that the boundary of the colorful union is covered by $O(n^{d-1})$ hyperplanes, and the colorful union is the union of $d+1$ star-shaped polyhedra. These properties lead to efficient data structures for point inclusion queries in the colorful union.

1 Introduction

Given a colored set S of n points in d-dimensional Euclidean space $\mathbb{R}^d$, a simplex is *colorful* if its vertices have pairwise distinct colors. The *simplicial depth* (resp., *colorful simplicial depth*) of a point $p \in \mathbb{R}^d$ is the number of full dimensional closed simplices (resp., colorful simplices) spanned by S and containing p. It is clear that set of all points of positive simplicial depth is the convex hull conv$S$. We call the set of all points of positive *colorful* simplicial depth the *colorful union* and denote it by U_S. It is the union of all colorful simplices, hence it is a polyhedron in $\mathbb{R}^d$. The study of colorful depth was pioneered by Bárány [9], who deduced a lower bound on the maximum simplicial depth by showing that a point lies in a colorful simplex for many random colorings of the point set.

The *colorful linear programming (CLP)* problem was proposed by Bárány and Onn [10]: For a colored set of n points in $\mathbb{R}^d$ and a query point $q \in \mathbb{R}^d$, find a colorful simplex that contains q or report that none exists. An important special case is that q lies in the *core* of the colored points, which is the intersection of the convex hulls of the color classes. In this special case, q is in the colorful union by the colorful Carathédory

---

* Supported by the German Research Foundation (DFG) under grant SCHU 2458/1-1.
** Supported in part by NSERC grant RGPIN 35586. Research by Tóth has been conducted at Tufts University, Medford, MA.

W. Wu and O. Daescu (Eds.): COCOA 2010, Part I, LNCS 6508, pp. 324–338, 2010.
© Springer-Verlag Berlin Heidelberg 2010

theorem [9], and so the CLP is guaranteed to be feasible. This case was thoroughly studied by Bárány and Onn [10,11] and Deza *et al.* [17]. The colorful Carathéodry theorem has recently been strengthened. Arocha *et al.* [7] and Holmsen *et al.* [22] have independently proved that the query point q lies in the colorful union already if q is contained in the convex hull of the union of any two color classes. However, very little in known about the general case that q is an arbitrary point in $\mathbb{R}^d$.

We design efficient data structures for a colored set S of n points in $\mathbb{R}^d$ that supports point inclusion queries for the colorful union U_S. For $d = 2$, it is easy to construct a data structure with $O(\log n)$ query time, $O(n)$ space and $O(n \log n)$ preprocessing time. Boissonnat *et al.* [12] proved that the union U_S of all colorful triangles for a set of n colored points in the plane is a simple polygon with at most $2n - 3$ vertices, which can be computed in $O(n \log n)$ time. Hence, a point location data structure of size $O(n)$ can support point inclusion queries for U_S. For an efficient data structure in higher dimensions one has to understand the combinatorial structure of the colorful union. We present the following results.

1. We show that for every $d \in \mathbb{N}$, the maximum combinatorial complexity of the colorful union for n colored points in $\mathbb{R}^d$ is between $\Omega(n^{(d-1)^2})$ and $O(n^{(d-1)^2} \log n)$ for every $d \in \mathbb{N}$. A tight worst case bound of $\Theta(n)$ has been known for $d = 2$, and we prove a stronger upper bound of $O(n^4 \alpha(n))$ for $d = 3$, where $\alpha(\cdot)$ is the extremely slowly growing inverse Ackermann function.
2. We show that U_S is the union of $d + 1$ star-shaped polyhedra, where the star-centers are the vertices of an arbitrary colorful simplex. This reduces point inclusion queries to ray-shooting queries, and leads to efficient data structures to support point inclusion queries in arbitrary fixed dimension. In particular, in $\mathbb{R}^d$, there is a data structure of size m, $n^{d-1} \leq m \leq n^{(d-1)^2}$, that supports point inclusion queries for U_S in $O(n^{d-1+\varepsilon}/m^{1/(d-1)})$ time for any $\varepsilon > 0$.
3. We show that the colorful union may have undesirable features already in $\mathbb{R}^3$. We construct colored sets S of n points in $\mathbb{R}^3$ with each of the following properties:
   (i) U_S is not star-shaped;
   (ii) U_S has a face whose edges are all adjacent to *reflex* dihedral angles;
   (iii) and the boundary of U_S contains a chain of $\Omega(n)$ reflex vertices.
   On the contrary, U_S is star-shaped and has no two consecutive reflex vertices in $\mathbb{R}^2$.

Related Work. The union of geometric objects in $\mathbb{R}^d$ has applications in constructive solid modeling, motion planning, proximity problems, and conflict-free colorings. We redirect the interested reader to the excellent survey [26] on unions of various geometric objects in Euclidean space. The maximum combinatorial complexity of the union of m full-dimensional simplices in $\mathbb{R}^d$ is $O(m^d)$, this bound is attained if the simplices are flat. Research efforts focused on finding families of simplices whose union have smaller complexity. One example is the family of *fat* simplices, where the dihedral angles of all simplices are at bounded from below by a constant $\delta > 0$, which is the *fatness parameter* of the family. The maximum complexity of the union of m fat triangles in $\mathbb{R}^2$ is known to be between $O(m 2^{\alpha(m)} \log^* m)$ and $\Omega(m \alpha(m))$, where $\alpha(\cdot)$ is the inverse Ackermann function [19,28,32] (the upper bound has recently been improved from $O(m \log \log m)$ [25]). Ezra and Sharir [20] proved that the complexity of m fat tetrahedra in $\mathbb{R}^3$ is $O(m^{2+\varepsilon})$ for every $\varepsilon > 0$. Our result about the complexity of the colorful union is another example: n colored points in $\mathbb{R}^d$ determine $m = O(n^{d+1})$ colorful simplices, yet the complexity of their union is only $O(n^{(d-1)^2} \log n) = O(m^{(d-1)(1-\frac{2}{d+1})} \log m)$.

The colorful simplices in a colored points set in $\mathbb{R}^d$ can be interpreted as a complete multipartite $(d+1)$-uniform geometric hypergraph. Alon and Akiyama [5] studied the number of pairwise disjoint simplices in such a hypergraph, Dey and Pach [15] studied the intersections of hyperedges, related to the higher dimensional analogues of the crossing number. To the best of our knowledge, the combinatorial complexity of the union of colorful simplices in $\mathbb{R}^d$ has not been considered before for dimensions $d \geq 3$.

2   Preliminaries

Let S be a colored set of $n \geq d + 1$ points in $\mathbb{R}^d$. We assume throughout this paper that every $d + 1$ points in S are affinely independent. For $k = 0, 1, \ldots, d$, a k-*simplex* is a subset $P \subseteq S$ of size $k + 1$. A d-simplex in $\mathbb{R}^d$ is also called a *simplex* for short. The convex hull of a subset $P \subseteq S$ is denoted by $\mathrm{conv} P$. For subsets with up to three elements, we use the shorthand notation $p = \{p\}$, $pq = \{p, q\}$, and $pqr = \{p, q, r\}$; if there is no danger of confusion, we also use the same notation for the convex hulls $pq = \mathrm{conv}\{p, q\}$ and $pqr = \mathrm{conv}\{p, q, r\}$. We say that a simplex P *contains* a point set $Q \subset \mathbb{R}^d$ if $Q \subseteq \mathrm{conv} P$.

The colors of the points in S are represented by positive integers. For a single point s, we denote by $\mathrm{color}(s)$ the color of s. For $S' \subseteq S$, we denote by $\mathrm{color}(S') \subset \mathbb{N}$ the set of colors that occur in S'. A k-simplex is *colorful* if its $k + 1$ vertices have pairwise distinct colors. We assume throughout that $|\mathrm{color}(S)| \geq d + 1$, hence there is at least one colorful simplex in S. The *colorful union* of S is the polyhedron

$$U_S = \bigcup \{\mathrm{conv} P : P \subseteq S, |P| = d + 1, \text{ and } |\mathrm{color}(P)| = d + 1\}.$$

The *combinatorial complexity* of a polyhedron U is the number of its k-faces for all $k = 0, 1, \ldots, d$. In $\mathbb{R}^3$, in particular, if a polyhedron U is simply connected, then its 0-, 1-, and 2-faces form a plane graph, and hence the combinatorial complexity of U is proportional to the number of vertices, edges, or faces.

Extremal sets and shells. A point $p \in S$ is *extremal* if there is a bounding halfspace H^- (supported by a hyperplane H that contains p) such that $\mathrm{color}(S \cap H^-) \subseteq \{\mathrm{color}(p)\}$. In general, a k-simplex $P \subset S$ is *extremal* if it is colorful and there is a hyperplane H containing P such that $\mathrm{color}(S \cap H^-) \subseteq \mathrm{color}(P)$, where H^- is again a halfspace bounded by H. For an extremal $(d - 2)$-simplex P, let the *wedge* $W(P)$ be the intersection of the closed halfspaces $\mathrm{cl}(H^-)$ for all hyperplanes H that witness that P is extremal. By definition, all points whose color is not in $\mathrm{color}(P)$ must lie in $W(P)$. The boundary of $W(P)$ consists of two half-hyperplanes, say H_1 and H_2, each containing P and one additional point of S, say $s_1, s_2 \in S$, respectively. The colors of s_1 and s_2 may be the same, but they differ from any color in $\mathrm{color}(P)$. We call the two $(d-1)$-simplices, $P \cup \{s_1\}$ and $P \cup \{s_2\}$, the *shells* of P. We also say that P is the *axis* of these two shells. It is clear that S determines $O(n^{d-1})$ extremal $(d-2)$-simplices, hence there are $O(n^{d-1})$ shells in $\mathbb{R}^2$.

Parity constraints. The following Lemma (without full proof) can be found in [16, Theorem 3.5]. For completeness, we give an easy proof using a variational approach.

Lemma 1. *Let S be a colorful point set in $\mathbb{R}^d$ such that each color class has even cardinality. If a point $q \in \mathbb{R}^d \setminus S$ does not lie on any hyperplane spanned by S, then q is contained in an even number of colorful simplices.*

Proof. Consider a continuous path γ from point q to an arbitrary point r in the exterior of conv$S$ avoiding all affine $(d-1)$-flats spanned by any d points in S. We follow γ from q to r, and keep track of the colorful simplicial depth. The colorful depth of r is zero. The colorful depth changes only if γ crosses a colorful $(d-1)$-simplex P spanned by S. Let $S_0 \subset S$ be the set of all points whose color is missing from color(P). By our assumption, the cardinality of S_0 is even. Denote by H the hyperplane spanned by P. When γ crosses conv$P$ from halfspace H^- to H^+, the colorful simplicial depth changes by $|S_0 \cap H^+| - |S_0 \cap H^-|$, which is even, since $|S_0| = |S_0 \cap H^+| + |S_0 \cap H^-|$ is even. $\square$

We show next that the boundary of the colorful union is covered by shells.

Lemma 2. *Every face of the colorful union U_S is contained in a shell.*

Proof. Let f be a face of the polyhedron U_S. Since U_S is the union
of colorful simplices, f lies on the
boundary of some colorful simplex
$P \subset S$. Hence, f is contained in a
colorful $(d-1)$-simplex $Q \subset P$. Assume that $P = Q \cup \{r\}$, and denote
by H the hyperplane spanned by Q
that induces the two open halfspaces
H^+ and H^-, such that $r \in H^-$. If
there is a point $s \in S \cap H^+$ with
color$(s) \notin$ color(Q), then f would

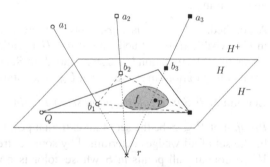

Fig. 1. Construction of Lemma 2

be in the interior of the union of two colorful simplices conv$(Q \cup \{r\}) \cup$conv$(Q \cup \{s\})$. Hence every point in $S \cap H^+$ is colored by some color from color(Q), that is, color$(S \cap H^+) \subseteq$ color(Q).

If color$(S \cap H^+) \subsetneq$ color(Q), then there is a colorful $(d-2)$-simplex $R \subset Q$ such that color$(S \cap H^+) \subseteq$ color(R). This means that R is an extremal $(d-2)$-simplex with axis R, and Q is a shell of R. Hence f is contained in shell Q, as required.

Now assume that color$(S \cap H^+) =$ color(Q). Let $A = \{a_1, \ldots, a_d\} \subset S \cap H^+$ be a set of d points with distinct colors (*i.e.*, color$(A) =$ color(Q)). For each a_i, let $b_i = a_i r \cap H$ (Fig. 2). Let $B = \{b_1, \ldots, b_d\}$, and color each b_i with the color of a_i. In the point set $Q \cup B$, each of the d color classes has cardinality 2. Pick a point p in the interior of face $f \subset H$ that does not lie on any hyperplane spanned by S. The point p is in the interior of the colorful d-simplex Q, so by Lemma 1, it is in the interior of some other colorful d-simplex Q' spanned by $Q \cup B$, which has at least one vertex in B. Let Q'' be the set obtained from Q' by replacing all points $b_i \in Q$ by their corresponding counterparts a_i. It can be easily checked that $Q'' \cup \{r\}$ is colorful and contains p in its interior. This contradicts our assumption that f (and p) are on the boundary of U_S.   □

Visibility within the colorful union. We define *visibility* with respect to the polyhedron U_S. We say that two points $p, q \in U_S$ are visible to each other if the line segment pq is disjoint from the exterior of U_S.

Theorem 1. *Let S be a colored point set in $\mathbb{R}^d$ and let $P \subset S$ be an arbitrary colorful simplex. Then every point in U_S is visible from some vertex of P.*

Proof. Let $q \in U_S$ be a point in the colorful union. If $q \in \mathrm{conv} P$, then any vertex of P sees q. Assume that $q \notin \mathrm{conv} P$. Since $q \in U_S$, there is a colorful simplex $Q \subseteq S$ that contains q. If P and Q have a common vertex, then it obviously sees q. Assume that $P \cap Q = \emptyset$. Note that P and Q are each colorful, but they do not necessarily have the same $d + 1$ colors. Successively pick a point in P and a point in Q whose colors are unique in $P \cup Q$, and recolor both points to a new color. The recoloring ensures that both P and Q remain colorful, and $P \cup Q$ has $d + 1$ colors. Clearly, if a simplex $R \subset P \cup Q$ is colorful in the new colors, then it was colorful in the original colors, too. Now $P \cup Q$ is a colored point set where every color class has size 2. Point p is contained in the colorful simplex $Q \subset P \cup Q$. By Lemma 1, p is contained in some other colorful simplex $R \subset P \cup Q$. Then R must have at least one common vertex with P, which is visible from p.   □

A polyhedron U in $\mathbb{R}^d$ is *star-shaped* if there is a point $p \in U$ such that every point in U is visible from p. Such a point $p \in U$ is called a *star center*. We show that in the plane, the colorful union U_S is star-shaped. In Section 5, however, we construct colored point configurations in $\mathbb{R}^3$ such that U_S is not star-shaped.

Lemma 3. *If S be a colored point set in $\mathbb{R}^2$, then U_S is star-shaped.*

Proof. Let $P \subseteq S$ be the set of all extremal points in S and let $\mathcal{W} = \{W(p) : p \in P\}$ be the set of all wedges determined by some extremal point in S. Recall that a wedge $W(p)$ contains all points in S whose color is not $\mathrm{color}(p)$. It follows that any three wedges in $\mathcal{W}$ have a non-empty intersection (in fact, their intersection contains a point of S). By Helly's theorem, all wedges in $\mathcal{W}$ have a common intersection point, say $o \in \mathbb{R}^2$. We show that U_S is star-shaped with star center o. It is enough to show that for every point $q \in U_S$, the line segment oq lies in U_S. Consider an arbitrary point $q \in U_S$, and let e be an arbitrary edge of U_S that intersects the ray $\overrightarrow{oq}$. By Lemma 2, edge e lies on the boundary of a wedge $W(p) \in \mathcal{W}$, where o lies in the interior of $W(p)$. So $\overrightarrow{oq}$ crosses e from the interior to the exterior of U_S. It follows that $\overrightarrow{oq}$ crosses the boundary of U_S at most once, and so $\overrightarrow{oq} \cap U_S$ is a line segment. Since both o and q are in $\overrightarrow{oq} \cap U_S$, segment oq lies in U_S, as required.   □

3   Efficient Data Structures for Point Inclusion Queries

Using Theorem 1, we can build a data structure for point inclusion queries in the colorful union U_S. Let S be a colored set of n points in $\mathbb{R}^d$, and let $G = \{g_1, \ldots, g_{d+1}\} \subset S$

be an arbitrary colorful simplex. For $i = 1, \ldots, d+1$, let $S_i = \{s \in S : \text{color}(s) \neq i\} \cup \{g_i\}$, and let U_i be the colorful union of S_i. Note that g_i is the only point of color i in S_i. By Theorem 1, $U_S = \cup_{i=1}^{d+1} U_i$. That is, for a query point $q \in \mathbb{R}^d$, we have $q \in U_S$ if and only if $q \in U_i$ for some $i = 1, \ldots, d+1$. It is easy to test $q \in U_i$ with a ray shooting query.

Lemma 4. *For every $i = 1, 2 \ldots, d+1$, we have $q \in U_i$ if and only if $q = g_i$ or the ray emitted from g_i in the direction of q passing through q before reaching a shell of $S_i \cup \{g_i\}$.*

Proof. Suppose that $q \neq g_i$. Recall that U_i is star-shaped with star-center g_i. If $q \in U_i$, then the ray $\overrightarrow{g_i q}$ passes trough q before reaching the boundary of U_i, and the boundary of U_i is a shell of S_i by Lemma 2. Conversely, suppose that the ray $\overrightarrow{g_i q}$ hits a shell convΔ of S_i. Since the ray starts from g_i, Δ is spanned by $S_i \setminus \{g_1\}$. If the ray passes though q before hitting convΔ, then q is contained in the colorful simplex $\Delta \cup \{g_i\}$. $\square$

Let T_i denote the set of shells of S_i. Since $|S_i| = O(n)$, we have $|T_i| = O(n^{d-1})$. A ray shooting query for T_i would report the *first* shell hit by a ray, not the *last* one. Nevertheless, the problem can be reduced to vertical ray shooting. If g_i is on the convex hull of S, then a projective transformation can map g_i to infinity such that rays emitted from g_i become vertical rays directed downwards. The *last* shell hit by a vertical downward ray passing through q is the *first* shell hit by a vertical upward ray starting from infinity. If g_i is not on the convex hull, we can partition $\mathbb{R}^d$ into two halfspaces by a hyperplane containing g_i, and build a ray shooting data structure for the set of shells in T_i clipped in each halfspace. The currently available data structures for ray shooting queries among a set of $(d-1)$-simplices in $\mathbb{R}^d$ are based on range spaces of finite VC-dimension, multi-level partition trees, and Megiddo's parametric search technique [2,3,14,29]. Since $O(d)$ vertical ray shooting data structures, each for $O(n^{d-1})$ shells in $\mathbb{R}^d$, can jointly answer a containment query for U_S, we have the following result.

Theorem 2. *For a set of n colored points in $\mathbb{R}^d$, there are data structures for answering point inclusion queries for the colorful union U_S. There is a data structure with $O(n^{(d-1)^2}\alpha(n))$ space and $O(\log n)$ query time. If the available space is reduced to m, $n^{d-1} \leq m \leq n^{(d-1)^2}$, then the query time increases to $O(n^{d-1+\varepsilon}/m^{1/(d-1)})$ for any $\varepsilon > 0$.*

Edelsbrunner [18] (see also [27,30]) proved that the maximum combinatorial complexity of the *upper envelope* of m possibly intersecting $(d-1)$-simplices in $\mathbb{R}^d$ is $\Theta(m^{d-1}\alpha(m))$. Since $U_S = \cup_{i=1}^{d+1} U_i$, then U_S is the union of $d+1$ star-shaped polyhedra, each of which has $O(n^{(d-1)^2}\alpha(n))$ combinatorial complexity. This, however, does not imply the same upper bound for the complexity of U_S.

4   The Combinatorial Complexity of the Colorful Union

The *minimum* combinatorial complexity of U_S for $n \geq d+1$ colored points in $\mathbb{R}^d$ is $\Theta(1)$. If the convex hull of S is a colorful simplex, then $U_S = \text{conv} S$ with $d+1$

vertices. If S is in convex position, then the minimum combinatorial complexity of U_S is $\Theta(n)$. This complexity is attained for a colored point set constructed recursively as follows. Start with the $d + 1$ vertices of a colorful simplex. In each step, choose an arbitrary (colorful) face Δ of the current convex hull, place a new point near the center of Δ in the exterior of the convex hull, and color it with a color that does not occur in color(Δ). In the remainder of this section, we present lower and upper bounds for the *maximum* combinatorial complexity of the colorful union of n colored points in $\mathbb{R}^d$.

4.1 Lower Bounds for the Maximum Combinatorial Complexity

Theorem 3. *For every integer $d \geq 2$, there are $(d + 1)$-colored point sets in $\mathbb{R}^d$ of size $n \geq d + 1$ such that the combinatorial complexity of the colorful union is $\Omega(n^{(d-1)^2})$.*

Proof. Let $d \geq 2$ be a fixed positive integer. For every $n \geq d + 1$, we construct a set S of n points of $d + 1$ colors in $\mathbb{R}^d$. We have one point of color d and $d + 1$ each. Let a (resp., b) be the point of color d (resp., $d + 1$) on the x_d-axis at N (resp., $N + 1$), for a sufficiently large N to be specified later.

The remaining $n - 2$ points are evenly distributed in the first $d - 1$ color classes. We construct the position of these point in three steps. Let $\mathbb{R}^{d-1}$ denote the subspace of $\mathbb{R}^d$ spanned by the first $d - 1$ coordinate axes. **Step 1.** For $i = 1, \ldots, d - 1$, place the points of color i in the interval $(0, 1)$ of the x_i-axis. Let S_1 denote the set of these points. **Step 2.** Perturb each point in S_1 in the subspace $\mathbb{R}^{d-1}$ by a sufficiently small $\delta_1 > 0$ such that the resulting point set S_2 is in general position in $\mathbb{R}^{d-1}$. **Step 3.** Perturb the x_d coordinate of each point in S_2 by a sufficiently small $\delta_2 > 0$ such that the resulting point set S_3 is in general position in $\mathbb{R}^d$. Our point set is $S = S_3 \cup \{a, b\}$.

The points in $S_1 \subset \mathbb{R}^{d-1}$ form $\Theta(n^{d-1})$ colorful $(d-2)$-simplices in $\mathbb{R}^{d-1}$. Since S_1 is in general position, the intersection of any $d - 1$ distinct $(d-2)$-simplices spanned by S_1 is either empty or a single point. Let $\mathcal{M}_1$ denote the $(d-1)$-tuples of colorful $(d-2)$-simplices of S_1 with a non-empty intersection. We have $|\mathcal{M}_1| = \Theta((n^{d-1})^{d-1}) = \Theta(n^{(d-1)^2})$ by the second selection theorem [6,24] which, in turn, follows from the colorful Tverberg theorem [31]. Let $\mathcal{M}_2$ and $\mathcal{M}_3$ denote the corresponding $(d - 1)$-tuples of $(d - 2)$-simplices of S_2 and S_3, respectively. After the first perturbation, the $(d - 1)$-tuples in $\mathcal{M}$ intersect in *distinct* points in $\mathbb{R}^{d-1}$. Denote by v_m the intersection point of a $(d - 1)$-tuple $m \in \mathcal{M}_2$, and let $V_{\mathcal{M}} = \{v_m : m \in \mathcal{M}\}$. Let $\varepsilon > 0$ be the minimum distance between the points in $V_{\mathcal{M}}$ in $\mathbb{R}^{d-1}$. For every point $v_m, m \in \mathcal{M}_2$, let $B_m \subset \mathbb{R}^d$ denote the d-dimensional ball of radius $\varepsilon/3$ centered at v_m. The balls $B_m, m \in \mathcal{M}_\varepsilon$, are pairwise disjoint.

After the second perturbation, S_3 is in general position in R_d, and so no point is contained in $d - 1$ distinct $(d-2)$-simplices. However, each colorful $(d-2)$-simplex in S_3 is extremal, with two almost vertical shells incident to points a and b, respectively. If $\delta_2 > 0$ is sufficiently small, the $d - 1$ pairs of shells whose axes are the $d - 1$ distinct $(d - 2)$-simplices in $m \in \mathcal{M}_3$ intersects in a unique ball B_m. If N is sufficiently large, then the lower-most intersection point of these $d - 1$ pairs of shells is a vertex of U_S. Since the balls B_m are pairwise disjoint, U_S has at least $\Theta(n^{(d-1)^2})$ vertices. $\qquad \square$

4.2 Upper Bounds for the Maximum Combinatorial Complexity

Boissonnat *et al.* [12] showed that the colorful union of a set S of n colored points in the plane is a simple polygon. They also showed polygon U_S has no two consecutive reflex vertices, and the convex vertices are points in S. It follows that U_S has at most $2n$ vertices in $\mathbb{R}^2$. By Lemma 2, the boundary of U_S is contained in $O(n^{d-1})$ shell simplices for every $d \geq 2$. Aronov and Sharir [8,30] proved that the combinatorial complexity of a single cell in the arrangement of m distinct $(d-1)$-simplices in $\mathbb{R}^d$ is $O(m^{d-1} \log m)$. The colorful union U_S has the same combinatorial complexity as the *outer face* in the arrangement of its shells, which is $O((n^{d-1})^{d-1} \log(n^{d-1})) = O(n^{(d-1)^2} \log n)$. We have shown the following.

Theorem 4. *For every $d \geq 2$, the combinatorial complexity of the union of colorful tetrahedra spanned by a set of n colored points in $\mathbb{R}^d$ is $O(n^{(d-1)^2} \log n)$.*

In the remainder of this section, we slightly improve this general bound for $d = 3$.

Theorem 5. *The combinatorial complexity of the union of colorful tetrahedra spanned by a set of n colored points in $\mathbb{R}^3$ is $O(n^4 \alpha(n))$, where $\alpha(\cdot)$ is the inverse of the Ackermann function.*

The proof builds on a the following lemma, which we prove in Section 4.4.

Lemma 5. *If S is a set of n colored points in $\mathbb{R}^3$, then the relative interior of every shell of S contains $O(n^2 \alpha(n))$ vertices of U_S.*

Proof of Theorem 5. It is enough to count the number of vertices of U_S. Every vertex of U_S is incident to at least three faces of U_S, which lie on shell triangles by Lemma 2. If a vertex v lies on the boundary of all incident shell triangles, then $v \in S$. So it is enough to count vertices lying in the (relative) interior of at least one shell triangle. A set of n colored points span $O(n^2)$ colorful edges. So there are $O(n^2)$ extremal colorful edges, hence $O(n^2)$ shells. By Lemma 5, U_S has at most $O(n^2 \alpha(n))$ vertices in the relative interior of each shell triangle. It follows that the total number of vertices is $O(n^4 \alpha(n))$.

4.3 Auxiliary Results in the Plane

Before the proof of Lemma 5, we present auxiliary results in the plane. The proof is deferred to the full version of the paper due to space constraints. We say that a line segment e is *fully visible* from a point $p \in U_S$ in $\mathbb{R}^2$, if $\text{conv}(e \cup \{p\}) \subseteq U_S$.

Lemma 6. *Let S be a colored point set in $\mathbb{R}^2$. Let $s_1, s_2 \in S$ with $\text{color}(s_1) = 1$ and $\text{color}(s_2) = 2$.*

(1) *If an edge e of U_S is not fully visible from both s_1 and s_2, then the points s_1, s_2, and the two endpoints of e are in convex position; both endpoints of e are in S; and e is an edge of $\text{conv} S$.*
(2) *U_S has at most two edges that are not fully visible from both s_1 and s_2.*
(3) *Any other edge e of U_S is incident to some extremal vertex $s \in S$ such that the convex hull of e and vertices $\{s_1, s_2\} \cap W(s)$ lies in U_S.*

4.4   Proof of Lemma 5

We are given a set S of n colored points in $\mathbb{R}^3$. Let $\Delta = \{v_1, v_2, v_3\}$ be a shell triangle with axis $\{v_1, v_2\}$. We want to show that the relative interior of $\text{conv}\Delta$ contains $O(n^2\alpha(n))$ vertices of U_S. Assume without loss of generality that $\text{color}(v_i) = i$ for $i = 1, 2, 3$. Let H be the plane spanned by Δ. As usual H defines two open halfspaces H^+ and H^-. Let H^- be the open halfspace containing all points of colors other than $\{1, 2, 3\}$. By the definition of shells, we have $\text{color}(S \cap H^+) \subseteq \{1, 2\}$. Let $U_\Delta = \text{cl}(\text{conv}\Delta \cap \text{int}(U_S))$ be the restriction of U_S to the triangle $\text{conv}\Delta$. It is enough to show that U_Δ has $O(n^2\alpha(n))$ vertices.

Definition of traces. By Lemma 2, every edge of U_Δ lies in the intersection of $\text{conv}\Delta$ and another shell triangle. Let T denote the set of all shell triangles t such that $\text{conv}\Delta \cap$ $\text{conv}\, t \neq \emptyset$ and $t \neq \Delta$. Since there are $O(n^2)$ shells, we have $|T| = O(n^2)$. We distinguish three *types* of triangles in T. A triangle $t \in T$ is of

type A if exactly one vertex of t is in $S \cap H^+$,
type B if exactly two vertices of t are in $S \cap H^+$ and an axis of t crosses H,
type C if exactly two vertices t are in $S \cap H^+$ and the axis of t is in H^+.

Denote by T_A, T_B, and T_C, respectively, the shell triangles of type A, B, and C. We have $T = T_A \cup T_B \cup T_C$. For every $t \in T$, the line segment $\text{conv}\Delta \cap \text{conv}\, t$ lies in U_Δ, and it may contain several collinear edges of U_Δ. Let $\text{trace}(t)$ be the convex hull of all edges of U_Δ along $\text{conv}\Delta \cap \text{conv}\, t$. We say that $\text{trace}(t)$ is of *type A* (resp., *type B* or *C*) if $t \in T_A$ (resp., T_B or T_C). A $\text{trace}(t)$ is called i-*visible*, for $i = 0, 1, 2$, if U_Δ contains the convex hull of $\text{trace}(t)$ and i vertices of Δ, and i is the maximum such integer. In particular, a i-visible trace is fully visible from i vertices of Δ, where visibility is understood with respect to the polygon U_Δ. If a 0-visible trace can be decomposed into two line segments which are each fully visible from a vertex of Δ, then fix one such decomposition, and call the two segments its *half-traces*. Every half-trace is 1- or 2-visible.

Outline of the proof of Lemma 5. Every vertex of U_Δ lying in the interior of $\text{conv}\Delta$ is at the intersection of two traces. The intersection is either the endpoint of one of the traces or the crossing point of the two traces (*i.e.*, lies in the relative interior of both traces). There are $O(n^2)$ traces, and so U_Δ has $O(n^2)$ vertices at endpoints of traces. It remains to show that U_Δ has $O(n^2\alpha(n))$ vertices at crossings of traces. Proposition 1 below shows that U_Δ has $O(n^2\alpha(n))$ vertices at the crossing of two traces which are fully visible from the *same* vertex of Δ.

For $j = 1, 2, 3$, the convex hull of v_j and a trace or half-trace fully visible from v_j forms a triangle lying in U_Δ. Let D_j denote the union of all $O(n^2)$ such triangles. It is clear that $D_j \subseteq U_\Delta$.

Proposition 1. *For $j = 1, 2, 3$, the set D_j has $O(n^2\alpha(n))$ vertices.*

Proof. By definition, D_j is the union of $O(n^2)$ triangles that lie in $\text{conv}\Delta$ and share vertex v_j. Apply a projective transformation that maps v_j to infinite, and maps the incident edges of Δ to vertical halflines pointing up. Every triangle incident to v_j and lying in $\text{conv}\Delta$ is mapped to a region vertically above a line segment. The number of vertices of D_j is the combinatorial complexity of the lower envelope of these segments. It is known that the lower envelope of $O(n^2)$ line segments has $O(n^2\alpha(n^2)) = O(n^2\alpha(n))$ vertices, which is the maximum length of a Davenport-Schinzel sequence of order 3 over $O(n^2)$ symbols [4].   □

To prove Lemma 5, it remains to bound the number of crossings of pairs of traces which are (1) either fully visible from different vertices of Δ, or (2) one of them is not fully visible from any vertex of Δ. For this, we study the properties of various types of traces in more detail. Some of the proofs are deferred to the full version of the paper.

Traces of type A. For every $p \in S \cap H^+$ and $q \in S \cap \mathrm{cl}(H^-)$, let $v_{pq} = pq \cap H$, that is, the intersection point of segment pq and plane H. For every $p \in S \cap H^+$, let U_p denote the union of all colorful tetrahedra spanned by p and three points in $S \cap \mathrm{cl}(H^-)$. It is clear that $U_p \subseteq U_S$. We also define the planar point set $S(p) = \{v_{pq} : q \in S \cap \mathrm{cl}(H^-), \mathrm{color}(q) \neq \mathrm{color}(p)\}$ and color each point v_{pq} with $\mathrm{color}(q)$. In particular, the three vertices of Δ are in $S(p)$, with their original colors. Observe that $H \cap U_p = U_{S(p)}$.

Proposition 2. *If* $t \in T_A$ *where* p *is the vertex of* t *in* $S \cap H^+$, *then* $\mathrm{trace}(t)$ *is contained in an edge of* U_p.

The proof of Proposition 2 as well as the proofs of the following propositions can be found in the full version of the paper.

By Lemma 6, at most two edges of U_p are not fully visible from any vertex of Δ, and at most one such edge intersects the interior of Δ. Therefore, there is at most one 0-visible trace on the boundary of each U_p.

Consider a segment pq with $p \in S \cap H^+$ and $q \in S \cap \mathrm{cl}(H^-)$. If v_{pq} is an extremal point of $S(p)$, then we define $w_{pq} = W(v_{pq})$ as the minimum wedge in H at apex v_{pq} that contains all points of $S(p)$ of colors different from v_{pq}. Note that w_{pq} might be different from the intersection of a 3-dimensional wedge $W(pq)$, defined for the entire point set S, and the plane H, since $W(pq)$ also contains the points in $S \cap H^+$ whose color is not in $\mathrm{color}(pq)$. By Proposition 2 and Lemma 2, every trace of type A lies on the boundary of some wedge w_{pq}. By Lemma 6(3), we can associate every $\mathrm{trace}(t)$ of type A with an adjacent wedge w_{pq} such that U_Δ contains the convex hull of $\mathrm{trace}(t)$ and all vertices of Δ lying in w_{pq}. Direct segment $\mathrm{trace}(t)$ toward the apex of the associated wedge. By Lemma 6(1), a 0-visible $\mathrm{trace}(t)$ lies on the boundary of two wedges, each of which contains a single vertex of Δ. We may associate a 0-visible trace to either of them, and so a 0-visible trace have two possible directions.

Traces of types B and C. We show that every trace of type B or C is 1- or 2-visible. Therefore, all 0-visible traces are of type A.

Proposition 3. *Every* $\mathrm{trace}(t)$ *of type B is*

– *2-visible, or*
– *1-visible and lies on the boundary of a wedge* w *such that (i) the apex of* w *lies on the axis of* t, *(ii)* $w \cap \mathrm{conv}\Delta \subset U_\Delta$, *and (iii)* w *contains either* v_1 *or* v_2.

For every extremal segment pr where $p \in H^+$ and $r \in H^-$, with $\mathrm{color}(r) = 3$, there is at most one wedge w described in Proposition 3. Note that if pr is an extremal segment in S, then v_{pr} is an extremal point in $S(p)$, and so wedge w_{pr} is defined. The wedges w_{pr} and w have the same apex, they both contain the same vertex of Δ. If both w_{pr} and w exist, let $\widehat{w}_{pr}$ be their union, otherwise let $\widehat{w}_{pr} = w_{pr}$. Then $\widehat{w}_{pr}$ is a wedge with apex v_{pr}, it contains only one vertex of Δ, and its boundary contains all traces lying on the boundaries of w_{pr} and w. Direct the traces on the boundary of $\widehat{w}_{pr}$ towards v_{pr}.

Proposition 4. *Every trace of type C is*

- *2-visible, or*
- *1-visible and lies on the boundary of a halfplane h^- such that $h^- \cap \text{conv}\Delta \subset U_\Delta$.*

2-visible traces. We show that 2-visible traces are incident to at most $O(n^2\alpha(n))$ vertices of U_Δ.

Proposition 5. *Let $t_1, t_2 \in T$ such that their traces intersect at point x and $\text{trace}(t_1)$ be 2-visible.*

(a) *If $\text{trace}(t_2)$ is 2-visible, then both traces are fully visible from a vertex of Δ.*
(b) *If $\text{trace}(t_2)$ is 1-visible, then both traces are fully visible from a vertex of Δ; or x is an endpoint of $\text{trace}(t_1)$ or $\text{trace}(t_2)$.*
(c) *If $\text{trace}(t_2)$ is 1-visible, then x is an endpoint of $\text{trace}(t_2)$; or $\text{trace}(t_2)$ can be decomposed into two half-traces, each fully visible from some vertices of Δ.*

It follows that all vertices of U_Δ that lie on some 2-visible trace must be an endpoint of a trace or a vertex of D_j for some $j \in \{1, 2, 3\}$. Observe also that if $\text{trace}(t)$ lies on the boundary of a halfplane h^- with $h^- \cap \text{conv}\Delta \subset U_\Delta$, then $\text{trace}(t)$ does not cross any other traces. We conclude that 2-visible traces and all traces of type C are involved in $O(n^2\alpha(n))$ vertices of U_Δ. It remains to consider the vertices of U_Δ at the crossings of 0- and 1-visible traces of type A or B. These traces lie on the boundaries of some wedges $\widehat{w}_{pq}$. Let X denote the set of vertices of U_Δ at crossings of 0- or 1-visible traces lying on the boundaries of wedges $\widehat{w}_{pq}$. It remains to show that $|X| = O(n^2\alpha(n))$.

Directed traces. Each vertex $x \in X$ is incident to two directed edges of U_Δ. Since 0-visible traces may have two possible directions, we define the *in-degree* of $x \in X$ as the number of ingoing edges along 1-visible traces. Let $Y \subseteq X$ be the set of vertices of in-degree 1 or 2. The following proposition implies that it is enough to prove that $|Y| = O(n^2\alpha(n))$.

Proposition 6. *We have $|X| \le O(|Y| + n^2\alpha(n))$.*

Crossing wedges. Consider a vertex $x \in Y$ of U_Δ. It is at the crossing of some 0- or 1-visible traces $\text{trace}(t_1)$ and $\text{trace}(t_2)$. Since the in-degree of x is 1 or 2, we may assume that $\text{trace}(t_1)$ is 1-visible and directed towards x, and $\text{trace}(t_2)$ is 0- or 1-visible directed arbitrarily. We also know that $\text{trace}(t_1)$ and $\text{trace}(t_2)$ lie on the boundaries of some wedges $\widehat{w}_{pr}$ and $\widehat{w}_{qs}$, respectively, each of which contains exactly one vertex of Δ. If $\text{trace}(t_2)$ is 1-visible, then $\widehat{w}_{pr}$ and $\widehat{w}_{qs}$ contain distinct vertices of Δ, otherwise the two traces would be visible from the same vertex of Δ. If $\text{trace}(t_2)$ is 0-visible, then it lies on the boundary of two wedges, each of which contains a distinct vertex of Δ. We may choose $\widehat{w}_{qs}$ such that $\widehat{w}_{pr}$ and $\widehat{w}_{qs}$ contain different vertices of Δ. The following proposition restricts how wedges $\widehat{w}_{pr}$ and $\widehat{w}_{qs}$ can intersect.

Proposition 7. *Let $p, q \in H^+$, and $r, s \in \mathrm{cl}(H^-)$ such that v_{pr} and v_{qs} are extremal in $S(p)$ and $S(q)$. Suppose that wedges $\widehat{w}_{pr}$ and $\widehat{w}_{qs}$ each contain only one vertex of Δ, denoted v_i and v_j, $i \neq j$, respectively. If two traces on the boundaries of $\widehat{w}_{pr}$ and $\widehat{w}_{qs}$ cross at point $x \in Y$, then we have $v_{pr} \in \widehat{w}_{qs}$ or $v_{qs} \in \widehat{w}_{pr}$.* $\square$

We return to study a crossing $x \in Y$ of traces $\mathrm{trace}(t_1)$ and $\mathrm{trace}(t_2)$, lying on the boundaries of wedges $\widehat{w}_{pr}$ and $\widehat{w}_{qs}$, respectively. We have assumed $\widehat{w}_{pr}$ and $\widehat{w}_{qs}$ each contain distinct vertices of Δ, say v_i and v_j, $i \neq j$, respectively. We also assumed that $\mathrm{trace}(t_1)$ enters the interior of $\widehat{w}_{qs}$ at x. By Proposition 7, we may assume that $\widehat{w}_{qs}$ contains the apex of $\widehat{w}_{pr}$, and so $\mathrm{trace}(t_1)$ remains in the interior of $\widehat{w}_{qs}$. However, x is not an endpoint of $\mathrm{trace}(t_1)$, it has to reach the boundary of U_Δ again, at its neighbor $x' \in \mathrm{trace}(t_1)$. Since $(\widehat{w}_{qs} \setminus w_{qs}) \cap \mathrm{conv}\Delta \subset U_\Delta$, $\mathrm{trace}(t_1)$ also enters wedge w_{qs}, and the polygon U_q, where $U_q \subset U_\Delta$. By Lemma 6, $\mathrm{trace}(t_1)$ can exit U_q through an edge of U_q fully visible from v_i. However, $\mathrm{trace}(t_i)$ is also fully visible from v_i. Therefore, segment $xx' \subset \mathrm{trace}(t_1)$ has to intersect the boundary of D_i at some vertex of D_i. We charge vertex x to this vertex of D_3 (even if it lies in the interior of U_Δ. Every vertex of $D_1, D_2,$ and D_3 is charged at most twice. By Proposition 1, we have $|Y| = O(n^2\alpha(n))$, as required. This completes the proof of Lemma 5.

5 Colored Point Configurations with Undesirable Features

A point configuration whose colorful union is not star-shaped. In contrast to Theorem 1 one (arbitrary) point is not sufficient to guard the interior of S. There are colored point sets in $\mathbb{R}^3$ whose colorful union is not star-shaped. One example is shown in Table 1 and Fig. 2(a).

Table 1. The coordinates of a point set whose colorful union is not star shaped

vertex	x	y	z	color		vertex	x	y	z	color
p_1	−0.9	−1	1	1		p_5	−1	−0.9	−1	3
p_2	−0.9	1	1	1		p_6	1	−0.9	−1	3
p_3	0.9	−1	1	2		p_7	−1	0.9	−1	4
p_4	0.9	1	1	2		p_8	1	0.9	−1	4

In the point set in Fig. 2, axes $p_1 p_3$ and $p_2 p_4$ have color $\{1, 2\}$. Every point in U_S that sees the relative interior of $p_1 p_3$ must lie in wedge $W(p_1, p_3)$. Similarly, the relative interior of $p_2 p_4$ can only be seen from wedge $W(p_2, p_4)$. Thus any point in U_S that sees both axes must lie in $W(p_2, p_4) \cap W(p_1, p_3)$. The intersection of these two wedges, however, is strictly below the xy-plane. A superset of the intersection is depicted in Fig. 2(b). Similarly, the relative interior of $p_5 p_7$ and $p_6 p_8$ is visible from the intersection of two wedges, which is a region strictly above the xy-plane (Fig. 2(b)). It follows that no point can see all these four edges of U_S.

A face bounded by edges with reflex dihedral angles. In the plane, every face of U_S is incident to a point in S (c.f. [12]). This property immediately implies that the combinatorial complexity of U_S is $O(n)$ in $\mathbb{R}^2$. In $\mathbb{R}^3$, however, there are colored point sets S such that a face of U_S is not incident to any colorful edge. See Fig. 3 for an example.

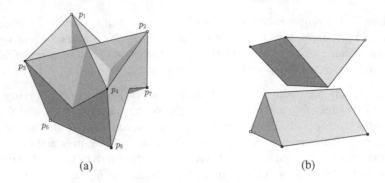

Fig. 2. (a) The colorful union of the point set S. (b) A superset of the visibility regions.

vertex	x	y	z	color
p_1	7	−10	−3	1
p_2	8	−9	0	1
p_3	0	−14	10	2
p_4	9	−15	6	2
p_5	4	−13	0	2
p_6	2	−10	−1	3
p_7	4	−6	0	3
p_8	5	−6	−4	4

Fig. 3. An example where a face is not incident to any axis

A chain of reflex vertices. In the plane, U_S has no two adjacent reflex vertices [12]. This no longer true in $\mathbb{R}^3$. Fig. 4 indicates a family of points sets in $\mathbb{R}^3$ where the boundary of a face may contain an arbitrary long chain of reflex vertices. Points of colors 1 and 2 are arranged along two parallel lines in the xy-plane, such that the complete bipartite graph between the first two color classes forms a convex chain of length $\Omega(n)$ (Fig. 4, left). Two points of color 3 and 4 are placed below the xy-plane on opposite sides of all vertical planes through edges of color $\{1, 2\}$. A small perturbation can make any 4 points affine independent.

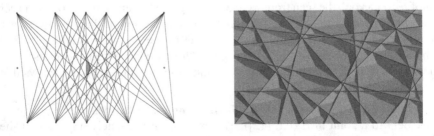

Fig. 4. A chain of reflex vertices of U_S along the boundary of a single face

6 Open Problems

We have tightened the gap between the lower and upper bound for the combinatorial complexity of the colorful union in $\mathbb{R}^3$. We do not know whether the term corresponding to the Ackerman function is necessary. It is an obvious open problem to simplify and extend our results to higher dimensions.

We have transformed inclusion queries for U_S into vertical ray shooting queries, which lead to the data structure proposed in Theorem 2. We build a ray shooting data structure for shell $(d-1)$-simplces spanned by the point set, that is, the ray shooting data structure ignores the fact the simplices are spanned by a *ground set* of only n points, and it also ignores the colors. It remains an interesting open question whether these two structural properties can be exploited to design a more efficient data structure.

Depth queries are a more general form of inclusion queries. Little is known about general colorful simplicial depth queries and it is desirable to come up with efficient data structures for this problem. For the monochromatic case, the simplicial depth can be computed in $\mathbb{R}^2$ in $O(n \log n)$ time [21,23], and in $\mathbb{R}^3$ in $O(n^2)$ time [13,23]. In higher dimensions no better strategy than the trivial $O(n^{1+d})$ brute force test seems to be known. Afshani and Chang [1] proposed a data structure for approximate simplicial depth queries. There are no similar approximate results for the colorful simplicial depth.

Acknowledgements. We thank Kushan Ahmadian, Boris Aronov, Gill Barequet, Marina Gavrilova, Andreas Holmsen, Mashhood Ishaque, Natan Rubin, Diane Souvaine, and Andrew Winslow for helpful conversations about these topics.

References

1. Afshani, P., Chan, T.M.: On approximate range counting and depth. Discrete Comput. Geom. 42(1), 3–21 (2009)
2. Agarwal, P.K.: Range searching. In: Goodman, J., O'Rourke, J. (eds.) Handbook of Discrete and Computational Geometry, ch. 36, pp. 809–838. CRC Press, Boca Raton (2004)
3. Agarwal, P.K., Matoušek, J.: Ray shooting and parametric search. SIAM J. Comput. 22, 794–806 (1993)
4. Agarwal, P.K., Sharir, M.: Davenport-Schinzel sequences and their geometric applications. In: Sack, J.-R., Urrutia, J. (eds.) Handbook of Computational Geometry, ch. 1, pp. 1–47. Elsevier, Amsterdam (2000)
5. Akiyama, J., Alon, N.: Disjoint simplices and geometric hypergraphs. In: Proceedings of the Third International Conference on Combinatorial Mathematics, pp. 1–3. Academy of Sciences, New York (1989)
6. Alon, N., Bárány, I., Füredi, Z., Kleitman, D.J.: Point selections and weak ε-nets for convex hulls. Combin. Probab. Comput. 1(3), 189–200 (1992)
7. Arocha, J.L., Bárány, I., Bracho, J., Fabila, R., Montejano, L.: Very colorful theorems. Discrete Comput. Geom. 42(2), 142–154 (2009)
8. Aronov, B., Sharir, M.: Castles in the air revisited. Discrete Comput. Geom. 12(1), 119–150 (1994)
9. Bárány, I.: A generalization of Carathéodory's theorem. Discrete Math. 40(2-3), 141–152 (1982)
10. Bárány, I., Onn, S.: Carathéodory's theorem, colourful and applicable. Bolyai Soc. Math. Stud., János Bolyai Math. Soc., Budapest 6, 11–21 (1997)

11. Bárány, I., Onn, S.: Colourful linear programming and its relatives. Math. Oper. Res. 22(3), 550–567 (1997)
12. Boissonnat, J.-D., Devillers, O., Preparata, F.P.: Computing the union of 3-colored triangles. Intern. J. Comput. Geom. Appl. 1, 187–196 (1991)
13. Cheng, A.Y., Ouyang, M.: On algorithms for simplicial depth. In: Canadian Conf. Comput. Geom., pp. 53–56 (2001)
14. de Berg, M.: Ray Shooting, Depth Orders and Hidden Surface Removal. LNCS, vol. 703. Springer, Heidelberg (1993); ch. 5, Ray shooting from a fixed point, pp. 53–65
15. Dey, T.K., Pach, J.: Extremal problems for geometric hypergraphs. Discrete Comput. Geom. 19, 473–484 (1998)
16. Deza, A., Huang, S., Stephen, T., Terlaky, T.: Colourful simplicial depth. Discrete Comput. Geom. 35(4), 597–615 (2006)
17. Deza, A., Huang, S., Stephen, T., Terlaky, T.: The colourful feasibility problem. Discrete Appl. Math. 156(11), 2166–2177 (2008)
18. Edelsbrunner, H.: The upper envelope of piecewise linear functions: Tight bounds on the number of faces. Discrete Comput. Geom. 4(1), 337–343 (1989)
19. Ezra, E., Aronov, B., Sharir, M.: Improved bound for the union of fat tetrahedra in three dimensions. In: Proc. 22nd Annual ACM-SIAM Symposium on Discrete Algorithms (SODA). ACM press, New York (to appear, 2011)
20. Ezra, E., Sharir, M.: Almost tight bound for the union of fat tetrahedra in three dimensions. In: Proc. 48th Sympos. on Foundations of Comp. Sci. (FOCS), pp. 525–535. IEEE, Los Alamitos (2007)
21. Gil, J., Steiger, W., Wigderson, A.: Geometric medians. Discrete Math. 108(1-3), 37–51 (1992)
22. Holmsen, A.F., Pach, J., Tverberg, H.: Points surrounding the origin. Combinatorica 28(6), 633–644 (2008)
23. Khuller, S., Mitchell, J.S.B.: On a triangle counting problem. Inf. Process. Lett. 33(6), 319–321 (1990)
24. Matoušek, J.: Lectures on Discrete Geometry. Graduate Texts in Math, vol. 212. Springer, Heidelberg (2002)
25. Matoušek, J., Pach, J., Sharir, M., Sifrony, S., Welzl, E.: Fat triangles determine linearly many holes. SIAM J. Comput. 23, 154–169 (1994)
26. Pach, J., Safruti, I., Sharir, M.: The union of congruent cubes in three dimensions. Discrete Comput. Geom. 30, 133–160 (2003)
27. Pach, J., Sharir, M.: The upper envelope of piecewise linear functions and the boundary of a region enclosed by convex plates: Combinatorial analysis. Discrete Comput. Geom. 4, 291–310 (1989)
28. Pach, J., Tardos, G.: On the boundary complexity of the union of fat triangles. SIAM J. Comput. 31, 1745–1760 (2002)
29. Pellegrini, M.: Ray shooting and lines in space. In: Goodman, J., O'Rourke, J. (eds.) Handbook of Discrete and Computational Geometry, ch. 36, pp. 809–838. CRC Press, Boca Raton (2004)
30. Tagansky, B.: A new technique for analyzing substructures in arrangements of piecewise linear surfaces. Discrete Comput. Geom. 16(4), 455–479 (1996)
31. Živaljević, R.T., Vrećica, S.T.: The colored Tverberg's problem and complexes of injective functions. J. Combin. Theory Ser. A 61(2), 309–318 (1992)
32. Wiernik, A., Sharir, M.: Planar realization of nonlinear Davenport Schinzel sequences by segments. Discrete Comput. Geom. 3, 15–47 (1988)

Compact Visibility Representation of 4-Connected Plane Graphs

Xin He, Jiun-Jie Wang[1],[*], and Huaming Zhang[2],[**]

[1] Department of Computer Science and Engineering
University at Buffalo, Buffalo, NY, 14260, USA
{xinhe,jiunjiew}@buffalo.edu
[2] Department of Computer Science
University of Alabama in Huntsville, Huntsville, AL, 35899, USA
hzhang@cs.uah.edu

Abstract. The *visibility representation* (VR for short) is a classical representation of plane graphs. The VR has various applications and has been extensively studied. A main focus of the study is to minimize the size of the VR. It is known that there exists a plane graph G with n vertices where any VR of G requires a size at least $\lfloor \frac{2n}{3} \rfloor \times (\lfloor \frac{4n}{3} \rfloor - 3)$. For upper bounds, it is known that every plane graph has a VR with size at most $\lfloor \frac{2}{3}n \rfloor \times (2n - 5)$, and a VR with size at most $(n - 1) \times \lfloor \frac{4}{3}n \rfloor$.

It has been an open problem to find a VR with both height and width simultaneously bounded away from the trivial upper bounds (namely of size $c_h n \times c_w n$ with $c_h < 1$ and $c_w < 2$). In this paper, we provide the first VR construction for a non-trivial graph class that simultaneously bounds both the height and the width. We prove that every 4-connected plane graph has a VR with height $\leq \frac{3n}{4} + 2\lceil \sqrt{n} \rceil + 4$ and width $\leq \lceil \frac{3n}{2} \rceil$. Our VR algorithm is based on an st-orientation of 4-connected plane graphs with special properties. Since the st-orientation is a very useful concept in other applications, this result may be of independent interests.

1 Introduction

Drawing plane graphs has emerged as a fast growing research area in recent years (see [1] for a survey). A *visibility representation* (VR for short) of a plane graph G is a drawing of G, where the vertices of G are represented by non-overlapping horizontal line segments (with integer end point coordinates), and each edge of G is represented by a vertical line segment touching the segments of its end vertices. Figure 1 shows a VR of a plane graph G. The problem of finding a compact VR is important not only in algorithmic graph theory, but also in practical applications. A simple linear time VR algorithm was given in [12,13] for 2-connected plane graphs. It uses an *st-orientation* of G and the corresponding st-orientation of its st-dual G^* to construct VR. Using this approach, the height of the VR is $\leq (n - 1)$ and the width of the VR is $\leq (2n - 5)$ [12,13].

---

[*] Research supported in part by NSF Grant CCR-0635104.
[**] Research supported in part by NSF Grant CCR-0728830.

W. Wu and O. Daescu (Eds.): COCOA 2010, Part I, LNCS 6508, pp. 339–353, 2010.
© Springer-Verlag Berlin Heidelberg 2010

As in many other graph drawing problems, one of the main concerns in VR research is to minimize the size of the representation. For the lower bounds, it was shown in [15] that there exists a plane graph G with n vertices where any VR of G requires a size at least $\lfloor \frac{2n}{3} \rfloor \times (\lfloor \frac{4n}{3} \rfloor - 3)$. Several papers have been published to reduce the height and width of the VR by carefully constructing special st-orientations. The following table summarizes related previous results.

The VR of plane graph G	The VR of 4-Connected plane graph G
Width $\leq (2n - 5)$ [12,13]	
Height $\leq (n - 1)$ [12,13]	
Width $\leq \lfloor \frac{3n-6}{2} \rfloor$ [5]	
Width $\leq \lfloor \frac{22n-42}{15} \rfloor$ [8]	Width $\leq (n - 1)$ [6]
Height $\leq \lfloor \frac{5n}{6} \rfloor$ [15]	
Width $\leq \lfloor \frac{13n-24}{9} \rfloor$ [16]	Height $\leq \lceil \frac{3n}{4} \rceil$ [14]
Height $\leq \lfloor \frac{4n-1}{5} \rfloor$ [17]	
Height $\leq \frac{2n}{3} + \lfloor 2\sqrt{n} \rfloor$ [4]	
Height $\leq \frac{2n}{3} + O(1)$ [18]	
Width $\leq \lfloor \frac{4n}{3} \rfloor - 2$ [3]	Height $\leq \lceil \frac{n}{2} \rceil + 2\lceil \sqrt{\frac{n-2}{2}} \rceil$ [2]

All these results concentrated on one dimension of the VR only. In the table above, the un-mentioned dimension is bounded by the trivial upper bound ($n-1$ for the height and $2n - 5$ for the width). In [10,11], heuristic algorithms were developed aiming at reducing the height and the width of VR simultaneously. It has been illusive to find a VR with both height and width simultaneously bounded away from the trivial upper bounds. In this paper, we prove that every 4-connected plane graph of n vertices has a VR with height $\leq \frac{3n}{4} + 2\lceil \sqrt{n} \rceil + 4$ and width $\leq \lceil \frac{3n}{2} \rceil$. The representation can be constructed in linear time.

The present paper is organized as follows. §2 introduces preliminaries. §3 presents the construction of the VR with the stated height and width bounds. §4 concludes the paper.

2   Preliminaries

In this section, we give definitions and preliminary results. Definitions not mentioned here are standard. A *planar graph* is a graph G such that the vertices can be drawn in the plane and the edges can be drawn as non-intersecting curves. Such a drawing is called a *plane embedding*. The drawing divides the plane into a number of connected regions. Each region is called a *face*. The unbounded face is the *exterior face*. Other faces are *interior faces*. A *plane graph* is a planar graph with a fixed embedding. A *plane triangulation* is a plane graph where every face is a triangle (including the exterior face). We abbreviate the words "counterclockwise" and "clockwise" as ccw and cw respectively.

When discussing VR, we assume G is a plane triangulation. (If not, we get a plane triangulation G' by adding dummy edges into G. After constructing a VR for G', a VR of G is obtained by deleting the vertical line segments for the dummy edges.)

A *numbering* $\mathcal{O}$ of a set $S = \{a_1, \ldots, a_k\}$ is a 1-1 mapping between S and the set $\{1, 2, \ldots, k\}$. We write $\mathcal{O} = \langle a_{i_1}, a_{i_2}, \ldots, a_{i_k} \rangle$ to indicate $\mathcal{O}(a_{i_1}) = 1$, $\mathcal{O}(a_{i_2}) = 2 \ldots$ A set S written this way is called an *ordered list*. For two elements a_i and a_j, if a_i is assigned a smaller number than a_j in $\mathcal{O}$, we write $a_i \prec_{\mathcal{O}} a_j$. Let S_1 and S_2 be two disjoint sets. If $\mathcal{O}_1$ is a numbering of S_1 and $\mathcal{O}_2$ is a numbering of S_2, their concatenation $\mathcal{O} = \langle \mathcal{O}_1, \mathcal{O}_2 \rangle$ is the numbering of $S_1 \cup S_2$ where $\mathcal{O}(x) = \mathcal{O}_1(x)$ for all $x \in S_1$ and $\mathcal{O}(y) = \mathcal{O}_2(y) + |S_1|$ for all $y \in S_2$.

An *orientation* of a (undirected) graph G is a digraph obtained from G by assigning a direction to each edge of G. Let $G = (V, E)$ be an undirected graph. A numbering $\mathcal{O}$ of V induces an orientation of G as follows: each edge of G is directed from its lower numbered end vertex to its higher numbered end vertex. The resulting digraph, denoted by $G_{\mathcal{O}}$, is called the *orientation derived from* $\mathcal{O}$ which, obviously, is acyclic. We use length($G_{\mathcal{O}}$) (or simply length($\mathcal{O}$) if G is clear from the context) to denote the length of the longest path in $G_{\mathcal{O}}$. (The length of a path is the number of edges in it.)

Let G be a 2-connected plane graph with an exterior edge (s, t). An orientation of G is called an *st-orientation* if the resulting digraph is acyclic with s as the only source and t as the only sink (also called an *st-graph*). For every 2-connected plane graph G and an exterior edge (s, t), there exists an *st*-orientation [7]. Properties of the *st*-orientations can be found in [9].

Let G be a 2-connected plane graph and (s, t) an exterior edge. An *st-numbering* of G is a one-to-one mapping $\xi : V \rightarrow \{1, 2, \ldots, n\}$, such that $\xi(s) = 1$, $\xi(t) = n$, and each vertex $v \neq s, t$ has two neighbors u, w with $\xi(u) < \xi(v) < \xi(w)$. Given an *st*-numbering ξ of G, the orientation of G derived from ξ is obviously an *st*-orientation of G. On the other hand, if $G = (V, E)$ has an *st*-orientation $\mathcal{O}$, we can define an 1-1 mapping $\xi : V \rightarrow \{1, \ldots, n\}$ by topological sort of $G_{\mathcal{O}}$. It is easy to see that ξ is an *st*-numbering and the orientation derived from ξ is $\mathcal{O}$. From now on, we will interchangeably use the term "an *st*-numbering" of G and the term "an *st*-orientation" of G.

Definition 1. Let G be a plane graph with an *st*-orientation $\mathcal{O}$, where (s, t) is an exterior edge drawn at the left on the exterior face of G. The *st-dual* graph G^* of G and the dual orientation $\mathcal{O}^*$ of $\mathcal{O}$ is defined as follows:

- Each face f of G corresponds to a node f^* of G^*. The unique interior face adjacent to the edge (s, t) corresponds to a node s^* in G^*, the exterior face corresponds to a node t^* in G^*.
- For each edge $e \neq (s, t)$ of G separating a face f_1 on its left and a face f_2 on its right, there is a dual edge e^* in G^* from f_1^* to f_2^*.
- The dual edge of the exterior edge (s, t) is directed from s^* to t^*.

Fig 1 (1) shows an *st*-graph G and its *st*-dual G^* (where circles and solid lines denote the vertices and the edges of G; squares and dashed lines denote the nodes and the edges of G^*). It is well known that the *st*-dual graph G^* defined above is an *st*-graph with source s^* and sink t^*. The following theorem was proved in [12,13]:

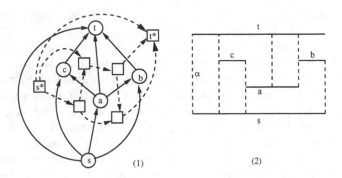

Fig. 1. (1) An st-graph G and its st-dual graph G^*; (2) A VR of G

Theorem 1. *Let G be a 2-connected plane graph with an st-orientation $\mathcal{O}$. Let $\mathcal{O}^*$ be the dual st-orientation of G^*. A VR of G can be obtained from $\mathcal{O}$ in linear time. The height of the VR is* length$(\mathcal{O}) \leq n - 1$. *The width of the VR is* length$(\mathcal{O}^*) \leq 2n - 5$ *(which is the number of nodes in G^*).*

Figure 1 (2) shows a VR of the graph G shown in Figure 1 (1). The width of the VR is length$(\mathcal{O}^*) = 5$. The height of the VR is length$(\mathcal{O}) = 3$.

Definition 2. A 4TP graph is a plane graph G satisfying the following two conditions: (1) G is 4-connected; (2) Every interior face of G is a triangle and the exterior face is a quadrangle.

The four exterior vertices of a 4TP graph will be denoted by v_S, v_W, v_N, v_E in cw order. Let H be a 4-connected plane triangulation and $e = (s,t)$ an exterior edge of G. If we delete e from H, the resulting graph $G = H - \{e\}$ is a 4TP graph. We label the exterior vertices of G so that $s = v_S$ and $t = v_N$. Our algorithm will construct a VR $\mathcal{D}$ of G so that the line segment l_s for s has the smallest y-coordinate, and the line segment l_t for t has the largest y-coordinate. From $\mathcal{D}$, we can obtain a VR $\mathcal{D}'$ of H as follows: Extend both l_s and l_t to the left by one unit, then add a vertical line segment α connecting them (see Figure 1 (2).) This operation does not change the height of $\mathcal{D}$ and increases the width of $\mathcal{D}$ by 1. From now on we will consider 4TP graphs only.

Definition 3. A *regular edge labeling* (REL for short) of a 4TP graph $G = (V, E)$ is a partition and orientation of the interior edges of G into two subsets E_{green}, E_{red} of directed edges such that the following hold:

1. For each interior vertex v, the edges incident to v appear in cw order around v as follows: a set of edges in E_{red} leaving v; a set of edges in E_{green} entering v; a set of edges in E_{red} entering v; a set of edges in E_{green} leaving v.
2. All interior edges incident to v_N are in E_{red} and entering v_N. All interior edges incident to v_W are in E_{green} and entering v_W. All interior edges incident to v_S are in E_{red} and leaving v_S. All interior edges incident to v_E are in E_{green} and leaving v_E.

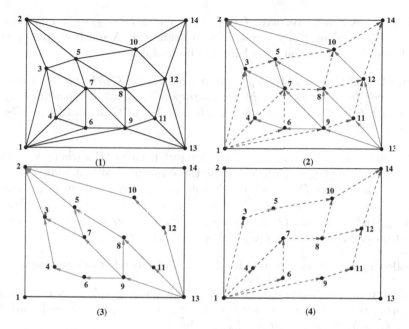

Fig. 2. (1) A 4TP graph G; (2) A REL $\mathcal{R}$ of G; (3) G_{green}; (4) G_{red}

Figure 2 (2) shows a REL of a 4TP graph G shown in Figure 2 (1). The green and red edges are drawn as solid and dashed lines respectively. Here $v_S = 1, v_W = 2, v_N = 14, v_E = 13$. (These 4 exterior vertices will always be drawn at the lower-left, upper-left, upper-right and lower-right corners, respectively). It was shown in [6] that every 4TP graph has a REL, constructible in linear time.

Let $\mathcal{R}$ be a REL of a 4TP graph G. Let $G_{\mathcal{R}}$ be the orientation of G obtained from $\mathcal{R}$ as follows. The interior edges are directed as in $\mathcal{R}$ (ignoring the colors). The exterior edges are oriented as: $v_S \rightarrow v_W$, $v_W \rightarrow v_N$, $v_S \rightarrow v_E$, $v_E \rightarrow v_N$. It was shown in [6] that $G_{\mathcal{R}}$ is an st-orientation of G with source v_S and sink v_N. $G_{\mathcal{R}}$ will be called the st-orientation derived from $\mathcal{R}$.

Let G_{green} (G_{red}, respectively) be the directed graph obtained from $G_{\mathcal{R}}$ by deleting all red (green, respectively) interior edges. Note that both G_{green} and G_{red} are st-graphs with source v_S and sink v_N. G_{green} and G_{red} are shown in Fig 2 (3) and (4).

Lemma 1. *Let $f(G_{green})$ and $f(G_{red})$ be the number of interior faces of G_{green} and G_{red}, respectively. Then $f(G_{green}) + f(G_{red}) = n - 1$.*

Proof. By Euler's formula, a 4TP graph G with n vertices has $m = 3n - 7$ edges. Both G_{green} and G_{red} contain n vertices. Let m_r and m_g be the number of edges in G_{red} and in G_{green} respectively. By Euler's formula, we have: $f(G_{green}) = m_g + 1 - n$ and $f(G_{red}) = m_r + 1 - n$. Each interior edge of G belongs to either G_{green} or G_{red}. The four exterior edges belong to both G_{green} and G_{red}. Thus $m_g + m_r = m + 4$. Therefore $f(G_{green}) + f(G_{red}) = m_g + m_r + 2 - 2n = m + 4 + 2 - 2n = n - 1$. □

In Fig 2, G has $n = 14$ vertices, $f(G_{green}) = 7$ and $f(G_{red}) = 6$. The following lemma was proved in [6] by a complicated argument. A much simpler proof is given below. The argument used here will be useful later.

Lemma 2. *Let G be a 4TP graph with a REL $\mathcal{R}$. Let $G_{\mathcal{R}}$ be the st-orientation of G derived from $\mathcal{R}$. Let $G_{\mathcal{R}}^*$ be the corresponding dual st-orientation of G^*. Then length($G_{\mathcal{R}}^*$) $\leq n - 1$. In other words, the VR of G obtained from $G_{\mathcal{R}}$ has width $\leq n - 1$.*

Proof. Let $P^* = \{e_{i_1}^*, e_{i_2}^*, \ldots, e_{i_k}^*\}$ be a longest path in $G_{\mathcal{R}}^*$, where $k = $ length $(G_{\mathcal{R}}^*)$. For each j ($1 \leq j \leq k$), let e_{i_j} be the edge in G corresponding to $e_{i_j}^*$. If e_{i_j} is red, then when P^* passes $e_{i_j}^*$, it enters a new red face in G_{red}. If e_{i_j} is green, then when P^* passes $e_{i_j}^*$, it enters a new green face in G_{green}. Because both G_{red} and G_{green} are plane st-graphs, each red or green face can be entered at most once. Therefore $k \leq f(G_{green}) + f(G_{red}) = n - 1$.   □

The following definition was used in [2,4] to find special st-orientations.

Definition 4. A *ladder graph* of order n is a plane graph $L = (A \cup B, E_L)$. The vertex set of L can be partitioned into $A = \{a_1, \ldots, a_{\lceil n/2 \rceil}\}$ and $B = \{b_1, \ldots, b_{\lfloor n/2 \rfloor}\}$. $E_L = L_A \cup L_B \cup L_{cross}$ where:

-   $L_A = \{(a_i, a_{i+1}) | 1 \leq i < \lceil n/2 \rceil\}$; $L_B = \{(b_j, b_{j+1}) | 1 \leq j < \lfloor n/2 \rfloor\}$.
-   L_{cross} consists of edges, (called *cross edges* of L), between a vertex $a_i \in A$ and a vertex $b_j \in B$; no two edges in L_{cross} cross each other; and the edges $(a_1, b_1), (a_{\lceil n/2 \rceil}, b_{\lfloor n/2 \rfloor}) \in L_{cross}$.

For a cross edge (a_i, b_j), define slope$(a_i, b_j) = j - i$. It is called a *level* (or *up* or *down*, respectively) edge if slope$(a_i, b_j) = 0$ (or > 0 or < 0, respectively).

Definition 5. An orientation $\mathcal{L}$ of a ladder graph L is *consistent* if the following hold: (1) For any i, the edge (a_i, a_{i+1}) is directed from a_i to a_{i+1} and the edge (b_i, b_{i+1}) is directed from b_i to b_{i+1}; (2) The edges in E_{cross} are oriented in a way such that $\mathcal{L}$ is acyclic.

From the definition, it is clear that a consistent orientation $\mathcal{L}$ is an st-orientation of L with source either a_1 or b_1, and sink either $a_{\lceil n/2 \rceil}$ or $b_{\lfloor n/2 \rfloor}$, depending on the orientations of (a_1, b_1) and $(a_{\lceil n/2 \rceil}, b_{\lfloor n/2 \rfloor})$.

Theorem 2. *Every ladder graph L of order n has a consistent orientation $\mathcal{L}$, constructible in linear time, such that the following hold:*
1. *a_1 is the only source and $b_{\lfloor n/2 \rfloor}$ is the only sink of $\mathcal{L}$.*
2. *length($\mathcal{L}$) $\leq \lceil n/2 \rceil + 2\lceil \sqrt{(n-2)/2} \rceil$.*

The essentially same theorem was originally proved in [4]. The theorem stated above is adapted from a slightly different version in [2]. It can be proved by a slight modification of the proof in [2].

3   Compact VR of 4-Connected Plane Graphs

In order to construct a VR of G with stated width and height, by Theorem 1, all we need is to find an st-orientation $\mathcal{O}$ of G so that both length$(\mathcal{O})$ and length$(\mathcal{O}^*)$ are not too large. The main difficulty of the construction is that these two goals often conflict. We will use a REL $\mathcal{R}$ of G to guide the construction of $\mathcal{O}$. (This is why we need the 4-connectivity: Only 4PT graphs have REL.)

Throughout this section, $G = (V, E)$ denotes a 4TP graph and $\mathcal{R}$ a REL of G. The basic idea of the construction is as follows: First, we use $\mathcal{R}$ to partition G into two subgraphs G_A and G_B of equal size. In the st-orientation $\mathcal{O}$, the orientations of the edges within G_A and G_B are the same as in $\mathcal{R}$. The edges of G between G_A and G_B form a ladder graph E_{cross}. The crux for constrcting $\mathcal{O}$ is to oricnt the edges in E_{cross} in order to bound both length$(\mathcal{O})$ and length$(\mathcal{O}^*)$.

3.1   Partition G into G_A, G_B and E_{cross}

Let $\mathcal{R}^{rev}$ be the orientation of G obtained from $\mathcal{R}$ by reversing the direction of green edges. Let $G_{\mathcal{R}^{rev}}$ be the orientation of G derived from $\mathcal{R}^{rev}$. Observe that if we flip G through a line that passes v_S and v_N, then $\mathcal{R}^{rev}$ is just a REL of G (with the roles of v_E and v_W switched). Thus $G_{\mathcal{R}^{rev}}$ is an st-orientation of G.

Let $\mathcal{P} = \{v_1, v_2, \ldots, v_n\}$ be a topological ordering of $G_{\mathcal{R}^{rev}}$. Then we have: $v_1 = v_S$, $v_2 = v_W$, $v_{n-1} = v_E$ and $v_n = v_N$. Partition V into two subsets: $A = \{v_1, v_2, \ldots, v_{\lceil n/2 \rceil}\}$ and $B = \{v_{\lceil n/2 \rceil + 1}, \ldots, v_n\}$. Let G_A (G_B, respectively) be the subgraph of G induced by the vertex set A (B, respectively). Let $G_{A\mathcal{R}}$ ($G_{B\mathcal{R}}$, respectively) denote the graph G_A (G_B, respectively) whose edges are partitioned and oriented according to $\mathcal{R}$.

Next, we order the vertex set of G_A as $A = \langle a_1, a_2, \ldots, a_{\lceil n/2 \rceil} \rangle$ by a topological sort of $G_{A\mathcal{R}}$. Note that $a_1 = v_S$ and $a_{\lceil n/2 \rceil} = v_W$. Similarly, we order the vertex set of G_B as $B = \langle b_1, b_2, \ldots, b_{\lfloor n/2 \rfloor} \rangle$ by a topological sort of $G_{B\mathcal{R}}$. Note that $b_1 = v_E$ and $b_{\lfloor n/2 \rfloor} = v_N$.

The edge set of G can be partitioned into three subsets: E_A is the edge set of G_A; E_B is the edge set of G_B; and E_{cross} is the set of edges between A and B. Let P_A be the path from a_1 to $a_{\lceil n/2 \rceil}$ on the exterior face of G_A. Let P_B be the path from b_1 to $b_{\lfloor n/2 \rfloor}$ on the exterior face of G_B. Let C be the region bounded by P_A, P_B and the edges (a_1, b_1) and $(a_{\lceil n/2 \rceil}, b_{\lfloor n/2 \rfloor})$. The faces of G in the region C are called the *cross faces* of G.

Lemma 3. *1. The numbering $\langle a_1, a_2, \ldots, a_{\lceil n/2 \rceil} \rangle$ is an st-numbering of G_A.*
2. The numbering $\langle b_1, b_2, \ldots, b_{\lfloor n/2 \rfloor} \rangle$ is an st-numbering of G_B.

Proof. We only prove (1). The proof of (2) is similar. Since $G_{\mathcal{R}}$ is acyclic and $G_{A\mathcal{R}}$ is a subgraph of $G_{\mathcal{R}}$, $G_{A\mathcal{R}}$ is acyclic. Clearly, $a_1 = v_S$ is a source and $a_{\lceil n/2 \rceil} = v_W$ is a sink of $G_{A\mathcal{R}}$. Consider any vertex $v = a_i$ $(1 < i < \lceil n/2 \rceil)$. We need to show a_i has two neighbors a_j and a_k with $j < i < k$.

Since v is an interior vertex of G, there is a red edge $e = u \to v$ in $\mathcal{R}$. e is oriented as $u \to v$ in $\mathcal{R}^{rev}$. Thus, in the topological ordering of $G_{\mathcal{R}^{rev}}$, u is

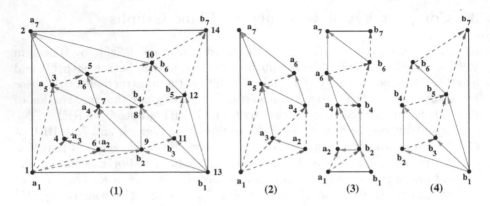

Fig. 3. (1) the numbering of the graph in Fig 2 (2). The vertices are numbered by a topological ordering of $G_{\mathcal{R}^{rev}}$. The labels $a_1, \ldots, a_7, b_1, \ldots, b_7$ indicate their numbering in $G_{A\mathcal{R}}$ and $G_{B\mathcal{R}}$ respectively; (2) $G_{A\mathcal{R}}$; (3) the edges in E_{cross} (with the paths P_A and P_B); (4) $G_{B\mathcal{R}}$. Here $P_A = \{a_1, a_2, a_4, a_6, a_7\}$ and $P_B = \{b_1, b_2, b_4, b_6, b_7\}$.

numbered before v. (Namely $u = v_p$ and $v = v_q$ with $p < q$). Hence $u \in A$. Since e is directed as $u \to v$ in $G_{A\mathcal{R}}$, u is numbered before v in the topological ordering of $G_{A\mathcal{R}}$. Namely $u = a_j$ for some $j < i$.

Since v is an interior vertex of G, there is a green edge $e' = v \to w$ in $\mathcal{R}$. e' is oriented as $w \to v$ in $\mathcal{R}^{rev}$. Thus, in the topological ordering of $G_{\mathcal{R}^{rev}}$, w is numbered before v. (Namely $w = v_r$ and $v = v_q$ with $r < q$). Hence $w \in A$. Since e' is directed as $v \to w$ in $G_{A\mathcal{R}}$, w is numbered after v in the topological ordering of $G_{A\mathcal{R}}$. Namely $w = a_k$ for some $i < k$. $\square$

Construct a ladder graph $L = (A \cup B, E_{cross})$ as follows: L contains a path $L_A = a_1 \to a_2 \ldots \to a_{\lceil n/2 \rceil}$, a path $L_B = b_1 \to b_2 \ldots \to b_{\lfloor n/2 \rfloor}$ and the edges in E_{cross}. (For the graph shown in Fig 3 (1), the corresponding ladder graph L can be obtained from the graph shown in Fig 3 (3) by inserting the vertex a_3 into the edge (a_2, a_4), a_5 into (a_4, a_6), b_3 into (b_2, b_4) and b_5 into (b_4, a_6)).

Definition 6. *Let $\mathcal{L}$ be a consistent orientation of L. $G_{\mathcal{RL}}$ denotes the orientation of G obtained as follows: The edges in E_A and E_B are oriented as in $G_{\mathcal{R}}$. The edges in E_{cross} are oriented as in $\mathcal{L}$.*

Lemma 4. *If $\mathcal{L}$ is consistent, then $G_{\mathcal{RL}}$ is an st-orientation of G.*

Proof. Since $G_{A\mathcal{R}}$ and $G_{B\mathcal{R}}$ are acyclic and $\mathcal{L}$ is a consistent orientation of L, $G_{\mathcal{RL}}$ is acyclic. Consider any interior vertex v of G. If $v \in A$, then v has two neighbors u, w in G_A such that $u \to v$ and $v \to w$ in $G_{A\mathcal{R}}$. If $v \in B$, then v has two neighbors u, w in G_B such that $u \to v$ and $v \to w$ in $G_{B\mathcal{R}}$. Depending on the orientation of the cross edge (a_1, b_1) in $\mathcal{L}$, either a_1 or b_1 is the unique source of $G_{\mathcal{RL}}$. Depending on the orientation of the cross edge $(a_{\lceil n/2 \rceil}, b_{\lfloor n/2 \rfloor})$ in $\mathcal{L}$, either $a_{\lceil n/2 \rceil}$ or $b_{\lfloor n/2 \rfloor}$ is the unique sink of $G_{\mathcal{RL}}$. So $G_{\mathcal{RL}}$ is an st-orientation of G. $\square$

$L_{\mathcal{L}}$ denotes the st-orientation of L derived from $\mathcal{L}$. Let $G_{\mathcal{RL}}$ be the st-orientation of G derived from $\mathcal{R}$ and $\mathcal{L}$, and $G^*_{\mathcal{RL}}$ the corresponding dual st-orientation of G^*.

Theorem 3. $\text{length}(G_{\mathcal{RL}}) \leq \text{length}(L_{\mathcal{L}})$.

Proof. Let P be a longest path in $G_{\mathcal{RL}}$. We transform P to a path P_L in $L_{\mathcal{L}}$ as follows. Consider any edge $e = u \to v$ in P. If e is a cross edge, we keep it in P_L. If e is an edge in G_A, then $u = a_i$ and $v = a_j$ for some $i < j$. We replace e by the sub-path in L_A from a_i to a_j. If e is in G_B, we replace it by a sub-path in L_B. After this operation is performed to all edges in P, we obtain a directed path P_L in $L_{\mathcal{L}}$. Thus: $\text{length}(G_{\mathcal{RL}}) = \text{length}(P) \leq \text{length}(P_L) \leq \text{length}(L_{\mathcal{L}})$. $\square$

Theorem 4. *Let P^* be a longest path in $G^*_{\mathcal{RL}}$. Then $\text{length}(P^*) \leq n - 1 + l$, where l is the number of cross faces of G passed by P^*.*

Proof. Because the way $G_{\mathcal{RL}}$ is oriented, P^* travels in one of the following ways:

I: (i) P^* first crosses the edge $(a_1, a_{\lceil n/2 \rceil})$ then travels some faces within G_A; (ii) crosses an edge in P_A and travels some l cross faces; (iii) crosses an edge in P_B, and travels some faces within G_B.

II: P^* first crosses the edge $(a_{\lceil n/2 \rceil}, b_{\lfloor n/2 \rfloor})$. It is similar to Case I, except the portion (i) is empty. We count the length of these sub-paths separately.

G_A has $n_a = \lceil n/2 \rceil$ vertices. Let o_a be the number of vertices on the exterior face of G_A. Let $i_a = n_a - o_a$ be the number of interior vertices of G_A. It is easy to show that the number of interior faces of G_A is $f_a = 2n_a - o_a - 2$. By Euler's formula, the number of edges in G_A is $m_a = n_a + f_a - 1 = 3n_a - o_a - 3$.

Let $G_{A,red}$ be the subgraph of G_A consisting of the exterior edges of G_A and its red interior edges. $G_{A,red}$ has n_a vertices. Let $m_{A,red}$ be the number of edges in $G_{A,red}$. By Euler's formula, the number on interior faces in $G_{A,red}$ is $f_{A,red} = m_{A,red} - n_a + 1$.

Let $G_{A,green}$ be the subgraph of G_A consisting of the exterior edges of G_A and its green interior edges. $G_{A,green}$ has n_a vertices, Let $m_{A,green}$ be the number of edges in $G_{A,green}$. By Euler's formula, the number on interior faces in $G_{A,green}$ is $f_{A,green} = m_{A,green} - n_a + 1$. Since each of the o_a exterior edges of G_A belongs to both $G_{A,red}$ and $G_{A,green}$, we have $m_{A,red} + m_{A,green} = m_a + o_a$. Thus:

$f_{A,red} + f_{A,green} = (m_{A,red} - n_a + 1) + (m_{A,green} - n_a + 1) = m_{A,red} + m_{A,green} - 2n_a + 2 = m_a + o_a - 2n_a + 2 = (3n_a - o_a - 3) + o_a - 2n_a + 2 = n_a - 1 = \lceil n/2 \rceil - 1$.

(For example, the graph G_A in Fig 3 (2) has $n_a = 7$ vertices. $i_a = 2$, $o_a = 5$, $m_a = 3n_a - o_a - 3 = 13$, and $f_a = 2n_a - o_a - 2 = 7$. $f_{A,red} = 3$ and $f_{A,green} = 3$).

Consider the sub-path of P^* when it travels within G_A. When P^* crosses a red edge, it enters a new face in $G_{A,red}$. When P^* crosses a green edge, it enters a new face in $G_{A,green}$. Since the edges in $G_{A\mathcal{R}}$ are oriented according to REL $\mathcal{R}$, each face in $G_{A,red}$ and $G_{A,green}$ can be entered at most once. Therefore the length of P^* within G_A is at most $f_{A,red} + f_{A,green} \leq \lceil n/2 \rceil - 1$.

Then P^* crosses an edge in P_A, and enters the first cross face. It continues to travel l cross faces. Then P^* crosses an edge in P_B (which adds 1 to the length of P^*) and enters the first face in G_B. By the same argument for the sub-path of P^* within G_A, the length of P^* within G_B is at most $n_b - 1 = \lfloor n/2 \rfloor - 1$. Hence length$(P^*) \le (\lceil n/2 \rceil - 1) + l + 1 + (\lfloor n/2 \rfloor - 1) = n - 1 + l$.   □

3.2 Orientation of E_{cross}

We next describe how to find a consistent orientation $\mathcal{L}$ for L so that the length of the longest path P in the st-orientation $G_{\mathcal{RL}}$ and the length of the longest path P^* in the dual st-orientation $G^*_{\mathcal{RL}}$ are not too large.

Let $k = |E_{cross}|$. Then G has $k - 1$ cross faces. Note that we always have $k \le (n - 1)$. (The equality holds when all vertices of G_A are in the exterior path P_A and all vertices of G_B are in the exterior path P_B. In this case, P_A has $\lceil n/2 \rceil - 1$ edges and P_B has $\lfloor n/2 \rfloor - 1$ edges. Since each cross face consumes one edge in either P_A or P_B, there are $\lceil n/2 \rceil + \lfloor n/2 \rfloor - 2 = n - 2$ cross faces).

Order the edges in E_{cross} from bottom up: $E_{cross} = \{e_1, e_2, \ldots, e_k\}$. Suppose that $e_t = (a_{i_t}, b_{j_t})$. In particular $e_1 = (a_1, b_1)$ and $e_k = (a_{\lceil n/2 \rceil}, b_{\lfloor n/2 \rfloor})$.

Consider any cross edge e_t. If e_t is an up edge (namely slope$(e_t) > 0$), it's natural to orient e_i as $a_{i_t} \to b_{j_t}$, because otherwise the length of P might increase. However, if we always orient cross edges this way, the length of P^* might be too large. (Consider a special case where all cross edges are up edges. If we orient all cross edges from A side to B side, then P^* may pass all cross faces). To avoid this, some up edge e_t might have to be oriented as $a_{i_t} \leftarrow b_{j_t}$. This, of course, might increase the length of P. The trick is to orient the cross edges in a way so that the lengths of P and P^* do not increase too much.

Let $p = \lfloor k/2 \rfloor$. Consider the edge $e_p = (a_{i_p}, b_{j_p})$. Without loss of generality, we assume slope$(e_p) = j_p - i_p \ge 0$. (If not, switch the roles of G_A and G_B).

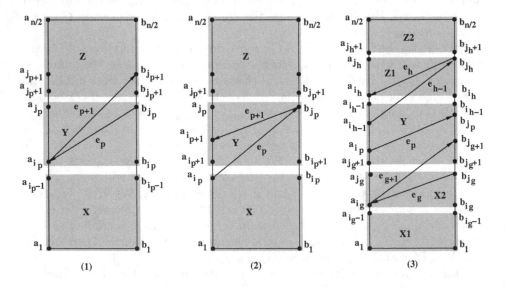

Fig. 4. (1) Case 1a; (2) Case 1b; (3) Case 2

Case 1: slope$(e_p) = j_p - i_p \leq n/4$.

Case 1a: $e_{p+1} = (a_{i_{p+1}}, b_{j_{p+1}})$ with $j_{p+1} \neq j_p$ and $i_{p+1} = i_p$ (see Fig 4 (1)).
   We divide L into three sub-ladder graphs:

- $X = (A_X \cup B_X, E_X)$ of order $x = 2(i_p - 1)$, $A_X = \{a_1, \ldots, a_{i_p-1}\}$ and $B_X = \{b_1, \ldots, b_{i_p-1}\}$. Let $\mathcal{L}_X$ be the consistent orientation of X in Theorem 2, with length$(\mathcal{L}_X) \leq \lceil x/2 \rceil + 2\lceil \sqrt{x-2)/2} \rceil$.
- $Y = (A_Y \cup B_Y, E_Y)$ of order $y = 2(j_p - i_p + 1)$, $A_Y = \{a_{i_p}, \ldots a_{j_p}\}$ and $B_Y = \{b_{i_p}, \ldots, b_{j_p}\}$. Define $\mathcal{L}_Y = \langle b_{i_p}, b_{i_p+1}, \ldots, b_{j_p}, a_{i_p}, a_{i_p+1}, \ldots, a_{j_p} \rangle$. Note that length$(\mathcal{L}_Y) = y - 1$ and $y/2 = j_p - i_p + 1 = $ slope$(e_p) + 1 \leq n/4 + 1$.
- $Z = (A_Z \cup B_Z, E_Z)$ of order $z = n - 2j_p$, $A_Z = \{a_{j_p+1}, \ldots, a_{\lceil n/2 \rceil}\}$ and $B_Z = \{b_{j_p+1}, \ldots, b_{\lfloor n/2 \rfloor}\}$. Let $\mathcal{L}_Z$ be the consistent orientation of Z in Theorem 2, with length$(\mathcal{L}_Z) \leq \lceil z/2 \rceil + 2\lceil \sqrt{(z-2)/2} \rceil$.

Define $\mathcal{L} = \langle \mathcal{L}_X, \mathcal{L}_Y, \mathcal{L}_Z \rangle$.

Case 1b: $e_{p+1} = (a_{i_{p+1}}, b_{j_{p+1}})$ with $j_{p+1} = j_p$ and $i_{p+1} \neq i_p$ (see Fig 4 (2)).
   We divide L into three sub-ladder graphs:

- $X = (A_X \cup B_X, E_X)$ of order $x = 2i_p$, $A_X = \{a_1, a_2, \ldots, a_{i_p}\}$ and $B_X = \{b_1, b_2, \ldots, b_{i_p}\}$. Let $\mathcal{L}_X$ be the consistent ordering for X in Theorem 2, with length$(\mathcal{L}_X) \leq \lceil x/2 \rceil + 2\lceil \sqrt{(x-2)/2} \rceil$.
- $Y = (A_Y \cup B_Y, E_Y)$ of order $y = 2(j_p - i_p)$, $A_Y = \{a_{i_p+1}, \ldots a_{j_p}\}$ and $B_Y = \{b_{i_p+1}, \ldots, b_{j_p}\}$. Define $\mathcal{L}_Y = \langle b_{i_p+1}, \ldots, b_{j_p}, a_{i_p+1}, \ldots, a_{j_p} \rangle$.
- $Z = (A_Z \cup B_Z, E_Z)$ of order $z = n - 2j_p$, $A_Z = \{a_{j_p+1}, \ldots, a_{\lceil n/2 \rceil}\}$ and $B_Z = \{b_{j_p+1}, \ldots, b_{\lfloor n/2 \rfloor}\}$. Let $\mathcal{L}_Z$ be the consistent ordering for Z in Theorem 2, with length$(\mathcal{L}_Z) \leq \lceil z/2 \rceil + 2\lceil \sqrt{(z-2)/2} \rceil$.

Define $\mathcal{L} = \langle \mathcal{L}_X, \mathcal{L}_Y, \mathcal{L}_Z \rangle$.

Case 2: slope$(e_p) = j_p - i_p > n/4$ (see Fig 4 (3)).
   Let $e_g = (a_{i_g}, b_{j_g})$ be the edge such that g is the largest index between 1 and $p-1$ with slope$(e_g) = j_g - i_g \leq \frac{n}{4}$. Let $e_h = (a_{i_h}, b_{j_h})$ be the edge such that h is the smallest index between $p+1$ and k with slope$(e_h) = j_h - i_h \leq \frac{n}{4}$. We divide L into five sub-ladder graphs:

- $X1 = (A_{X1} \cup B_{X_1}, E_{X1})$ of order $x_1 = 2(i_g - 1)$, $A_{X1} = \{a_1, \ldots, a_{i_g-1}\}$ and $B_{X1} = \{b_1, \ldots, b_{i_g-1}\}$. Let $\mathcal{L}_{X1}$ be the consistent orientation for $X1$ in Theorem 2, with length$(\mathcal{L}_{X1}) \leq \lceil x_1/2 \rceil + 2\lceil \sqrt{(x_1-2)/2} \rceil$.
- $X2 = (A_{X2} \cup B_{X2}, E_{X2})$ of order $x_2 = 2(j_g - i_g + 1)$, $A_{X1} = \{a_{i_g}, a_{i_g+1}, \ldots, a_{j_g}\}$ and $B_{X2} = \{b_{i_g}, b_{i_g+1}, \ldots, b_{j_g}\}$.
   Define $\mathcal{L}_{X2} = \langle b_{i_g}, b_{i_g+1}, \ldots, b_{j_g}, a_{i_g}, a_{i_g+1}, \ldots, a_{j_g} \rangle$.
- $Y = (A_Y \cup B_Y, E_Y)$ of order $y = 2(i_h - j_g - 1)$, $A_Y = \{a_{j_g+1}, a_{j_g+2}, \ldots, a_{i_h-1}\}$ and $B_Y = \{b_{j_g+1}, b_{j_g+2}, \ldots, b_{i_h-1}\}$.
   Define $\mathcal{L}_Y = \langle a_{j_g+1}, a_{j_g+2}, \ldots, a_{i_h-1}, b_{j_g+1}, b_{j_g+2}, \ldots, b_{i_h-1} \rangle$.
- $Z1 = (A_{Z_1} \cup B_{Z1}, E_{Z1})$ of order $z_1 = 2(j_h - i_h + 1)$, $A_{Z1} = \{a_{i_h}, \ldots, a_{j_h}\}$ and $B_{Z1} = \{b_{i_h}, \ldots, b_{j_h}\}$. Define $\mathcal{L}_{Z1} = \langle b_{i_h}, b_{i_h+1}, \ldots, b_{j_h}, a_{i_h}, a_{i_h+1}, \ldots, a_{j_h} \rangle$.

– $Z2 = (A_{Z2} \cup B_{Z2}, E_{Z2})$ of order $z_2 = n - 2j_h$, $A_{Z2} = \{a_{j_h+1}, \ldots, a_{\lceil n/2 \rceil}\}$ and $B_{Z2} = \{b_{j_h+1}, \ldots, b_{\lfloor n/2 \rfloor}\}$. Let $\mathcal{L}_{Z2}$ be the consistent orientation for $Z2$ in Theorem 2, with length($\mathcal{L}_{Z2}$) $\leq \lceil z_2/2 \rceil + 2\lceil \sqrt{(z_2 - 2)/2} \rceil$.

Define $\mathcal{L} = \langle \mathcal{L}_{X1}, \mathcal{L}_{X2}, \mathcal{L}_Y, \mathcal{L}_{Z1}, \mathcal{L}_{Z2} \rangle$.
We need the following lemmas for the analysis of Case 2.

Lemma 5. *There are at most $\lfloor n/2 \rfloor$ cross faces between e_g and e_h.*

Proof. Consider the cross edge e_{g+1}. By the choice of e_g, we have slope(e_{g+1}) $> n/4$. So $e_{g+1} = (a_{i_g}, b_{j_{g+1}})$ for some $j_{g+1} > j_g$ and $j_{g+1} - i_g > n/4$. This implies $j_{g+1} > n/4 + i_g \geq n/4 + 1$.

Consider the cross edge e_{h-1}. By the choice of e_h, we have slope(e_{h-1}) $> n/4$. Thus $e_{h-1} = (a_{i_{h-1}}, b_{j_h})$ for some $i_{h-1} < i_h$ and $j_h - i_{h-1} > n/4$. This implies $i_{h-1} < j_h - n/4 \leq \lfloor n/2 \rfloor - n/4$.

Each cross face between e_{g+1} and e_{h-1} consumes either one edge in P_A between a_{i_g} and $a_{i_{h-1}}$, or one edge in P_B between $b_{j_{g+1}}$ and b_{j_h}. Hence the number of these cross faces is at most: $w \leq (j_h - j_{g+1}) + (i_{h-1} - i_g) < (\lfloor n/2 \rfloor - n/4 - 1) + (\lfloor n/2 \rfloor - n/4 - 1) = 2\lfloor n/2 \rfloor - n/2 - 2 \leq \lfloor n/2 \rfloor - 2$.

Thus the number of cross faces between e_g and e_h is at most $w + 2 \leq \lfloor n/2 \rfloor$. $\square$

Lemma 6. *Let $U = X2 \cup Y \cup Z1$. Let $\mathcal{L}_U = \langle \mathcal{L}_{X_2}, \mathcal{L}_Y, \mathcal{L}_{Z1} \rangle$. Then*

$$\text{length}(\mathcal{L}_U) \leq (x_2 + y + z_1)/2 + n/4 + 1.$$

Proof. Let $S = \{e_{g+1}, e_{g+2}, \ldots, e_{h-2}, e_{h-1}\}$. Note that for any $e_t \in S$, slope(e_t) $> n/4$. Let P be a longest path in $\mathcal{L}_U$. Since b_{i_g} is the source and a_{j_h} is the sink in $\mathcal{L}_U$, P must start at b_{i_g} and end at a_{j_h}. The following are three different ways that P may achieve the maximum length. (In the following, the symbol $\overset{l}{\Longrightarrow}$ means a sub-path of length l. The symbol $\rightarrow$ means a single edge).

– $Q_1: b_{i_g} \overset{x_2/2-1}{\Longrightarrow} b_{j_g} \rightarrow a_{i_g} \overset{j_h-i_g}{\Longrightarrow} a_{j_h}$.
  Then length(Q_1) $= (x_2/2 - 1) + 1 + (j_h - i_g) \leq (x_2 + y + z_1)/2 + n/4$. (Here we use the facts: $j_h - i_g = (x_2 + y + z_1)/2 - 1$ and $x_2/2 = j_g - i_g + 1 =$ slope(e_g) $+ 1 \leq n/4 + 1$).
– $Q_2: b_{i_g} \overset{j_h-i_g}{\Longrightarrow} b_{j_h} \rightarrow a_{i_h} \overset{z_1/2-1}{\Longrightarrow} a_{j_h}$.
  By the same argument, we can show length(Q_2) $\leq (x_2 + y + z_1)/2 + n/4$.
– $Q_3: b_{i_g} \overset{x_2/2-1}{\Longrightarrow} b_{j_g} \rightarrow a_{i_g} \overset{i_t-i_g}{\Longrightarrow} a_{i_t} \overset{e_t}{\rightarrow} b_{j_t} \overset{j_h-j_t}{\Longrightarrow} b_{j_h} \rightarrow a_{i_h} \overset{j_h-i_h}{\Longrightarrow} a_{j_h}$.
  Then: length(Q_3) $= (x_2/2 - 1) + 1 + (i_t - i_g) + 1 + (j_h - j_t) + 1 + (j_h - i_h) = (x_2/2 + 2) + (j_h - i_g) + [(j_h - i_h) - (j_t - i_t)]$.
  Note that $x_2/2 \leq n/4 + 1$, slope(e_h) $= j_h - i_h \leq n/4$ and slope(e_t) $= j_t - i_t > n/4$. Because slope(e_h) and slope(e_t) are integers, $[(j_h - i_h) - (j_t - i_t)] \leq -1$. Thus:
  length(Q_3) $\leq n/4 + 3 + (x_2 + y + z_1)/2 - 1 - 1 = (x_2 + y + z_1)/2 + n/4 + 1$. $\square$

3.3 Analysis

Let $\mathcal{L}$ be the orientation for the ladder graph L constructed above.

Lemma 7. $\mathcal{L}$ *is a consistent orientation of* L.

Proof. We prove the lemma for Case 2. The proof for Cases 1a and 1b are similar. All we have to do is to show $\mathcal{L}$ is acyclic. The sub-orientations $\mathcal{L}_{X1}, \mathcal{L}_{Z2}$ are acyclic by Theorem 2. The sub-orientations $\mathcal{L}_{X2}, \mathcal{L}_{Y}, \mathcal{L}_{Z1}$ are acyclic by the construction. Since $\mathcal{L}$ is the concatenation of $\mathcal{L}_{X1}, \mathcal{L}_{X2}, \mathcal{L}_{Y}, \mathcal{L}_{Z1}, \mathcal{L}_{Z2}$, the orientations of the edges whose end vertices belong to different sub-ladder graphs do not create cycles. Hence $\mathcal{L}$ is acyclic. □

Note that in all cases, a_1 is the source and $b_{\lfloor n/2 \rfloor}$ is the sink of $\mathcal{L}$. Let $G_{\mathcal{RL}}$ be the st-orientation of G derived from $\mathcal{R}$ and $\mathcal{L}$. Let $G^*_{\mathcal{RL}}$ be the corresponding dual st-orientation of G^*.

Lemma 8. $\text{length}(G^*_{\mathcal{RL}}) \le \lceil \frac{3n}{2} \rceil - 1$.

Proof. Let P^* be a longest path in $G^*_{\mathcal{RL}}$. Let l be the number of cross faces passed by P^*. By Theorem 4, it is enough to show $l \le \lceil n/2 \rceil$.

Case 1a: Because the cross edges e_p and e_{p+1} are oriented in opposite direction, P^* can pass the cross faces either in the region above e_p or in the region below e_{p+1}, but not both. Because each of these two regions has at most $\lceil (k-1)/2 \rceil + 1 \le \lceil (n-2)/2 \rceil + 1$ cross faces, we have $l \le \lceil n/2 \rceil$.

Case 1b: Similar to Case 1a.

Case 2: Note that e_g is oriented as $a_{i_g} \leftarrow b_{j_g}$ (see Fig 4 (3)). By the choice of e_g, slope$(e_{g+1}) > n/4$. Hence $e_{g+1} = (a_{i_g}, b_{j_{g+1}})$ for some $j_{g+1} > j_g$. So e_{g+1} is oriented $a_{i_g} \to b_{j_{g+1}}$ in $\mathcal{L}$. Similarly, we can show e_h is oriented as $a_{i_h} \leftarrow b_{j_h}$, and e_{h-1} is oriented as $a_{i_{h-1}} \to b_{j_h}$ for some $i_{h-1} < i_h$. Because of the orientations of the cross edges e_g, e_{g+1}, e_{h-1} and e_h, the path P^* can pass cross faces in only one of the following three regions:

- The region below the edge e_{g+1}. The number of cross faces in this region is at most $\lceil n/2 \rceil$ because this region is below e_p.
- The region between e_g and e_h. The number of cross faces in this region is at most $\lceil n/2 \rceil$ by Lemma 5.
- The region above e_{h-1}. The number of cross faces in this region is at most $\lceil n/2 \rceil$ because this region is above e_p. □

Lemma 9. $\text{length}(G_{\mathcal{RL}}) \le \frac{3n}{4} + 2\lceil \sqrt{n} \rceil + 4$.

Proof. Let P be a longest path in $L_{\mathcal{L}}$. By Theorem 3, it's enough to show $\text{length}(P) \le \frac{3n}{4} + 2\lceil \sqrt{n} \rceil + 4$.

Case 1: Let P_X, P_Y and P_Z be the sub-paths of P in the sub-ladder graphs X, Y and Z, respectively. By Theorem 2, $\text{length}(P_X) \le \lceil x/2 \rceil + 2\lceil \sqrt{(x-2)/2} \rceil$ and $\text{length}(P_Z) \le \lceil z/2 \rceil + 2\lceil \sqrt{(z-2)/2} \rceil$. Since Y contains y vertices, $\text{length}(P_Y) \le y - 1$. The edges connecting these three sub-paths add 2 to the length of P. Noting the facts: $x + y + z = n$ and $y/2 \le n/4 + 1$, we have:

$$\text{length}(P) = \text{length}(P_X) + \text{length}(P_Y) + \text{length}(P_Z) + 2$$
$$\leq (\lceil x/2 \rceil + 2\lceil \sqrt{(x-2)/2} \rceil) + (y-1) + (\lceil z/2 \rceil + 2\lceil \sqrt{(z-2)/2} \rceil) + 2$$
$$\leq (x/2 + 1/2 + 2\lceil \sqrt{(x-2)/2} \rceil) + (y-1) +$$
$$(z/2 + 1/2 + 2\lceil \sqrt{(z-2)/2} \rceil) + 2$$
$$= n/2 + y/2 + 2\lceil \sqrt{(x-2)/2} \rceil + 2\lceil \sqrt{(z-2)/2} \rceil + 2$$

Let $f(x,z) = 2\lceil \sqrt{(x-2)/2} \rceil + 2\lceil \sqrt{(z-2)/2} \rceil$. Since $x+z \leq n$, it is easy to check $f(x,z)$ reaches the maximum value when $x = z = n/2$: $f(n/2, n/2) \leq 2\lceil \sqrt{n} \rceil$. Hence: $\text{length}(P) \leq 3n/4 + 2\lceil \sqrt{n} \rceil + 3$.

Case 1b: Similar to Case 1a.

Case 2: Let $U = X2 \cup Y \cup Z1$ and $\mathcal{L}_U = \langle \mathcal{L}_{X_2}, \mathcal{L}_Y, \mathcal{L}_{Z1} \rangle$ (as in Lemma 6).

Let P_{X1}, P_U, P_{Z2} be the sub-paths of P in the sub-ladder graphs $X1$, U, and $Z2$ respectively. By Theorem 2, $\text{length}(P_{X1}) \leq \lceil x_1/2 \rceil + 2\lceil \sqrt{(x_1-2)/2} \rceil$ and $\text{length}(P_{Z2}) \leq \lceil z_2/2 \rceil + 2\lceil \sqrt{(z_2-2)/2} \rceil$. By Lemma 6, $\text{length}(P_U) \leq \text{length}(\mathcal{L}_U)$ $\leq (x_2 + y + z_1)/2 + n/4 + 1$. The edges connecting these 3 sub-paths add 2 to $\text{length}(P)$. Noting the facts that: $x_1 + x_2 + y + z_1 + z_2 = n$, $x_1 + x_2 \leq n$, we have:

$$\text{length}(P) = \text{length}(P_{X1}) + \text{length}(P_U) + \text{length}(P_{Z2}) + 2$$
$$\leq (\lceil x_1/2 \rceil + 2\lceil \sqrt{(x_1-2)/2} \rceil) + (x_2 + y + z_1)/2 + n/4 + 1 +$$
$$(\lceil z_2/2 \rceil + 2\lceil \sqrt{(z_2-2)/2} \rceil) + 2$$
$$\leq n/2 + n/4 + 2\lceil \sqrt{(x_1-2)/2} \rceil + 2\lceil \sqrt{(z_2-2)/2} \rceil + 4$$
$$\leq 3n/4 + 2\lceil \sqrt{n} \rceil + 4 \qquad \square$$

Theorem 5. *Every 4-connected plane graph with n vertices has a VR, constructible in linear time, with height $\leq \frac{3n}{4} + 2\lceil \sqrt{n} \rceil + 4$ and width $\leq \lceil 3n/2 \rceil$.*

Proof. We have the following algorithm.

1. Delete an exterior edge $e = (v_S, v_N)$ from H. The resulting graph G is a 4TP graph.
2. Find a REL $\mathcal{R}$ of G in $O(n)$ time [6].
3. Partition G into G_A and G_B by a topological sort of $G_{\mathcal{R}^{rev}}$.
4. Construct the ladder graph L and find the orientation $\mathcal{L}$ as in §3.2.
5. Constructed the st-orientation $G_{\mathcal{RL}}$. Find a VR $\mathcal{D}'$ for G in linear time by Theorem 1.
6. Add a vertical line for the deleted edge, we get a VR $\mathcal{D}$ of H.

By Lemma 8 and 9, the VR $\mathcal{D}'$ for G has height $\leq \frac{3n}{4} + 2\lceil \sqrt{n} \rceil + 4$ and width $\leq \lceil 3n/2 \rceil - 1$. The last step increases the width of $\mathcal{D}$ by 1. So the height and the width of $\mathcal{D}$ satisfy the stated bounds. All steps of the algorithm can be done in linear time. So the total run time is $O(n)$. $\qquad \square$

4   Conclusion

In this paper, we present a VR construction for 4-connected plane graphs, which simultaneously bounds height $\leq \frac{3n}{4} + 2\lceil \sqrt{n} \rceil + 4$ and width $\leq \lceil 3n/2 \rceil$. This is the first VR construction that simultaneously bounds the height by $c_h n$ and the width by $c_w n$ where $c_h < 1$ and $c_w < 2$. It would be interesting to find such VR for broader classes of plane graphs.

References

1. di Battista, G., Eades, P., Tammassia, R., Tollis, I.: Graph Drawing: Algorithms for the Visualization of Graphs. Princeton Hall, Princeton (1998)
2. Chen, C.Y., Hung, Y.F., Lu, H.I.: Visibility Representations of Four Connected Plane Graphs with Near Optimal Heights. In: Tollis, I.G., Patrignani, M. (eds.) GD 2008. LNCS, vol. 5417, pp. 67–77. Springer, Heidelberg (2009)
3. Fan, J.H., Lin, C.C., Lu, H.I.: Width-optimal visibility representations of plane graphs. In: Tokuyama, T. (ed.) ISAAC 2007 LNCS, vol. 4835, pp. 160–171. Springer, Heidelberg (2007)
4. He, X., Zhang, H.: Nearly optimal visibility representations on plane graphs. In: Bugliesi, M., Preneel, B., Sassone, V., Wegener, I. (eds.) ICALP 2006. LNCS, vol. 4051, pp. 407–418. Springer, Heidelberg (2006)
5. Kant, G.: A more compact visibility representation. International Journal of Computational Geometry and Applications 7, 197–210 (1997)
6. Kant, G., He, X.: Regular edge labeling of 4-connected plane graphs and its applications in graph drawing problems. TCS 172, 175–193 (1997)
7. Lempel, A., Even, S., Cederbaum, I.: An algorithm for planarity testing of graphs. In: Proc. of an International Symposium on Theory of Graphs, Rome, pp. 215–232 (July 1967)
8. Lin, C.-C., Lu, H.-I., Sun, I.-F.: Improved compact visibility representation of planar graph via Schnyder's realizer. SIAM J. Disc. Math. 18, 19–29 (2004)
9. Ossona de Mendez, P.: Orientations bipolaires. PhD thesis, Ecole des Hautes Etudes en Sciences Sociales, Paris (1994)
10. Papamanthou, C., Tollis, I.G.: Applications of Parameterized st-Orientations in Graph Drawing Algorithms. In: Healy, P., Nikolov, N.S. (eds.) GD 2005. LNCS, vol. 3843, pp. 355–367. Springer, Heidelberg (2006)
11. Papamanthou, C., Tollis, I.G.: Parameterized st -Orientations of Graphs: Algorithms and Experiments. In: Kaufmann, M., Wagner, D. (eds.) GD 2006. LNCS, vol. 4372, pp. 220–233. Springer, Heidelberg (2007)
12. Rosenstiehl, P., Tarjan, R.E.: Rectilinear planar layouts and bipolar orientations of planar graphs. Discrete Comput. Geom. 1, 343–353 (1986)
13. Tamassia, R., Tollis, I.G.: An unified approach to visibility representations of planar graphs. Discrete Comput. Geom. 1, 321–341 (1986)
14. Zhang, H., He, X.: Canonical Ordering Trees and Their Applications in Graph Drawing. Discrete Comput. Geom. 33, 321–344 (2005)
15. Zhang, H., He, X.: Visibility Representation of Plane Graphs via Canonical Ordering Tree. Information Processing Letters 96, 41–48 (2005)
16. Zhang, H., He, X.: Improved Visibility Representation of Plane Graphs. Computational Geometry: Theory and Applications 30, 29–39 (2005)
17. Zhang, H., He, X.: An Application of Well-Orderly Trees in Graph Drawing. In: Healy, P., Nikolov, N.S. (eds.) GD 2005. LNCS, vol. 3843, pp. 458–467. Springer, Heidelberg (2006)
18. Zhang, H., He, X.: Optimal st-orientations for plane triangulations. J. Comb. Optim. 17(4), 367–377 (2009)

Some Variations on Constrained Minimum Enclosing Circle Problem

Arindam Karmakar[1], Sandip Das[1],
Subhas C. Nandy[1], and Binay K. Bhattacharya[2]

[1] ACM Unit, Indian Statistical Institute, Kolkata - 700108, India
[2] School of Computing Science, Simon Fraser University, Canada - V5A 1S6

Abstract. Given a set P of n points and a straight line L, we study three important variations of minimum enclosing circle problem. The first problem is on computing k circles of minimum (common) radius with centers on L which can cover the members in P. We propose three algorithms for this problem. The first one runs in $O(nk \log n)$ time and $O(n)$ space. The second one runs in $O(nk + k^2 \log^3 n)$ time and $O(n \log n)$ space assuming that the points are sorted along L, and is efficient where $k << n$. The third one is based on parametric search and it runs in $O(n \log n + k \log^4 n)$ time. The next one is on computing the minimum radius circle centered on L that can enclose at least k points. The time and space complexities of the proposed algorithm are $O(nk)$ and $O(n)$ respectively. Finally, we study the situation where the points are associated with k colors, and the objective is to find a minimum radius circle with center on L such that at least one point of each color lies inside it. We propose an $O(n \log n)$ time algorithm for this problem.

1 Introduction

Geometric facility location problem is an important area of algorithmic research that deals with the identification of appropriate resource locations for serving a set of demands efficiently. Many variations of this problem has come up depending on practical applications. A typical facility location problem is the k-center problem. Here a set of points $P = \{p_1, p_2, \ldots, p_n\}$ is given in $I\!\!R^d$ as clients. The objective is to identify k positions in $I\!\!R^d$ for placing the facilities such that the maximum distance of a client from its nearest facility is minimized. We shall restrict ourself to $d = 2$. The *1-center problem* (or, the *minimum enclosing circle problem*) can be solved in $O(n)$ time [18]. Hurtado et al. [12] considered a variation of the 1-center problem where the center of the smallest enclosing circle of P is constrained to lie inside a given convex polygon of size m; the proposed algorithm runs in $O(n + m)$ time. Bose et al. [5] considered the generalized version of the problem where the center of the smallest enclosing circle of P is constrained to lie inside a given simple polygon of size m. The worst case time complexity of the proposed algorithm is $O((n + m) \log(n + m))$ time. The online query versions of the 1-center problem are also studied, where the objective is to preprocess the points in P so that given any arbitrary line or a line segment, the

W. Wu and O. Daescu (Eds.): COCOA 2010, Part I, LNCS 6508, pp. 354–368, 2010.

optimal location of the center can be identified efficiently. Two algorithms are available for this problem. The older one runs in $O(n \log n)$ preprocessing time using $O(n)$ space, and the query answering time is $O(\log^2 n)$ [22]. The recent one reduces the query time to $O(\log n)$ with an expense of $O(n^2)$ preprocessing time and space [14].

For the *2-center problem* the best known algorithm was proposed by Chan [6]. He suggested two algorithms; the first one is a deterministic algorithm, and it runs in $O(n \log^2 n (\log \log n)^2)$ time, and the second one is a randomized algorithm that runs in $O(n \log^2 n)$ time with high probability. A variation of this problem is the *discrete two-center problem*, where the objective is to find two closed disks whose union covers the point set P and whose centers are a pair of points in P. If all the points of P are in convex position, Kim et al.'s [15] $O(n \log^2 n)$ time algorithm is the best known result. For general position of P, the best result of discrete 2-center problem is $O(n^{\frac{4}{3}} \log^5 n)$ time complexity by Agarwal et al. [2].

The *k-center problem* in $\mathbb{R}^d$ is known to be NP-complete if $d \geq 2$ [16]. In its decision version, a radius r is given, and the problem is to determine whether k circles of radius r can cover the points in P. Hwang et al. [10] proposed an $n^{O(\sqrt{k})}$ time algorithm for the k-center problem in $\mathbb{R}^2$. Therefore it makes sense to search for efficient approximation algorithms and heuristics for the general version [13,20]. Recently, Brass et al. [4] studied several interesting variations of the constrained k center problem, where the centers of the circles lie on a line L. For the case where the line L is given, the proposed algorithm uses parametric search, and runs in $O(n \log^2 n)$ time. If the orientation of the line is given and one can choose L to minimize the radius, then their proposed algorithm runs in $O(n^2 \log^2 n)$ time. If no constraint on choosing L is given their algorithm can choose L and report k circles of minimum radius in $O(n^4 \log^2 n)$ time. Alt et al. [3] studied several variations of circle covering problem with circles of arbitrary radii and centers lying on a straight line. Their objective function is to minimize the sum/sum-of-square of radii of the covering circles. They proposed optimal algorithms for the above two optimization problems with time complexity $O(n^2 \log n)$ and $(n^4 \log n)$ respectively. They also proposed constant factor approximation algorithms for these problems that run in $O(n \log n)$ time.

In the colored variation of the k enclosing circle problem, each point is assigned a color in $\{1, 2, \ldots, k\}$ and the objective is to find a minimum radius circle containing at least one point of each color. Alt et al. [1] introduced the concept of the farthest color Voronoi diagram, which can be used to compute the smallest color spanning circle in $O(kn)$ time. The farthest color Voronoi diagram can be computed in $O(kn \log n)$ time [11].

In this paper, we study several constrained versions of the k-center problem. We first consider the *fixed radius covering problem (FRCP)*, where a real number r and a straight line L are given along with the point set P. The objective is to find the minimum number of circles of radius r centered on L that can cover the points in P (if at all possible). We show that, if the points are sorted with respect to their projections on L, then this problem can be solved in $O(n)$ time.

If $O(n \log n)$ preprocessing time and space are allowed, then the query version of the FRCP problem can be solved in $O(k \log^2 n)$ time, where the radius r is the input to the query. Next, we shall consider the *minimum radius covering problem (MRCP)*, where the number of circles k is given as input. The objective is to cover all the members in P by k circles of equal radii and centered on the given line L. Our aim is to minimize the radius of the circles. We propose two implementable algorithms for this problem. Given the set P and an integer k, the first one computes the optimum radius in $O(nk \log n)$ time and $O(n)$ space. The second one is efficient if $k << n$; we apply $O(k \log^2 n)$ time *FRCP* algorithm to solve the *MRCP* in $O(n \log n + nk + k^2 \log^3 n)$ time. Finally, we show that there exists an $O(n \log n + k \log^4 n)$ time algorithm for the *MRCP* using parametric searching technique [17]. It needs to be mentioned that Brass et al. [4] proposed an $O(n \log^2 n)$ time algorithm for the *MRCP* using the slope selection problem [7,17], which in turn uses the parametric searching technique [18]. Next, we address the *minimum radius k-enclosing circle problem*, where the set P and an integer k is given as input. The objective is to compute a circle with center on the line L which can cover k points in P and its radius is minimum among all the k points enclosing circles with centers on the line L. We propose an easy to implement algorithm for this problem that runs in $O(n \log n + nk)$ time and $O(n)$ space. Finally, we study the situation where each point in P is associated with one of the k given colors, and the objective is to find a minimum radius circle with center on L such that at least one point of each color lies inside it. We propose an $O(n \log n)$ time algorithm for this problem.

2   FRCP: Fixed Radius Covering Problem

Here the line L and radius r are given along with the points in P. The objective is to find the minimum number of circles of radius r centered on L that covers all the points in P, provided at least one such a solution exists. We consider a slab S bounded by a pair of lines parallel to L and both at distance r from L. If each member of P lies inside the slab, then there exists a feasible solution to the problem. We now explain the method of getting the optimum solution for this problem.

Without loss of generality, we assume that the line L is the x-axis, and the members in P are above the line L. If a point p lies below L and a circle C passes through p, then C also passes through the point p', where p' is the mirror image of p with respect to L. We also assume that the points in P are sorted from left to right. We use $C(p, r)$ to denote the circle of radius r centered at the point p. Since $p_i \in S$, the intersection of $C(p_i, r)$ and L is an interval $I_i = [a_i, b_i]$ where $a_i \leq b_i$. Let $I = \{I_1, I_2, \ldots, I_n\}$ be the set of intervals on the line L. Note that, (i) for any point $q \in I_i$, $C(q, r)$ contains p_i, and (ii) if the intervals in a subset $J \subseteq I$ overlap, then a circle $C(q, r)$ centered at a point q on the overlapping region $I^* = \cap_{j \in J} I_j$ contains all the points $\{p_j, j \in J\}$ (see Figure 1). Thus, we have the following result.

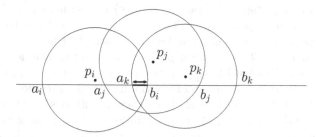

Fig. 1. Three intervals are overlapping on the interval $[a_k, b_i]$

Lemma 1. *If the end-points of the intervals in I are presorted along the line L, then the minimum number of circles of a given radius and centered on the line L required to enclose P, can be obtained in $O(n)$ time.*

Proof. The minimum number of circles of radius r centered on ℓ that are required to enclose the points in P is equal to the cardinality of the minimum clique cover of the interval graph formed by the intervals in I. If the points in P are sorted with respect to their x-coordinates, then computing the minimum clique cover of the intervals in I needs $O(n)$ time [9].   ⊓

We now improve the worst case time complexity of solving the FRCP problem to $O(k \log^2 n)$, where k is the size of the output.

Lemma 2. *If $\mathcal{C} = \{C_1, C_2, \ldots, C_k\}$ be the set of circles in the optimal solution, then the leftmost circle of $C_1 \in \mathcal{C}$ must enclose the leftmost point in P.*

Proof. It is easy to observe that I_1 contains the projection π_1 of the point p_1 on L. Let χ_1 be the leftmost maximal clique of the interval graph formed by the intervals in I. We need to prove that $\chi_1 \subseteq I_1$. Assume the contrary, let $\chi_1 \cap I_1 = \emptyset$. Let $p \ (\neq p_1)$ be a point whose corresponding interval I_p on L contributes to χ_1 (i.e., $I_p \cap \chi_1 \neq \emptyset$). As mentioned earlier, I_p must contain the projection π of the point p on L. Since π_1 is to the left of π, I_p must contain π_1. This is true for all the points that contribute to χ_1. Thus, π_1 lies in χ_1, and we have a contradiction.   □

Lemma 3. *Suppose the intersection of the intervals $I_1, I_2, \ldots I_{m-1}$ on the line L is non-empty (say $\chi = I_1 \cap I_2 \cap \ldots \cap I_{m-1}$), and I_m does not overlap on χ. If $I_{m+j} \ (j \geq 1)$ overlaps on χ, then I_{m+j} also overlaps on I_m.*

Proof. Similar to the proof of Lemma 2.   □

Lemmata 2 and 3 lead to the following procedure: (i) identify the leftmost interval $I_m = [a_m, b_m]$, such that $a_m \notin I_1$; (ii) compute the center of the circle (of radius r) on L that covers $p_1, p_2, \ldots, p_{m-1}$, and (iii) repeat steps (i) and (ii) assuming p_m as p_1. This again needs $O(n)$ time. In order to expedite the algorithm, we need to identify m without inspecting all the intervals $I_2, I_3, \ldots I_m$.

As a preprocessing step, we create a (binary) tournament tree $\mathcal{T}$ whose leaf nodes correspond to $p_1, p_2, \ldots, p_n$. The leaf level is considered to be the 0-th level of the tree $\mathcal{T}$. The i-th node at level 1 is attached with a pair of points (p_{2i+1}, p_{2i+2}). If $n - 2\lfloor \frac{n}{2} \rfloor = 1$, then p_n is also moved in this level. Thus, the number of nodes at this level is $\lceil \frac{n}{2} \rceil$. In the i-th node of the j-th level of $\mathcal{T}$, the set of points attached are the union of the points attached to the $(2i+1)$-th and $(2i+2)$-th nodes of the $j-1$-th level. The process continues up to the root of $\mathcal{T}$. We attach a secondary structure V_v at each node v of $\mathcal{T}$. It is an array containing the intersection points of the furthest point Voronoi diagram of the set of points attached with the node v and the line L (see Figure 2). Each intersection point θ can serve as the center of the minimum enclosing circle of all the points attached to this node. The radius ρ_θ of the corresponding circle is also attached with θ. It can be shown that the radii attached to these intersection points in order form a unimodal sequence [22].

The number of points attached to a node in the j-th level is at most 2^j, and computing the secondary structure at this node is $O(2^j)$. Since the number of nodes in the j-th level is at most $\frac{n}{2^j}$, the total time spent at the j-th level is $O(n)$. Thus, the construction of $\mathcal{T}$ needs $O(n \log n)$ time and space.

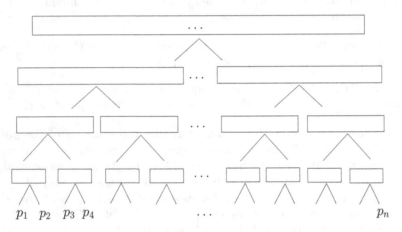

Fig. 2. Demonstration of the secondary structures at each node of $\mathcal{T}$

We now compute the maximum number of consecutive points starting from p_i that can be covered by a circle of radius r centered on L as follows. We consider two sets of points A and B, where the points in B are all to the right of the rightmost point in A, and it is known that the points in A can be covered by a circle of radius r centered on L. We use θ to denote the rightmost point on L such that a circle of radius r centered on θ can enclose the points in A. We try to append maximum number of points from B which can also be covered (along with A) with a circle of radius r centered on L.

We start execution initializing the set A by p_i, and θ be a point on L at distance r from p_i. Let v be the node of $\mathcal{T}$ representing the point p_i. At any stage, let u be the least common ancestor of v and its inorder successor in $\mathcal{T}$. Let u' be the right child of u. Let B be the set of points attached to the node u'. Let ρ^* be the minimum radius in the secondary structure $V_{u'}$ attached to u' and the corresponding center be θ^*. Here the following cases may arise:

Case 1: θ is to the left of θ^* in the array $V_{u'}$.

If the distance of the furthest neighbor of θ among the points in B is less than r, we set $A = A \cup B$, and repeat the same step with $v = u$, without changing the value of θ.

If the distance of the furthest neighbor of θ among the points in B is greater than r, then the minimum enclosing circle of the points in B with center at θ can not cover all the points in B. So, we need to move to the left child of u' to condider a subset of B. If the left child of u' is non-null, then set $u' =$ left child of the present u', and repeat the same process.

Case 2: θ is to the right of θ^* in the array $V_{u'}$.

If the radius attached to θ^* in $V_{u'}$ is greater than r, then a circle of radius r centered on L can not even cover the points in B. As in the second part of Case 1, here also we move to the left child of u'.

If the radius attached to θ^* in $V_{u'}$ is less than r, then we identify the furthest neighbor p of θ among the points in B. If the distance of p from θ is less than r, $C(p_i, r)$ with center at θ can enclose the points in B along with A. So, do not change the value of θ.

Otherwise, we again perform binary search in the array $V_{u'}$ to locate a point θ' on L to the right of θ^* such that a circle of radius r centered at θ' can contain all the points in B. Note that, there may exist some more points to the right of the points in B which can also be enclosed by the circle $C(p_i, r)$. So, in both the cases (i) we set $A = A \cup B$, (ii) set $v = u$, v' $=$ inorder successor of u, and then (iv) repeat the same process setting $u =$ least common ancestor of v and v'.

Thus finding the center of each circle, we may need to visit $O(\log n)$ nodes in $\mathcal{T}$, and in each node, we need to perform binary search in the secondary structure attached to that node to update θ. Since we need k such circles (the size of the output) to cover all the points in P, we have the following result.

Theorem 1. *The time complexity of the FRCP problem is $O(k \log^2 n)$ with an $O(n \log n)$ time preprocessing step. Here k is the minimum number of circles of radius r centered on L that are required to cover the points in P.*

3   MRCP: Minimum Radius Covering Problem

Here the number of circles k is given, and our objective is to find the minimum value of the radius r such that k circles of radius r centered on L can cover the members in P.

Let $\mathcal{C} = \{C_1, C_2, \ldots, C_k\}$ be the solution of the MRCP. Assume that the centers of the members in $\mathcal{C}$ are ordered from left to right on the line L. Each member in the set C must pass through at least one point in P. If the circle passes through a single point p_i then its center will be the perpendicular projection of p_i on the L. In case, a circle passes through at least two points $p_i, p_j \in P$ then its center is at the intersection point of L and the perpendicular bisector of the line segment $[p_i, p_j]$. A trivial $O(n^2 \log n)$ time and $O(n^2)$ space algorithm for the *MRCP* is as follows: (i) Compute the $\binom{n}{2}$ possible radii of the circles as discussed above and sort them in an array R; (ii) perform a binary search in R to find the minimum radius r_{opt} such that k circles of equal radius r_{opt} can cover all the points in P. At each step of this binary search, we need to invoke the $O(n)$ time algorithm FRCP (see Lemma 1) to decide whether k circles of the given radius is sufficient to cover P.

We now present three algorithms for the problem using Lemma 2. The first two are easy to implement, and is based on the following greedy strategy. Let r_{opt} be the optimum radius. Our algorithm selects a point of maximum index i such that $P_i = \{p_1, p_2, \ldots, p_i\}$ are all covered by the circle C_1. The remaining points will be covered by $k-1$ circles of radius r_{opt} centered on L. Similarly, C_2 can also be defined, and so on. The third one is based on parametric searching, and the objective is to demonstrate that the time complexity of the *MRCP* can be reduced to $O(n \log n)$ where k is a small constant.

3.1   Algorithm 1

Lemma 4. *If r' is the radius of the minimum radius circle C_1 containing $\{p_1, p_2, \ldots, p_i\}$, and r'' is the radius of the minimum radius circle containing $\{p_1, p_2, \ldots, p_i, p_{i+1}\}$ then $r' \leq r_{opt} < r''$.*

The index i, that defines C_1, can be computed by performing a binary search among the points in P. At each step of this binary search, we select an index α. Next we compute the radius r of the minimum enclosing circle with center on L that covers the points $\{p_1, p_2, \ldots, p_\alpha\}$, and then invoke FRCP with radius r. After executing $O(\log n)$ steps, we get two radii r' and r'' such that more than k circles of radius r'' are needed to cover P and less than or equal to k circles of radius r'' are needed to cover P. The circle of radius r' determines the index i.

Lemma 5. *The worst case time complexity for computing i (or equivalently C_1) is $O(n \log n)$.*

Proof. Follows from the fact that each step of the binary search needs $O(n)$ time (see (i) [18], and (ii) Lemma 1). □

We consider a circle (say C_1') of radius r' with center on L that covers $P_i = \{p_1, p_2, \ldots, p_i\}$. Thus, p_{i+1} is the leftmost point in $P \setminus P_i$. We repeat the same procedure with $P \setminus P_i$ of size $n - i$. This again returns a pair of radii r^* and r^{**}, where $r^* \leq r_{opt} < r^{**}$. We update $r' = \max(r', r^*)$ and $r'' = \min(r'', r^{**})$. This process may iterate k times. At the last step, we have a set of points $\{p_a, p_{a+1}, \ldots, p_n\}$. We compute the minimum enclosing circle of these points. If the radius r of this circle is greater than r'', then $r_{opt} = r''$, otherwise $r_{opt} = r'$.

Thus, each iteration of the procedure MRCP fixes one covering circle from left to right, and k such iterations are required. The time complexity of one iteration is $O(n \log n)$ in the worst case (see Lemma 5). Also, the extra space required by our algorithm is the tree T and the working space for computing the minimum enclosing circle of a set of points with center on a given line, which is $O(n)$. Thus, we have the following result.

Lemma 6. *The time and space complexities of Algorithm 1 for the MRCP are $O(kn \log n)$ and $O(n)$ respectively.*

3.2   Algorithm 2

We now present a more efficient algorithm where $k <.<. n$, by avoiding the binary search in computing C_1. Since C_1 passes through p_1, the center of C_1 is either the projection of p_i on the line L or the point of intersection of the perpendicular bisector of the line segment $[p_1, p_i]$ with the line L for some i. We compute n possible radii as stated above. In order to compute r' and r'', we choose the median of the radii, say r_{med}, and invoke the improved algorithm for FRCP. If it returns "NO" (resp. "YES"), we identify the radii which are greater (resp. less) than r_{med} and execute the same procedure. We need to iterate $O(\log n)$ steps to get r' and r''.

Lemma 7. *The time and space complexities of Algorithm 2 for the MRCP are $O(kn + n \log n + k^2 \log^3 n)$ and $O(n \log n)$ respectively.*

Proof. The preprocessing of the FRCP needs $O(n \log n)$ time and space. During the computation of C_1, we may need to compute median and invoke the algorithm for FRCP $O(\log n)$ time. Note that, after each step the number of elements is reduced to half of that of the previous step. So, the total time required for the median computation is $O(n) + O(\frac{n}{2}) + +O(\frac{n}{4}) + \ldots = O(n)$. Since the time complexity of the improved algorithm for FRCP is $O(k \log^2 n)$, the time complexity of computing r' and r'' (or equivalently C_1) is $O(n + k \log^3 n)$. The time complexity result follows from the fact that, in order to determine r_{opt}, we need to iterate the same procedure k times.   □

3.3   Algorithm 3

Our approach to solve MRCP is to run FRCP algorithm parametrically without knowing the value of optimal radius r^* a priori. For a parameter value r (may not be optimum), the FRCP solution takes $O(n)$ time. We are assuming that the projection of the points on the given line L are already sorted. Surely the optimum value of the parameter r^* is within some interval, say $\Lambda = [a, b]$. Initially, we start with $\Lambda = (0, \infty)$. Let $I_i(r) = [a_i(r), b_i(r)]$ be the intersection of the circle $C(p_i, r)$ with the given line L. Assuming that the line L is the x axis, we have $a_i(r) = x_i - \sqrt{r^2 - y_i^2}$ and $b_i(r) = x_i + \sqrt{r^2 - y_i^2}$. We thus have $2n$ endpoints $a_i(r)$ and $b_i(r)$, $1 \le i \le n$. We now determine the relative order of the endpoints of the intervals $I_i(r^*)$, $1 \le i \le n$ on L. Note that, the relative order of two intervals $[a_i(r^*), b_i(r^*)]$ and $[a_j(r^*), b_j(r^*)]$ can be determined by performing the feasibility test $FRCP$ with $r = r'$, where r' is the euclidean distance between p_i (or p_j) and the intersection point (α) of L and the perpendicular bisector of p_i and p_j. For a pair of points p_i and p_j, r' can be obtained in $O(1)$ time. If $r^* \leq r'$, either one of the intervals $I_i(r')$ and $I_j(r')$ is completely contained in the other, or the two intervals do not overlap, but they may touch. In the other hand, if $r^* > r'$ then the two intervals properly overlap. After the feasibility test with r' we will be able to find the relative order of the endpoints $a_i(r^*), b_i(r^*), a_j(r^*), b_j(r^*)$, and the interval $\Lambda = [a, b]$ containing r^* is now reduced to either $[a, r']$ or $[r', b]$, depending on whether $r^* \leq r'$ or $r^* > r'$ respectively. The sorting step of the end-points of the intervals with unknown r^* can be performed by solving $O(\log^2 n)$ feasibility tests [18]. However, the number of feasibility tests can be reduced to $O(\log n)$ (see Cole [8]). Let the reduced interval be $\Lambda = [\mu, \nu]$ which contains r^* after the sorting step. The relative order of the endpoints remains the same for any point in Λ. Thus, we have μ as the smallest radius. Now we have the following theorem.

Theorem 2. *The parametric searching based algorithm solves MRCP in* $O(n \log n)$ *time. The storage space requirement is* $O(n)$.

4 k-MRCP: Minimum Radius k-Enclosing Circle Problem

A circle is said to be k-enclosing if it encloses at least k points. In this section, we study the problem of finding the k-enclosing circle of minimum radius whose center is constrained to lie on a given line L. A brute-force $O(n^2)$ time algorithm is easy to design. We propose an $O(n \log n + nk)$ time algorithm for this problem.

Our algorithm starts with partitioning the given plane into $\lceil \frac{n}{k} \rceil$ slabs $S = \{S_1, S_2, \ldots, S_{\frac{n}{k}}\}$ perpendicular to the line L such that each slab (excepting the last one) contains exactly k points. As in the earlier problem, we may assume that L is the x-axis, and all the points in P are above the line L. The slabs in S are defined by the vertical lines $\mathcal{L} = \{\ell_1, \ell_2, \ldots \ell_{\frac{n}{k}-1}\}$, where ℓ_i intersects L at the point a_i. The members in the set $\mathcal{A} = \{a_1, a_2, \ldots, a_{\frac{n}{k}-1}\}$ are ordered with respect to their x-coordinate. The entire task can be done in $O(n \log \frac{n}{k})$ time using the recursive median finding algorithm. Let P_i denote the points in the slab S_i. We use $KNR(p)$ (resp. $KNL(p)$) to denote the set of k points nearest to a point p among those who lie to the right (resp. left) side of the vertical line at p. For each a_i, we store $KNL(a_i)$ and $KNR(a_i)$. The minimum radius k-enclosing circle centered on a_i encloses k points among the $2k$ points in $KNL(a_i) \bigcup KNR(a_i)$. We use $KN(a_i)$ to denote the set of points inside this circle. Now we have the following observations.

Lemma 8. *For all* $a_i \in \mathcal{A}$, $KN(a_i)$ *can be computed in* $O(n)$ *time.*

Proof. $KNL(a_1)$ and $KNR(a_{\lceil \frac{n}{k} \rceil - 1})$ can be computed in $O(k)$ time since there are only k candidate points. For $1 < i < \lceil \frac{n}{k} \rceil - 1$, each member in the set

$KNR(a_i)$ either lies in the slab S_{i+1} or a member of the set $KNR(a_{i+1})$. If $KNR(a_{i+1})$ is available, then $KNR(a_i)$ can be computed in $O(k)$ time. Thus the sets $KNR(a_{\lceil \frac{n}{k} \rceil - 1}), KNR(a_{\lceil \frac{n}{k} \rceil - 2}), \ldots, KNR(a_1)$ can be computed in $O(n)$ time. Similarly, the sets $KNL(a_1), KNL(a_2), \ldots, KNL(a_{\lceil \frac{n}{k} \rceil - 1})$ can also be computed in $O(n)$ time. In order to compute $KN(a_i)$, we compute the distances of a_i from the members in $KNL(a_i) \bigcup KNR(a_i)$, and then choose the median among these distances in $O(k)$ time. The result follows by adding the time complexities for computing $KN(a_i)$ for all $i = 1, 2, \ldots, \lceil \frac{n}{k} \rceil - 1$.   □

We now describe two algorithms for computing the minimum radius k-enclosing circle whose center is on the line segment $[a_i, a_{i+1}]$. The first one is relatively simple, and it takes $O(k^2)$ time and space. A different approach is adopted to reduce the space complexity to $O(k)$ keeping the time complexity invariant.

4.1   Algorithm 1

Note that, the minimum radius k-enclosing circle with center on the line segment $[a_i, a_{i+1}]$ contains k points among the members in the set $P_i \bigcup KN(a_i) \bigcup KN(a_{i+1})$. The cardinality of this set is at most $3k$. It either passes through one or two points in the above set. If it passes through only a single point $p \in P_i$, then the center and radius of the circle will be determined by the perpendicular projection of p on $[a_i, a_{i+1}]$. If it passes through two points $p, q \in P_i \bigcup KN(a_i) \bigcup KN(a_{i+1})$, then the bisector of the line segment $[p, q]$ must intersect $[a_i, a_{i+1}]$, and this determines the center and radius of the corresponding circle. Our desired circle will have one among the $\binom{k}{2}$ possible radii. A simple way to solve this problem is to sort the $O(k^2)$ radii and then perform binary search to choose the desired r^*. At each step of the binary search (with a radius r) we test whether radius r is enough for getting a k-enclosing circle by executing the following steps.

Step 1: Draw circular arcs of radius r centered at each point of $P_i \bigcup KN(a_i)$ $\bigcup KN(a_{i+1})$. This generates an interval graph of at most $3k$ intervals on the line segment $[a_i, a_{i+1}]$.
Step 2: Compute the maximum clique of this graph.
Step 3: If the size of the maximum clique is less than k, then we need to increase the radius; if it is greater or equal to k, we need to decrease the radius.

The search terminates with a radius r^* such that the corresponding interval graph has a k-clique. But, if we choose the next smaller radius from the sorted list, the size of the maximum clique of the corresponding interval graph is less than or equal to $k - 1$.

For a given radius r, the generation of $O(k)$ intervals and finding the maximum clique of that interval graph needs $O(k \log k)$ time. Since, we need to consider at most $\log k$ radii, the overall time needed for Step 2 is $O(k \log^2 k)$. Thus, the time complexity of this algorithm is dominated by the sorting of $O(k^2)$ radii, which is $O(k^2 \log k)$.

The time complexity can be reduced to $O(k^2)$ by avoiding the sorting as follows. We first find the median r of the aforesaid $O(k^2)$ radii in less than αk^2 time, where α is a constant. The testing of r for getting a k-enclosing circle needs $O(k \log k)$ time. For the next iteration of the binary search, we can get the required set of radii in another k^2 comparisons with r. The selection of middle-most radius in that part is done by executing the median find algorithm in $\alpha \frac{k^2}{2}$ time. But, testing time complexity remains $O(k \log k)$. Proceeding similarly, at the i-th stage the median finding needs $\alpha \frac{k^2}{2^i}$ time. Thus, the total time complexity for finding median at different stages is $O(k^2)$ in the worst case. The time spent for the testing of k-enclosing property of the desired circle at different stages is $O(k \log^2 k)$. Thus we have the following result:

Theorem 3. *The time and space complexities of Algorithm 1 for computing the minimum radius k-enclosing circle are $O(nk)$ and $O(\max(n, k^2))$ respectively.*

4.2   Algorithm 2

We now adopt a different approach (adapted from Brass et al. [4]) for reducing the space complexity of the problem . Let us consider a point $p = (\alpha, \beta)$ on the x-y plane, and consider the equation of the circle of radius r $(r \geq \beta)$ centered at p, which is $(x - \alpha)^2 + (y - \beta)^2 = r^2$. By putting $y = 0$, we get the segment intercepted by this circle on the x-axis, which is $\Psi(p) = r^2 - (x - \alpha)^2 = \beta^2$. This is the equation of a hyperbola in the x-r plane (where both r and x vary). Let $\ell(p)$ be a line parallel to the x-axis, which is the tangent of $\Psi(p)$. The point of contact of $\ell(p)$ and $\Psi(p)$ is called the vertex of $\Psi(p)$, and its r-coordinate gives the minimum radius required to enclose point p by a circle with center on the x-axis. For a pair of points p and q in the x-y plane, the minimum radius required to enclose p and q by a circle with center on the x-axis is obtained as follows: if the vertex of one hyperbola, say $\Psi(p)$, is inside the other hyperbola $\Psi(q)$, then the r-coordinate of the vertex of the hyperbola $\Psi(p)$ gives the minimum radius for enclosing both p and q (see Figure 3(a)), otherwise, the r coordinate of the intersection point θ of $\Psi(p)$ and $\Psi(q)$ gives the minimum radius for enclosing both p and q (see Figure 3(b)). Similarly, the minimum radius required to enclose a set of k points with center on the x-axis is the point having minimum r-coordinate in the intersection region of the hyperbolas corresponding these k points in the x-r plane. It may be a vertex of a hyperbola or an intersection point of a pair of hyperbolas.

Thus, in order to find the minimum radius k-enclosing circle with center on $[a_i, a_{i+1}]$, we need to consider $O(k)$ hyperbolas $\Psi_i = \{\Psi(p) | p \in P_i \bigcup KN(a_i) \bigcup KN(a_{i+1})\}$, and have to choose a point with minimum r-coordinate where exactly k hyperbolas in the set Ψ_i overlap. We need to inspect the two types of events (i) vertices of the members in Ψ_i, and (ii) the pairwise intersection points of the members in Ψ_i. In other words, we need to compute the k-th layer in the arrangement of hyperbolas. The r-coordinate of all the type (i) and type (ii) event points in this level need to be checked. Vahrenhold [23] recently described an inplace algorithm for computing the intersections of m line segments

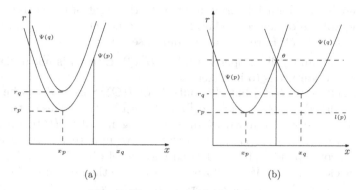

Fig. 3. Illustration of Algorithm 2

in $O(m \log^2 m + M)$ time and $O(m)$ space, where M is the number of intersections. We can use that algorithm for computing all the event points in $O(k^2)$ time and $O(k)$ space. Among these $O(k^2)$ points, we need to identify a point having minimum r value and which appears in the kth level of the arrangement of these $O(k)$ hyperbolas. Thus, we have the following result.

Theorem 4. *Given a set of n points, the minimum radius k-enclosing circle centered on a given line L can be obtained in $O(nk)$ time and $O(n)$ space.*

5   Color-Spanning Circle with Center on a Query Line

Here each point in the set $P = \{p_1, p_2, \ldots, p_n\}$ is associated with a color from a set of k colors, namely $\{1, 2, \ldots, k\}$. We use $\sigma(p_j)$ to denote the color of the point p_j, and P_i to denote the set of points in the color class i. The objective is to find the minimum radius circle C^* with center on the given line L such that it contains at least one point of each P_i. If $k = n$, the minimum enclosing circle of P with center on line L solves the problem.

Without loss of generality, we assume that the points in P are in general position. We compute the nearest point Voronoi diagram for points in P_i for each $i = 1, 2, \ldots, k$. Let NV_i denote the intersection points of the Voronoi diagram for points in P_i and the line L. The number of such intersection points is $O(|P_i|)$. Thus, NV_i, $i = 1, 2, \ldots, k$ splits L into $\sum_{i=1}^{k} |NV_i|$ intervals, which is $O(n)$ in the worst case. For each interval, the nearest point in each color class is known from the respective Voronoi diagrams. We process these intervals from left to right. We start with the left-most interval, and consider the nearest point of each color class in this interval. Let Q denote the set of these k points (of distinct colors). We compute the convex hull $CH(Q) = \{q_1, q_2, \ldots, q_m\}$ ($m \le k$) of the points in Q in $O(k \log k)$ time and then compute the points of intersection of the farthest point Voronoi diagram of $CH(Q)$ and the line L. We use FV to denote this set of points. We can compute the minimum enclosing circle of $CH(Q)$ with center on L by inspecting the members in FV. This needs $O(k)$ time [22].

While moving to the neighboring interval to the right of the current interval, the existing point q of one color class leaves from Q and a point q' in the same color class enters in Q. Here one of the four cases may arise:

Case 1: Both q and q' lie properly inside $CH(Q)$. $CH(Q)$ is not updated, and hence no action needs to be taken.

Case 2: q is inside $CH(Q)$ but q' is outside $CH(Q)$. Surely q' will be a vertex of the updated $CH(Q)$. Let $[q', q_a]$ and $[q', q_b]$ be two edges of the updated $CH(Q)$. These are obtained by drawing tangents of $CH(Q)$ from q'. The vertices $\{q_{a+1}, q_{a+2}, \ldots, q_{b-1}\}$ are deleted. We update FV by deleting the elements corresponding to the deleted vertices of $CH(Q)$ and adding one or two new elements in FV as follows. We compute the perpendicular bisectors of $[q_a, q']$ and $[q', q_b]$. If they intersect prior to reaching L, then the point of intersection of perpendicular bisectors of $[q_a, q_b]$ and L is added in FV. But if they intersect after reaching L, then the intersections of both these line with L are added in FV. The total time needed for this operation is proportional to $(b - a)$ (the number of vertices of the existing $CH(Q)$ that goes inside the updated $CH(Q)$).

Case 3: q is a vertex of $CH(Q)$ and q' is inside $CH(Q)$. Here, the entry corresponding to q in FV is deleted. Some new entries corresponding to the new vertices in the updated $CH(Q)$ are added in FV. These are obtained using a method similar to Case 2, and the time taken for this opration is proportional to the number of vertices that are added to $CH(Q)$.

Case 4: q is a vertex of $CH(Q)$ and q' is outside $CH(Q)$. We process this case by deleting q from $CH(Q)$ and updating FV as in Case 3. Next, we add q' in $CH(Q)$ and update FV as we did in Case 2. The total time needed is $O(m_1 + m_2)$, where m_1 and m_2 are respectively the number of vertices added to and deleted from $CH(Q)$.

Theorem 5. *The time and space complexities of computing the minimum radius color spanning circle with center on a given line L are $O(n \log n)$ and $O(n)$ respectively.*

Proof. The nearest neighbor Voronoi diagram for the points with color i can be computed in $O(|P_i| \log |P_i|)$, and then computing NV_i needs another $O(|P_i|)$ time. Thus, the total time required for splitting L into intervals by the Voronoi diagram of all color classes is $O(n \log n)$ where $\sum_{i=1}^{k} |P_i| = n$. While processing the first interval the time required for computing the convex hull and the corresponding FV array is $O(k \log k)$.

While computing the subsequent intervals, updating the convex hull needs $O(\log k + m)$ time, where m vertices of the existing convex hull are deleted to get the updated convex hull for the current interval. Once a point is deleted in Case 2, it may appear again in the Case 3 while processing some other point. Thus, a point can appear at most twice, once as a case 2 event and once as a case 3 event.

After getting the updated convex hull, the updated FV array can be computed in $O(k)$ time. Thus, the amortized time complexity for processing all the intervals is $O(n \log n)$. The space complexity result is easy to follow.   □

References

1. Abellanas, M., Hurtado, F., Icking, C., Klein, R., Langetepe, E., Ma, L., Sacriston, V.: The farthest color voronoi diagram and related problems. In: 17th European Workshop on Computational Geometry (2001)
2. Agarwal, P.K., Sharir, M., Welzl, E.: The discrete 2-center problem. In: 13th Annual Symposium on Computational Geometry, pp. 147–155 (1997)
3. Alt, H., Arkin, E.M., Bronnimann, H., Erickson, J., Fekete, S.P., Knauer, C., Lenchner, J., Mitchell, J.S.B., Whittlesey, K.: Minimum-cost Coverage of point sets by disks. In: Proc. 22nd Annual ACM Symposium on Computational Geometry, pp. 449–458 (2006)
4. Brass, P., Knauer, C., H.-S. Na, C.-S. Shin, Vigneron, A.: Computing k-centers on a line Technical Report CoRR abs/0902.3282 (2009)
5. Dose, P., Wang, Q.: Facility location constrained to a polygonal domain. In: Rajsbaum, S. (ed.) LATIN 2002. LNCS, vol. 2286, pp. 153–164. Springer, Heidelberg (2002)
6. Chan, T.M.: More planar two-center algorithms. Computational Geometry: Theory and Applications 13, 189–198 (1999)
7. Cole, R., Salowe, J., Steiger, W., Szemeredi, E.: An optimal-time algorithm for slope selection. SIAM Journal on Computing 18, 792–810 (1989)
8. Cole, R.: Slowing down sorting networks to obtain faster sorting algorithms. J. ACM 34, 200–208 (1987)
9. Gupta, U.I., Lee, D.T., Leung, J.Y.-T.: Efficient algorithms for interval graphs and circular-arc graphs. Networks 12, 459–467 (1982)
10. Hwang, R.Z., Chang, R.C., Lee, R.C.T.: The searching over separators strategy To solve some NP-hard problems in subexponential time. Algorithmica 9, 398–423 (1993)
11. Huttenlocher, D.P., Kedem, K., Sharir, M.: The upper envelope of Voronoi surfaces and its applications. Discrete Computational Geometry 9, 267–291 (1993)
12. Hurtado, F., Sacristan, V., Toussaint, G.: Facility location problems with constraints. Studies in Locational Analysis 15, 17–35 (2000)
13. Hochbaum, D.S., Shmoys, D.: A best possible heuristic for the k- center problem. Mathematics of Operations Research 10, 180–184 (1985)
14. Karmakar, A., Roy, S., Das, S.: Fast computation of smallest enclosing circle with center on a query line segment. Information Processing Letters 108, 343–346 (2008)
15. Kim, S.K., Shin, C.S.: Efficient algorithms for two-center problems for a convex polygon. In: Du, D.-Z., Eades, P., Sharma, A.K., Lin, X., Estivill-Castro, V. (eds.) COCOON 2000. LNCS, vol. 1858, pp. 299–309. Springer, Heidelberg (2000)
16. Marchetti-Spaccamela, A.: The p-center problem in the plane is NP-complete. In: Proc. 19th Allerton Conf. on Communication, Control and Computing, pp. 31–40 (1981)
17. Matousek, J.: Randomized optimal algorithm for slope selection. Information Processing Letters 39, 183–187 (1991)
18. Megiddo, N.: Linear-time algorithms for linear programming in R^3 and related problems. SIAM Journal Comput. 12, 759–776 (1983)
19. Olariu, S., Schwing, J.L., Zhang, J.: Optimal parallel algorithms for problems modeled by a family of intervals. IEEE Transactions on Parallel and Distributed Systems 3, 364–374 (1992)

20. Plesnik, J.: A heuristic for the p-center problem in graphs. Discrete Applied Mathematics 17, 263–268 (1987)
21. Sharir, M.: A near-linear algorithm for the planar 2-center problem. Discrete Computational Geometry 18, 125–134 (1997)
22. Roy, S., Karmakar, A., Das, S., Nandy, S.C.: Constrained minimum enclosing circle with center on a query line Segment. In: Královič, R., Urzyczyn, P. (eds.) MFCS 2006. LNCS, vol. 4162, pp. 765–776. Springer, Heidelberg (2006)
23. Vahrenhold, J.: Line-segment intersection made in-place. Computational Geometry: Theory and Applications 38, 213–230 (2007)

Searching for an Axis-Parallel Shoreline

Elmar Langetepe

University of Bonn, Department of Computer Science I, Bonn, Germany

Abstract. We are searching for an unknown horizontal or vertical line in the plane under the competitive framework. We design a framework for lower bounds on all *cyclic* and *monotone* strategies that result in two-sequence functionals. For optimizing such functionals we apply a method that combines two main paradigms. The given solution shows that the combination method is of general interest. Finally, we obtain the current best strategy and can prove that this is the best strategy among all cyclic and monotone strategies which is a main step toward a lower bound construction.

Keywords: Search games, online algorithm, competitive analysis, combinatorial optimization, two-sequence functionals.

1 Introduction

Let us assume that we are lost at sea without sight and we are searching for an unknown shoreline in the competitive sense. That is, we compare the length of our search path until arriving at the shoreline to the length of the shortest path to the shoreline if it was known in advance. The logarithmic spiral conjecture says that the best strategy is a logarithmic spiral and the best spiral achieves a competitive ratio of $13.81113\ldots$, see [2,3,5,6]. This old fundamental search problem of searching for a line in the plane has recently attracted new attention.

On the one hand it was recently shown that spiral search is optimal for the *searching-for-point-in-the-plane* scenario [16]. This result gives hope that it will be possible to prove that the logarithmic spiral conjecture is also true if we are searching for a line.

On the other hand there is a new upper bound on the competitive ratio for finding an axis-parallel shoreline [11]. The given strategy makes use of a special representation that has to be optimized. Finally, the strategy achieves a ratio of $12.5406\ldots$

In this paper we make use of a unique representation of the lower bounds of all cyclic and monotone strategies in the axis-parallel shoreline setting. The representation is shown in detail in Sect. 3. The main benefit is that the problem results in the optimization of two-sequence functionals and finally gives an optimal cyclic and monotone strategy. For optimizing two-sequence functionals we can apply a combination of two main paradigms. This approach was introduced in [12,17] and the given example shows that the generic approach is of general interest. We easily achieve the same ratio as in [11] by considering a special

W. Wu and O. Daescu (Eds.): COCOA 2010, Part I, LNCS 6508, pp. 369–384, 2010.

case of our representation. Furthermore, we can slightly improve the above ratio and we can also state that for *cyclic and monotone* strategies there is definitely no hope for further improvements. The main open question is whether there is always a cyclic and monotone optimal strategy.

2   Preliminaries

We are searching for an unknown line l that is parallel to one of the axes. Since l can be everywhere in the plane any search path Π finally has to visit all possible lines l. Consider the length of a search path Π from the origin o to the first point p_l where some line l is met. Let $\Pi_o^{p_l}$ denote this path and $|\Pi_o^{p_l}|$ its distance. Competitive analysis compares $|\Pi_o^{p_l}|$ to the length of the shortest path from o to l, denoted by $|ol_\perp|$. The worst-case location of l gives the competitive ratio of the strategy which means $C = \sup_l \frac{|\Pi_o^{p_l}|}{|ol_\perp|}$. Then we ask for a strategy Π that attains the smallest possible constant C.

Competitive analysis was introduced by Sleator and Tarjan [22], and used in many settings since then, see for example the survey by Fiat and Woeginger [4] or, for the field of online robot motion planning, see the surveys [10,19]. Note that we assume that the unknown line is at least one step away from the origin. Otherwise we have to introduce a fixed additive constant. Both interpretations are equivalent, see [22].

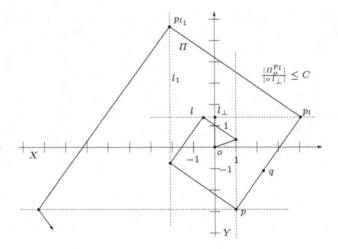

Fig. 1. The strategy of Jeż and Łopuszański [11] is cyclic and monotone. The current worst case occurs at kinks of the strategy, see the points p_l and p.

Somehow it seems to be reasonable that a strategy visits the four possible directions, east, north, west and south and so on in a cyclic order. Additionally, if a corresponding direction was already visited at distance d from the start, the next visit should be beyond d. In this sense we define cyclic and monotone strategies as follows. The strategy Π can be parameterized in polar-coordinates $(\phi, d(\phi))$, the angle ϕ is monotonically increasing and we have $d(\phi) \leq d(\phi + 2\pi)$.

In [11] such a strategy is given as a polygonal chain defined by a sequence of points (x_i, y_i) for $i = 1, 2, 3, \ldots$ These points are visited in the given order. More precisely Jeż and Łopuszański define a strategy by $x_{2k+1} = -\alpha\, x_{2k-1}$, $x_{2k} = x_{2k-3}$, $y_{2k+2} = -\alpha\, y_{2k}$ and $y_{2k+1} = y_{2k-2}$ with an expanding factor α. The reason for this special formulation is that they would like to let the projections onto the axes expand by a factor α. In Fig. 1 there is an example of the strategy for the best α with starting values $x_1 = 1$ and $y_2 = \sqrt{a}$. For $\alpha = 2.03 \ldots$ the strategy achieves a ratio of $12.5406 \ldots$, the worst case is attained at the points (x_i, y_i).

While a strategy is running there is a current depth up to which all lines in a given direction already have been visited. The smallest current depth among all directions will be responsible for a worst case ratio in the very near future. For example in Fig. 1 at point q the smallest current depth among all directions up to point q is given at line l. At p_l a *local* worst case ratio is achieved. Thus we have some discrete points where a worst case ratio in a given direction is achieved. In general such local worst case situation will at least define a lower bound on the competitive ratio.

When a local worst case situation for a given direction is attained, the strategy might expand this direction for a while. For example in Fig. 1 from p_l to p_{l_1} the given strategy *expands* into the north until the next local worst case situation is achieved in the west at p_{l_1}. In our case the ratio for all lines to the north which are newly met during this movement is smaller than the ratio previously attained at p_l. Of course it is not clear that an optimal strategy has to behave in this fashion.

Furthermore, the strategy presented in [11] and illustrated in Fig. 1 behaves in a very special way. The local worst case situation for the strategy in Fig. 1 is always attained exactly at a kink at coordinate $x_{2k} = x_{2k-3}$ or $y_{2k+1} = y_{2k-2}$. Additionally, the strategy ends its expansion there.

3   A Lower Bound Design

We would like to make use of a more general representation of a cyclic and monotone strategy. Finally, we would like to find a unique lower bound design for all local worst case situations. It will turn out that this gives a strategy that achieves the best local worst case ratio as its overall ratio. Thus the strategy is an optimal cyclic and monotone strategy. The example of a cyclic and monotone strategy in Fig. 2(i) shows its general behaviour and the local worst case situations. In the following let d_i denote an axis parallel line and for simplicity d_i also denotes its orthogonal distance to the origin. While the strategy is moving around, local worst case situations for the ratio occur at points p_i on axis-parallel lines d_i, see Fig. 2(i). The corresponding line d_i was visited *one round before* at point q_i. Before p_i is visited, exactly the lines with distance $\leq d_i$ in the specified direction already have been detected. Closely behind d_i and after p_i was visited

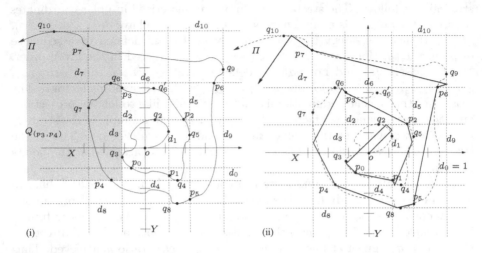

Fig. 2. (i) A cyclic and monotone strategy successively visits local worst case situations at discrete points p_i. Between p_i and p_{i+1} the strategy can only *expand* the directions of d_i and d_{i-1}. (ii) For the original strategy in (i) we replace the movements between local worst case situations p_i and p_{i+1} accordingly. The ratio closely behind the points p_i will never be greater than the ratio at those points in (i).

a line at distance $d_i + \varepsilon$ is met. This gives a local worst case for the ratio and in turn a lower bound for the overall ratio. In the very beginning there is a local worst case ratio for the last line that is visited at distance 1, see for example p_0 in Fig. 2(i). The strategy proceeds in a cyclic manner and therefore such local worst case situation occur in a cyclic order east, north, south, west and so on.

Between two local worst case situation p_i and p_{i+1} the strategy might further *expand* a given direction. This will help for future local worst case situations. For example in Fig. 2(i) between the local worst case situations at p_4 (south) and p_5 (east) the strategy expands up to distance d_8 at point q_8 to the south. This will later give a new local worst case situation for d_8.

The expansion between two local worst case situations need not be unique. For example between p_1 (east) and p_2 (north) the cyclic strategy expands a bit more to the south at q_4 and then to the east up to q_5 before visiting p_2 in the north. Another interesting situation occurs between the local worst case situations at p_2 and p_3. The direction north is expanded up to point q_6'. But later on between p_3 and p_4 the point q_6' is subsumed by a visit of q_6. The corresponding line d_6 is farther away from the start than a line that runs through q_6'.

In general for a cyclic and monotone strategy we can characterize the following behaviour. Between two local worst case situations p_i and p_{i+1} the strategy can only expand the direction of d_i and the direction of the preceeding worst case situation d_{i-1}. This is always true since the corresponding lines d_i and d_{i+1} of p_i and p_{i+1} *span* a quadrant where the strategy has to run in. In this quadrant only the distances of d_i and of d_{i-1} can be expanded to d_{i+4} and d_{i+3}, respectively. For example in Fig. 2(i) between p_3 and p_4 the strategy runs in the quadrant indicated by the shaded rectangle $Q_{(p_3,p_4)}$. Here the strategy expands to the south up to d_6 at q_6 and to the west up to d_7 at q_7. Between p_3 and p_4 the strategy has to move inside the quadrant indicated by $Q_{(p_3,p_4)}$.

3.1 Polygonal Path between p_i and p_{i+1}

As mentioned in the beginning we would like find a lower bound on all local worst case situation. Therefore we replace the movement from p_i to p_{i+1} by a polygonal path as follows. For the strategy in Fig. 2(i) the corresponding replacement is shown in Fig. 2(ii).

Polygonal path from p_i to p_{i+1} and its expansions

1. A segment from p_i to p_{i+1}, if the maximal expansion of d_i and d_{i-1} to d_{i+4} and d_{i+3} occurs outside the path from p_i to p_{i+1}. See for example the path between p_2 and p_3 in Fig. 2(i) and Fig. 2(ii) where neither d_6 nor d_5 is visited.
2. A path from p_i to p_{i+1} that makes a specular reflection at d_{i+4}, if the maximal expansion for d_i to d_{i+4} occurs between p_i to p_{i+1} and the maximal expansion for d_{i-1} to d_{i+3} occurs before p_i is met. See for example the path between p_4 and p_5 in Fig. 2(i) and Fig. 2(ii) where d_8 is visited but not d_7.
3. A path from p_i to p_{i+1} that makes a specular reflection at d_{i+3}, if the maximal expansion for d_{i-1} to d_{i+3} occurs between p_i to p_{i+1} and the maximal expansion for d_i to d_{i+4} occurs before after p_{i+1} is met. See for example the path between p_6 and p_7 in Fig. 2(i) and Fig. 2(ii) where d_9 is visited but not d_{10}.
4. A path from p_i to p_{i+1} that makes a specular reflection at d_{i+3} and d_{i+4} if the maximal expansion for d_i to d_{i+4} and for d_{i-1} to d_{i+3} occurs between p_i to p_{i+1}. See for example the path between p_3 and p_4 in Fig. 2(i) and Fig. 2(ii) where d_7 and d_6 is visited.
5. In the beginning we move from the start to p_0 and visit the last shoreline at distance $d_0 = 1$ with some specular reflections at d_1, d_2 and d_3. See for example the path from the start to p_0 in Fig. 2(i) and Fig. 2(ii) where d_1, d_2 and d_3 is visited.

By using specular reflections we have constructed the shortest path from p_i and p_{i+1} that also fulfills the expansion into the specified directions. This means that in the above reformulation the ratio at the local worst case situations at points p_i will never be greater than the ratio at such points in the original strategy. Additionally, in comparison to the strategy of Jeż and Łopuszański presented in Fig. 1 our reformulation is more general. We allow additional kinks between the points p_i and p_{i+1}.

For any cyclic and monotone strategy we can do the replacement of the path between p_i and p_{i+1} as indicated above. The next step is that we would like to find a unique representation of the paths between p_i and p_{i+1}. Obviously any cyclic and monotone strategy defines an infinite sequence of positive values $(d_0, d_1, d_2, d_3, \ldots)$. The remaining task is that we have to fix the points p_i on d_i.

Instead of fixing all p_i we would like to find a unique representation for the length of the polygonal chain from p_i and p_{i+1}. We will make use of a second infinite sequence of values $(\beta_0, \beta_1, \beta_2, \beta_3, \ldots)$. Our goal is to use this second sequence, so that the length of the path between any p_i and p_{i+1} is uniquely determined by

$$\sqrt{(d_{i+1} + \beta_i \, d_i)^2 + (2d_{i+4} - \beta_{i+1} \, d_{i+1} - d_i)^2} \tag{1}$$

regardless of which of the first four cases in the enumeration above occurs.

The starting round from the start to p_0 with some specular reflections at d_1, d_2 and d_3 (item 5. in the enumeration above) should also fit into our representation. Therefore, for technical reasons we will make use of $d_{-1} := -2d_1$. Furthermore, we will extend β by $\beta_{-1} := -\frac{d_2}{d_1}$. One can verify that in this case

$$\sqrt{(d_0 + \beta_{-1} \, d_{-1})^2 + (2d_3 - \beta_0 \, d_0 - d_{-1})^2} \tag{2}$$

exactly represents the first three reflections. We omit the details.

If we can guarantee that the length for any polygonal path between p_i and p_{i+1} is given by (1) and β_{-1} and d_{-1} choosen adequately as motivated above, it is easy to verify that the supremum of all local worst case situations can be defined by

$$\sup_k \frac{\sum_{i=0}^{k} \sqrt{(d_i + \beta_{i-1} \, d_{i-1})^2 + (2d_{i+3} - \beta_i \, d_i - d_{i-1})^2}}{d_k}. \tag{3}$$

The numerator gives a lower bound to the length of the path at the local worst case situation at p_k. The local worst case shoreline is detected closely behind d_k.

Fortunately, (3) represents a functional defined by two infinite sequences $(d_{-1}, d_0, d_1, d_2, d_3, \ldots)$ and $(\beta_{-1}, \beta_0, \beta_1, \beta_2, \beta_3, \ldots)$. A method for minimizing such functionals was recently presented in [17]. In the next sections we will show how to find the optimal sequences for minimizing (3).[1] Finally, this gives the minimal lower bound for all cyclic and monotone strategies and will also define a discrete strategy that attains the corresponding optimal ratio.

3.2   Unique Representation of the Path from p_i to p_{i+1}

The remaining task of this section is to interpret the values $(\beta_{-1}, \beta_0, \beta_1, \beta_2, \beta_3, \ldots)$. so that the movement between p_i and p_{i+1} always has length (1). The idea is that we translate the specular reflections from p_i to p_{i+1} to a single segment of the same length, see Fig. 3(i). If there is only a single reflection between p_i and p_{i+1}, we reflect either p_i or p_{i+1} on d_{i+3} or d_{i+4}, respectively. For example in Fig. 3(i) p_5 is reflected on d_8 to p'_5 for the path between p_4 and p_5 and the segment $p_4 p'_5$ has the same length as the path between p_4 and p_5. Analogously, p_6 is reflected on d_9 to p'_6 for the path between p_6 and p_7. Here $p'_6 p_7$ has the same length as the polygonal path between p_6 and p_7.

---

[1] In the following we will always fix the values $d_{-1} := -2d_1$ and $\beta_{-1} := -\frac{d_2}{d_1}$.

Similarily, if there is a specular reflection on d_{i+3} and d_{i+4} between p_i and p_{i+1} we project p_i on d_{i+3} and p_{i+1} on d_{i+4}. Thus we obtain p'_i and p'_{i+1} and the length of the segment $p'_i p'_{i+1}$ equals the length of the path from p_i and p_{i+1}. For example see the path from p_3 to p_4 which can be replaced by the line segment $p'_3 p'_4$.

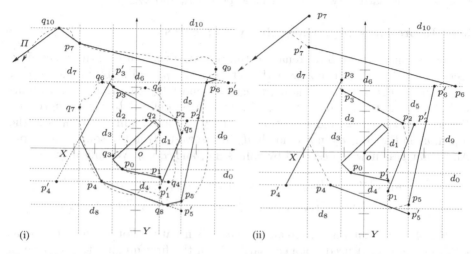

Fig. 3. (i) The polygonal path from p_i to p_{i+1} is replaced by a segment if we *resolve* the reflections. (ii) After renaming, the polygonal path from the original p_i to the original p_{i+1} is always represented by the segment $p_i p'_{i+1}$.

In general the rule is that we reflect p_i on d_{i+3} and obtain p'_i. For a general representation we would like to translate the path between p_i and p_{i+1} by the segment $p_i p'_{i+1}$. The problem is that there might be two reflection between p_i and p_{i+1} and we have to consider the segment $p'_i p'_{i+1}$ instead of $p_i p'_{i+1}$, compare $p'_3 p'_4$ in Fig. 3(i). Similarily, the polygonal path between p_i and p_{i+1} should be translated to $p'_i p_{i+1}$ instead of $p_i p'_{i+1}$, compare $p'_6 p_7$ in Fig. 3(i). We solve this problem by renaming.

3.3   Let $p_i p'_{i+1}$ Always Represent the Path from p_i and p_{i+1}

In order to use a unique representation we make use of the following trick. If $p_i p'_{i+1}$ does not represent the polygonal path between p_i and p_{i+1}, we allow to rename p_i by p'_i or p_{i+1} by p'_{i+1} and vice versa. The overall rule is that if the expansion of p_i to d_{i+4} does not occur between p_i and p_{i+1} (and therefore has to occur between p_{i+1} and p_{i+2}), we will rename p_{i+1} by p'_{i+1} and vice versa. For example, in Fig. 3(ii) we have renamed the reflection points p_1, p_2, p_3, p_6 and p_7 from Fig. 3(i) in this sense.

This renaming is always consistent for the neighboring paths. If the expansion of p_i on d_{i+4} does occur between p_{i+1} and p_{i+2}, we have the following situation. The segment for p_i and p_{i+1} should end at p_{i+1} and the segment for p_{i+1} and

p_{i+2} should start at p'_{i+1}. Therefore the renaming of p_{i+1} by p'_{i+1} and vice versa guarantees that the polygonal paths are finally interpreted by $p_i p'_{i+1}$ and $p_{i+1} p'_{i+2}$. For example, in Fig. 3(i) the expansion of p_5 to d_9 occurs between p_6 and p_7. Therefore we rename p_6 and p'_6, see Fig. 3(ii). Thus the path $p_5 p'_6$ represents the length of the path between the original points p_5 and p_6 and $p_6 p'_7$ represents the path between the original points p_6 and p_7.

3.4 Represent $p_i p'_{i+1}$ Uniquely by Sequences

Let us assume that we have renamed the points in the sense of the preceeding section. Finally, we will represent the set of points p_i and p'_i by making use of two infinite sequences $(d_{-1}, d_0, d_1, d_2, d_3, \dots)$ and $(\beta_{-1}, \beta_0, \beta_1, \beta_2, \beta_3, \dots)$. This will show that the segment $p_i p'_{i+1}$ always can be defined to have length (1). W.l.o.g. for the cyclic and monotone strategy we can assume that d_0 is in the south, d_1 in the east, d_2 in the north, d_3 in the west, d_4 in the south and so on. Any point p_i is now represented by values $\pm d_i$ and $\pm \beta_i d_i$. We uniquely set

$$p_{4k} := (-\beta_{4k} d_{4k}, -d_{4k}), \quad p_{4k+1} := (d_{4k+1}, -\beta_{4k+1} d_{4k+1}),$$
$$p_{4k+2} := (\beta_{4k+2} d_{4k+2}, d_{4k+2}), \quad p_{4k+3} := (-d_{4k+3}, \beta_{4k+3} d_{4k+3}),$$

see Fig. 4. Note that we have to allow that β_i is negative. For example the point $(\beta_{4k+2} d_{4k+2}, d_{4k+2})$ need not necessarily lie in the first quadrant, it can be in the first or second quadrant. More precisely, the point $p_6 = (\beta_6 d_6, d_6)$ in Fig. 4 could have been located in the second quadrant, if β_6 is negative. In principle this is allowed. For convenience and for abusing confusing intersections we did not use negative β_i in our examples, in general this would make no difference.

In any case the reflection points p'_i are always given by a reflection of p_i on d_{i+3}. For example, if we reflect $p_{4k+2} := (\beta_{4k+2} d_{4k+2}, d_{4k+2})$ on d_{4k+5} in order to obtain p'_{4k+2}, we have the same Y-coordinate d_{4k+2} but the X-coordinate of p'_{4k+2} is $d_{4k+5} + (d_{4k+5} - \beta_{4k+2} d_{4k+2}) = 2d_{4k+5} - \beta_{4k+2} d_{4k+2}$. See for example $p'_6 = (2d_9 - \beta_6 d_6, d_6)$ in Fig. 4. Note, that the reflection construction from p_i to p'_i also holds, if β_i is negative.

In general the reflection points can be now uniquely defined be the following representation:

$$p'_{4k} := (-2d_{4k+3} + \beta_{4k} d_{4k}, -d_{4k}), \quad p'_{4k+1} := (d_{4k+1}, -2d_{4k+4} + \beta_{4k+1} d_{4k+1})$$
$$p'_{4k+2} := (2d_{4k+5} - \beta_{4k+2} d_{4k+2}, d_{4k+2}), \quad p'_{4k+3} := (-d_{4k+3}, 2d_{4k+6} - \beta_{4k+3} d_{4k+3}).$$

Altogether, we can now express the length of $p_i p'_{i+1}$ by a unique formula using the correct definition of p_i and p'_i. Surprisingly, it turns out that $p_i p'_{i+1}$ has always length (1). For example, for $i = 4k + 1$ the Euclidean length of the segment from $p_{4k+1} := (d_{4k+1}, -\beta_{4k+1} d_{4k+1})$ to $p'_{4k+2} := (2d_{4k+5} - \beta_{4k+2} d_{4k+2}, d_{4k+2})$ is given by $\sqrt{(d_{4k+2} + \beta_{4k+1} d_{4k+1})^2 + (2d_{4k+5} - \beta_{4k+2} d_{4k+2} - d_{4k+1})^2}$. One can easily check that $p_i p'_{i+1}$ has length (1) for any i.

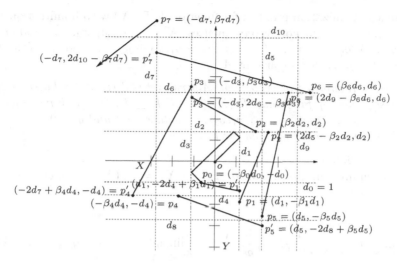

Fig. 4. After renaming the segment $p_i p'_{i+1}$ can be uniquely represented by (1).

4   Optimization of Single Sequence Functionals

We will briefly repeat the methods presented in [17] and start with a single sequence functional. There are two main paradigms for computing optimal strategies by optimizing single sequence functionals. We exemplify the application of the first approach to the functional

$$F_k(d_{-1}, d_0, d_1, \ldots) := \frac{\sum_{i=0}^k \sqrt{(d_{i+3} - d_{i-1})^2 + (d_{i+2} + d_i)^2}}{d_k}. \tag{4}$$

More precisely, we are searching for an infinite strategy $D = d_{-1}, d_0, d_1 \ldots$ so that

$$\inf_Y \sup_k F_k(Y) = C \quad \text{and} \quad \sup_k F_k(D) = C.$$

For our situation the above functional can be interpreted as follows. If we let $\beta_i d_i$ always be equal with d_{i+3}, we will obtain a strategy that only makes kinks at the local worst case situations, compare Fig. 3(ii) and Fig. 1.

The length of the segment $p_i p'_{i+1}$ in (1) is given by $\sqrt{(d_{i+4} - d_i)^2 + (d_{i+3} + d_{i+1})^2}$ and (3) equals (4). Again we can adapt d_{-1} so that the first round is expressed by the first summand of the numerator in (4).

Altogether, (4) defines a strategy that behaves like the strategy of Jeż and Łopuszański presented in Fig. 1. We can prove that without knowing it they found the best strategy among all such strategies.

4.1   Optimality of the Exponential Function

The following theorem states that the supremum of a functional is minimized by an exponential function, if certain properties are fulfilled. The given problem results in an optimization problem for functionals $F_k(X)$ with infinite sequences $X = (x_0, x_1, x_2, \ldots)$. For two sequences $X = (x_0, x_1, x_2, \ldots)$ and $Y = (y_0, y_1, y_2, \ldots)$ let $X + Y := (x_0 + y_0, x_1 + y_1, x_2 + y_2, \ldots)$ and $A \cdot X := (A \cdot x_0, A \cdot x_1, A \cdot x_2, \ldots)$ for a constant A.

Theorem 1. *(adapted from Gal [7,6], Alpern and Gal [1] and Schuierer [21])*
Given a sequence of functional $F_k(X)$ for all $k \geq k_0$ and infinite sequences
$X = (x_0, x_1, x_2, \ldots)$ and $Y = (y_0, y_1, y_2, \ldots)$ with $x_i > 0$ and $y_i > 0$.
  If the following conditions hold for F_k:

(i) F_k is continous,
(ii) F_k is unimodal, which means: $F_k(A \cdot X) = F_k(X)$ and $F_k(X + Y) \leq \max\{F_k(X), F_k(Y)\}$,
(iii)

$$\liminf_{a \to \infty} F_k\left(\frac{1}{a^k}, \frac{1}{a^{k-1}}, \ldots, \frac{1}{a}, 1\right) = \liminf_{\varepsilon_k, \varepsilon_{k-1}, \ldots, \varepsilon_1 \to 0} F_k(\varepsilon_k, \varepsilon_{k-1}, \ldots, \varepsilon_1, 1),$$

(iv)

$$\liminf_{a \to 0} F_k\left(1, a, a^2, \ldots, a^k\right) = \liminf_{\varepsilon_k, \varepsilon_{k-1}, \ldots, \varepsilon_1 \to 0} F_k(1, \varepsilon_1, \varepsilon_2, \ldots, \varepsilon_k,),$$

(v) $F_{k+1}(x_0, \ldots, x_{k+2}) \geq F_k(x_1, \ldots, x_{k+2})$.

then

$$\sup_k F_k(X) \geq \inf_a \sup_k F_k(A_a)$$

with $A_a = a^0, a^1, a^2, \ldots$ und $a > 0$. The supremum of the functional is minimized
by an exponential function.

We are able to prove that the functional $F_k(d_{-1}, d_0, \ldots, d_k)$ fulfills the conditions
of Theorem 1 and we conclude

$$\sup_k F_k(X) \geq \inf_a \sup_k F_k(A_a)$$

where $A_a = a^0, a^1, a^2, \ldots$ and $a > 0$. We can substitute d_i by a^i and can make
use of a geometric serie. We omit some simple analytic details and only present
the results. The problem now can be solved by

$$\inf_a \frac{\sum_{i=0}^k \sqrt{(a^{i+3} - a^{i-1})^2 + (a^{i+2} + a^i)^2}}{a^k} = \min_a \frac{\sqrt{(a^4 - a^2 + 1)(a^2 + 1)^2}}{a - 1}$$

Thus we have found a simple function that has to be minimized over a. Optimizing the last function in a by analytic means gives $a = 1.425421\ldots$ and exactly
the ratio $12.54064\ldots$ presented in [11]. The ratio is attained asymptotically at
the local worst case situations, for the first round there is some freedom.

  Altogether, this analysis shows that there is definitely no room for improvements, if we choose a cyclic and monotone strategy that always makes a kink at
the local worst case situation.

4.2   Equality Approach

On the other hand some authors [9,15,18,13,20] suggest to adjust an optimal strategy $X = x_0, x_1, x_2 \ldots$ with $F_k(X) \leq C$ to an optimal strategy $X' = x'_0, x'_1, x'_2 \ldots$ with $F_k(X') = C$ where C is the (probably unknown) best achievable factor. Then one will try to retrieve a recurrence for the values of X' from the equation $F_k(X') = C = F_{k+1}(X')$ and find the smallest C that fulfills this recurrence with positive values. In this section we do not apply this method to the functional (4) because this seems to be difficult. Instead we use the functional $F_k(X) := \frac{\sum_{i=0}^{k} x_i}{x_k}$, that stems from the 2-ray search problem, see [2,14].

It can be shown that for the 2-ray search problem such a strategy X with $F_k(X) = C = F_{k+1}(X)$ exists. How will we find the optimal strategy in this case? One will try to retrieve a recurrence for the values of X from the equation $F_k(X) = C = F_{k+1}(X)$.

For the 2-ray search problem we assume that $X = x_0, x_1, x_2 \ldots$ achieves equality in every step. We conclude $\sum_{i=0}^{k+2} x_i = C x_{k+1}$ and $\sum_{i=0}^{k+1} x_i = C x_k$. Subtracting both sides gives the recurrence $x_{k+2} = C(x_{k+1} - x_k)$ for $k = 0, 1, 2, \ldots$ Obtaining positive solutions for recurrences can be solved by analytic means, see [8]. It can be shown that for $C < 4$ there is no positive sequence that fulfills the given recurrence $x_{k+2} = C(x_{k+1} - x_k)$. Furthermore, for $x_i := (i+1)2^i$ we have $x_{k+2} = (k+3)2^{k+1} = 4(x_{k+1} - x_k) = (3k+4)2^{k+2} - (k+1)2^{k+2}$. This means, that there is a positive sequence, $x_i := (i+1)2^i$ that attains the competitive ratio $C = 4$. This means that $C = 4$ is optimal.

Altogether, we have two different approaches for obtaining optimal strategies stemming from different paradigms. In the following we will combine both paradigms in order to solve two-sequence functionals, see also [17].

5   Optimizing Two-Sequence Functionals

We would like to find the best cyclic strategy and have to optimize ratio (3). Our task is to find optimal sequences $\beta = (\beta_{-1}, \beta_0, \beta_1, \ldots)$ and $D = (d_{-1}, d_0, d_1, d_2, \ldots)$ that minimizes the supremum of the following functional for all k.

$$F_k(\beta, D) := \frac{\sum_{i=0}^{k} \sqrt{(d_i + \beta_{i-1} d_{i-1})^2 + (2d_{i+3} - \beta_i d_i - d_{i-1})^2}}{d_k} \tag{5}$$

Let us first assume that the sequence β is fixed. We omit the details but we can apply Theorem 1 to ratio (5). This means that a strategy $d_i = a^i$ will optimize $\sup_k F_k(D, \beta)$. But there is still a second sequence β that has to be optimized and a simple function for finding a and β is not given. Therefore we suggest to apply the second paradigm (equality approach) of Sect. 4.2 first. We would like to reduce the complexity of the problem, see also the example in [17].

Lemma 1. *For the functional (5) and the optimal sequences D and β with $\sup_k F_k(D, \beta) = C$ there always exists sequences D' and β' so that $F_l(D', \beta') = C$ is fulfilled for all $l \geq 0$.*

Proof. The proof works by induction on k. It is non-constructive because an optimal strategy is not known so far and we use a limit process for the inductive step.

Let us assume that an optimal cyclic strategy that minimizes $\sup_k F_k(D, \beta) = C$ is given. We would like to show by induction that for every k there is always a strategy D' and β' so that $F_l(D, \beta) = C$ is fulfilled for all $0 \leq l \leq k$ and $F_l(D, \beta) \leq C$ for $l > k$.

First, we show that we can let a single d_k shrink a bit. Let us assume that $F_k(D, \beta) < C$ holds. Then we can let d_k decrease to d'_k and adjust β_k to β'_k so that $F_k(D', \beta') = C$ and $F_l(D', \beta') \leq C$ for all $l \neq k$ holds.

The adjustment works as follows and is motivated in Fig. 5. If we move the line of d_k toward the origin then three *movements* of the strategy are concerned. First, the movement between the local worst case points on d_{k-1} to d_{k+1} that reflects on d_k changes. This part of the strategy will always get shorter, if we move d_k towards to the origin. If one of the corresponding segments become horizontal or vertical, we proceed by moving the corresponding segments toward the origin, also.

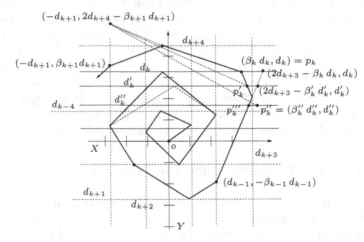

Fig. 5. If we move x_k downwards to d'_k closer to the origin, we decrease all distances which are concerned. This remains true, if we move beyond the reflection at d_{k+3}, compare d''_k. The ratio $F_k(D', \beta')$ increases, all other ratios decrease or remain the same.

Additionally, d_k and $\beta_k d_k$ defines a point p_k. The movement from p_k to the next local worst case point on d_{k+1} that might reflect on d_{k+4} and/or d_{k+3} and the movement of p_k toward the local worst case point of d_{k-1} that might reflect on d_{k+3} and/or d_{k+2} could be concerned, if we move d_k to the origin. In Fig. 5 the path from p_k to p_{k+1} reflects on d_{k+4} and the path from p_k to p_{k-1} reflects on d_{k+3}.

Let the line for d_k move towards the origin. Let us first assume that the kink happens on d_{k+3} before p_k is met as in Fig. 5. The other case will be handled later.

Let the intersection point p'_k with the original strategy define the new local worst case point. In turn it defines the value of β'. Thus, the length of the path will not increase. In fact by triangle inequality the path length decrease a bit because the kink appears a bit *earlier*, see the dashed path in Fig. 5.

Until reaching the reflection point at d_{k+3} using the adapted values d'_k and β'_k will always decrease the overall path length as indicated above. Fortunately, we can even move further on without changing the description of the ratio or functional, respectively.

We simply let the point p''_k move beyond d_{k+3} which means that $\beta''_k d''_k > d_{k+3}$ holds, see Fig. 5. The coordinates of the corresponding point still depends on d''_k and $\beta''_k d''_k$ in the same way. Similar to the renaming in Sect. 3.3 here p''_k and its reflection on d_{k+3}, namely p'''_k, change the role.

The movement from p_{k-1} to p'''_k and the movement from p''_k to p_{k+1} is described by the same formulas as before. More precisely, by

$$\sqrt{(d_{k+1} + \beta''_k d''_k)^2 + (2d_{k+4} - \beta_{k+1} d_{k+1} - d''_k)^2} \text{ and}$$

$$\sqrt{(d''_k + \beta_{k-1} d_{k-1})^2 + (2d_{k+3} - \beta''_k d''_k - d_{k-1})^2}.$$

Although the reflection on d_{k+3} now appears after p''_k we have the same description of the strategy. That is we can move d'_k closer to the origin to d''_k. The length of the overall path further decrease. We simply adjust β''_k accordingly.

While decreasing d_k the ratio $F_k(D, \beta)$ finally has to increase since there are always elements of the path that will not be concerned. In the induction proof we will see that finally there is some $d'_k > 0$ that let the ratio increase to $F_k(D', \beta') = C$. We will also see that we never have to move d_k below d_{k-4} in order to obtain this equality.

Therefore finally we will have $F_k(D', \beta') = C$ and $F_l(D', \beta') \leq C$ for all $l \neq k$.

The full proof works by induction. First, we let d_0 shrink so that $F_0(D', \beta') = C$ and $F_l(D', \beta') \leq C$ for all $l \geq 1$ holds. This is always possible because some parts of the path to the first point on d_0 remains the same.

The induction hypothesis says that for index k there is always an adjustment of a strategy so that D' and β' exists with $F_l(D', \beta') = C$ holds for all $0 \leq l \leq k$ and $F_l(D', \beta') \leq C$ for $l > k$. Additionally, the adjustment let the values of D' shrink but they will never get to zero.

Now we adjust d_{k+1} and let it shrink to d'_{k+1}. By induction hypothesis $F_{k-3}(D', \beta') = C$ holds before d_{k+1} and β_{k+1} is adjusted. This means that we will attain equality for $F_{k+1}(D', \beta') = C$ for d'_{k+1} and β'_{k+1} before d'_{k+1} reaches d'_{k-3}.

Now we have $F_{k+1}(D', \beta') = C$ and $F_l(D', \beta') \leq C$ for $l \neq k+1$. We apply the induction hypothesis again for the first k values of the new D'. We can repeat this process. This means that we will have shrinking values $(d'_0, d'_1, d'_2, \ldots, d'_{k+1})$ but they will never get to zero. Finally, they have to run into a limit that gives $F_l(D', \beta') = C$ for $0 \leq l \leq k+1$.

Thus, the inductive step is true. For all k there is an optimal strategy so that $F_l(D', \beta') = C$ for $0 \leq l \leq k$.   □

Altogether, we can now apply the idea of Sect. 4.2. We will make use of the fact that there is an optimal cyclic strategy with $F_k(D, \beta) = C = F_{k-1}(D, \beta)$. This means

$$C\,(d_k - d_{k-1}) = \sqrt{(d_k + \beta_{k-1}\,d_{k-1})^2 + (2d_{k+3} - \beta_k\,d_k - d_{k-1})^2}$$

and we obtain a new functional

$$G_k(\beta, D) := \frac{\sqrt{(d_k + \beta_{k-1}\,d_{k-1})^2 + (2d_{k+3} - \beta_k\,d_k - d_{k-1})^2}}{d_k - d_{k-1}} \tag{6}$$

and the new task is to find sequences D and β so that $\sup_k G_k(D, \beta)$ is minimal.

Fortunately, for fixed β we can again apply Theorem 1 to $G_k(D, \beta)$. We omit the details here. This means that there is an optimal strategy for (6) with $d_i := a^i$ for $a > 1$.

Substituting d_i by a^i and using some simple transformation shows that

$$G_k(A, \beta) = \frac{\sqrt{(a + \beta_{k-1})^2 + (2a^2 - \beta_k\,a - 1)^2}}{a - 1}. \tag{7}$$

Let us assume that we have found the best a. Since β_{k-1} takes over the role of β_k if we consider G_{k+1}, the best we can do is let β_i be a constant for all i. A similar statement was shown with more details in [17]. This means that we have to optimize a function $f(a, b) := \frac{\sqrt{(a+b)^2 + (2a^4 - ba - 1)^2}}{a-1}$. We would like to find the minimum of $f(a, b)$ by analytic means and do not present all details. The derivative of $f(a, b)$ in b gives $\frac{b + 2a - 2a^5 + ba^2}{\sqrt{(a+b)^2 + (2a^4 - ba - 1)^2}(a-1)}$. It is zero if and only if $b := 2(a^2 - 1)a$ holds. So for all a we have to use this b for minimization. Finally, we only have to optimize the function $f(a) := \frac{\sqrt{(a^2+1)(2a^2-1)^2}}{a-1}$. For $a > 1$ this function has a unique minimum of $12.53853842\ldots$ for $a = 1.431489\ldots$ and we have $b = 2(a^2 - 1)a = 3.0037344\ldots$

Note that this is only a very small improvement on the strategy of Jeż and Łopuszański but in comparison to the former result we can state that this is the best strategy that visits the directions in a cyclic order and in a monotone manner.

Theorem 2. *An optimal cyclic and monotone strategy that finds a horizontal or vertical shoreline can be described by $d_i = a^i$ and $\beta_i\,d_i = b\,a^i$ in the sense of Sect. 3.4 and obtains an optimal competitive ratio of $12.53853842\ldots$ for $a = 1.431489\ldots$ and $b = 2(a^2 - 1)a = 3.0037344\ldots$*

The strategy is shown in Fig. 6. The values d_i and $\beta_i\,d_i = b\,a^i$ has to be interpreted in the sense of Sect. 3.4. Note that we have $\beta_i\,d_i > d_{i+3}$ and the strategy first visits p_i and then expands the direction of d_{i-1} to d_{i+3}. There is a specular reflection on d_{i+3} between p_i and p_{i+1}. Interestingly, there is also a very slight kink at p_i itself.[2]

---

[2] We have also checked this behaviour analytically. An optimal strategy that behaves in the same way but has kinks only at d_{i+3} attains a greater ratio.

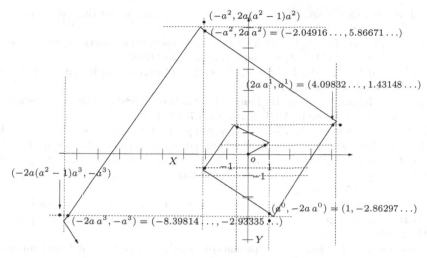

Fig. 6. The optimal cyclic strategy defined by $d_i = a^i$ and $\beta_i d_i = b a^i$ with $a = 1.431489\ldots$ and $b = 2a = 2.826979\ldots$ There is also a small kink whenever a local worst case is met.

Furthermore, the strategy presented above attains its worst case ratio exactly at the local worst case situations. Additionally, it optimizes the local worst case situations for all cyclic and monotone strategies and therefore the strategy is the optimal cyclic and monotone strategy.

6   Conclusion

In this paper we found the current best competitive strategy for searching a horizontal or vertical shoreline. We formalized a lower bound on all cyclic and monotone strategies and by optimizing two-sequence functionals we slightly improve the current best strategy. The main open question is how to find a general (tight) lower bound. It remains to show that there is always an optimal strategy that visits the directions in a cyclic order and a monotone manner.

References

1. Alpern, S., Gal, S.: The Theory of Search Games and Rendezvous. Kluwer Academic Publications, Dordrecht (2003)
2. Baeza-Yates, R., Culberson, J., Rawlins, G.: Searching in the plane. Inform. Comput. 106, 234–252 (1993)
3. Chrobak, M.: A princess swimming in the fog looking for a monster cow. SIGACT News 35(2), 74–78 (2004)
4. Fiat, A., Woeginger, G. (eds.): On-line Algorithms: The State of the Art. LNCS, vol. 1442. Springer, Heidelberg (1998)
5. Finch, S.R., Zhu, L.-Y.: Searching for a shoreline (2005)

6. Gal, S.: Search Games. Mathematics in Science and Engineering, vol. 149. Academic Press, New York (1980)
7. Gal, S., Chazan, D.: On the optimality of the exponential functions for some minmax problems. SIAM J. Appl. Math. 30, 324–348 (1976)
8. Graham, R.L., Knuth, D.E., Patashnik, O.: Concrete Mathematics, 2nd edn. Addison-Wesley, Reading (1994)
9. Hipke, C., Icking, C., Klein, R., Langetepe, E.: How to find a point on a line within a fixed distance. Discrete Appl. Math. 93, 67–73 (1999)
10. Icking, C., Kamphans, T., Klein, R., Langetepe, E.: On the competitive complexity of navigation tasks. In: Hager, G.D., Christensen, H.I., Bunke, H., Klein, R. (eds.) Dagstuhl Seminar 2000. LNCS, vol. 2238, pp. 245–258. Springer, Heidelberg (2002)
11. Jeż, A., Łopuszański, J.: On the two-dimensional cow search problem. Inf. Process. Lett. 109(11), 543–547 (2009)
12. Kamphans, T., Langetepe, E.: On optimizing multi-sequence functionals for competitive analysis. In: Abstracts 21st European Workshop Comput. Geom., pp. 111–114 (2005)
13. Kamphans, T., Langetepe, E.: Optimal competitive online ray search with an error-prone robot. In: Nikoletseas, S.E. (ed.) WEA 2005. LNCS, vol. 3503, pp. 593–596. Springer, Heidelberg (2005)
14. Kao, M.-Y., Reif, J.H., Tate, S.R.: Searching in an unknown environment: An optimal randomized algorithm for the cow-path problem. Inform. Comput. 133(1), 63–79 (1996)
15. Langetepe, E.: Design and Analysis of Strategies for Autonomous Systems in Motion Planning. PhD thesis, Dep. of Comp. Science, University of Hagen (2000)
16. Langetepe, E.: On the optimality of spiral search. In: SODA 2010: Proc. 21st Annu. ACM-SIAM Symp. Disc. Algor., pp. 1–12 (2010)
17. Langetepe, E.: Optimizing two-sequence functionals for competitive analysis. Accepted for Computer Science – Research and Development (2010)
18. López-Ortiz, A., Schuierer, S.: The ultimate strategy to search on m rays? Theor. Comput. Sci. 261(2), 267–295 (2001)
19. Rao, N.S.V., Kareti, S., Shi, W., Iyengar, S.S.: Robot navigation in unknown terrains: introductory survey of non-heuristic algorithms. Technical Report ORNL/TM-12410, Oak Ridge National Laboratory (1993)
20. Schuierer, S.: Searching on m bounded rays optimally. Technical Report 112, Institut für Informatik, Universität Freiburg, Germany (1998)
21. Schuierer, S.: Lower bounds in on-line geometric searching. Comput. Geom. Theory Appl. 18, 37–53 (2001)
22. Sleator, D.D., Tarjan, R.E.: Amortized efficiency of list update and paging rules. Commun. ACM 28, 202–208 (1985)

Bounded Length, 2-Edge Augmentation of Geometric Planar Graphs

Evangelos Kranakis[1,*], Danny Krizanc[2],
Oscar Morales Ponce[3,**], and Ladislav Stacho[4,***]

[1] School of Computer Science, Carleton University, Ottawa, ON, K1S 5B6, Canada
[2] Department of Mathematics and Computer Science, Wesleyan University,
Middletown CT 06459, USA
[3] School of Computer Science, Carleton University, Ottawa, ON, K1S 5B6, Canada
[4] Department of Mathematics, Simon Fraser University, 8888 University Drive,
Burnaby, British Columbia, Canada, V5A 1S6

Abstract. Algorithms for the construction of spanning planar subgraphs of Unit Disk Graphs (UDGs) do not ensure connectivity of the resulting graph under single edge deletion. To overcome this deficiency, in this paper we address the problem of augmenting the edge set of planar geometric graphs with straight line edges of bounded length so that the resulting graph is planar and 2-edge connected. We give bounds on the number of newly added straight-line edges and show that such edges can be of length at most 3 times the max length of the edges of the original graph; also 3 is shown to be optimal. It is shown to be NP-hard to augment a geometric planar graph to a 2-edge connected geometric planar with the minimum number of new edges of a given bounded length. Further, we prove that there is no local algorithm for augmenting a planar UDG into a 2-edge connected planar graph with straight line edges.

Keywords and Phrases: Augmentation, Deletion, 2-edge connected, Geometric, Local, Minimum number of edges, Planar, UDG.

1 Introduction

In several network applications it is desired to construct a spanning subgraph of a given unit disk graph graph with "robust connectivity", in the sense that the spanning subgraph remains connected under edge deletion. The usual graph parameter quantifying this robustness is called k-connectivity: a graph G is called k-edge (respectively, k-vertex) connected if it remains connected despite the deletion of any $k-1$ edges (respectively, vertices). k-connectivity is an important property because it implies fault tolerance under either edge or vertex deletions.

The main question arising is given a UDG on a set of sensors how to construct a k-edge (respectively, vertex) connected spanning graph respecting the

---

* Supported in part by NSERC and MITACS grants.
** Supported in part by CONACyT and NSERC grants.
*** Supported in part by NSERC grant.

W. Wu and O. Daescu (Eds.): COCOA 2010, Part I, LNCS 6508, pp. 385–397, 2010.

UDG and that has good planarity and connectivity characteristics as well as edge lengths which are a constant multiple of the unit radius of the UDG. Two approaches (or a combination thereof) can be considered. In the first one, called *edge deletion*, existing edges of a given graph are removed to obtain the desired spanning planar subgraph of the given graph. In this case, the main issue arising is whether or not such a spanning subgraph exists and how to attain it with a minimum number of deletions. In the second one, called *edge augmentation*, a planar spanning graph is obtained from the original graph by adding new edges. In this case, the main issues arising are the number of edges being added, as well as the length of the augmented edges (since these represent the ranges of the corresponding sensors) which should be bounded by a constant independent of the size of the network.

1.1 Related Work

Both edge deletion and edge augmentation problems have been considered in the literature. Characterizations on the number of edges for augmenting a graph to a 2-edge connected graph as well as weighted versions (shown to be NP-complete) can be found in [4]. For any integer $k > 1$, [5] and [13] give an algorithm for the minimum number of edges for augmenting any graph G to a k-edge connected graph in polynomial time. In [2] it is proved that given a 2-edge connected graph there is an algorithm running in time $O(mn)$ which finds a 2-edge connected spanning subgraph whose number of edges is $17/12$ times the optimal, where m is the number of edges and n the number of vertices of the graph. An improvement is provided in [12] in which a $4/3$ approximation algorithm is given. Later, Jothi et al. [6] provided a $5/4$-approximation algorithm. However in these papers the resulting spanning subgraph are not guaranteed to be planar.

The problem changes significantly when we restrict our attention to planar graphs. In fact, [8] proved it is NP-hard to determine the minimum number of edges required to be added to augment a given planar graph into a 2-vertex connected planar graph. [11] proved that it is also NP-hard to augment a geometric planar graph to a 2-edge connected geometric planar with the minimum number of new straight line edges (but the newly added edges may be of unbounded length). Also [7] considers the case of outerplanar graphs. Planar augmentation results for geometric graphs can be found in [1]. They show that $2n/3$ additional edges are required in some cases and $6n/7$ edges are always sufficient for augmenting a planar graph into a 2-edge connected planar graph. For the case of trees these bounds become $n/2$ and $2n/3$, respectively. Although the planar graphs constructed in [1] are geometric the edge lengths of the augmented edges are not bounded.

Very little is known for UDGs. [3] considers the edge deletion problem in the context of UDGs and describes two simple algorithms that find subgraphs with maximal node degree of 10 and 6 that ensure both 2-edge and 2-node connectivity, respectively. However the resulting graphs are not planar.

1.2   Contributions and Outline of the Paper

In this paper we relate the lengths of the edges of the resulting augmented graph to the original UDG. More specifically, in this paper we address the problem of augmenting the edge set of planar geometric graphs with straight line edges of bounded length so that the resulting graph is planar (no crossing edges except in the end points) and 2-edge connected. In Section 2 we give bounds on the number of newly added straight-line edges and show that such edges can be of length at most three times the max length of the edges of the original graph; the number of newly added edges is shown not to exceed the number of cut edges of the original graph. If the original graph is a tree with n nodes and max degree Δ then we prove that at most $n(1 - 1/2\Delta)$ edges are sufficient, while for MSTs at most $5n/6$ edges are shown to be sufficient. All these algorithms are linear in the number of sensors. In addition, in Subsection 2.2 we indicate how to extend the NP-completeness proof of [11] in order to show that it is NP-hard to augment a geometric planar graph to a 2-edge connected geometric planar graph with the minimum number of new edges of bounded length. In Section 3, we prove that there is no local algorithm (i.e. a distributed algorithm that finishes in constant time by using only information at constant distance assuming each link takes one time unit to traverse) for augmenting a planar UDG into a 2-edge connected planar graph with straight line edges.

2   Augmentation with Bounded Length Edges

In this section we consider the augmentation problem for planar graphs. First we give an upper bound on the number of edges for arbitrary planar graphs and then consider the special cases of MSTs and general geometric trees. We point out that a zig-zag path with n vertices of a convex polygon [1] requires $\lceil (n - 2)/2 \rceil$ additional edges to augment it into a two-edge connected planar graph as Figure 1 depicts. This bound is valid regardless of the length of the edges.

All the algorithms given in this section are linear in the number of sensors. Later we prove the NP-completeness for computing the minimum number of augmented edges of bounded length for the planar 2-edge augmentation problem.

All planar graphs considered can be drawn in the plane with straight line edges. Moreover, edges can be intersected only at their endpoints. A two edge connected planar graph $G = (V, E)$ consists of a set of cycles, i.e. every edge

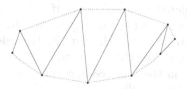

Fig. 1. A graph of n vertices that requires $\lceil (n - 2)/2 \rceil$ new edges to augment it into a two-edge connected planar graph

$\{u, v\} \in E$ is in at least one cycle. We say that $\{u, w\}$ is an immediate neighbour of $\{u, v\}$ if $\angle wuv < \pi$ and u does not have any other neighbour inside the angle $\angle wuv$. Observe that an edge can have up to 4 immediate neighbours.

2.1   Upper Bounds

Theorem 1. *Let* $G = (V, E)$ *be a connected geometric graph with* $|V| \geq 3$, *vertices in general position in the plane (no three points are collinear),* b *cut edges and maximum edge length* ≤ 1. *Then,* G *can be augmented to a 2-edge connected planar graph* $G' = (V, E \cup E')$ *with at most* b *additional edges of length at most* 3 *in time* $O(|V|)$.

Proof. In the proof below we indicate how to add new edges leading to the construction of the new graph G'. Firstly, we introduce some notation that we require for the proof. We classify edges of G' in the following categories:

- E_1. All the original edges of G or additional edges of length at most one. Thus, $E \subseteq E_1$.
- E_2. Additional edges of length at most two.
- E_3. Additional edges of length at most three.

The proof is constructive and in each step an additional edge of length at most three is added to create a cycle which includes at least one cut edge of G. The intuitive idea is to create a cycle by joining one cut edge e with one of its immediate neighbours called pivot of e which in turn forms a triangle. Thus, an additional edge has always associated exactly one cut edge in E_1. We present the details in Algorithm 1.

The invariant throughout the proof is that additional edges belong to at least one cycle of G' having length at most three and each edge $e \in E_3$ is always incident in its triangle to one edge in E_2. Moreover, each additional edge is incident to an original cut edge.

Let $e = \{u, v\} \in E$ be any cut edge of G' (If it exists, otherwise G' is already two-edge connected planar graph.) Let $e_1 = \{u, w\}$ be the immediate neighbour of e in G' with min length such that if $e_1 \notin E_1$ then e_1 is in a sector formed by the angle between e and an immediate neighbour of e in G, i.e. in a convex sector formed by original consecutive edges. e_1 exists since $|V| \geq 3$. Thus, a priority is given to the pivot e_1 in the following order: E_1, E_2 and E_3 (we refer this as priority order). Three cases can occur:

Case 1 $e_1 \in E_1$. Consider the triangle wuv, if it is empty, add $\{v, w\} \in E_2$ to G' and form a cycle with e_1 and e; see Figure 2a. Otherwise, there must exist a vertex x inside the triangle wvu such that the triangle xvu is empty (If there exists more than one then choose the closest to u.) Consider the two components of $G - e$. If v and x are in the same component, then add $\{u, x\} \in E_1$ to G'; see Figure 2b. This creates a cycle with $\{u, v\}$. Otherwise, add $\{x, v\} \in E_2$ to G'; see Figure 2c. This creates a cycle with e. Thus, exactly one edge is added to G'. Observe that any cut edge e' incident to x (if it exists) will never choose

---

Algorithm 1: Augmenting the connectivity of an arbitrary planar graph with bounded length straight line edges

---

    input : $G = (V, E)$; G is a connected geometric planar with max length 1.
    output: $G' = (V, E \cup E_1 \cup E_2 \cup E_3)$; G' is 2-edge connected planar geometric

1   Let B be the set of bridges of G;
2   **while** $B \neq \emptyset$ **do**
3   Remove the first element $e = \{u, v\}$ from B;
4   Let $e_1 = \{u, w\}$ be the immediate neighbour of e in G' in priority order;
5   **if** $e_1 \in E_1$ **or** $e_1 \in E_2$ **then**
6   Let $i \in \{1, 2\}$ be the integer such that $e_1 \in E_i$;
7   Find the vertex x such that $\triangle(xuv)$ is empty and x is closest to u;
8   **if** $x = w$ **then** $E_{i+1} \longleftarrow E_{i+1} \cup \{w, v\}$;
9   **else if** x *and* v *are in the same component of* $G - e$ **then**
             $E_i \longleftarrow E_i \cup \{x, u\}$;
10   **else** $E_{i+1} \longleftarrow E_{i+1} \cup \{x, v\}$;
11   **end**
12   **else**
13   Let e' be the cut edge associated to e_1;
14   Remove e_1 of E_3;
15   Let e_2 be the immediate neighbour of e in G';
16   Remove e_2 of E_2;
17   Add the cut edges to B in the chain from e' to e resulting from the removal of e_1 and e_2.
18   **end**
19 **end**

$\{x, v\}$ as its pivot since $\{x, v\}$ is in a concave angle formed by two consecutive neighbours of e'.

Case 2 $e_1 \in E_2$. This case only occurs when e is either a leaf or when all its immediate neighbours in G' are in E_2. Similar to the previous case, consider the triangle wuv if it is empty, then add $\{w, v\} \in E_3$ to G'. Otherwise, let x be a vertex inside the triangle wvu such that the triangle xvu is empty (If there exists more than one then choose the closest to u.) Consider the two components of $G - e$. If x and v are in the same component, then add $\{u, x\} \in E_2$ to G'. Otherwise, then add $\{v, x\} \in E_3$ to G'. In both cases e is part of a cycle and exactly one edge is added to G'. Similar to the previous case any cut edge e' incident to x (if it exists) will never choose the new edge as its pivot since it is in a concave angle formed by two consecutive neighbours of e'.

Case 3 $e_1 \in E_3$. This case also occurs only when e is either a leaf or all its immediate neighbours in G' are in E_3. Since an edge of length greater than three is not allowed, some added edges must be removed to reconfigure G' and be able to employ the previous two cases. We will show that exactly two edge removals of previously added edges is always sufficient. Let $e' \in E_1$ be the cut edge associated to e_1. Clearly e' exists and is not an immediate neighbor of e in G', otherwise e' would have chosen e as its pivot. Therefore, the third edge

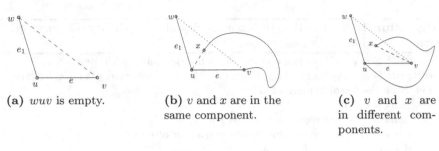

(a) *wuv* is empty. **(b)** v and x are in the same component. **(c)** v and x are in different components.

Fig. 2. $e, e_1 \in E_1$

$e_2 \in E_2$ of the triangle formed by e_1 and e' is incident to e. Similarly, let $e'' \in E_1$ be the cut edge associated to e_2 which always exists. We will prove that e'' and e_2 form a triangle with another edge in E_1. If e'' is incident to e then the third edge, say e_3, of the triangle formed by e_2 and e'' is also in E_1, otherwise e'' would have chosen e as its pivot. On the other hand, if e'' is not incident to e then e'' is incident to e' (otherwise e' would have never chosen e_2 as its pivot since e_2 would have been in a concave sector of e') and e_2 is incident to e and e_3 and therefore e_3 is also in E_1. Thus, by removing e_2 and e_1 from G' and processing in order the cut edges in the chain from e' to e will add edges only in E_1 and E_2; see Figure 3.

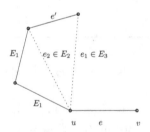

Fig. 3. The removal of e_1 and e_2 leaves a chain of four edges in E_1

It is easy to see that the number of additional edges is at most b since every additional edge is associated to exactly one cut edge.

The correctness of the Algorithm 1 comes from the proof above. Regarding the complexity, line 1 can be done in linear time by traversing the faces since there are $O(|V|)$ edges and each edge is visited twice. The while statement takes linear time on the number of bridges. Further, every step inside the while statement can be done in constant time by doing some preprocessing steps in linear time. Observe that the removal of added edges only affects processed cut edges at constant distance. Therefore, each cut edge is considered only a constant number of times. This completes the proof of Theorem 1. ∎

The upper bound 3 on the length of the augmented edges proved in Theorem 1 cannot be improved further as indicated by the example below.

Fig. 4. A planar graph that can be augmented to a 2-edge connected planar graph only by adding an edge of length $3 - \epsilon$

Example 1. It is easy to see that edge length 3 is sometimes necessary for augmenting a planar graph to a 2-edge connected planar graph with straight line edges. Figure 4 depicts a graph indicating that this can be done only by adding the new edge ux of length $3 - \epsilon$. It is easy to see that any other possibility will create a crossing and therefore create a non-planar graph.

Example 2. If we allow crossings, the planar graph depicted in Figure 5 can be augmented to a 2-edge connected graph by adding two edges of length $2 - \epsilon$.

Fig. 5. If we allow crossings, it can be augmented to a 2-edge connected graph by adding two edges of length $2 - \epsilon$

Theorem 2. *Let T be an Euclidean MST on a set P of n points, $|P| \geq 3$ in general position. Then, it can be augmented to a 2-edge connected planar graph G with at most $\lfloor 5n/6 \rfloor$ additional edges of length at most 3 times the maximum edge length of T in time $O(n)$.*

Proof. Let T be an Euclidean MST and l be its number of leaves. We may assume that T has max degree five (Although T can have max degree six, it is trivial to find an MST T' with max degree five with the same weight.) Observe that in any MST, two consecutive vertices v, w with common neighbour u form an empty triangle uvw. Coloring the vertices of T with two colours, say 1 and 2. Let C be the set of vertices in the chromatic class, say 1, with less number of leaves. It is easy to see that internal vertices of T with color, say 1, are stars with center of color 1 and leaves of color 2. Further, a star with center of color 1 does not share any edge with any other stars with center of the same chromatic class. Let S_u be a star with center $u \in C$ and $u_0, u_1, \ldots, u_{d(u)-1}$ be the neighbours of u in clockwise order around u such that the widest angle is formed by $u_{d(u)-1}$ and u_0. If $d(u) = 2k$, let

$$S'_u = S_u \cup \bigcup_{i=0}^{k-1} \{u_{2i}, u_{2i+1}\}.$$

Otherwise, if $d(u) = 2k + 1$, let

$$S'_u = S_u \cup \bigcup_{i=0}^{k-1} \{u_{2i}, u_{2i+1}\} \cup \{u_{2(k-1)+1}, u_{2k}\}.$$

Let

$$G' = T \cup \bigcup_{\forall u \in C} S'_u.$$

Observe that G' has at most $\lfloor l/2 \rfloor$ cut edges which correspond to leaves of T. By Theorem 1, $\lfloor l/2 \rfloor$ additional edges are enough to augment G to a 2-ECP graph. To be more precise we present the Algorithm 2 that summarizes these steps.

Algorithm 2: Augmenting of the connectivity of an MST with bounded length straight line edges

---

    input : Euclidean $MST(V, E)$ with max length 1.
    output: $G = (V, E \cup E_2)$; G is 2-edge connected planar graph with length
            bounded by 3
1 Find a two-coloration of T;
2 Let C be the chromatic class with the minimum number of leaves;
3 **foreach** $u \in C$ such that $d(u) > 1$ **do**
4   Let $u_0, u_1, \ldots, u_{d(u)-1}$ be the neighbours of u such that $\angle u_0 u_{d(u)-1}$ is the widest angle;
5   **if** $d(u) = 2k$ **then** $E_2 \leftarrow E_2 \cup \bigcup_{i=0}^{k-1} \{u_{2i}, u_{2i+1}\}$;
6   **else** $E_2 \leftarrow E_2 \cup \bigcup_{i=0}^{k-1} \{u_{2i}, u_{2i+1}\} \cup \{u_{2(k-1)+1}, u_{2k}\}$
7 **end**
8 Run Algorithm 1 with $G(V, E \cup E_2)$;

Let V_i denote the number of vertices in C of degree i. Thus, $V_1 = l/2 \le n/2$. Observe that

$$n - 1 = \sum_{\forall u \in C} d(u) = \sum_{i=1}^{5} i V_i.$$

Therefore, the number of additional edges is

$$V_1 + V_2 + 2V_3 + 2V_4 + 3V_5 \le V_1 + \frac{2}{3}(n - 1 - V_1)$$
$$\le \frac{V_1}{3} + \frac{2}{3}(n - 1)$$
$$< \frac{n}{6} + \frac{2n}{3}$$
$$= \frac{5n}{6}$$

The correctness of Algorithm 2 comes from the proof above and the running time is easily seen to be linear in the number of vertices. This completes the proof of Theorem 2. ∎

Theorem 3. *Let T be any arbitrary planar tree T, with $n \ge 3$ vertices in general position and max degree Δ. Then, it can be augmented to a 2-edge connected planar graph G with at most $n(1 - \frac{1}{2\Delta})$ additional edges of length at most 3 times the maximum edge length of T in time $O(n)$.*

Proof. Color vertices of T with two colours, say 1 and 2. Let C be the set of vertices in the chromatic class with the minimum number of leaves, say 1. Each internal vertex of C is a star with center of color 1 and leaves of color 2. Moreover, stars with center in C do not have any common edge. However, a triangle formed with two consecutive vertices of a vertex u is not necessarily empty. We will show how to create cycles for each independent star of C. Let u be any vertex in C such that $d(u) \geq 2$ and u_0 and u_1 be two consecutive neighbours of u forming an angle less than π. Consider triangle uu_0u_1. If it is empty, then add a cycle with $\{u_0, u_1\}$. Otherwise, consider the concave region of all the vertices inside the triangle uu_0u_1 including u_0 and u_1. There must exist two adjacent vertices v and w in the concave region such that v and w are in different components of $T \setminus u$. Moreover, by Theorem 1 at most $d - 2$ additional edges are enough to create cycles with the remaining $d - 2$ neighbours of u. Thus, each star with center u can be augmented with $d(u) - 1$ additional edges. The Algorithm 3 summarizes this process.

Algorithm 3: Augmenting of the connectivity of a arbitrary tree with bounded length straight line edges

---

    input : $T(V, E)$; T is a tree embedded in the plane such that the max length is 1.

    output: $G = (V, E \cup E_2)$; G is 2-edge connected planar graph with length bounded by 3

1 Find a two-coloration of T;

2 Let C be the chromatic class with the minimum number of leaves;

3 **foreach** $u \in C$ *such that* $d(u) > 1$ **do**

4   Let u_0, u_1 two consecutive neighbours of u such that $\angle u_0uu_1 < \pi$;

5   Find two vertices x, y inside the triangle u_0uu_1 such that x and y are in different component of $T - u$;

6      $E_2 \leftarrow E_2 \cup \{x, y\}$;

7 **end**

8 Run Algorithm 1 with $G(V, E \cup E_2)$;

Let V_i denote the number of vertices in C of degree i. It is easy to see that $n - 1 = \sum_{\forall u \in C} d(u) = \sum_{i=1}^{\Delta} iV_i$. Hence, the number of additional edges is

$$V_1 + \sum_{i=2}^{\Delta}(i - 1)V_i \leq V_1 + \frac{\Delta - 1}{\Delta}\sum_{i=2}^{\Delta} iV_i$$

$$= V_1 + \frac{\Delta - 1}{\Delta}(n - 1 - V_1)$$

$$\leq l/2 + \frac{\Delta - 1}{\Delta}(n - 1 - l/2)$$

$$< n/2 + \frac{\Delta - 1}{\Delta}(n - 1 - n/2)$$

$$\leq n\left(1 - \frac{1}{2\Delta}\right)$$

Algorithm 3 can be implemented in linear time by doing some preprocessing in linear time and the correctness comes from the proof above. This proves Theorem 3.   ∎

2.2   NP Completeness

In this subsection we prove that it is NP-hard to augment a geometric planar graph to a 2-edge connected geometric planar with the minimum number of new edges of any given bounded length. The proof is in fact a simple modification of the proof given in [11] to ensure that new edges are always bounded. Below we indicate the theorem which is necessary in order to derive the result.

Theorem 4. *Augmenting a geometric planar graph G into 2-edge connected geometric planar graph with the minimum number of new straight line of length r times the longest edge of G is NP-hard.*

Proof. (Outline) Take an instance of a planar 3-SAT and consider the construction of the graph G given in [11][Theorem 1]. Let l be the longest edge of G and c the longest candidate edge to be added. If $c > rl$, then add a cycle at any given arbitrary vertex in the convex hull of G in such a way that it does not interfere with G and has at least one edge of length $l' \geq c/r$. Thus, every new edge will be bounded by rl'. The theorem follows by applying the same arguments given in [11][Theorem 1].   ∎

3   Impossibility of Local Algorithm for Augmentation

In this section we prove that there is no local algorithm for augmenting a planar UDG to a 2-edge connected planar graph. We prove the following theorem.

Theorem 5. *There is a unit disk graph G with n nodes located in the plane in such a way that the following hold.*

1. *Any distributed algorithm for augmenting G into a 2-edge connected planar graph with straigh line edges requires $\Omega(\log^* n)$ rounds.*
2. *There is a distributed algorithm for augmenting G into a 2-edge connected planar graph with straight line edges which takes $O(\log^* n)$ rounds.*

In particular, such a planar augmentation of the unit disk graph cannot be done locally.

Proof. Consider the unit disk graph depicted in Figure 6. Nodes are placed in the plane in such a way that for each node i $(0 < i < n)$, the angle $\angle((i-1)i(i+1))$ formed by node i with its two neighbours $i-1$ and $i+1$ is $\pi - \epsilon$, where $\epsilon > 0$ is sufficiently small. Observe that as a consequence of this geometric representation of the graph, for $i < j$ and $i' < j'$ the straight lines joining vertices i, j and i', j' intersect if either $i < i' < j$ or $i' < i < j'$. The vertices form a *line graph* and have arbitrary distinct identities, namely for all i, the i-th node has identity id_i,

Fig. 6. n nodes arranged in a line. For each i, node i has identity id_i.

for $i = 0, 1, \ldots, n - 1$. Further, identifiers are picked from the range $\{1, \ldots, n\}$ and form an arbitrary permutation of this set.

We assume a standard distributed computing model (see [10]) whereby every node has two ports (one for each of its two neighbours) except for the two endpoint nodes id_0, id_{n-1} which have only one port and assume that they are consistently oriented to form the communication model of a line graph.

First we prove the impossibility result in Part 1. Assume there is a distributed algorithm, say $\mathcal{A}$, for augmentation of the graph into a 2-edge connected planar graph terminating in T rounds. Consider n consecutive points on a line. In order to form a planar 2-edge connected graph, new edges are added to the line graph. This is done by informing each vertex u of the vertex u' with which it forms a new edge. We call two such vertices a *pair* (note that some vertices may not be paired with any other vertex). Therefore by time T and after execution of the algorithm $\mathcal{A}$ every node u either

1. is paired with another node u' so that $\{u, u'\}$ forms a new edge, or
2. it is not paired with any other node of the line graph,

but the graph resulting by augmenting the line graph with the new pairs $\{u, u'\}$ is planar and 2-edge connected. Now the main proof is in two cases.

Case 1. Observe that all the new edges of the augmented planar graph are above the line graph as Figure 7 depicts. It is clear that u and u' of each pair $\{u, u'\}$ are at distance at most T in the line graph since the running time of the algorithm is T and a message takes one time unit to traverse an edge. Now we can give an algorithm for coloring the vertices of the graph that has running time $O(T)$. The algorithm is in two phases.

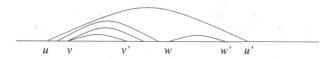

Fig. 7. Pairing nodes of the line graph so as to form a 2-edge connected planar augmentation

Phase 1: Maximum Intervals Discovery Algorithm. In Phase 1, we calculate maximum intervals between paired nodes as follows. In executing the algorithm $\mathcal{A}$ nodes can remember the port to which they received the messages. Therefore it is easy to see that if nodes v, v' are paired then they can execute an *interval* algorithm to discover the identities of all the nodes in the interval $I(v, v')$ delimited by the nodes v, v' in the line graphs. Clearly, the running time of this discovery algorithm is T. Further, every node v which is paired with node v' runs an additional *maximal interval* algorithm to discover the maximum interval $I_{u,u'}$ such that $I_{v,v'} \subseteq I_{u,u'}$. It is clear that since the augmented graph is planar this maximum interval is unique and well-defined. Moreover, the length of the maximum interval $I_{u,u'}$ is at most T since the running time of the algorithm $\mathcal{A}$ is T.

Phase 2: Coloring Algorithm. This Phase is executed only by pairs u, u' of nodes whose interval $I_{u,u'}$ is maximum. First of all such nodes receive the same color, say a. Both u, u' can easily remember the sequence of identities as they are being received in their ports and color the nodes in the interval $I_{u,u'}$ consistently using exactly two colors, say b, c. Clearly, the resulting coloring is consistent and has three colors. However, it is well known that any deterministic algorithm for 3 coloring a ring of n nodes requires $\Omega(\log^* n)$ rounds (see [10][page 89]). Therefore $T \in \Omega(\log^* n)$, as desired. This completes the proof of Part 2 and hence also the proof of Theorem 5.   ∎

The impossibility theorem just proved hinges on the fact that edges are drawn as straight lines. If we drop this requirement then we can show that a local algorithm is possible even for augmentation to a 2-vertex connected planar spanning graph.

Theorem 6. *There is a distributed, local algorithm which takes two rounds for augmenting a line graph into a 2-vertex connected, planar (non-geometric) graph.*

Proof. To prove the theorem, consider the case whereby the vertices form a line graph and have arbitrary distinct identities, namely for all i, the i-th node has identity id_i, for $i = 0, 1, \ldots, n - 1$. Further, identifiers are picked from the range $\{1, \ldots, n\}$ and form an arbitrary permutation of this set (see [10]). Now we give the local algorithm. Each node forms two new edges with nodes two hops away from it (one, if it is a node of degree 1 at an endpoint). In particular, for each $i \geq 2$ the following edges $\{i, i+2\}, \{i-2, i\}$ are formed. The resulting graph has a planar representation (as depicted in Figure 8 by appropriately drawing the edges above and below the line) and it is also 2-vertex connected. This completes the proof of Theorem 6.   ∎

Fig. 8. Planar representation of the line graph augmented with new edges

Observe that the proof of Theorem 6 was based on an underlying line graph. Moreover the newly added edges that were used to form the augmented graph were not straight lines. We note that augmentation of a planar graph to a 2-vertex connected planar graph may not even be possible, in general. This is easily seen from the unit disk graph depicted in Figure 6. Clearly, a 2-vertex connected planar spanning graph is possible only by connecting the two endpoints $0, n-1$ with a straight line edge, which requires n communication steps.

Example 3. The impossibility result given in Theorem 5 is also due to the fact that the identities are not ordered. For example, if $id_i = i$, where $i = 0, 1, \ldots, n - 1$, then the planar augmentation problem is easy. E.g., execute a distributed algorithm that draws new edges between nodes with even identities. This algorithm clearly works if n is odd. If n is even then the augmented graph resulting after execution of this algorithm will not have the rightmost edge $\{n - 2, n - 1\}$ in a cycle. For this reason we add instead the edge $\{n - 3, n - 1\}$ which is of length 3.

4   Conclusion

In this paper we focused on the problem of constructing 2-edge connected geometric planar spanning graphs respecting an existing UDG. Such graphs are fault tolerant under edge deletion and because of their planarity can also be used to implement geometric routing with guaranteed delivery [9].

References

1. Abellanas, M., García, A., Hurtado, F., Tejel, J., Urrutia, J.: Augmenting the connectivity of geometric graphs. Comp. Geom. Theory Appl. 40(3), 220–230 (2008)
2. Cheriyan, J., Sebö, A., Szigeti, Z.: An Improved Approximation Algorithm for Minimum Size 2-Edge Connected Spanning Subgraphs. In: Bixby, R.E., Boyd, E.A., Ríos-Mercado, R.Z. (eds.) IPCO 1998. LNCS, vol. 1412, pp. 126–136. Springer, Heidelberg (1998)
3. Dong, Q., Bejerano, Y.: Building Robust Nomadic Wireless Mesh Networks Using Directional Antennas. In: IEEE INFOCOM 2008, The 27th Conference on Computer Communications, pp. 1624–1632 (2008)
4. Eswaran, K.P., Tarjan, R.E.: Augmentation problems. SIAM J. Comput. 5, 653–665 (1976)
5. Jackson, B., Jordán, T.: Independence free graphs and vertex connectivity augmentation. J. Comb. Theory Ser. B 94(1), 31–77 (2005)
6. Jothi, R., Raghavachari, B., Varadarajan, S.: A 5/4-Approximation Algorithm for Minimum 2-Edge-Connectivity. In: SODA 2003: Proceedings of the Fourteenth Annual ACM-SIAM Symposium on Discrete Algorithms, pp. 725–734. Society for Industrial and Applied Mathematics, Philadelphia (2003)
7. Kant, G.: Augmenting outerplanar graphs. J. Algorithms 21(1), 1–25 (1996)
8. Kant, G., Bodlaender, H.: Planar graph augmentation problems. In: Workshop on Algorithms and Data Structures, pp. 286–298 (1991)
9. Kranakis, E., Singh, H., Urrutia, J.: Compass routing on geometric networks. In: Proc. 11th Canadian Conference on Computational Geometry (1999)
10. Peleg, D.: Distributed computing: a locality-sensitive approach. Society for Industrial and Applied Mathematics, Philadelphia (2000)
11. Rutter, I., Wolff, A.: Augmenting the connectivity of planar and geometric graphs. Electronic Notes in Discrete Mathematics 31, 53–56 (2008)
12. Vempala, S., Vetta, A.: Factor 4/3 Approximations for Minimum 2-Connected Subgraphs. In: Jansen, K., Khuller, S. (eds.) APPROX 2000. LNCS, vol. 1913, pp. 262–273. Springer, Heidelberg (2000)
13. Watanabe, T., Nakamura, A.: Edge-connectivity augmentation problems. J. Comput. Syst. Sci. 35(1), 96–144 (1987)

Scheduling Packets with Values and Deadlines in Size-Bounded Buffers

Fei Li*

Department of Computer Science
George Mason University
Fairfax, VA 22030, USA
lifei@cs.gmu.edu

Abstract. Motivated by providing quality-of-service differentiated services in the Internet, we consider buffer management algorithms for network switches. We study a *multi-buffer model*. A network switch consists of multiple size-bounded buffers such that at any time, the number of packets residing in each individual buffer cannot exceed its capacity. Packets arrive at the network switch over time; they have values, deadlines, and designated buffers. In each time step, at most one pending packet is allowed to be sent and this packet can be from any buffer. The objective is to maximize the total value of the packets sent by their respective deadlines. A 9.82-competitive online algorithm (Azar and Levy. SWAT 2006) and a 4.73-competitive online algorithm (Li. AAIM 2009) have been provided for this model, but no offline algorithms have yet been described. In this paper, we study the offline setting of the multi-buffer model. Our contributions include a few optimal offline algorithms for some variants of the model. Each variant has its unique and interesting algorithmic feature.

1 Introduction

Motivated by providing quality-of-service differentiated services in the Internet, we consider buffer management algorithms for network switches. We study a *multi-buffer model*. A network switch consists of m size-bounded buffers Q_1, Q_2, ..., Q_m and their sizes are denoted as B_1, B_2, ..., B_m respectively. At any time, the number of packets residing in each individual buffer Q_i cannot exceed its capacity B_i. Time is discretized into time steps. Packets arrive at the network switch over time and each packet p has an integer arriving time (release time) $r_p \in \mathbb{Z}^+$, a non-negative value $v_p \in \mathbb{R}^+$, an integer deadline $d_p \in \mathbb{Z}^+$, and a designated buffer $b_p \in \{Q_1, ..., Q_m\}$ that it can reside in. The deadline d_p specifies the time by which the packet p should be sent. This model is preemptive such that the packets already existing in the buffers can be dropped at any time before they are transmitted. A dropped packet cannot be delivered any more. In each time step, at most one pending packet is allowed to be sent

---

* Research is partially supported by NSF grant CCF-0915681.

W. Wu and O. Daescu (Eds.): COCOA 2010, Part I, LNCS 6508, pp. 398–407, 2010.

and this packet may be from any buffer. The objective is to maximize *weighted throughput*, which is defined as the total value of the packets transmitted by their respective deadlines. A network switch consisting of either a single buffer or multiple buffers for transmitting packets is illustrated in Figure 1.

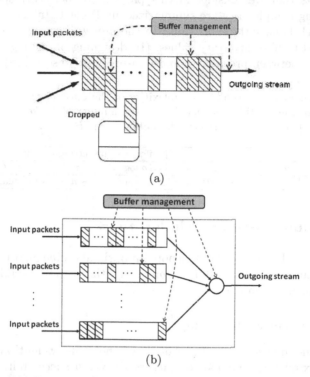

Fig. 1. Buffer management for (a) a singe buffer and (b) multiple buffers

The first QoS buffer management model is introduced in [1]. Since then, quite a few researchers have studied this model as well as other variants, mostly in the online settings [7][6][3][9][5]. A well-studied model is called the *bounded-delay model*. In this model, there is only one buffer. Packets have integer release time, integer deadlines, and non-negative values. The objective is to maximize the total value of the packets sent by their deadlines. An implicit assumption on this model is the buffer's sufficiently large size. All released packets can be stored in the buffer before they are delivered or they get to expire. For the bounded-delay model, an optimal offline algorithm running in $O(n^2)$ time has been proposed in [7], where n is the number of packets released. We call the bounded-delay model a *bounded-buffer model* in case the buffer size is enforced to be finite. The bounded-buffer model generalizes the bounded-delay model, if we allow the buffer size to be larger than any packet's *slack time*. (A packet's slack time is defined as the difference between its deadline and release time.) The bounded-buffer model is one variant of the *multi-buffer model* proposed by Azar and Levy [2]. A 9.82-competitive online algorithm [2] and a 4.73-competitive online algorithm [8] have been provided for this model, but no offline algorithms yet been described. In this paper, we study the offline setting of the multi-buffer model. Our contributions include a few optimal offline algorithms for some variants of the model. Each variant has its unique and interesting algorithmic feature.

The variants that we consider in this paper and their corresponding algorithms' running complexities are summarized in Table 1. In the *uniform-value* setting, all packets have the same value. In the *non-uniform-value* setting, packets are allowed to have arbitrary values. (In designing offline algorithms, there is no difference between preemptive and non-preemptive settings.)

Table 1. Summary of the running complexities of the optimal offline algorithms for some variants of the multi-buffer model. n is the number of packets in the input sequence. For the bounded-buffer model, the buffer size is $B \in \mathbb{Z}^+$.

	uniform-value setting	non-uniform-value setting
$m = 1$	$\Theta(n \log \min\{B,\ n\})$	$O(n^2)$
$m > 1$ (packets sharing a common deadline)	$O(n \max\{m,\ \log n\})$	$O(n^2 \max\{m,\ \log n\})$

2   The Bounded-Buffer Model, $m = 1$

There is only one buffer. Let OPT denote an optimal offline algorithm. Without loss of generality, we assume OPT is *non-idling*, that is, OPT sends a packet as long as the buffer is non-empty.

2.1   The uniform-value Setting

In the uniform-value setting, all packets have the same 'weight' and the objective is to maximize the number of packets delivered successfully. An optimal algorithm called EDF (which stands for 'Earliest-Deadline-First') works simply as follows. Actually, EDF is an *online* algorithm.

Algorithm 1. An algorithm for the uniform-value setting with $m = 1$

1: All packets in the buffer are organized by their deadlines using an augmented red-black tree [4].
2: Upon each new arrival, we insert it into the packet queue in increasing order of deadlines.
   {Let the current time be t.}
3: If the buffer is full or if more than $t' - t$ packets are to be sent by some deadline t' (we call these cases '*tight*'), we drop the packet with the earliest deadline.
4: In each time step, the earliest-deadline packet in the buffer is sent.

Lemma 1. *For the bounded-buffer model in the uniform-value setting, there exists an optimal offline algorithm running in $O(n \log \min\{B,\ n\})$ time, where n is the number of packets released and B is the buffer size.*

Proof. We first prove EDF's correctness using a loop invariant. The loop invariant is: At any time, there exists a one-to-one mapping (injection) from each packet q in OPT's buffer to a packet j in EDF's buffer such that $d_q \leq d_j$. Without loss of generality, we align the mappings such that an earlier-deadline packet in OPT's buffer maps to an earlier-deadline packet in EDF's buffer. For example, assume q_1 and q_2 in OPT's buffer map to j_1 and j_2 in EDF's buffer respectively. If $d_{q_1} < d_{q_2}$ but $d_{j_1} \geq d_{j_2}$, we swap the mappings and let q_1 map to j_2 and q_2 map to j_1. Note $d_{q_1} \leq d_{q_2} \leq d_{j_2} \leq d_{j_1}$.

This invariant holds before any packet is released. Let us assume it holds at time t. Consider a new arrival p accepted by OPT. Recall that all packets are with the same value. The packet p is either accepted by EDF or there exists a packet j which is not mapped yet by any packet in OPT's buffer and having a deadline $d_j \geq d_p$. (In this case, we can map p in OPT's buffer to j in EDF's buffer.) Otherwise, we can drop j and accept p or OPT's buffer is 'tight' as well and OPT rejects p. In each time step, both OPT and EDF send one packet as long as their buffers are non-empty. Without loss of generality, we can assume OPT sends the earliest-deadline packet in its buffer. Thus, the loop invariant still holds after each step's deliveries. The loop invariant implies the correctness of the algorithm.

For each new arrival, it takes $O(\log \min\{B, n\})$ to insert p into or drop p out of the packet queue in EDF's buffer. The algorithm has an upper bound of running time $O(n \log \min\{B, n\})$. The proof is completed.   □

The following instance shows that no algorithm has a running complexity asymptotically better than $\Omega(n \log \min\{B, n\})$.

Example 1. Assume $B \geq n$. All packets are released at the same time 0. To identify whether all packets can be delivered successfully, we have to sort them by deadlines such that packets can be delivered in an earliest-deadline-first (EDF) manner. The lower bound of comparison-based sorting n numbers takes $\Omega(n \log n)$ [4].

Corollary 1. *Consider the bounded-buffer model in the uniform-value setting. If packets' deadlines are weakly increasing along with their release time, EDF is an optimal algorithm running in linear time $O(n)$.*

2.2   The non-uniform-value Setting

If $B \geq n$, the optimal offline algorithm [7] for the bounded-delay model applies on the bounded-buffer model and has a running time of $O(n \log n)$. We assume $B < n$. Fix an input sequence $\mathcal{I}$. We have the following algorithm.

Lemma 2. *For the bounded-buffer model in the non-uniform-value setting, there exists an optimal offline algorithm running in $O(n^2)$ time, where n is the number of packets released.*

Algorithm 2. An algorithm for the non-uniform-value setting with $m = 1$

1: Sort all packets in $\mathcal{I}$ in non-increasing value order. For packets with the same value, sort them in decreasing order of deadlines.
2: Start from a set of packets $S = \emptyset$. For each packet $j \in (\mathcal{I} \setminus S)$, pick up j in EDF order and run EDF to examine whether all packets in $S \cup \{j\}$ can be delivered successfully by their respective deadlines. That is, send packets in EDF order and if one selected packet in S cannot be put in the buffer with size B, then S cannot be delivered successfully.
   {Actually, we can start from the time r_j to run EDF over the packets $S \cup \{j\}$ instead of from scratch; though this does not help to reduce the asymptotic running complexity.}
3: If 'yes', update S with $S \cup \{j\}$.
4: For each examined packet j, no matter whether we insert j into S or not, drop it out of $\mathcal{I}$.
5: Examine all packets in $\mathcal{I}$ in order until $\mathcal{I}$ gets empty.

Proof. We first claim that the model is a matroid. A matroid is an ordered pair $M = (S, \varPi)$ satisfying the following conditions [4].

1. S is a finite set.
2. $\varPi$ is a nonempty family of subsets of S such that if $B \in \varPi$ and $A \subseteq B$, then $A \in \varPi$. $\varPi$ is called *hereditary* if it satisfies this property. The empty set $\emptyset$ is necessarily a member of $\varPi$.
3. If $A \in \varPi$, $B \in \varPi$, and $|A| < |B|$, then there exists some element $x \in B \setminus A$ such that $A \cup \{x\} \in \varPi$. We say that M satisfies the *exchange property*.

In our case, we have the following observations on our algorithm.

1. The set of packets that we consider is finite.
2. Consider a set of packets that can be delivered successfully by their deadlines in an EDF manner. Its any subset can be delivered successfully as well. Thus, the heredity property is satisfied.
3. Let two sets of packets be A and B with $|A| < |B|$ that can be successfully sent by an algorithm. We show that there exists a packet $j \in B \setminus A$ such that $A \cup \{j\}$ can be scheduled successfully as well. Otherwise, if for any packet $j \in B$ such that if j is scheduled, then one (and only one) packet in A cannot be sent by its deadline, then we can modify A to B by inserting each packet j from $B \setminus A$ to A and drop the packet $i \in A \setminus B$ that cannot be sent by its deadline. At the end of this procedure, we conclude that $|B| \leq |A|$.

We then claim that the schedule of S we finally get from Algorithm 2 has the maximum total value. Note that Algorithm 2 is a greedy algorithm. As the model is a matroid, Algorithm 2 is optimal.

Let $|\mathcal{I}| = n$. Sorting packets in $\mathcal{I}$ takes $O(n \log n)$ time. The buffer has at most B packets at any time, thus, each packet insertion (in increasing deadline order) takes $O(\log B)$ time. Running EDF over a set of packets $S \cup \{j\}$ takes time $|S| + 1 \leq n$. For each packet j, examining $S \cup \{j\}$ of being successfully sent takes time $O(\log B + n)$. Thus, the total running time of the algorithm is $O(n \log n + n(n + \log B)) = O(n^2)$. Thus, our algorithm has a running time of $O(n^2)$. The proof is completed.   □

3   Scheduling Packets with a Common Deadline or without Deadlines, $m > 1$

In this Section, we consider the cases with multiple buffers. Let OPT denote an optimal offline algorithm. Without loss of generality, we assume OPT is non-idling.

In scheduling packets without deadlines, we assume all packets have a common deadline $r_{\max} + n$, where $r_{\max}$ is the largest release time of a packet. We also note that when there are no new arrivals, all packets already in the buffers can be sequentially delivered successfully.

Let $P_i(t)$ denote the set of packets released at time t designated to the buffer Q_i. Since each buffer Q_i cannot accommodate more than B_i packets at any time, we assume that for each Q_i, at any release time t, $|P_i(t)| \leq B_i$. Let $Q_i(t)$ and $|Q_i(t)|$ denote the packet queue in the buffer Q_i and its size, respectively. Let $r_{\max}^i$ denote the largest release time of a packet designated to the buffer Q_i. Let D be the common deadline.

3.1   The uniform-value Setting

In the uniform-value setting, all packets have the same 'weight' and the objective is to maximize the number of packets delivered successfully. Note that an algorithm with the minimum number of packets unsent achieves the maximum throughput since the total number of packets released is a fixed number. Thus, instead of considering maximizing the total number of packets delivered in designing algorithms, we tackle with this variant from the perspective of minimizing the number of packets dropped.

For each buffer, our idea is to calculate the number of buffer slots that we have to reserve in order to accept future arrivals (that is, minimizing the number of packets dropped due to 'packet overflow'). This value indicates to us the latest time that we have to deliver a packet from a buffer. Based on this idea, we design an algorithm described in Algorithm 3 for this variant.

Theorem 1. *In scheduling packets with the same value and same deadline, there exists an optimal offline algorithm running in $O(n \max\{m, \log n\})$ time, where n is the number of packets released.*

Proof. Let us call our algorithm 3 TS (standing for 'Tight Schedule'). We first show the correctness of Algorithm 3 using the exchange argument.

Remember that all packets are with the same value and same deadline and TS accepts packets in a greedy manner for each buffer, thus, as long as OPT and TS schedule packets from the same buffer in each time step, they achieve the same throughput. Let $\mathcal{O}$ denote the set of packets sent by OPT. Let t be

---

Algorithm 3. An algorithm for the uniform-value setting with $m > 1$

1: For each buffer Q_i, consider $P_i(t)$ in decreasing order of release time t.
2: Define a variable $Z_i(t)$ to denote the number of buffer slots that are needed from the buffer Q_i to accommodate packets released at/after time t.
3: Initially, set $Z_i(r^i_{\max}) = \max\{|P_i(r^i_{\max})|,\ D - r^i_{\max}\}$.
4: In the reverse order of release time, calculate $Z_i(t) = \min\{B_i,\ Z_i(t') + |P_i(t)| - (t' - t)\}$, where t' is the immediate next release time (of packets) after time t for Q_i.
5: For each new arrival, if its designated buffer is full, drop the packet. Otherwise, append the packet at the end of the queue.
6: In each time step t, send any packet from the buffer Q_i if $Z_i(\tilde{t}) + |Q_i(t)| \geq B_i$, where $\tilde{t}$ is the immediate next release time of packets for the buffer Q_i. Ties are broken arbitrarily.
7: If all the buffers Q_i have $Z_i(\tilde{t}) + |Q_i(t)| < B_i$, choose any packet to send.
8: Switch to another buffer to send a packet only if this buffer is empty or if another buffer Q_i satisfies $Z_i(\tilde{t}) + |Q_i(t)| \geq B_i$ at time t.

---

the first time step in which OPT and TS deliver packets from different buffers. Assume OPT sends a packet q_1 from a buffer Q_1 and TS sends a packet p_1 from a buffer Q_2 where $Q_1 \neq Q_2$. At time t, we use $\hat{t}$ and $\tilde{t}$ to differentiate the two (possibly) distinct next release time of packets designated to buffers Q_1 and Q_2 respectively.

If $p_1 \notin \mathcal{O}$, then OPT can be modified by sending p_1 in this time step and dropping q_1 out of its packet sending sequence such that $\mathcal{O}$ is updated with $\mathcal{O} \cup \{p_1\} \setminus \{q_1\}$. The updated OPT has no a no less total value. Here, we assume $p_1 \in \mathcal{O}$. Since we choose Q_2 to send a packet, one of the following cases must happen.

1. Assume $Z_1(\hat{t}) + |Q_1(t)| < B_1$ and $Z_2(\tilde{t}) + |Q_2(t)| < B_2$. In this case, delivering either p_1 or q_1 will not result in packet overflow for both buffers Q_1 and Q_2. Thus, OPT can be changed to choose Q_2 to send a packet.
2. Assume $Z_1(\hat{t}) + |Q_1(t)| < B_1$ and $Z_2(\tilde{t}) + |Q_2(t)| \geq B_2$. In this case, if TS does not choose Q_2 to send a packet, one packet released at time $\tilde{t}$ or later will not be delivered successfully. Let this packet be p. Then, among all the packets in Q_2's current buffer and those packets released later designated to Q_2, one of them must not be in $\mathcal{O}$. Otherwise, OPT will choose Q_2 to send a packet to avoid Q_2's packet overflow. Assume the packet sending sequence since time t for OPT is $q_1, \ldots, p_1, \ldots$. We modify the sequence for OPT as $p_1, \ldots, p, \ldots$ and update $\mathcal{O}$ as $\mathcal{O} \cup \{p\} \setminus \{q_1\}$. Since p_1 is delivered in this time step, there exists an extra buffer slot (compared with that of the unmodified OPT which does not send p_1 for step t) to accommodate p in the buffer Q_2 and thus, the new packet sequence is schedulable. After our modification, OPT's total gain is not reduced and OPT chooses the same queue as TS does to send a packet in this time step.

3. Assume $Z_1(\hat{t})+|Q_1(t)| \geq B_1$ and $Z_2(\hat{t})+|Q_2(t)| \geq B_2$. In this case, delivering either p_1 or q_1 will result in packet overflow for the other buffer. Thus, with the same analysis as the above case, OPT can be changed to choose Q_2 to send a packet.

We then show the running time of Algorithm 3. Sorting all distinct release time for each buffer takes $O(n \log n)$ time. Calculating the variables $Z_i(t)$ takes linear time $O(n)$. For each time t, we identify the buffer to send a packet and this takes time $O(m)$. The overall time of identifying packets to send from the buffers is $O(n \cdot m)$. In total, the running complexity of our algorithm is $O(n \max\{m, \log n\})$. The proof is completed.   □

The proof of Theorem 1 immediately implies the following corollary.

Corollary 2. *In scheduling packets with the same value and same deadline, Algorithm 3 provides a way of identifying whether a set of packets can be delivered successfully.*

3.2   The non-uniform-value Setting

We realize that when each buffer size is large enough, the multi-buffer model is the same as the bounded-delay model since all arriving packets can be accommodated in the buffers. Hence, we have two trivial results on the non-uniform-value setting.

Lemma 3. *For the multi-buffer model, if all the buffers have their sizes larger than the maximum slack of a packet designated to them, the multi-buffer model is same as the bounded-delay model. An optimal offline algorithm running in time $O(n^2)$ exists, where n is the number of packets released.*

Corollary 3. *Consider the multi-buffer model at a time t. There exists an optimal offline algorithm sending all the packets in the current buffers, running in $O(n \log n)$ time, where n is the number of packets pending in the current buffers.*

In scheduling weighted packets sharing a common deadline, our idea is to combine Algorithm 2 and Algorithm 3. We note that this variant is a matroid as well (this claim can be verified easily as that in the proof of Lemma 2). Then a greedy algorithm scheduling packets with more values is optimal. Let S be a set of packets we decide to send. Initially, S is empty. We order packets in decreasing order of values. Then, we examine packets one by one, as long as the new one and those already selected packets can be delivered by the common deadline, we add this new packet into S. Otherwise, we drop this newly considered packet. There is a question unsolved: How do we identify whether a set of selected packets can be delivered since they belong to multiple buffers at different times? We apply the idea of Algorithm 3, specifically, the result of Corollary 3. The algorithm is described in Algorithm 4.

Algorithm 4. An algorithm for the non-uniform-value setting with $m > 1$

1: Fix an input instance $\mathcal{I}$. Sort all packets in $\mathcal{I}$ in non-increasing value order.
2: Start from a set of packets $S = \emptyset$. For each packet $j \in (\mathcal{I} \setminus S)$, pick up j in order and examine whether all packets in $S \cup \{j\}$ can be delivered successfully. (See Algorithm 5.)
3: If 'yes', update S with $S \cup \{j\}$.
4: For each examined packet j, no matter whether we insert j into S or not, drop it out of $\mathcal{I}$.
5: Examine all packets in $\mathcal{I}$ in order till $\mathcal{I}$ gets empty.

Algorithm 5. Identifying whether a set of packets can be delivered successfully

---
1: Let $P_i'(t)$ denote a subset of selected packets (S) which are released at time t designated to the buffer Q_i.
$\{P_i'(t) = P_i(t) \cap S.\}$
2: For each buffer Q_i, consider $P_i(t)$ in decreasing order of release time t.
3: In reverse order of release time, we calculate $Z_i(t) = \min\{B_i, Z_i(t') + |P_i(t)| - (t' - t)\}$, where t' is the immediate next release time (of packets) after time t for Q_i.
4: For each new arrival, if its designated buffer is full, drop the packet and return 'no'. Otherwise, append the packet to the queue.
5: In each time step t, send any packet from the buffer Q_i if $Z_i(\tilde{t}) + |Q_i(t)| \geq B_i$, where $\tilde{t}$ is the immediate next release time of packets for the buffer Q_i. Ties are broken arbitrarily.
6: If all the buffers Q_i have $Z_i(\tilde{t}) + |Q_i(t)| < B_i$, choose any packet to send.
7: We switch to another buffer to send a packet only if this buffer is empty or if another buffer Q_i satisfies $Z_i(\tilde{t}) + |Q_i(t)| \geq B_i$ at time t.

Theorem 2. *In scheduling packets with the same deadline, there exists an optimal offline algorithm running in $O(n^2 \log n)$ time, where n is the number of packets released.*

Proof. The correctness of Algorithm 4 depends on the matroid property of this variant and Corollary 3.

We then show the running time of Algorithm 4. Sorting all distinct release time for each buffer takes $O(n \log n)$ time. Calculating the variables $Z_i(t)$ takes linear time $O(n)$. For each time t, we identify the buffer to send a packet and this takes time $O(m)$. In total, the running complexity of our algorithm in examining one packet is $O(n \max\{m, \log n\})$. Thus, the total running time of Algorithm 4 is $O(n^2 \max\{m, \log n\})$. The proof is completed.   □

4   Conclusion

In this paper, we design offline algorithms for some variants of the multi-buffer model. We show that if the number of buffers is restricted to 1 or if all packets share a common deadline, some efficient offline algorithms can be developed. However, for the general case of the multi-buffer model, the constraints from the buffer sizes, packets' deadlines and packets' values complicate this packet scheduling problem. An optimal offline algorithm for the general multi-buffer model is being developed.

References

1. Aiello, W., Mansour, Y., Rajagopolan, S., Rosen, A.: Competitive queue policies for differentiated services. Journal of Algorithms 55(2), 113–141 (2005)
2. Azar, Y., Levy, N.: Multiplexing packets with arbitrary deadlines in bounded buffers. In: Arge, L., Freivalds, R. (eds.) SWAT 2006. LNCS, vol. 4059, pp. 5–16. Springer, Heidelberg (2006)
3. Chrobak, M., Jawor, W., Sgall, J., Tichy, T.: Improved online algorithms for buffer management in QoS switches. ACM Transactions on Algorithms 3(4), Article number 50 (2007)
4. Cormen, T.H., Leiserson, C.E., Rivest, R.L., Stein, C.: Introduction to Algorithms, 3rd edn. MIT Press, Cambridge (2009)
5. Englert, M., Westermann, M.: Considering suppressed packets improves buffer management in QoS switches. In: Proceedings of the 18th Annual ACM-SIAM Symposium on Discrete Algorithms (SODA), pp. 209–218 (2007)
6. Hajek, B.: On the competitiveness of online scheduling of unit-length packets with hard deadlines in slotted time. In: Proceedings of 2001 Conference on Information Sciences and Systems (CISS), pp. 434–438 (2001)
7. Kesselman, A., Lotker, Z., Mansour, Y., Patt-Shamir, B., Schieber, B., Sviridenko, M.: Buffer overflow management in QoS switches. SIAM Journal on Computing (SICOMP) 33(3), 563–583 (2004)
8. Li, F.: Improved online algorithms for multiplexing weighted packets in bounded buffers. In: Goldberg, A.V., Zhou, Y. (eds.) AAIM 2009. LNCS, vol. 5564, pp. 265–278. Springer, Heidelberg (2009)
9. Li, F., Sethuraman, J., Stein, C.: An optimal online algorithm for packet scheduling with agreeable deadlines. In: Proceedings of the 16th Annual ACM-SIAM Symposium on Discrete Algorithms (SODA), pp. 801–802 (2005)

Transporting Jobs through a Processing Center with Two Parallel Machines

Hans Kellerer[1], Alan J. Soper[2], and Vitaly A. Strusevich[2]

[1] Institut für Statistik und Operations Research, Universität Graz,
Universitätstraße 15, A-8010, Graz, Austria
hans.kellerer@uni-graz.at
[2] School of Computing and Mathematical Sciences, University of Greenwich,
Old Royal Naval College, Park Row, Greenwich, London SE10 9LS, U.K.
{A.J.Soper,V.Strusevich}@greenwich.ac.uk

Abstract. In this paper, we consider a processing system that consists of two identical parallel machines such that the jobs are delivered to the system by a single transporter and moved between the machines by the same transporter. The objective is to minimize the length of a schedule, i.e., the time by which the completed jobs are collected together on board the transporter. The jobs can be processed with preemption, provided that the portions of jobs are properly transported to the corresponding machines. We establish properties of feasible schedule, define lower bounds on the optimal length and describe an algorithm that behaves like a fully polynomial-time approximation scheme (FPTAS).

Keywords: scheduling with transportation; parallel machines; FPTAS.

1 Introduction

Integrating scheduling and logistics decision-making into a single model can be seen as one of the current trends of scheduling theory. In these enhanced models it is required to combine typical scheduling decisions with various logistics decisions, normally related to inventory control, machine breakdowns, maintenance, and various transportation issues.

In this paper, we consider a processing system that consists of two identical parallel machines. The jobs are delivered to the system by a single transporter, moved between the machines by that transporter, and on their completion are transported away.

In the scheduling literature there are several approaches that address the issue of scheduling with transportation. Normally, transportation occurs between the processing stages, and therefore more often than not the processing system is a multi-stage or shop system, e.g., the flow shop and the open shop. Recall that for two machines, e.g., denoted by A and B, in the case of the flow shop each job is first processed on machine A and then on machine B, while for the open shop, the processing route of each job is not known in advance. In both shop models, each job is seen as consisting of two operations, and the operations of

W. Wu and O. Daescu (Eds.): COCOA 2010, Part I, LNCS 6508, pp. 408–422, 2010.
© Springer-Verlag Berlin Heidelberg 2010

the same job are not allowed to overlap. As a rule, for the problems considered and reviewed in this paper the objective is to minimize the completion time of all jobs on all machines.

In our study, we focus on approximability issues, which have been a topic of considerable interest in the area. A polynomial-time algorithm that creates a schedule with a objective function value that is at most $\rho \geq 1$ times the optimal value is called a $\rho-approximation$ algorithm; the value of ρ is called a *worst-case ratio bound.* A family of $\rho-$approximation algorithms is called a *fully polynomial-time approximation scheme (FPTAS)* if $\rho = 1 + \varepsilon$ for any $\varepsilon > 0$ and the running time is polynomial with respect to both the length of the problem input and $1/\varepsilon$.

Reviews of four known types of scheduling models with a transportation component can be found in [8] and [9]. Two of those types (the robotic cells and the transportation networks) appear to be less relevant to this study and are not discussed below.

The model with *Transportation Lags* is the most studied among those that combine scheduling with transportation. Here it is assumed that there is a known time lag between the completion of an operation and the start of the same job on the machine that is next in the processing route. These lags can be interpreted as *transportation times* needed to move a job between the machines, provided that the transportation device is always available. A detailed review of the complexity results on open shop and flow shop scheduling with transportation lags is given in [1]. For the general case with job-dependent transportation lags, the two-machine open shop problem is unary $NP-$hard even if for any job the durations of its operations are equal. A $\frac{3}{2}-$ approximation algorithm for this problem is due to [12]. The two-machine flow shop problem is unary NP-hard even if all processing times are unit, see [13]; several 2-approximation algorithms are given in [2]. A polynomial-time approximation scheme for the classical flow shop problem developed in [3] can be modified to handle the transportation lags.

The model that we study in this paper belongs to the class of *Models with Interstage Transporters.* For the two-machine flow shop the general model of this type is introduced in [7]. Assume that there are v transporters each capable of carrying c jobs between the machines. The transportation time from A to B is equal to τ, while the travel time of an empty transporter from B to A is equal to σ. The problem with $c = 1$ and $\sigma = 0$ is shown to be unary NP-hard in [4]. The problem with $v = 1$ and $c \geq 3$ is unary NP-hard, while the case of $c = 2$ is open, see [7]. The open shop version of this problem is addressed in [8] and [9].

There are no known approximation results for these models, apart from the two-machine flow shop and open shop with a single uncapacitated transporter, i.e., $v = 1$ and $c \geq n$. For both the flow shop and open shop models, it is assumed that the jobs are brought by the transporter to one of the machines, moved between the machines in batches, and when the processing is over, the transporter collects the jobs together and carries them away. For this model, the objective is to minimize the time by which all completed jobs are collected on board the transporter.

The classes of heuristic flow shop schedules in which the jobs are split in at most b batches on each machine are studied in [8] for $b = 2$ and in [11] for $b = 3$, and $\frac{b+1}{b}$ −approximation algorithms are designed, these ratios being the best possible as long as a heuristic schedule contains at most b batches. For the open shop counterpart of the above problem with $\tau = \sigma$ a $\frac{7}{5}$ − approximation algorithm is developed in [9].

The problem that we study in this paper belongs to the same family, and its main features are as follows. There are two identical parallel machines, M_1 and M_2. The processing time of a job $j \in N = \{1, 2, \ldots, n\}$ is equal to p_j. At time zero, the jobs are brought to the system by a transporter. Each job is either processed without preemption on one of the machines or its processing is split into several portions, to be performed on different machines. In the latter case, the total duration of the portions of a job j is equal to p_j. For a job to be (partially) processed on a machine it must be delivered there by the transporter. A move of the transporter between the machines takes τ time units, and the number of jobs transferred by a move can be arbitrary. Extending the notation adopted in [8], we call this problem $TP2|v = 1, c \geq n|K_{\max}$, where $K_{\max}$ is the time by which all completed jobs are collected together on board the transporter.

We are aware of only one other study that combines scheduling on parallel machines with transportation. This is the paper by Qi [10]. As in our case, Qi's model also involves two parallel identical machines. However, there are several points of difference between the two models. First, the type of transportation used by Qi is essentially a transportation lag. Second, there is no preemption allowed in Qi's model. Third, the jobs are known to be assigned to the machines in advance and are moved only to be processed on the other machine and returned to the originally assigned machine.

The remainder of this paper is organized as follows. In Sect. 2 we establish properties of schedules that are optimal for problem $TP2|v = 1, c \geq n|K_{\max}$ and derive lower bounds on the optimal length. Section 3 describes and analyzes an algorithm that creates a schedule with an even number (either two or four) of moves of the transporter, while the case of the schedules with three moves is considered in Sect. 4. Some concluding remarks and contained in Sect. 5.

2   Feasible Schedules: Properties, Structure and Lower Bounds

In this section, we describe properties of optimal schedules for problem $TP2|v = 1, c \geq n|K_{\max}$, identify their structures and establish lower bounds on the optimal value of the objective function.

Recall that the jobs of a set $N = \{1, 2, \ldots, n\}$ have to be processed on any of two identical processing machines, M_1 and M_2. The processing time of a job $j \in N$ is equal to p_j. Since the machines are identical, we may assume that all jobs are brought by the transporter to machine M_1 at time zero, so that the first move of the transporter is made from machine M_1 to machine M_2. Sometimes we refer to machine M_1 as the *top* machine and to machine M_2 as *bottom* machine;

also the transporter will be said to move down if it moves from M_1 to M_2, and to move up otherwise. On their arrival, some of the jobs will be left at M_1 to be processed and totally completed on that machine. Some other jobs will be moved to machine M_2 to be processed and totally completed on that machine. There may be jobs that are partly processed on M_1 and partly on M_2; however each such job has to be transported to the corresponding machine for (partial) processing. The transporter can move any number of jobs at a time, and the length of a move in either direction is equal to τ time units. While a job is being transported it cannot be processed on either machine; besides, it is not allowed to process a job on both machines simultaneously. The objective is to minimize the *length* of a schedule, i.e., the time $K_{\max}$ by which all completed jobs are collected together on board the transporter.

For problem $TP2|v = 1, c \geq n|K_{\max}$, let S^* denote an optimal schedule, i.e., $K_{\max}(S^*) \leq K_{\max}(S)$ for all feasible schedules S. There are three possible types of a feasible schedule S:

Type 0: all jobs are processed on one machine M_1;
Type 1: the number of moves of the transporter in S^* is odd (upon their completion the jobs are collected at machine M_2);
Type 2: the number of moves of the transporter in S^* is even (upon their completion the jobs are collected at machine M_1).

In what follows, we assume that a Type 0 schedule is not optimal; otherwise, the problem is trivial.

Definition 1. *For problem $TP2|v = 1, c \geq n|K_{\max}$, a class of feasible schedules in which (i) the first move of the transporter starts at time zero, and (ii) the last move does not transfer any jobs that need to be completed is called Class $\mathcal{S}$.*

The statement below describes a possible structure of an optimal schedule and can be proved by ruling out all dominated structures.

Theorem 1. *For problem $TP2|v = 1, c \geq n|K_{\max}$, the search for an optimal schedule can be limited to schedules of Class $\mathcal{S}$ with two, three or four moves.*

Given a schedule S, let $Q(S)$ denote the set of jobs processed with preemption, partly on machine M_1 and partly on machine M_2; we call these jobs *fractional*. If in a schedule S some job j is fractional, i.e., $j \in Q(S)$, then we denote by x_j and y_j the total length of the time intervals during which job j is processed on machine M_1 and machine M_2, respectively. Obviously, $x_j + y_j = p_j$. The set of *whole* jobs that are processed without preemption on machine M_i is denoted by $Z_i(S)$, where $i \in \{1, 2\}$. If no confusion arises, we may drop the reference to a schedule and write Q, Z_1 and Z_2. For a non-empty set $H \subseteq N$ of jobs, denote $p(H) = \sum_{j \in H} p_j$. The pieces of notation $x(H)$ and $y(H)$ are used analogously. We now derive lower bounds on the length of a feasible schedule.

Lemma 1. *For any schedule S that is feasible for problem $TP2|v = 1, c \geq n|K_{\max}$, the lower bound*

$$K_{\max}(S) \geq T, \tag{1}$$

holds, where

$$T = \frac{p(N)}{2} + \tau. \tag{2}$$

Proof. For schedule S, let X denote the total load on machine M_1, i.e., the sum of processing times of the jobs and portions of jobs processed on M_1. If S is a Type 1 schedule then $K_{\max}(S) \geq X + \tau$ and $K_{\max}(S) \geq p(N) - X + \tau$. Otherwise, if S is a Type 2 schedule then $K_{\max}(S) \geq X$ and $K_{\max}(S) \geq p(N) - X + 2\tau$. In any case, the required lower bound (1) follows immediately.   □

It is not possible that there is exactly one move in an optimal schedule, since the length of such a schedule would be equal to $p(N) + \tau$, which is larger than $p(N)$, the length of a Type 0 schedule. Thus, there are at least three moves in any Type 1 schedule S, so that the lower bound

$$K_{\max}(S) \geq p_j + \tau \tag{3}$$

holds for each job $j \in N$. Indeed, if a job p_j is completed on the top machine M_1, it has to be moved to machine M_2, where the schedule terminates. If it is completed on machine M_2, it has to be brought there before its (possibly, partial) processing may start on that machine. This implies the bound (3).

In each Type 2 schedule the first move of the transporter is made from the top machine M_1, the last move is made from the bottom machine M_2, and the schedule terminates when all completed jobs are collected on board the transporter at the top machine M_1. For any Type 2 schedule S a lower bound

$$K_{\max}(S) \geq p_j \tag{4}$$

holds for each job $j \in N$. Besides, if in a Type 2 schedule S job j or its part is processed on machine M_2, then

$$K_{\max}(S) \geq p_j + 2\tau. \tag{5}$$

To see this, notice that job j must be brought to M_2 and returned to M_1, and while it is processed on M_2 or being moved it cannot be processed on M_1.

Lemma 2. *For problem $TP2|v = 1, c \geq n|K_{\max}$, the search for the best schedule with two moves can be limited to the non-preemptive schedules.*

Proof. If for a schedule $S \in \mathcal{S}$ with two moves set $Q(S)$ of fractional jobs is not empty, then each job $j \in Q(S)$ is brought to machine M_2 by the first move of the transporter that starts at time zero, and has to be completed subsequently on M_1, which is impossible for schedules of Class $\mathcal{S}$. See Fig. 1.   □

Lemma 3. *For problem $TP2|v = 1, c \geq n|K_{\max}$, let $S_3 \in \mathcal{S}$ be a schedule with three moves in which some jobs are fractional. Then without increasing the value of the objective function, schedule S_3 can be transformed into either a non-preemptive schedule (see Fig. 2) or into a schedule $S_3(k)$ in which only some job $k \in Q(S_3)$ remains fractional (see Fig. 3). In the latter case, the lower bound*

$$K_{\max}(S_3(k)) \geq p_k + 3\tau \tag{6}$$

holds.

Fig. 1. A schedule with two moves

Fig. 2. A non-preemptive schedule with three moves

The lemma can be proved by performing appropriate transformations of an original schedule that do not increase the length of the schedule.

In the proofs and algorithms presented in this paper, we will often use the following splitting procedure.

Procedure Split(X, Y, γ)
Input: Two sets X and $Y = \{\sigma(1), \ldots, \sigma(y)\}$ of jobs and a bound γ such that $|X| \geq 0$, $|Y| = y > 1$, and $p(X) < \gamma$, $p(X) + p(Y) > \gamma$
Output: A set $Y' \subset Y$ and a job $u \in Y$ that is split in two portions $p_u = x_u + y_u$, so that $p(X) + p(Y') + x_u = \gamma$

Step 1: Considering the jobs of set Y in the order given by the list σ, find the position v, where $1 \leq v \leq y$ such that

$$p(X) + \sum_{j=1}^{v-1} p_{\sigma(j)} < \gamma, \quad p(X) + \sum_{j=1}^{v} p_{\sigma(j)} \geq \gamma.$$

Step 2: Output $Y' := \{\{\sigma(1), \ldots, \sigma(v-1)\}\}$, $u := \sigma(v)$, $x_u := \gamma - p(X) - p(Y')$, $y_u = p_u - x_u$.

It is clear that the running time of Procedure Split is linear in $|X \cup Y|$.

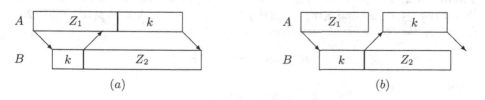

Fig. 3. A preemptive schedule $S_3(k)$ with three moves

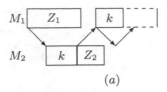

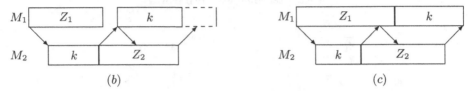

Fig. 4. A preemptive schedule $S_4(k)$ with four moves

Definition 2. *A job $j \in N$ is said to have rank r, $r \in \{0, 1, 2, 3\}$ if*

$$p_j + r\tau \geq T. \tag{7}$$

It can be shown that a non-preemptive schedule with four moves cannot be better than that with two moves; thus, we assume that any schedule with four moves contains at least one fractional job.

Lemma 4. *Let A and B be disjoint sets of jobs such that $p(A) \leq T$ and $p(A) + p(B) > T$. Suppose that Procedure $Split(A, B, T)$ is run and a set $B' \subset B$ and a job $k \in B\backslash B'$ are found such that $p(A) + p(B') + x_k = T$, where $0 < x_k < p_k$. Define $Z_1 = A \cup B'$ and $Z_2 = N\backslash(Z_1 \cup \{k\})$. Let $S_4(k)$ be a Type 2 schedule shown in Fig. 4. Then either (i) job k is a rank 2 job, or (ii) $K_{\max}(S_4(k)) = T$, provided that $p(Z_2) \geq 2\tau$.*

Proof. In schedule $S_4(k)$ only job k is fractional, and $p(Z_1) + x_k = p(Z_2) + y_k + 2\tau = T$. If $y_k + 2\tau > T - x_k = p(Z_1)$, then job k is a rank 2 job, and the structure of schedule $S_4(k)$ is as shown in Fig. 4(a) or (b). Otherwise, job k is delivered to machine M_1 before all jobs of set Z_1 are completed on that machine, and, due to the condition $p(Z_2) \geq 2\tau$, the next downward move of the transporter is finished no later than all jobs are completed on machine M_2. Thus, the structure of schedule $S_4(k)$ is as shown in Fig. 4(c) and $K_{\max}(S_4(k)) = T$.   □

Theorem 2. *For problem $TP2|v = 1, c \geq n|K_{\max}$, let W_2^* be a set of jobs such that $p(W_2^*) \leq T - 2\tau$ and $p(W_2^*) \geq p(W)$ for any set $W \subseteq N$ with $p(W) \leq T - 2\tau$. Let S_4^* be the best schedule with four moves. Then*

$$K_{\max}(S_4^*) \geq y^* + 4\tau, \tag{8}$$

where

$$y^* = \frac{p(N) - p(W_2^*)}{2} - 2\tau. \tag{9}$$

Table 1. Instances of the problem

Instance	p_1	p_2	p_3	T	$K_{\max}(S)$	Tight lower bound	Moves in S^*
1	10	7	5	15	15	(1,5)	2
2	10	7	3	14	14	(1)	3
3	9	2	9	14	14	(1,6)	3
4	10	8	8	17	17	(1)	4
5	10	7	7	16	16.5	(8)	4

Proof. Suppose that in schedule S_4^* machine M_2 processes a set W_2 of whole jobs, while $Q = Q(S_4^*)$ is the set of fractional jobs.

For any schedule $S_4 \in S$ with four moves no fractional job is processed on machine M_2 between the second and the third moves of the transporter. This implies that $K_{\max}(S_4^*) \geq y(Q) + 4\tau$.

The length of schedule S_4^* cannot be shorter than the total processing of the jobs assigned to machine M_1, i.e., $K_{\max}(S_4^*) \geq p(N) - p(W_2) - y(Q)$.

For a fixed set W_2, let y be the root of the equation $p(N) - p(W_2) - y(Q) = y(Q) + 4\tau$, i.e.,

$$y = \frac{p(N) - p(W_2)}{2} - 2\tau.$$

Then by definition of set W_2^* the lower bound (8) holds.   □

The examples below exhibit instances of problem $TP2|v = 1, c \geq n|K_{\max}$ for which an optimal schedule includes two, three or four moves of the transporter. The examples also demonstrate that all established lower bounds are tight. In all five listed instances the transportation time τ is equal to 4. Optimal schedules for Instances 1 and 2 have no preemption; in the other schedules exactly one job is fractional. For Instances 1-4 the value of T is a tight lower bound, possibly together with another bound. For Instance 5, the optimal length of the schedule is strictly larger than T. Here $T = 16$, so that either $W_2^* = \{2\}$ or $W_2^* = \{3\}$, and $p(W_2^*) = 7 < T - 2\tau = 8$, see Theorem 2. Thus, the value of $y^* = 0.5$ is computed in accordance with (9).

In this paper, in our analysis of various algorithms we use the following statement; see [5] and Lemma 4.6.1 in [6].

Theorem 3. *Consider the subset-sum problem of the form*

$$\max \sum_{j \in H} p_j x_j$$
$$\sum_{j \in H} p_j x_j \leq c \qquad (10)$$
$$x_j \in \{0,1\}, \; j \in H \subseteq N,$$

This problem admits an FPTAS that for a given positive ε a solution $x_j^\varepsilon \in \{0,1\}$, $j \in H$, that is either optimal, provided that

$$\sum_{j \in H} p_j x_j^\varepsilon < (1 - \varepsilon)c$$

or

$$(1 - \varepsilon)c \leq \sum_{j \in H} p_j x_j^\varepsilon \leq c.$$

Such an FPTAS requires no more than $O(n/\varepsilon)$ time.

We split our further consideration in accordance with the number of moves in a schedule.

3   Finding the Best Schedule with an Even Number of Moves

In this section, we present an algorithm that behaves as an FPTAS, provided that the number of moves in an optimal schedule is either 2 or 4.

Using polynomial reductions of a well-known problem PARTITION to the decision versions of problem $TP2|v = 1, c \geq n|K_{\max}$ we can prove that the following problems are binary NP-hard: (i) finding the best (non-preemptive) schedule with two moves, and (ii) finding the best (preemptive) schedule with four moves, provided that $p(N) \geq 4\tau$.

The following algorithm outputs a schedule problem $TP2|v = 1, c \geq n|K_{\max}$ and uses an FPTAS for the subset-sum problem as a subroutine.

Algorithm MoveEven

Step 1. Compute T in accordance with (2).

Step 2. Given an $\varepsilon > 0$, run an FPTAS for the subset-sum problem

$$\max \sum_{j \in N} p_j x_j$$
$$\sum_{j \in N} p_j x_j \leq T \qquad (11)$$
$$x_j \in \{0, 1\}, \ j \in N.$$

Determine $Z_1' = \{j \in N | x_j^\varepsilon = 1\}$ and $Z_2' = \{j \in N | x_j^\varepsilon = 0\}$ and create schedule S_2' shown in Fig. 1 with $Z_1 = Z_1'$ and $Z_2 = Z_2'$. If $p(Z_1') \geq (1 - \varepsilon)T$, then go to Step 7.

Step 3. Given an $\varepsilon > 0$, run an FPTAS for the subset-sum problem

$$\max \sum_{j \in N} p_j x_j$$
$$\sum_{j \in N} p_j x_j \leq T - 2\tau \qquad (12)$$
$$x_j \in \{0, 1\}, \ j \in N,$$

Determine $Z_2'' = \{j \in N | x_j^\varepsilon = 1\}$ and $Z_1'' = \{j \in N | x_j^\varepsilon = 0\}$. Create schedule S_2'' shown in Fig. 1 with $Z_1 = Z_1''$ and $Z_2 = Z_2''$. If $p(Z_2'') \geq (1 - \varepsilon)(T - 2\tau)$, then go to Step 7.

Step 4. Find set $B_2 = \{j \in N | p_j + 2\tau > T\}$ of the jobs of rank 2. If either $p(Z_1'') \leq 4\tau$ or $p(B_2) \geq T$, then go to Step 7; otherwise go to Step 5.

Step 5. If $p(Z_2'') \geq 2\tau$, select an arbitrary job $k \in Z_1'' \backslash B_2$, define $y_k = (T - 2\tau) - p(Z_2'')$ and $x_k = p_k - y_k$, make a schedule S_4' with four moves as shown in Fig. 4(c) with $Z_1 = Z_1'' \backslash \{k\}$, $Z_2 = Z_2''$ and go to Step 7; otherwise go to Step 6.

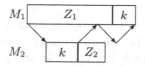

Fig. 5. Schedule S_4'' with four moves and $y_k = y^*$

Step 6. Compute y^* by formula (9) with $W_2^* = Z_2''$. Select an arbitrary job $k \in Z_1'' \backslash B_2$, define $y_k = y^*$ and $x_k = p_k - y_k$, make a schedule S_4'' with four moves as shown in Fig. 5 with $Z_1 = Z_1'' \backslash \{k\}$, $Z_2 = Z_2''$ and go to Step 7.
Step 7. Output the best of the found schedules as schedule S^ε.

Algorithm MoveEven requires $O(n/\varepsilon)$ time. Below we analyze its performance.

Theorem 4. *For problem $TP2|v = 1, c \geq n|K_{\max}$, Algorithm MoveEven behaves as an FPTAS, provided that there exists an optimal schedule S^* either with two or with four moves.*

Proof. Suppose first that in an optimal schedule S^* the transporter makes two moves.

Consider schedule S_2' found in Step 2. Notice that $p(Z_1') \leq T$. We have that $K_{\max}(S_2') = \max\{p(Z_1'), p(Z_2') + 2\tau\}$. If $p(Z_1') \geq (1-\varepsilon)T$, then we derive $p(Z_2') + 2\tau \leq 2T - (1 - \varepsilon)T = (1 + \varepsilon)T$. Thus, due to (1) we obtain that $K_{\max}(S_2') \leq (1 + \varepsilon)T \leq (1 + \varepsilon)K_{\max}(S^*)$.

Similarly, for schedule S_2'' found in Step 3, notice that $p(Z_2'') + 2\tau \leq T$. We have that $K_{\max}(S_2'') = \max\{p(Z_1''), p(Z_2'') + 2\tau\}$. If $p(Z_2'') \geq (1 - \varepsilon)(T - 2\tau)$, then we derive $p(Z_1'') = 2(T - \tau) - p(Z_2'') \leq (1 + \varepsilon)T - 2\varepsilon\tau$. Thus, due to (1) we obtain that $K_{\max}(S_2'') \leq (1 + \varepsilon)T \leq (1 + \varepsilon)K_{\max}(S^*)$.

If for schedule S_2' found in Step 2 the inequality $p(Z_1') < (1 - \varepsilon)T$ holds, then due to Theorem 3, the solution found by the FPTAS is an optimal solution to problem (11). This means that the value $p(Z_1')$ cannot be enlarged and the value $p(Z_2') + 2\tau$ cannot be reduced, as long as we require $p(Z_1') \leq T$, i.e., $K_{\max}(S^*) \geq p(Z_2') + 2\tau$.

Similarly, if for schedule S_2'' found in Step 3 the inequality $p(Z_2'') < (1 - \varepsilon)(T - 2\tau)$ holds, then due to Theorem 3, the solution found by the FPTAS is an optimal solution to problem (12). The algorithm outputs schedule S^ε such that $K_{\max}(S^\varepsilon) \leq \min\{K_{\max}(S_2'), K_{\max}(S_2'')\} = \min\{p(Z_2'') + 2\tau, p(Z_1'')\}$, which is optimal.

The conditions of Step 4 describe situations in which the algorithm still outputs a two-move schedule that is optimal. If $p(Z_1'') \leq 4\tau$ then no schedule with four moves can be shorter than $K_{\max}(S_2'') = p(Z_1'')$, i.e., for an optimal schedule with an even number of moves the equality $K_{\max}(S^*) = \min\{K_{\max}(S_2'), K_{\max}(S_2'')\}$ holds. For the set B_2 of rank 2 jobs found in Step 4 suppose that $p(B_2) \geq T$. If in schedule S^* each job of set B_2 is processed only on machine M_1, then $K_{\max}(S^*) \geq p(B_2)$. In this case, schedule S_2'' found in Step 3 is optimal. If in S^* a job of rank 2 is processed, even partly, on machine M_2 then due to (5) we deduce that $K_{\max}(S^*) \geq p_u + 2\tau$, where $u \in B_2$ is the shortest rank 2 job, so that schedule S_2' found in Step 2 is optimal.

In the rest of the proof, it is assumed that there are four moves in an optimal schedule S^*. We arrive at Step 5 if $p(Z_1'') > 4\tau$ and $p(B_2) < T$. Consider the sets Z_1'' and Z_2'' found in Step 3. Since no rank 2 job belongs to set Z_2'', we deduce that $B_2 \subset Z_1''$ and there exists a job $k \in Z_1'' \backslash B_2$ such that $p_k + 2\tau < T$.

If $p(Z_2'') > 2\tau$, then due to Lemma 4, we can create schedule S_4' shown in Fig. 4(c) with $K_{\max}(S_4') = T$, which means this schedule is optimal; see Step 5.

On the other hand, if $p(Z_2'') \leq 2\tau$, then $p(Z_1'') = p(N) - p(Z_2'') \geq (2T - 2\tau) - 2\tau = 2T - 4\tau$. Recall that the FPTAS in Step 3 solves the corresponding subset-sum problem optimally, so that for the value y^* found by formula (9) with $W_2^* = Z_2''$ the lower bound $K_{\max}(S^*) \geq y^* + 4\tau$ holds. It can be seen that $y^* > 0$. Since the solution found by the FPTAS is an optimal solution to problem (12), it follows that for any job $k \in Z_1''$ the inequality $p(Z_2'') + p_k > T - 2\tau$ holds, which is equivalent to $p(Z_1'') - p_k < T$. Thus, we derive from $p(Z_1'') \geq 2T - 4\tau$ that $p_k > y^*$.

In Step 6, the algorithm selects a job $k \in Z_1'' \backslash B_2$ and assigns it to be processed on machine M_2 for y^* time units. For job $k \in Z_1'' \backslash B_2$, we derive

$$p(Z_1'') - p_k = \frac{p(Z_1'')}{2} + \frac{p(Z_1'')}{2} - p_k \geq \frac{p(Z_1'')}{2} + (T - 2\tau) - (T - 2\tau) = \frac{p(Z_1'')}{2}.$$

In schedule S_4'' job k is delivered back to machine M_1 at time $2\tau + y^* = \frac{p(Z_1'')}{2} < p(Z_1'') - p_k$, so that the structure of schedule S_4'' is as shown in Fig. 5 with $Z_1 = Z_1'' \backslash \{k\}$. Since $K_{\max}(S_4'') = y^* + 4\tau = p(Z_1'') - y^* = \frac{p(Z_1'')}{2} + 2\tau$, this schedule is optimal due to (8).   □

4   Finding the Best Schedule with Three Moves

In this section, we present an algorithm for problem $TP2|v = 1, c \geq n|K_{\max}$ that for many instances finds the best schedule with three moves, but in general behaves as an FPTAS.

As follows from Lemma 3, the search for the best schedule with three moves can be limited to (i) non-preemptive schedules with the closest possible loads on the machines (see Fig. 2), or (ii) preemptive schedules with a single fractional job (see Fig. 3).

Below, we refer to a rank 3 job as *long*; otherwise, if a job does not satisfy (7) for $r = 3$, it is called *short*. As seen from the consideration below, the presence of the short jobs is crucial for fast finding an exact solution to the problem.

It can be proved that for problem $TP2|v = 1, c \geq n|K_{\max}$ with no short jobs and $p(N) \geq 4\tau$, finding the best non-preemptive schedule as well as the best preemptive schedule with three moves is binary NP-hard.

Algorithm Move3 presented below finds a schedule with three moves for problem $TP2|v = 1, c \geq n|K_{\max}$ that is either optimal or arbitrarily close to the optimal one.

Algorithm Move3

Step 1. Compute T in accordance with (2).

Step 2. Find the set A of all short jobs and the set B of all long jobs. If set A is not empty, go to Step 3; otherwise, go to Step 9.

Step 3. If $p(A) \geq T - \tau$, then run Procedure Split$(\emptyset, A, T - \tau)$ to find a set $A' \subset A$ and a job $u \in A$ such that $p(A') + x_u = T - \tau$. If $x_u = p_u$ then create a non-preemptive schedule S_3 shown in Fig. 2 with $Z_1 = A' \cup \{u\}$ and $Z_2 = N \backslash Z_1$. If $x_u < p_u$, then define $Z_1 = A'$ and $Z_2 = N \backslash (Z_1 \cup \{u\})$. Create a Type 1 schedule $S_3(u)$ shown in Fig. 3(a) with $k = u$. Go to Step 12.

Step 4. Compute $\varepsilon_0 = p(A)/(T - \tau)$, take $\varepsilon = \varepsilon_0$ and run an FPTAS for the subset-sum problem

$$\max \sum_{j \in H} p_j x_j$$
$$\sum_{j \in H} p_j x_j \leq T - \tau \qquad (13)$$
$$x_j \in \{0, 1\}, \ j \in H,$$

where $H = B$. Define $H^{(1)} = \{j \in H | x_j^\varepsilon = 1\}$. If $p(H^{(1)}) \geq (1 - \varepsilon)(T - \tau)$, then go to Step 5, otherwise go to Step 6.

Step 5. Run Procedure Split$(H^{(1)}, A, T - \tau)$ to find a set $A' \subset A$ and a job $u \in A$ such that $p(H^{(1)}) + p(A') + x_u = T - \tau$. If $x_u = p_u$ then create a non-preemptive schedule S_3 shown in Fig. 2 with $Z_1 = H^{(1)} \cup A' \cup \{u\}$. If $x_u < p_u$, then define $Z_1 := H^{(1)} \cup A$ and $Z_2 := N \backslash (Z_1 \cup \{u\})$ and create a schedule $S_3(u)$ shown in Fig. 3(a) with $k = u$. Go to Step 12.

Step 6. Create a schedule S_3 shown in Fig. 2 with $Z_1 = H^{(1)} \cup A$.

Step 7. Renumber the jobs in such a way that $B = \{1, 2, \ldots, h\}$ and $p_1 \leq p_2 \leq \ldots \leq p_h$. Define $\underline{k} := 1$ and $\overline{k} := h$.

(a) If $\underline{k} = \overline{k}$, then go to Step 8; otherwise, compute $k = \lceil (\overline{k} + \underline{k})/2 \rceil$.

(b) Define $B_k = B \backslash \{k\}$. Compute $\varepsilon_k = p_k/(T - \tau)$, take $\varepsilon = \varepsilon_k$ and apply an FPTAS to the subset-sum problem of the form (13) with $H = B_k$. Define $H^{(1)} = \{j \in H | x_j^\varepsilon = 1\}$.

(c) If $p(H^{(1)}) + p(A) \geq (1 - \varepsilon_k)(T - \tau)$, then define $\overline{k} := k$; otherwise define $\underline{k} := k$. Go to Step 7(a).

Step 8. For the current value of k, define $Z_1 := H^{(1)} \cup A$ and $Z_2 := N \backslash (Z_1 \cup \{k\})$. Create a schedule $S_3(k)$ shown in Fig. 3(b). Go to Step 12.

Step 9. If $p(N) < 4\tau$ then take $\varepsilon = \varepsilon_0 = (2\tau - p(N)/2)/(T - \tau)$; otherwise, take a given $\varepsilon > 0$.

Step 10. With the chosen ε, run an FPTAS for the problem (13) with $H = N$. Define $H^{(1)} = \{j \in H | x_j^\varepsilon = 1\}$. Create a non-preemptive schedule S_3 as shown in Fig. 2 with $Z_1 = H^{(1)}$. If either $p(N) < 4\tau$ or $p(H^{(1)}) \geq (1 - \varepsilon)(T - \tau)$, then go to Step 12, otherwise go to Step 11.

Step 11. Find a schedule $S_3(k)$ as described in Steps 7-8 above with $B = N$ and $A = \emptyset$.

Step 12. Output the best of all found schedules as the candidate schedule S^ε and Stop.

The statements below analyze the performance of Algorithm Move3.

Lemma 5. *For problem $TP2|v = 1, c \geq n|K_{\max}$ with at least one short job, Algorithm Move3 finds the best candidate schedule with three moves in $O(n \log n)$ time.*

Proof. In this lemma, we analyze Steps 1–8 of Algorithm Move3.

Suppose first that $p(A) \geq T - \tau$. If for job u found in Step 3 we have that $x_u = p_u$, then $p(A) + p_u = T - \tau$ and we have achieved equal loads on the machines without splitting a job. Therefore, $K_{\max}(S_3) = \max\{T, 3\tau\}$, and S_3 is an optimal Type 1 schedule. If $x_u < p_u$, then job u is a short fractional job. It follows from $p_u + 3\tau < T$ that $y_u + 2\tau < p(Z_1)$ and $x_u + 2\tau < p(Z_2)$. Thus, schedule $S_3(u)$ is as shown in Fig. 3(a), so that $K_{\max}(S_3(u)) = T$ and this schedule is optimal. In the remainder of this proof, assume that $p(A) < T - \tau < p(B)$.

In Step 4, since $A \neq \emptyset$, we have that $\varepsilon_0 > 0$. The analysis of schedules found in Step 5 is similar to that performed above for schedules found in Step 3: either we have a non-preemptive schedule with equal loads or a preemptive schedule with a short fractional job. The fact that a short fractional job exists follows from the observation that $p\left(H^{(1)}\right) + p(A) \geq (1 - \varepsilon_0)(T - \tau) + \varepsilon_0(T - \tau) = T - \tau$.

If $p\left(H^{(1)}\right) < (1 - \varepsilon_0)(T - \tau)$ then due to Theorem 3, the value $p\left(H^{(1)}\right)$ is optimal for the corresponding subset-sum problem of the form (13). Besides, $p\left(H^{(1)}\right) + p(A) < T - \tau$, i.e., there can be no short fractional job. Due to optimality of the value $p\left(H^{(1)}\right)$, for schedule S_3 found in Step 6 set $Z_1 = H^{(1)} \cup A$ forms an optimal solution of the subset-sum problem (13) with $H = N$, so that for set $Z_2 := N \backslash Z_1$ the value $p(Z_2)$ cannot be reduced. We have that $K_{\max}(S_3) = \max\{3\tau, \tau + p(Z_2)\}$, which is the length of the best non-preemptive schedule with three moves.

We arrive at Step 7 when $p\left(H^{(1)}\right) + p(A) < T - \tau$, knowing that in the best preemptive schedule $S_3(k)$ the fractional job k is going to be long and $K_{\max}(S_3(k)) = p_k + 3\tau$; see Fig. 3(b). Thus, we need to find the shortest long job k that can be fractional in schedule $S_3(k)$. This is done by the binary search procedure in Step 7 that takes at most $O(\log h)$ iterations.

The best of all found schedules is the best candidate schedule with three moves, provided that the instance of the problem contains short jobs. The overall running time of Algorithm Move3 in the presence of short jobs is $O(n \log n)$. □

Lemma 6. *For problem $TP2|v = 1, c \geq n|K_{\max}$ with no short jobs, Algorithm Move3 finds the best schedule with three moves in $O(n)$ time, provided that $p(N) < 4\tau$.*

In this lemma, we analyze Step 10 of Algorithm Move3, where $\varepsilon = \varepsilon_0$ is as defined in Step 9. We can prove that a non-preemptive schedule S_3 with $Z_1 = H^{(1)}$ found in Step 10 is optimal, irrespective the sign of the difference $p(Z_1) - (1 - \varepsilon_0)(T - \tau)$.

Lemma 7. *For finding the best candidate schedule with three moves for the instances of problem $TP2|v = 1, c \geq n|K_{\max}$ with no short jobs and $p(N) \geq 4\tau$, Algorithm Move3 behaves as an FPTAS that requires at most $O(n \log n + n/\varepsilon)$ time.*

Proof. In this lemma, we analyze Steps 10 and 11 of Algorithm Move3. If for set $H^{(1)}$ found in Step 10 the inequality $p(Z_1) = p(H^{(1)}) \geq (1 - \varepsilon)(T - \tau)$ holds, then $p(Z_2) = p(N \backslash H^{(1)}) \leq (1 + \varepsilon)(T - \tau)$. Since $\max\{p(Z_1), p(Z_2)\} \geq p(N)/2 \geq 2\tau$, we derive that $K_{\max}(S_3) = \max\{3\tau, p(Z_1) + \tau, \tau + p(Z_2)\} \leq (1 + \varepsilon)(T - \tau) + \tau \leq (1 + \varepsilon)K_{\max}(S^*)$.

If $p(Z_1) < (1 - \varepsilon)(T - \tau)$, then the only schedule that may be better than schedule S_3 is a preemptive schedule $S_3(k)$ with three moves such that $K_{\max}(S_3(k)) = p_k + 3\tau < \tau + p(Z_2)$, where k is a long job. Such a schedule is found in Step 11.

Running Step 10 requires $O(n/\varepsilon)$ time, Step 11 is essentially a binary search algorithm that takes $O(n \log n)$ time.   □

Lemmas 5–7 can be summarized to completely describe the behavior of Algorithm Move3.

Theorem 5. *For problem $TP2|v = 1, c \geq n|K_{\max}$ Algorithm Move3 behaves as an FPTAS, provided that there exists an optimal schedule S^* with three moves.*

For the instances with no short jobs and $p(N) \geq 4\tau$ the problem is NP-hard and the algorithm runs as an FPTAS that takes $O(n \log n + n/\varepsilon)$ time. In all other cases, the best schedule with three moves can be found in $O(n \log n)$ time.

5   Conclusion

The general algorithm for handling problem $TP2|v = 1, c \geq n|K_{\max}$ involves the following stages:

1. Create a schedule S_0 in which all jobs are processed on machine M_1.
2. Run Algorithm MoveEven.
3. Run Algorithm Move3.
4. Output the best of all found schedules.

As shown in Sect. 3 and 4, for some instances of the problem the final algorithm will find an optimal schedule, while for others it will behave as an FPTAS. The running time of the algorithm does not exceed $O(n \log n + n/\varepsilon)$.

If we limit our search to the schedules in which each job is processed on a machine with no preemption, then our algorithm can be modified to behave as a $(4/3)$−approximation algorithm, and a worst-case bound of $4/3$ cannot be improved in this class, i.e., there are instances of the problem for which the length of the best non-preemptive schedule can be arbitrarily close to $4/3$ times the length of the optimal preemptive schedule.

References

1. Brucker, P., Knust, S., Cheng, T.C.E., Shakhlevich, N.V.: Complexity results for flow-shop and open-shop scheduling problems with transportation delays. Ann. Oper. Res. 129, 81–106 (2004)

2. Dell'Amico, M.: Shop problems with two machines and time lags. Oper. Res. 44, 777–787 (1996)
3. Hall, L.A.: Approximability of flow shop scheduling. Math. Progr. B 82, 175–190 (1998)
4. Hurink, J., Knust, S.: Makespan minimization for flow-shop problems with transportation times and a single robot. Discrete Appl. Math. 112, 199–216 (2001)
5. Kellerer, H., Mansini, R., Pferschy, U., Speranza, M.G.: An efficient fully polynomial approximation scheme for the Subset-Sum Problem. J. Comput. Syst. Sci. 66, 349–370 (2003)
6. Kellerer, H., Pferschy, U., Pisinger, D.: Knapsack Problems. Springer, Berlin (2004)
7. Lee, C.-Y., Chen, Z.-L.: Machine scheduling with transportation times. J. Schedul. 4, 3–24 (2001)
8. Lee, C.-Y., Strusevich, V.A.: Two-machine shop scheduling with an uncapacitated interstage transporter. IIE Trans. 37, 725–736 (2005)
9. Lushchakova, I.N., Soper, A.J., Strusevich, V.A.: Transporting jobs through a two-machine open shop. Naval. Res. Log. 56, 1–18 (2009)
10. Qi, X.: A logistic scheduling model: scheduling and transshipment for two processing centers. IIE Trans. 38, 609–618 (2006)
11. Soper, A.J., Strusevich, V.A.: An improved approximation algorithm for the two-machine flow shop scheduling problem with an interstage transporter. Int J. Found. Computer Sci. 18, 565–591 (2007)
12. Strusevich, V.A.: A heuristic for the two-machine open-shop scheduling problem with transportation times. Discrete Appl. Math. 93, 287–304 (1999)
13. Yu, W., Hoogeveen, H., Lenstra, J.K.: Minimizing makespan in a two-machine flow shop with delays and unit-time operations is NP-hard. J. Schedul. 7, 333–348 (2004)

Author Index

Aisu, Hideyuki II-131

Balamohan, Balasingham II-58
Ballinger, Brad II-1
Bazgan, Cristina I-237
Beletska, Anna I-104
Belotti, Pietro I-65
Benbernou, Nadia II-1
Bhattacharya, Binay K. I-354
Bollig, Beate II-16
Bose, Prosenjit II-1
Busch, Arthur H. II-207

Cafieri, Sonia I-65
Cai, Zhipeng I-85
Chebotko, Artem II-97
Chen, Danny Z. I-270
Chen, Hong II-46
Chen, Wenping II-281
Chen, Xujin II-31
Chen, Zhixiang I-309
Cheng, Eddie I-222
Crespelle, Christophe I-1

Daescu, Ovidiu I-41
Damaschke, Peter II-117
Damian, Mirela II-1, II-181
D'Angelo, Gianlorenzo II-254
Das, Sandip I-354
Das, Shantanu I-11
Demaine, Erik D. II-1
Ding, Wei II-243, II-268
Dinh, Heiu I-184
Di Stefano, Gabriele II-254
Dobrev, Stefan II-72
Dragan, Feodor F. II-207
Du, Hongwei I-252
Duan, Zhenhua II-374
Dujmović, Vida II-1

Eppstein, David I-128

Fan, Hongbing II-292
Fan, Neng I-170
Fekete, Sándor I-21

Feldmann, Andreas Emil I-11
Flatland, Robin II-1
Flocchini, Paola II-58
Fu, Bin I-309, II-97

Gao, Yong II-332
Gonzalez-Hernandez, Loreto I-51
Goodrich, Michael T. I-128
Gray, Chris I-21
Gu, Qian-Ping II-107

Harutyunyan, Ararat I-31
Hasan, Maryam I-85
Hashim, Mashitoh II-195
He, Jing II-160
He, Xin I-339
Hu, Xiaodong II-31
Hurtado, Ferran II-1

Iacono, John II-1
Italiano, Giuseppe F. I-157

Jia, Xiaohua II-107
Ju, Wenqi I-41

Karmakar, Arindam I-354
Kellerer, Hans I-408
Kiyomi, Masashi II-362
Kranakis, Evangelos I-385, II-72, II-303
Krizanc, Danny I-385, II-72, II-303
Kröller, Alexander I-21
Kundeti, Vamsi I-184

Lambadaris, Ioannis II-303
Langetepe, Elmar I-369
Latapy, Matthieu I-1
Laura, Luigi I-157
Lee, Jon I-65
Lee, Wonjun I-252
Li, Deying II-46, II-281
Li, Fei I-398
Li, Zheng II-46, II-281
Liang, Hongyu II-160
Liberti, Leo I-65
Lin, Guohui I-85, II-243

Liu, Guizhen   II-170
Liu, Jin-Yi   I-300
Liu, Qinghai   I-212
Lubiw, Anna   II-1
Luo, Jun   I-41

Ma, Weidong   II-31
Ma, Wenkai   II-46, II-281
Marek, Palkowski   I-104
Marzban, Marjan   II-107
Matsuhisa, Takashi   I-77
Memar, Julia   I-142
Miao, Zhengke   I-114
Miri, Ali   II-58
Misiołek, Ewa   I-270
Morin, Pat   II-1
Muhammad, Azam Sheikh   II-117
Mukhopadhyay, Asish   II-401

Nagamochi, Hiroshi   II-347
Nandy, Subhas C.   I-354
Narayanan, Lata   II-303
Nastos, James   II-332
Navarra, Alfredo   II-254
Nguyen, Dung T.   I-197
Nguyen, Thanh Qui   I-1
Nguyen, Viet Hung   I-260, II-144

Olsen, Martin   II-87
Opatrny, Jaroslav   II-72, II-303
Otsuki, Tomoshi   II-131

Pan, Xuejun   II-170
Pardalos, Panos M.   I-170
Park, Haesun   I-252
Phan, Thi Ha Duong   I-1
Ponce, Oscar Morales   I-385, II-72

Qiu, Ke   I-222

Rajasekaran, Sanguthevar   I-184
Rangel-Valdez, Nelson   I-51
Raudonis, Kristin   II-181

Sacristán, Vera   II-1
Saitoh, Toshiki   II-362
Santaroni, Federico   I-157
Santoro, Nicola   II-58
Sarker, Animesh   II-401
Schulz, André   I-324
Schuurmans, Dale   I-85
Shen, Yilin   I-197

Shen, Zhi Zhang   I-222
Shi, Yi   I-85
Shu, Jinlong   I-114
Singh, Gaurav   I-142
Soper, Alan J.   I-408
Souvaine, Diane   II-1
Sritharan, R.   II-207
Stacho, Ladislav   I-385, II-72
Strash, Darren   I-128
Strusevich, Vitaly A.   I-408
Sun, Jonathan Z.   II-170
Switzer, Tom   II-401

Takaoka, Tadao   II-195
Tan, Jinsong   II-317
Tanaka, Toshiaki   II-131
Thai, My T.   I-197
Tian, Cong   II-374
Tomasz, Klimek   I-104
Torres-Jimenez, Jose   I-51
Tóth, Csaba D.   I-324
Toubaline, Sonia   I-237
Trott, Lowell   I-128

Uehara, Ryuhei   II-1, II-362

Vanderpooten, Daniel   I-237
Viglas, Anastasios   II-87

Wang, Jiun-Jie   I-339
Wang, Yuexuan   I-252
Widmayer, Peter   I-11
Wlodzimierz, Bielecki   I-104
Wu, Bang Ye   II-219
Wu, Yu-Liang   II-292

Xiao, Mingyu   II-387
Xue, Guoliang   II-243, II-268

Yamakami, Tomoyuki   I-285
Yang, Boting   II-228
Ye, Qiang   I-252
Yu, Guanglong   I-114
Yu, Zhihua   I-212

Zhang, Huaming   I-339
Zhang, Zhao   I-212
Zhong, Jioafei   I-252
Zhuang, Bingbing   II-347
Zinder, Yakov   I-142
Zvedeniouk, Ilia   II-87